ERCOFTAC Series **31**

ERCOFTAC (European Research Community on Flow, Turbulence and Combustion) was founded as an international association with scientific objectives in 1988. ERCOFTAC strongly promotes joint efforts of European research institutes and industries that are active in the field of flow, turbulence and combustion, in order to enhance the exchange of technical and scientific information on fundamental and applied research and design. Each year, ERCOFTAC organizes several meetings in the form of workshops, conferences and summer schools, where ERCOFTAC members and other researchers meet and exchange information. The *ERCOFTAC* series publishes the proceedings of ERCOFTAC meetings, which cover all aspects of fluid mechanics. The series comprises proceedings of conferences and workshops, and of textbooks presenting the material taught at summer schools. The series covers the entire domain of fluid mechanics, which includes physical modelling, computational fluid dynamics including grid generation and turbulence modelling, measuring techniques, flow visualization as applied to industrial flows, aerodynamics, combustion, geophysical and environmental flows, hydraulics, multi-phase flows, non-Newtonian flows, astrophysical flows, laminar, turbulent and transitional flows.

Indexed by SCOPUS, Google Scholar and SpringerLink.

Cristian Marchioli · Maria Vittoria Salvetti ·
Manuel Garcia-Villalba · Philipp Schlatter
Editors

Direct and Large Eddy Simulation XIII

Proceedings of DLES13

Springer

Editors
Cristian Marchioli
Department of Engineering and Architecture
University of Udine
Udine, Italy

Manuel Garcia-Villalba
TU Wien
Institute of Fluid Mechanics and Heat
Transfer
Wien, Austria

Maria Vittoria Salvetti
Department of Civil and Industrial
Engineering
University of Pisa
Pisa, Italy

Philipp Schlatter
Institute of Fluid Mechanics
Friedrich Alexander University
Erlangen-Nuremberg, Germany

ISSN 1382-4309 ISSN 2215-1826 (electronic)
ERCOFTAC Series
ISBN 978-3-031-47027-1 ISBN 978-3-031-47028-8 (eBook)
https://doi.org/10.1007/978-3-031-47028-8

This Springer imprint is published by the registered company Springer Nature Switzerland AG
The registered company address is: Gewerbestrasse 11, 6330 Cham, Switzerland

Contents

Internal and External Aerodynamics

LES Modeling and Hybrid LES-RANS

Numerical Methods

Turbulent Flows

Bluff Bodies and Separated Flows

Dynamics of Turbulent Kinetic Energy Advection in a Turbulent Boundary Layer Under Unsteady Pressure Gradients

F. Ambrogi(✉), U. Piomelli, and D. E. Rival

Department of Mechanical and Materials Engineering, Queen's University, Kingston, ON K7L 3N6, Canada
{f.ambrogi,ugo,d.e.rival}@queensu.ca

1 Introduction

Pressure-induced flow separation, which occurs in the presence of a strong adverse pressure gradient, is very common in many physical systems, ranging from the medical field, to engineering applications, to the natural sciences. In most of the practical applications the pressure gradient varies both spatially and temporally giving rise to a transient separation process often associated with a drop in performance of many systems (e.g. wings, turbomachinery). For instance, dynamic stall on an airfoil at large angles of attack, blood flow through valves and stenosis, and the dynamics of fish maneuvering, are all examples of cases in which unsteady flow separation plays a crucial role. Although a large effort has been invested to shed light onto the physics of separated flows, many questions still remain unanswered [1]. In a recent investigation we performed wall-resolved large eddy simulations (LES) of a spatially developing turbulent boundary layer (TBL) over a flat plate with unsteady flow separation [2]. The unsteadiness was generated by an oscillating free-stream wall-normal velocity profile following the approach by [3]. Results were analyzed in terms of mean and phase-averaged flow quantities and we found that the oscillation frequency greatly affects the transient separation cycle. Separation was observed at every considered frequency though the size of the separation bubble greatly changed as the frequency was decreased. The lowest frequency showed a clear trend towards a quasi-steady state in which the dimensions of the separation bubble (height and length) match very well a steady calculation. Even more interesting was the finding of a frequency at which the separation region is advected downstream and periodically washed-out of the domain. This behaviour is consistent with that of dynamic stall on an airfoil, pulsating flow over a backward facing step, and a stalled diffuser [1,4,5]. The purpose of the present paper is to shed light on questions that arose during our previous analysis: (i) what is the effect of the oscillation frequency on higher order statistics, (ii) what is the effect of the "washing-out" behaviour on the flow physics, and (iii) what are the causes behind the "washing-out" behaviour. In the following sections the problem formulation will be explained in more details, followed by some results and concluding remarks.

C. Marchioli et al. (Eds.): DLES 2023, ERCO 31, pp. 3–8, 2024.
https://doi.org/10.1007/978-3-031-47028-8_1

2 Problem Definition

Simulations are performed using the LES technique and the incompressible Navier-Stokes equations, mass and momentum, are solved for filtered quantities (here indicated with an overline):

$$\frac{\partial \overline{u}_i}{\partial x_i} = 0 \tag{1}$$

$$\frac{\partial \overline{u}_i}{\partial t} + \frac{\partial}{\partial x_j}\left(\overline{u}_i \overline{u}_j\right) = -\frac{\partial \overline{p}}{\partial x_i} + \nu \nabla^2 \overline{u}_i - \frac{\partial \tau_{ij}}{\partial x_j} \tag{2}$$

where $i = 1, 2,$ and 3 (or x, y, and z) are the streamwise, wall-normal and spanwise directions, respectively, u_i are the velocity components, P is the density-normalized hydrodynamic pressure, ν is the kinematic viscosity, and τ_{ij} is the subfilter-scale stress tensor. In this work, τ_{ij} is modelled using the Vreman eddy-viscosity model [6]. We use a well-validated finite-difference code that has previously been applied to similar cases [7–9]. The dimensions of the computational domain are $L_x = 600\delta_o^*$, $L_y = 64\delta_o^*$, and $L_z = 55\delta_o^*$. δ_o^* is the boundary layer displacement thickness at the inflow plane. The Reynolds number based on δ_o^* and the freestream velocity U_o at the inflow (which is constant in time) is $Re_{\delta_o^*} = 1,000$. For more information about computational domain, grid size, and grid convergence study, the interested reader is referred to [2]. A periodic boundary condition is used in the spanwise direction, a convective boundary condition at the outflow plane, and a velocity obtained from a parallel auxiliary simulation is used at the inflow.

Following the approach in [3], flow separation is achieved by prescribing a suction-blowing wall-normal velocity profile V_∞ at the freestream. In our case, however, V_∞ is both space and time dependent, and follows the relation $V_\infty(x,t) = V_o(x)\sin(2\pi t/T)$, where T is the oscillation period, and V_o a streamwise distribution of wall-normal velocity, chosen to match the case studied by [3]. The streamwise freestream velocity U_∞ is obtained by imposing a zero vorticity condition [3,9,10].

The main parameter that characterizes the unsteadiness is the reduced frequency $k = \pi f L_{PG}/U_o$, where $f = 1/T$ is the imposed frequency, L_{PG} is a characteristic length (we use the length over which the pressure gradient varies), and U_o is the freestream velocity at the inflow plane. The oscillation period has been divided in 20 equally-spaced phases, and phase averaging (here indicated with a $\langle\cdot\rangle$) has been carried out for every quantity. Starting at phase $\Phi = 0°$ in which the boundary layer experiences a zero pressure gradient (ZPG), Φ increases and the flow moves into a region in which a favorable pressure gradient (FPG) is followed by an adverse one (APG), then it goes back to a ZPG again at $\Phi = 180°$, then into a region in which an APG is followed by a FPG, until the cycle repeats itself. Finally, simulations were performed for a wide range of reduced frequencies k (0.2, 1, and 10); for the sake of conciseness only results for $k = 0.2$ and $k = 1$ will be presented in the following section.

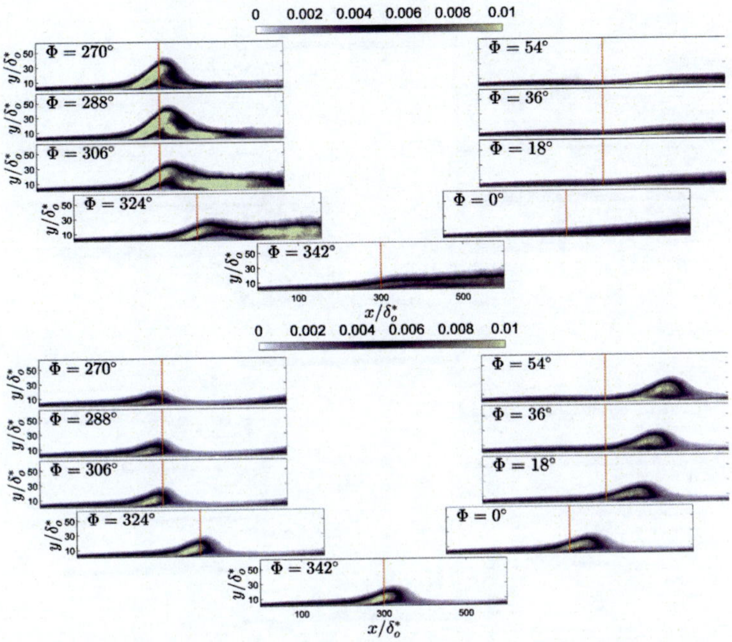

Fig. 1. Contours of phase-averaged turbulent kinetic energy for the $k = 0.2$ case (top) and $k = 1$ case (bottom). Red line denotes the center of the computational domain (Color figure online).

3 Results

In our recent characterization of unsteady flow separation in a turbulent boundary layer [2], we described how the transient separation process is affected by the reduced frequency k using mean and phase-averaged first-order quantities. As mentioned above, many aspects of the flow dynamic remain to be addressed. In the present work we use the phase-averaged turbulent kinetic energy TKE as a metric to investigate the dynamics of unsteady flow separation with the final goal to answer the aforementioned questions and shed light into the complexity of separated flows.

Figure 1 (top) shows contours of phase-averaged turbulent kinetic energy (TKE) from the extreme phase $\Phi = 270°$ (maximum APG-FPG) for 9 consecutive phases at the lowest frequency ($k = 0.2$). First, we notice that an advection-like phenomenon of TKE plays a significant role even at low frequency, however its dynamic behaviour is completely different from the $k = 1$ case. At phase $\Phi = 270°$, the high TKE region on the front of the separation bubble is generated due to the separated shear layer. This high TKE region grows and shrinks in the wall-normal direction but does not move in space. At phase $\Phi = 288°$, the high TKE region after the separation is due to flow impingement on the wall, and we notice a clear transfer of TKE from the front to the back of the

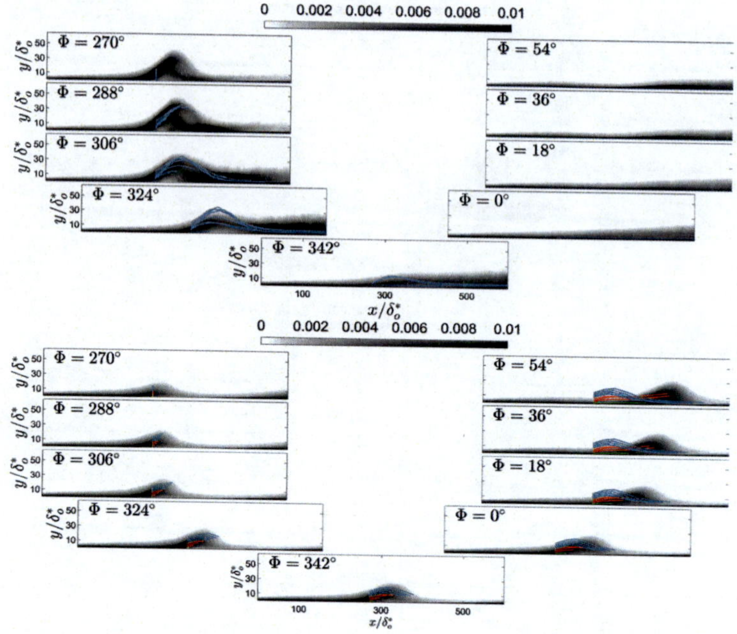

Fig. 2. Contours of phase-averaged turbulent kinetic energy overlapped with averaged particle pathlines for the $k = 0.2$ case (top) and $k = 1$ case (bottom). Red lines denoting particles that move upwards in the $k = 1$ case (Color figure online).

bubble and its advection out of the domain. This dynamic process takes place in 5 phases, before the freestream forcing changes sign. At this frequency, the time-evolution of TKE is in-phase with the freestream forcing. As the reduced frequency increases, however, the convective timescale of the flow and the unsteady imposed one are comparable, and a lag in response to the free stream forcing is generated and is present throughout the entire cycle [2]. In Fig. 1 (bottom), we first notice that the high TKE region due to the separated shear layer grows in the wall-normal direction in the first 5 phases ($\Phi = 270°$ to $\Phi = 342°$) but it does not start moving downstream until the freestream condition changes sign around $\Phi = 0°$. At this point, the high TKE region is advected downstream at a constant velocity of $0.6U_\infty$, and the most interesting aspect is that it does not deform while moving out of the domain.

To understand the causes of such behaviours we analyzed averaged particles pathlines defined as:

$$\frac{d\langle x_p(t)\rangle}{dt} = \langle u(x_p(t), t)\rangle \tag{3}$$

$$x_p\big|_{t=0} = x_{p,o} \tag{4}$$

where x_p is the particle position as a function of time and $x_{p,o}$ is the particle initial position. We chose a set of particles positions along the wall-normal direction and seeded them inside the separation region at phase $\Phi = 270°$ for both the low and intermediate frequency cases. The goal was to observe the averaged path of a particle during the advection process. Figure 2 (top) shows contours of the phase-averaged TKE for the low frequency case overlapped with particle pathlines. We notice that at this frequency, pathlines follow what the freestream does; suction corresponds to upward motion and blowing to downward motion. Therefore, pathlines closely resemble streamlines for the $k = 0.2$ case which reinforce the conclusion that this frequency is approaching a quasi-steady state [2]. Moreover, all particles are out of the domain in the first 5 phases, before the freestream changes sign. When performing the same analysis for the $k = 1$ case, we observe a different behaviour (Fig. 2 - bottom). After being seeded at $\Phi = 270°$ particles start behaving according to the freestream forcing and suction corresponds to upwards motion. At this frequency the period is much shorter than the $k = 0.2$ case, and when the freestream forcing changes sign, around $\Phi = 0°$, the particles are still in the domain and the flow response to the perturbation can be divided in 3 parts: (1) the green particles, very close to the wall, do not feel the change in the freestream, they slow down due to the separation region and accelerate again as in an oscillating boundary layer, (2) the blue particles are fast enough that when the freestream changes sign they already reached a position very close to the wall and are not affected by the suction, (3) the red particles are in a region of the flow that is fast enough to escape the wall-effect, but is slow enough that as soon as the freestream changes, these particles are picked up and start moving downstream exactly corresponding at the center of the high TKE region.

4 Conclusions

An LES of a spatially developing turbulent boundary layer with unsteady separation was performed to answer some questions arose during our previous mean and phase-averaged flow analysis in [2]: (i) what is the effect of the oscillation frequency on turbulent kinetic energy (TKE) and higher order statistics, (ii) what is the effect of the "washing-out" behaviour (observed for $k = 1$) on the flow physics, and (iii) what are the causes behind the "washing-out" behaviour. We showed that the time evolution of TKE highlight peculiar differences between the $k = 0.2$ and $k = 1$ cases. An advection-like phenomenon plays a significant role also at very low frequency but its dynamic is completely different compared to the $k = 1$ case, in which the high TKE region is advected downstream while keeping its shape. The reason for this behavior was found analyzing averaged particles pathlines and observing that while at low frequency they closely resemble streamlines, for the $k = 1$ case they cross each other and the flow response to the freestream is divided in three separate region based on particle position along the wall-normal direction.

Acknowledgements. Research supported by the Natural Sciences and Engineering Council of Canada (NSERC). This research was enabled by computational support provided by Digital Research Alliance of Canada.

References

1. Simpson, R.L.: Turbulent boundary-layer separation. Ann. Rev. Fluid Mech. **21**, 205–232 (1989)
2. Ambrogi, F., Piomelli, U., Rival, D.E.: Characterization of unsteady separation in a turbulent boundary layer: mean and phase-averaged flow. J. Fluid Mech. **945**, A10 (2022)
3. Na, Y., Moin, P.: Direct numerical simulation of a separated turbulent boundary layer. J. Fluid Mech. **374**, 379–405 (1998)
4. Mullin, T., Greated, C.A., Grant, I.: Pulsating flow over a step. Phys. Fluids **23**(4), 669–674 (1980)
5. Leishman, G.J.: Principles of Helicopter Aerodynamics. Cambridge University Press, Cambridge (2006)
6. Vreman, A.W.: An eddy-viscosity subgrid-scale model for turbulent shear flow: algebraic theory and applications. Phys. Fluids **16**(10), 3670–3681 (2004)
7. Keating, A., Piomelli, U., Bremhorst, K., Nešić, S.: Large-eddy simulation of heat transfer downstream of a backward-facing step. J. Turbul. **5**(1), 020 (2004)
8. Yuan, J., Piomelli, U.: Numerical simulation of a spatially developing accelerating boundary layer over roughness. J. Fluid Mech. **780**, 192–214 (2015)
9. Wu, W., Piomelli, U.: Effects of surface roughness on a separating turbulent boundary layer. J. Fluid Mech. **841**, 552–580 (2018)
10. Lund, T.S., Wu, X., Squires, K.D.: Generation of turbulent inflow data for spatially-developing boundary layer simulations. J. Comput. Phys. **140**, 233–258 (1998)

DNS of the Flow About a 5:1 Rectangular Body with Sharp Corners

R. Corsini, A. Cimarelli, and E. Stalio(✉)

Dipartimento di Ingegneria "Enzo Ferrari",
Università degli Studi di Modena e Reggio Emilia, Modena, Italy
enrico.stalio@unimore.it

1 Introduction

The turbulent flow about blunt, rectangular bodies is often encountered in civil and wind engineering applications as well as in land transport vehicles. Relevant examples include deck slabs of long-span bridges and high-rise buildings of rectangular section and the problem of the aerodynamics of vehicles of simple shape like trucks and trains. The flow about an elongated rectangular cylinder with sharp corners involves several critical aerodynamics features like fixed-position separation in laminar conditions, reattachment and vortex shedding in the wake. In the applications, this gives place to vortex-induced vibration effects, affects convective heat transfer between the atmospheric air and the body as well as pollutant dispersion in the urban environment.

A considerable research effort has been expended since the past 20 years in the investigation of the aerodynamics of a two-dimensional rectangular cylinder with aspect ratio 5:1, especially by research groups based in Europe [1,2,4,11,13] and this in turn has made this a canonical case in aerodynamics, of acronym BARC, which stands for "Benchmark on the Aerodynamics of a two-dimensional Rectangular Cylinder".

Despite the context of abundance of research and publications, both numerical predictions and experimental data are characterized by a strong dependency upon the details of the setup and the quest for the "correct solution" at different Reynolds numbers is still underway. This is true also for Direct Numerical Simulations [6], where the difficulty in the numerical representation of the physical phenomenon does not include the modeling of turbulence.

In this work we present results from two Direct Numerical Simulation around the BARC body. The Reynolds numbers of the simulations are 3000 and 14000 respectively. Simulations are carried out using NEK5000 [7], which combines the geometric flexibility of finite element method with the high-order typical of spectral methods.

The result section includes a description of the flow configuration, the mean velocity and pressure field as well as pressure fluctuations on the wall. The Reynolds number effect on the above characteristics is investigated. Also passive scalar transport is shortly presented.

C. Marchioli et al. (Eds.): DLES 2023, ERCO 31, pp. 9–16, 2024.
https://doi.org/10.1007/978-3-031-47028-8_2

2 Numerical Methodology

The numerical code used for this study is NEK5000 . It implements the spectral element method. In each hexahedral element, velocity and passive scalar interpolants are polynomials of order N, pressure is treated with order $N - 2$. Simulations are run at $N = 7$. Time integration is performed treating implicitly the viscous terms by a third-order backward differentiation scheme while for the advective term, a third-order extrapolation scheme is used. Dirichlet boundary conditions are set for the uniform velocity field at the inlet. The open boundary condition is imposed at outflow and along the crossflow directions. The rectangular cylinder of aspect ratio 5:1 and height D is represented in a computational domain of extension $L_x \times L_y \times L_z = 80D \times 31D \times 5D$, where x, y and z directions are the streamwise, cross-flow and spanwise directions.

A two-dimensional structured grid consisting of quadrilateral conforming elements is generated and then extruded along the plane normal direction, corresponding to the spanwise direction. The grid is designed in order to fully resolve turbulence, details of the mesh employed at the two Reynolds numbers is given in Table 1.

Table 1. Details of the spatial discretization employed at the two Reynolds numbers considered. The minimum spacings are located at the leading-edge corner. The overbar indicates a streamwise average over the rectangle length.

Re	3000	14000
$\Delta x^+_{\max}, \overline{\Delta x^+}$	$4.3, 1.7$	$4.1, 2.3$
$\Delta y^+_{w\,\max}, \overline{\Delta y^+_w}$	$0.47, 0.23$	$0.66, 0.41$
$\Delta z^+_{\max}, \overline{\Delta z^+}$	$3.9, 1.9$	$5.1, 3.1$
$(\Delta x/\eta)_{\max}$	3.9	4.2
$(\Delta y/\eta)_{\max}$	3.3	4.6
$(\Delta z/\eta)_{\max}$	3.4	6.3

3 The Question of Spatial Discretization

The flow about the rectangular cylinder BARC is characterized by the mean flow topology displayed in Fig. 1. The laminar flow separates at the upstream angle of the body and reattaches on the side. Reattachment is turbulent, the time-averaged reattachment length depends upon *Re*. After reattachment, the fluid moving in the upstream direction gives place to a second, counter rotating mean vortex (secondary vortex). The fluid which is not included in the main separation bubble finally detaches at the downstream corners of the rectangular cylinder giving place to a mean vortex at the end of the body (wake vortex).

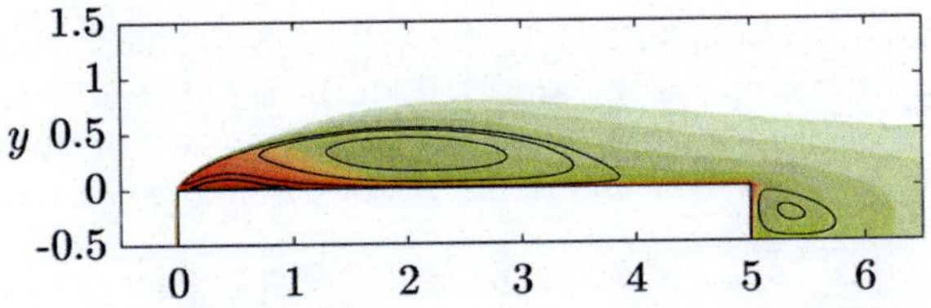

Fig. 1. Sketch of the large time-averaged structures

Table 2. Comparison between spatial discretization on the present DNS against Ref. [5] and Ref. [3]

	Present	Ref. [5]	Ref. [3]
$L_x \times L_y \times L_z$	$80 \times 31 \times 5$	$112 \times 50 \times 5$	$62.5 \times 42 \times 5$
$n_x \times n_y \times n_z$	$1247 \times 477 \times 316$	$- \times - \times 144$	$1776 \times 942 \times 150$
Spatial DoF	187 962 804	15×10^6	250 948 800
$\Delta x_{\min}, \Delta x_{\max}, \overline{\Delta x}$	0.004, 0.475, 0.064	0.004, –, –	0.0015, –, 0.035
$\Delta y_{\min}, \Delta y_{\max}, \overline{\Delta y}$	0.004, 0.503, 0.065	0.004, –, –	0.0015, –, 0.044
Δz	0.016	0.035	0.033

Direct numerical simulations carried out by different groups using different numerical algorithms and spatial discretization calculate statistics that – although qualitatively very similar– are quantitatively different. This is to be ascribed primarily to the different spatial discretization employed, see Ref. [6] for this discussion. Table 2 displays the spatial discretization used for $Re = 3000$ in the present work, in Ref. [5] and Ref. [3]. Table 3 reports the centers of the most important structures of the mean flow: primary, secondary and wake vortex. It is observed that for increasing spatial discretization the structures of the mean flow move downstream and that the convergence to a fully resolved solution is still in question. Notice that all the DNSs considered were originally meant for providing grid independent and method independent results.

Table 3. Coordinates of the main flow structures at $Re = 3000$: comparison between present results and results in Ref. [5] and Ref. [3]. In the first column x_r is the reattachment length; (x_c, y_c) are the coordinates of the vortex center; x_s is the start of the secondary vortex, x_e indicates the end of the secondary and wake vortices.

	Present	Ref. [5]	Ref. [3]
Primary vortex			
x_r	4.17	3.65	3.955
(x_c, y_c)	(2.67, 0.33)	(2.04, 0.35)	(2.357, 0.33)
Secondary vortex			
x_s	1.14	0.4	0.63
x_e	1.93	1.4	1.59
x_r	0.79	1.0	0.96
(x_c, y_c)	(1.57, 0.037)	-	(1.2, 0.041)
Wake vortex			
x_e	5.88	6.2	5.975
(x_c, y_c)	(5.38, −0.25)	(5.5, −0.27)	(5.415, −0.25)

4 Results

Results of the simulations at two different Reynolds numbers ($Re = 3000$ and $Re = 14000$) are presented in the following. The text is organized in two subsections. Subsection 4.1 is about the fluid flow and pressure fields, Subsect. 4.2 describes the main outcomes of a scalar transport study.

4.1 Velocity and Pressure Fields

The full DNS dataset at two different Reynolds numbers provide a wealth of quantitative detail on the case. The flow configuration described above is well reproduced by the isocontours of the time-averaged streamfunction in Fig. 2 for both the Reynolds numbers considered. In the Reynolds number range investigated, the reattachment position progressively moves downstream while the secondary vortex migrates upstream because of the larger energy content of the counter-flow boundary layer and its consequent increased capability of overcoming adverse pressure gradients. The same Fig. 2 reports also the distribution of the time-averaged turbulent kinetic energy $\langle k \rangle$. The regions of peak $\langle k \rangle$ are associated to shear layer instability. For increasing Reynolds numbers, when the flow is more prone to transition, the regions of peak turbulent kinetic energy are located more upstream. The same tendency is also observed by Moore et al. [11], at a higher range of Reynolds numbers. Also transition to turbulence is shifted upstream for increasing Reynolds numbers. Figure 2 shows that locations of peak $\langle k \rangle$ correspond to the regions of production of turbulent kinetic energy.

It can be observed from the $\langle c_p \rangle$ profiles on the solid walls in Fig. 3 that due to the displacement of the structures of the mean flow, also the mean pressure

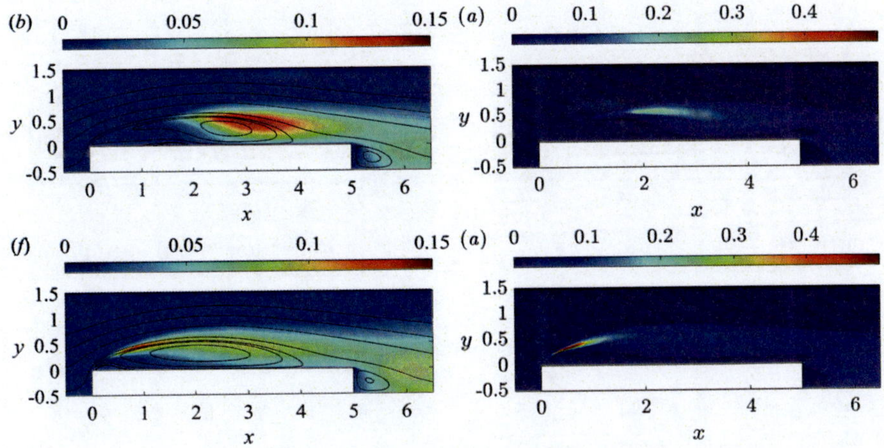

Fig. 2. Contours of turbulent kinetic energy $\langle k \rangle$ (left panels) and of turbulent kinetic energy production $\mathcal{P}_k$ (right panels) at $Re = 3000$ and 14000.

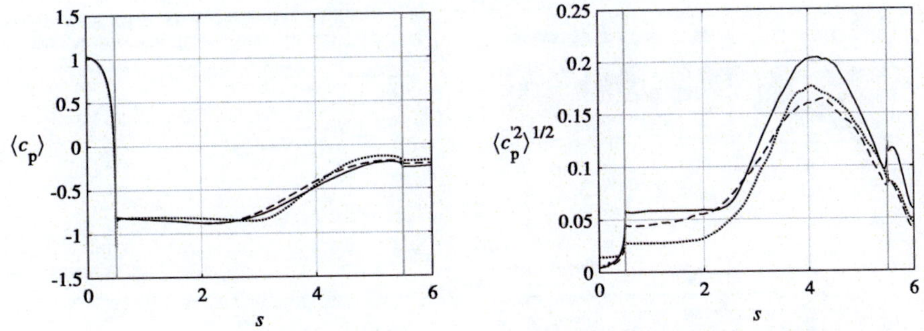

Fig. 3. Pressure coefficient profiles and standard deviations of c_p: dotted line, $Re = 3000$; dashed line, $Re = 8000$; and solid line, $Re = 14000$.

distribution depends upon the Reynolds number. Pressure fluctuations displayed in Fig. 3 are seen to peak close to the reattachment of the primary separation. This is to be ascribed to the recurrent lengthening and shortening of the primary vortex.

4.2 Passive Scalar Transport

Maps of the mean passive scalar φ at the two Reynolds numbers investigated are displayed in Fig. 4, together with passive scalar fluctuations. Low φ regions are observed in the secondary vortex region for both Reynolds numbers. The flow recirculating region prevents scalar transport between the free stream and the wall. This lack in scalar transport between side walls and outer flow will be explained later in this text from the discussion on scalar fluxes. Scalar field

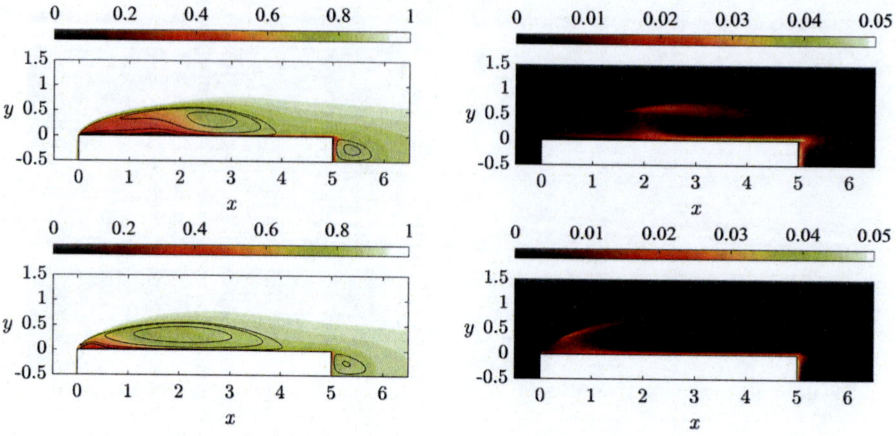

Fig. 4. Mean passive scalar field and its variance at $Re = 3000$ (top panels) and $Re = 14000$ (bottom panels).

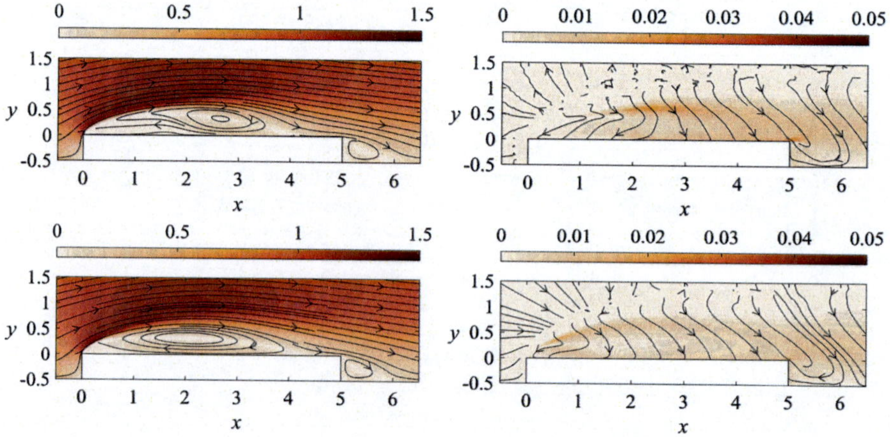

Fig. 5. Lines of heat fluxes with isocontours of their magnitude at $Re = 3000$ (top panels) and $Re = 14000$ (bottom panels). Left panels show heat fluxes due to the mean field and right panels show heat fluxes due to the fluctuating field.

fluctuations show very similar patterns as the turbulent kinetic energy, peak scalar fluctuations fall in the regions of peak turbulent kinetic energy.

Figure 5 shows that mean scalar fluxes do not cross the primary vortex and have the tendency to transport the passive scalar downstream or to form idle closed loops in the circulating regions. Turbulent heat fluxes instead can transport the scalar between the outer flow and the wall. In case the scalar is interpreted as internal energy or temperature, this appear to be the only mechanism which allows for heat to be exchanged at the wall. The different magnitudes of the two types of fluxes suggests that at these Reynolds numbers, passive scalar exchange between the solid wall region and the free stream is tiny.

5 Conclusions

This numerical study about a 5:1 rectangular body with sharp corners (BARC) does not make use of turbulence models and is characterized by a high-resolution representation. The two simulations carried out ($Re = 3000$ and $Re = 14000$) span the low to moderate Reynolds number range, where the Strouhal number of the vortex shedding still depends upon Re.

The flow configuration about the BARC body consists in the laminar separation at the upstream corner, a transition process occurring in the separated shear-layer and the reattachment of the turbulent flow. Two boundary layers are formed starting from the reattachment position: one which leads to the final separation at the downstream corner and the shedding of vortices in the wake, and a second boundary layer developing upstream which finally generates a smaller, counter-rotating, mean spanwise vortex.

By comparison of different DNS studies at $Re = 3000$ it appears that the location and the shape of the mean-flow structures depend upon the specific methodology employed. The spatial discretization level is likely to affect results more than other details like the type of spatial schemes and time advancement. Differences observed are still non-negligible and the question of spatial discretization of the BARC geometry is still open at $Re = 3000$.

While the general topology of the flow is not affected by Re, within the range of Reynolds numbers investigated, the reattachment position progressively moves downstream for increasing Reynolds numbers. The secondary vortex instead migrates upstream. This has an effect in the mean pressure field and its fluctuations.

Regarding scalar transport it appears that the flow recirculating region at the sides of the BARC prevents mean scalar fluxes to transport φ between freestream and wall. Turbulent fluxes, which are insted directed perpendicular to the solid walls, have a magnitude which is only a few percent of the total fluxes.

Acknowledgements. This work was granted access to the HPC resources FROM TGCC-GENCI: Projet A0122A12037. The authors acknowledge PRACE for awarding access to HPC resources on Joliot-Curie (KNL) hosted by GENCI at CEA, France.

References

1. Bruno, L., Fransos, D., Coste, N., Bosco, A.: 3D flow around a rectangular cylinder: a computational study. J. Wind Eng. Ind. Aerodyn. **98**(6–7), 263–276 (2010)
2. Brusiani, F., De Miranda, S., Patruno, L., Ubertini, F., Vaona, P.: On the evaluation of bridge deck flutter derivatives using rans turbulence models. J. Wind Eng. Ind. Aerodyn. **119**, 39–47 (2013)
3. Chiarini, A., Quadrio, M.: The turbulent flow over the BARC rectangular cylinder: a DNS study. Flow Turbul. Combust. **107**, 875–899 (2021)

4. Cimarelli, A., Leonforte, A., Angeli, D.: On the structure of the self-sustaining cycle in separating and reattaching flows. J. Fluid Mech. **857**, 907–936 (2018)
5. Cimarelli, A., Leonforte, A., Angeli, D.: Direct numerical simulation of the flow around a rectangular cylinder at a moderately high reynolds number. J. Wind Eng. Ind. Aerodyn. **174**, 38–49 (2018)
6. Corsini, R., Angeli, D., Stalio, E., Chibbaro, S., Cimarelli, A.: Flow solutions around rectangular cylinders: the question of spatial discretization. Wind Struct. Int. J. **34**, 151–159 (2022)
7. Fischer, P.F., Lottes, J.W., Kerkemeier, S.G.: "nek5000 web page" (2008). https://nek5000.mcs.anl.gov
8. Hourigan, K., Thompson, M., Tan, B.T.: Self-sustained oscillations in flows around long blunt plates. J. Fluids Struct. **15**, 387–398 (2001)
9. Mannini, C., Schewe, G.: Numerical study on the three-dimensional unsteady flow past a 5:1 rectangular cylinder using the DES approach. In: Proceedings of the Thirteenth International Conference on Wind Engineering, Amsterdam, The Netherlands (2011)
10. Mariotti, A., Siconolfi, L., Salvetti, M.V.: Stochastic sensitivity analysis of large-eddy simulation predictions of the flow around a 5:1 rectangular cylinder. Eur. J. Mech. B. Fluids **62**, 149–165 (2017)
11. Moore, D.M., Letchford, C.W., Amitay, M.: Energetic scales in a bluff body shear layer. J. Fluid Mech. **875**, 543–575 (2019)
12. Nakamura, Y., Ohya, Y., Ozono, S., Nakayama, R.: Experimental and numerical analysis of vortex shedding from elongated rectangular cylinders at low reynolds numbers 200–103. J. Wind Eng. Ind. Aerodyn. **65**, 301–308 (1996)
13. Nguyen, D.T., Hargreaves, D.M., Owen, J.S.: Vortex-induced vibration of a 5:1 rectangular cylinder: a comparison of wind tunnel sectional model tests and computational simulations. J. Wind Eng. Ind. Aerodyn. **175**, 1–16 (2018)
14. Schewe, G.: Reynolds-number-effects in flow around a rectangular cylinder with aspect ratio 1:5. J. Fluids Struct. **39**, 15–26 (2013)

In-Situ Analysis of Backflow Events and Their Relation to Separation in Wings Through Well-Resolved LES

F. Mallor[1(✉)], J. Liu[2], A. Peplinski[1], R. Vinuesa[1], R. Örlü[1,3], T. Weinkauf[2], and P. Schlatter[1,4]

[1] FLOW, Engineering Mechanics, KTH Royal Institute of Technology, Stockholm, Sweden
mallor@kth.se

[2] Division of Computational Science and Technology, KTH Royal Institute of Technology, Stockholm, Sweden

[3] Department of Mechanical, Electronic and Chemical Engineering OsloMet – Oslo Metropolitan University, Oslo, Norway

[4] Institute of Fluid Mechanics (LSTM), Friedrich-Alexander-Universität (FAU) Erlangen-Nürnberg, Erlangen, Germany

1 Introduction

Wall-bounded turbulent flows as those occurring in transportation (*e.g.* aviation) or industrial applications (*e.g* turbomachinery), are usually subjected to pressure gradients (PGs). The presence of such PGs affects greatly the development and physics of the turbulent boundary layer (TBL), making it an open research area. An important phenomena associated with the presence of strong adverse PGs (APGs) as appearing in wings, is the separation of the boundary layer, which can lead to stall. Moreover, recent and upcoming environmental restrictions have made drag reduction crucial in aviation. With a high proportion of the drag associated to the TBLs developing over the wet surfaces of the aircraft, their study and understanding is of paramount importance.

Initial research on TBLs in wings was conducted experimentally in wind tunnels, such as the nominal study of the turbulent flow past a NACA 4412 wing profile [4]. In recent years, developments in high-performance computing (HPC) architectures and numerical methods have enabled the study of these complex flows through accurate numerical simulations. Some examples of these are the large-eddy simulations (LES) of the turbulent flow around wing profiles [6,7,11]. However, the chord-based Reynolds numbers (Re_c) and angles of attack (AoA) achieved so far are still moderate, and the computational domains small (particularly in the spanwise direction, a fact that can affect separation physics greatly).

One of the most prevalent features of TBLs under strong APGs is flow separation. Moreover, even in zero pressure-gradient (ZPG) TBLs, instantaneous structures of reversed flow (referred to interchangeably as backflow events or

C. Marchioli et al. (Eds.): DLES 2023, ERCO 31, pp. 17–22, 2024.
https://doi.org/10.1007/978-3-031-47028-8_3

features in short) have been reported. Previous work has characterised backflow events for a variety of flow cases [3] and pressure-gradient conditions (up to moderate PGs) and found that both the length ($20l^+$) and life time ($2t^+$) of these events scales in inner viscous units. Instead, merely the frequency of these events was found to increase with increasing pressure-gradient conditions [10]. With diminishing mean wall-shear stress, when approaching mean separation, it is, however, known that the inner scaling will eventually not be valid. Hence, a structural change in the size and life time of these events is anticipated when approaching strong pressure-gradient conditions and mean-flow separation. This also explains the larger sizes and life times observed in stronger APG TBLs [2].

The current project leverages both the increase in availability of HPC resources, as well as the improvements in simulation methods (such as the introduction of adaptive mesh refinement and non-conformal meshing into the spectral-element method code Nek5000), in order to increase the achievable Reynolds number and angle of attack, while using a sufficiently large computational domain. Moreover, Nek5000 is coupled to the visualization software Catalyst, allowing for the in-situ analysis of backflow events and the turbulent structures associated with them. The high-fidelity simulation data is then used to study the characteristics of such events, and their relation with APGs and flow separation.

2 Methodology

Well-resolved (quasi-DNS) large-eddy simulations (using a relaxation-term filtering, analogous to the one described in Ref. [8]) are carried out using the massively-parallel spectral-element method code Nek5000 [5]. The case considered in this work is the turbulent flow past a NACA 4412 wing profile, as in Refs. [4,6,11]. The Reynolds number based on the chord and inflow velocity ($Re = cU_\infty/\nu$) is equal to 400,000, and the angle of attack 11°. The chosen angle of attack chosen is close to the maximum-lift condition, and the airfoil exhibits mean-flow separation near the trailing edge (starting at $x/c = 0.85$, where x is the chord-wise distance from the leading edge), as seen in the left subplot of Fig. 1. The introduction of adaptive mesh refinement (AMR) and non-conformal meshing allows to overcome some of the limitations faced in previous computational studies (such as in Ref. [11]). Firstly, the domain size (L_x; L_y; L_z) is considerably larger: (50c; 40c; 0.6c) in the AMR case and (6c; 4c; 0.2c) in previous conformal cases. Secondly, the boundary conditions in conformal cases have to be obtained from a precursor Reynolds-averaged Navier-Stokes (RANS) simulation due to the proximity of the domain boundaries to the wing. As this is not an issue in the AMR case, a uniform Dirichlet boundary condition is imposed in the inlet, top and bottom boundaries. The larger extent in the spanwise dimension is critical in the case presented here, as the separated region exhibits structures with large spanwise correlation lengths. Moreover, having a large distance to the boundaries is important, as possible interferences with the wake are avoided. Another advantage of using AMR is that the elements

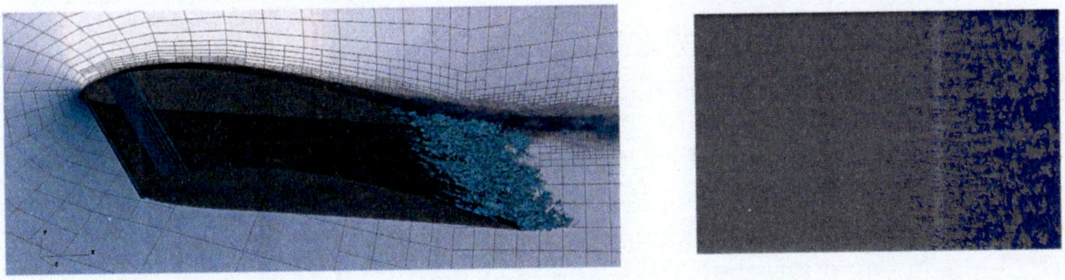

Fig. 1. Instantaneous contours of negative streamwise velocity, with the spectral element mesh in the background (left), and instantaneous separated spots at the wall, in the wall-aligned reference frame (right)

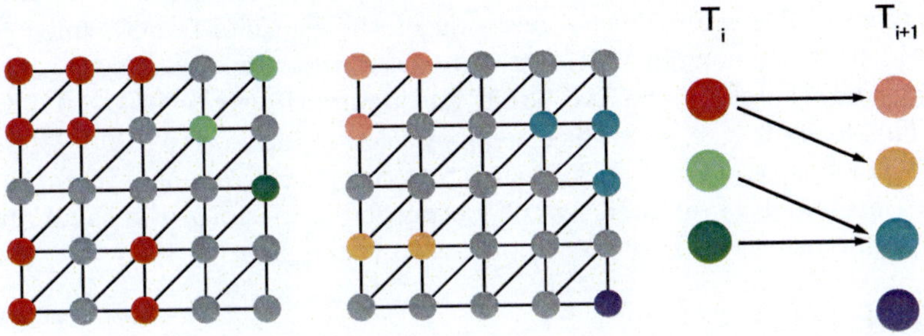

Fig. 2. Tracking of backflow structures at the wall nodes. Identified structures at an initial (left) and successive times (center) appear colored, while the rest of nodes are grey. The tracking result is shown in the rightmost sketch (Color figure online)

are concentrated in areas in which a higher resolution is needed (based on a spectral error indicator as described in Ref. [9]), allowing to reduce the overall computational cost of the simulation.

In-situ tracking of backflow events is carried out using the Catalyst adaptor for Nek5000 [1]. The use of in-situ data analysis is paramount in this work: the high level of refinement around the wing surface and the TBL, together with the time resolution needed to track backflow events makes the traditional offline post-processing methods unfeasible. In this work, the focus is on separated regions at the wall. In order to identify those negative skin-friction ($\tau_w = \mu du/dy$) regions, the wall-normal derivative of the streamwise velocity component is evaluated in Nek5000 (derivatives are computed and then a tensor rotation is performed, obtaining a local wall-aligned reference frame, as shown in the right panel of Fig. 1). The value of the skin friction (together with the coordinates of the wall nodes) are then used in order to identify the separated structures.

In order to find connected nodes forming single backflow structures, a triangulated mesh is built from the wall nodes. This mesh, as in the simulation, is periodic in the spanwise direction. Nodes with negative scalar values (*i.e.* $\tau_w < 0$) are considered feature nodes. A Breadth-first search (BFS) algorithm is performed on these feature nodes to identify connected feature regions, each of which is given a unique ID (as illustrated in the left panel of Fig. 2). The same processes is applied to the data in the next time step (middle panel of Fig. 2). Subsequently, a tracking arc is created between two feature regions if they overlap in consecutive time steps (right panel of Fig. 2). This approach allows for the identification and tracking of complex-shape structures in-situ. Moreover, thanks to the built-in capabilities of VTK, the area (size) and centroid of the features is easily computed, which, together with the tracking of features across different time steps, allows us to get a complete picture on how reversed flow appears and evolves in time and space. The in-situ tracking described in the previous section is applied every 4×10^{-4} flow-over times (*i.e.* c/U_{∞}, less than 0.5 inner-scaled time units t^{+}). The main data processing of the features is performed in-situ (feature identification, area and centroid calculation and tracking to previous time step) and, therefore, the individual grid points (in excess of 1×10^{9}) are not stored. This leads to the total disk memory used being less than 200 MB. This is a stark contrast to the 44 GB needed to store in disk a single instantaneous flow-field (around 1000 of them would be needed to perform the same analysis offline), highlighting the importance of in-situ processing.

3 Results

Close to 3 million individual areas of negative wall-shear stress are identified in total in the suction side of the wing (with the area of interest starting at $x/c = 0.25$ and continuing up to the trailing edge, in order to avoid having influence of the TBL tripping used at $x/c = 0.1$). The distribution of size and position of all identified features is shown as histograms in Fig. 3. Already in the area close to the leading edge a large number of backflow events are identified. Their number increases rapidly as the flow moves towards the trailing edge. In fact, the growth in number of backflow events is exponential after the point of maximum thickness of the airfoil at $x/c = 0.4$. This is closely related to the growing adverse pressure gradient seen in this region, as indicated by the growth of the Rotta-Clauser pressure-gradient parameter β. Nevertheless, the increase in the number of individual features does not continue all the way to the trailing edge, not even up to the mean-flow separation point: there is a region in between $x/c = 0.67$ and 0.77 in which a maximum is reached. Afterwards, their frequency decreases sharply (exponentially as well, but with a larger slope than in the previous build-up phase). Therefore, we can identify two clearly distinct regions from this analysis: A first part (from the leading edge up to around $x/c = 0.7$) in which the flow is attached but individual backflow events become more prevalent (*i.e.* their frequency increases) driven by the increasing APG, followed by a rapid transition to a second region in which the flow detaches and the number of individual backflow events decreases sharply (only a few, but large regions of separated flow fill most of the wing span). This behaviour can be also observed in Fig. 1, in which an instantaneous snapshot of the backflow events is shown.

Feature tracking allows to study this transition in more detail. In order to do that, newly-appearing backflow events (those not related to any other event in a

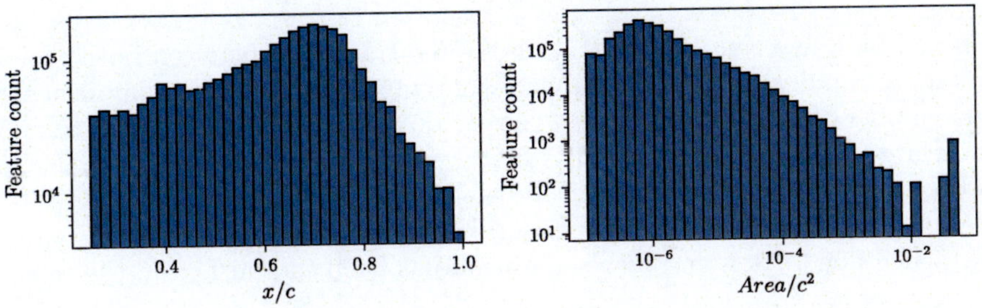

Fig. 3. Histograms of the position (left) and area (right) of all the individual features identified

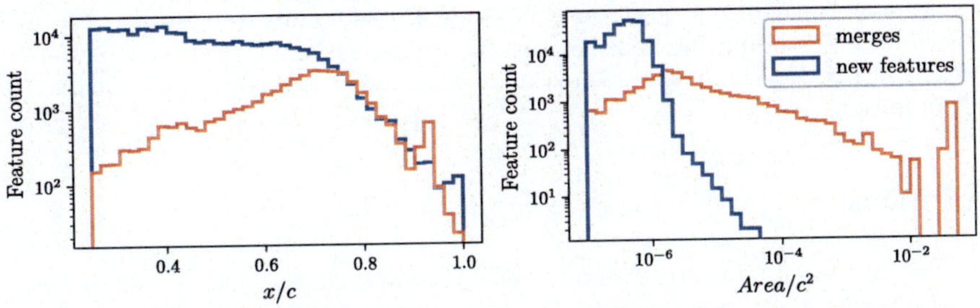

Fig. 4. Histograms of the position (left) and area (right) of all newly-appeared features and those formed after the merge of 2 or more features

previous time step) and merged events (those formed from two or more features in the previous time step) are identified and their distributions in position and size studied (shown in Fig. 4). It is clear that individual events spawn in the region closer to the leading edge, and that they are less likely to appear as one moves towards the separated flow region. The newly-spawned events are all small in size, as seen in the right plot of Fig. 4. As the flow moves downstream, merging activity increases, and with it, the size of the backflow regions. This continues until just before the point where mean-flow separation occurs, in which merging activity reaches its maximum. Later on, merging activity is also reduced, as the flow becomes completely dominated by a few massive structures.

4 Conclusions

In this work, the relationship between backflow events and mean-flow separation on a NACA 4412 wing profile has been studied by means of performing an in-situ tracking analysis of backflow events. The use of AMR in Nek5000 allowed to carry out a very well-resolved LES in which even the smallest backflow events appearing at the wall were captured and tracked. Two clearly differentiated regions were found: A first region in which small individual backflow events spawn (with their frequency increasing as the APG increases), and a second one, starting just before the point of mean flow separation, in which the flow is dominated by a few, very large structures. The transition between both regions occurs quite rapidly and is driven by the merging of individual backflow events. As the number of mergings increases, the regions of instantaneous separated flow become larger, leading to more individual events being merged into larger ones. This takes place consecutively until a few (very large) structures dominate the whole span. Therefore, a clear connection between individual backflow events in the attached flow region and flow separation has been established in the present work.

Acknowledgements. The simulations were performed on resources provided by both the Swedish National Infrastructure for Computing (SNIC) at the PDC Center for High Performance Computing, in KTH (Stockholm), and by the European High-Performance Computing Joint Undertaking (EuroHPC JU) project EHPC-REG-2021R0088 in LUMI (Finland). This research is funded by the Knut and Alice Wallenberg Foundation.

References

1. Atzori, M., et al.: In situ visualization of large-scale turbulence simulations in Nek5000 with ParaView catalyst. J. Supercomput. **78**(3), 3605–3620 (2022)
2. Bross, M., Fuchs, T., Kähler, C.J.: Interaction of coherent flow structures in adverse pressure gradient turbulent boundary layers. J. Fluid Mech. **873**, 287–321 (2019)
3. Chin, R.C., et al.: Backflow events under the effect of secondary flow of Prandtl's first kind. Phys. Rev. Fluids **5**(7), 074606 (2020)
4. Coles, D., Wadcock, A.J.: Flying-hot-wire study of flow past an NACA 4412 airfoil at maximum lift. AIAA J. **17**(4), 321–329 (1979)
5. Fischer, P., Kruse, J., Mullen, J., Tufo, H., Lottes, J., Kerkemeier, S.: Nek5000: Open source spectral element CFD solver. Argonne National Laboratory, Mathematics and Computer Science Division, Argonne, IL (2008)
6. Frère, A., Hillewaert, K., Chatelain, P., Winckelmans, G.: High Reynolds number airfoil: from wall-resolved to wall-modeled LES. Flow Turbul. Combust. **101**(2), 457–476 (2018)
7. Sato, M., Asada, K., Nonomura, T., Kawai, S., Fujii, K.: High Reynolds number airfoil: Large-eddy simulation of NACA 0015 airfoil flow at Reynolds number of 1.6×10^6, *AIAA J.* **55 (2)**, 673–679 (2017)
8. Schlatter, P., Stolz, S., Kleiser, L.: LES of transitional flows using the approximate deconvolution model. Int. J. Heat Fluid Flow **25**(3), 549–558 (2004)
9. Tanarro, A., Mallor, F., Offermans, N., Peplinski, A., Vinuesa, R., Schlatter, P.: Enabling adaptive mesh refinement for spectral-element simulations of turbulence around wing sections. Flow Turbul. Combust. **105**, 415–436 (2020)
10. Vinuesa, R., Örlü, R., Schlatter, P.: Characterisation of backflow events over a wing section. J. Turbulence **18**, 170–185 (2017)
11. Vinuesa, R., Negi, P.S., Atzori, M., Hanifi, A., Henningson, D.S., Schlatter, P.: Turbulent boundary layers around wing sections up to Re_c= 1,000,000. Int. J. Heat Fluid Flow **72**, 86–99 (2018)

Large-Eddy Simulations of the Accelerating Flow Around a Square Cylinder

A. Mariotti[1(✉)], S. Brusco[2], G. Lunghi[1], G. Piccardo[2], and M. V. Salvetti[1]

[1] Dipartimento di Ingegneria Civile e Industriale, University of Pisa, Pisa, Italy
{alessandro.mariotti,maria.vittoria.salvetti}@unipi.it,
gianmarco.lunghi@phd.unipi.it

[2] Dipartimento di Ingegneria Civile, Chimica e Ambientale, Polythechnic School, University of Genova, Genova, Italy
stefano.brusco@edu.unige.it, giuseppe.piccardo@unige.it

1 Introduction

The present work aims at investigating the high-Reynolds accelerating flow around a cylinder with a square cross-section. This is a classic shape of interest for wind-engineering applications, typical, e.g., of skyscrapers and towers. Despite the simple geometry, the flow features around square cylinders or prisms show a significant complexity. In particular, the flow is characterized by the shear-layer separation at the upstream edges. The separated shear layers undergo to Kelvin-Helmholtz instability, but they do not lose coherence until they form the von Karman vortex street in the wake. In terms of mean flow features, no reattachment is found on the lateral surface of the cylinder. Square cylinders with constant-velocity inflow conditions have been widely investigated in the literature, and no significant changes in flow topology and dynamics are found within the range $Re \simeq 10^3$–10^6 (e.g., [1,2,6,10,11,15–18,20]). However, in the last twenty years, the accelerating and decelerating flow conditions, typical, e.g., of thunderstorm outflows, have received significant attention since they may lead to serious damages of civil buildings and to economical losses. When high accelerations are present, transient effects become important (see, e.g., [4]) in the flow dynamics and in the consequent mean and fluctuating aerodynamic forces. Also the shear-layer dynamics and the vortex-shedding frequency are sensitive to the variation of the inflow velocity. Usually, the effect of wind acceleration has been experimentally and numerically investigated starting at rest [5,21]. However, typical conditions present flow accelerations/decelerations starting from a non-zero velocity. The latter case has been studied in the framework of the ERC THUNDERR project [19] by performing experiments at the multiple-fan wind tunnel of the Tamkang University in Taipei [3]. In particular, the effect of gaussian-type inflow accelerations/decelerations on a square cylinder in the range $Re = 1.72 \times 10^4$–6.34×10^4 have been analyzed. The main outcome was the measuring of constant-frequency time cells in the vortex-shedding.

C. Marchioli et al. (Eds.): DLES 2023, ERCO 31, pp. 23–28, 2024.
https://doi.org/10.1007/978-3-031-47028-8_4

In the present research activity, a Large-Eddy Simulation is carried out to study the effect of inflow accelerations on the flow dynamics around a square cylinder and on the vortex-shedding frequency. The same inflow acceleration and Reynolds-number range as in [3] are considered for cross-validation. This simulation may be considered a first step towards the characterization of the dynamic loads on civil buildings due to thunderstorm outflows, and to better understand the limits of the predictions for structure design obtained under the assumption of steady wind.

2 Problem Definition and Numerical Methodology

We consider the incompressible flow around a square cylinder at zero angles of attack. A Large-Eddy Simulation is performed by using the open-source code Nek5000, based on a high-order spectral element method. Each spectral element is rectangular or a suitable coordinate mapping of a rectangle. The basis functions inside the elements are Legendre polynomials of order N for velocity and $N-2$ for pressure in each direction; $N=6$ has been used in this work, as in [7,8,14]. A third-order backward finite-difference scheme based on the high-order splitting method is used for time advancing.

The computational domain is sketched in Fig. 1a, together with the cylinder cross section and the reference system in Fig. 1b. The cylinder center is located at $x/D = y/D = 0$, being D the width of the square cylinder, and the computational domain spans the following dimensions: $-75 \leq x/D \leq 125$, $-75 \leq y/D \leq 75$ and $0 \leq z/D \leq 5$. The spectral element size and distribution are the same used in [7,8,14]. In particular, the element size in the streamwise and lateral directions is $\Delta x/D = \Delta y/D = 0.125$ near the cylinder, while the element size is uniform in the spanwise direction, $\Delta z/D = 0.558$. A quadratic filter is applied to the three highest spectral modes. The effect of the filter can be considered as a subgrid scale dissipation (see, e.g., [8]).

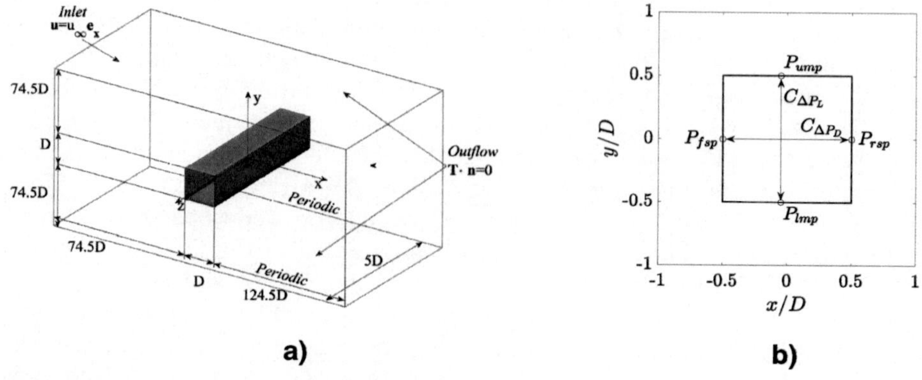

Fig. 1. a. Sketch of the computational domain and boundary conditions. b. Position of the pressure taps (cross section $z/D = 0$).

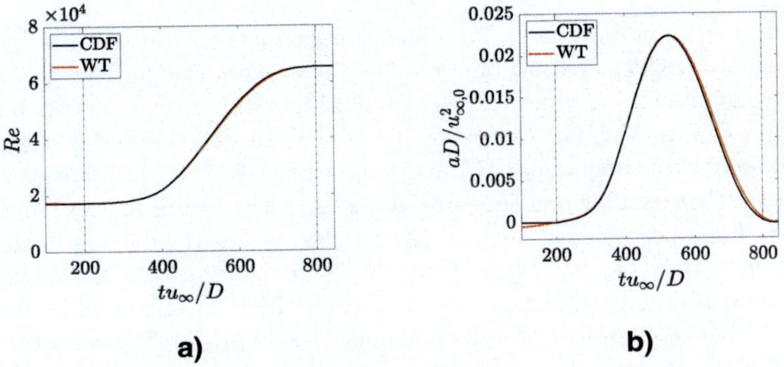

Fig. 2. a. Inflow Reynolds number (experiment from [3] in red). b. Inflow acceleration (experiment from [3] in red) (Color figure online).

A uniform unsteady velocity profile with no turbulence (smooth flow) is imposed at the inlet. The same unsteady flow conditions as in the experiments in [3] are considered for cross-comparison. The time histories of the inflow Reynolds number, *Re*, and of the acceleration made non-dimensional by using the inlet velocity at time $t = 0$, $aD/u^2_{\infty,0}$, are reported in Fig. 2a and Fig. 2b, respectively. Two dotted vertical lines at $tu_\infty/D = 200$ and $tu_\infty/D = 800$ delimit the acceleration region. The inflow Reynolds number varies from $Re = 1.720 \times 10^4$ to $Re = 6.536 \times 10^4$ and the maximum non-dimensional acceleration is equal to 2.35×10^{-2}. It should be noted that the acceleration can be made non-dimensional also considering the instantaneous velocity value, in this case the maximum value is $aD/u^2_\infty = 7.6 \times 10^{-3}$. Moreover, no-slip condition is imposed at the body surface, and traction-free boundary conditions for both the outflow and the upper and lower boundaries of the domain. Finally, periodicity is imposed in the spanwise direction.

3 Results and Discussion

We first compare the time-histories of differential-pressure coefficients in streamwise and crossflow directions, shown in Fig. 3a and Fig. 3b, respectively. Referring to Fig. 1b, both in experiments [3] and LES simulations they are defined as:

$$C_{\Delta P_D} = \frac{P_{fsp} - P_{rsp}}{1/2\rho u^2_\infty}, \qquad C_{\Delta P_L} = \frac{P_{ump} - P_{lmp}}{1/2\rho u^2_\infty} \tag{1}$$

being P_{fsp}, P_{rsp}, P_{ump}, P_{lmp} the instantaneous pressures evaluated on the body surface at the front stagnation point, the rear stagnation point, the upper-side midpoint, and the lower-side midpoint, respectively; ρ is the fluid density and u_∞ the instantaneous velocity at the inlet.

As discussed in [3], the two differential-pressure coefficients are a rough estimation of the drag and lift coefficients and they present the same frequency content. It is evident how the numerical simulation provides results in overall good agreement with the experiments in [3]. The streamwise differential-pressure coefficient has moving-average values slightly higher than the one of the experiments. The standard deviation of $C_{\Delta P_D}$ is significantly higher in simulations rather than in experiments [3]. Conversely, the crossflow differential-pressure coefficient present oscillations that are almost identical in numerical and experimental results. The oscillation cycles have time-periods decreasing in time with the increase of the velocity. Since these pressure fluctuations are related with the vortex-shedding dynamics, it is clear that the vortex-shedding frequency increases with the increase of inlet velocity. Its behavior is characterized by using a time-frequency analysis based on the continuous wavelet transform (for details, please refer to the comprehensive discussion in [9,13]). In this context wavelet are preferred to Fourier and Welch procedures (see, e.g., [12]) because they permit to obtain a clearer and mathematically-rigorous definition of the frequency peaks in time [13]. For our purpose, we used a complex Morlet wavelet with a central frequency of 6π and we derived from the wavelet energy maps the ridges indicating the vortex-shedding frequency, n^*, at each time window. The results of the wavelet ridges for $C_{\Delta P_L}$ are shown in Fig. 4a.The same discrete changes in the vortex-shedding frequency as in the experiments in [3] are found in the numerical simulations. Constant-frequency time cells are clearly visible in Fig. 4a. The related Strouhal number, $St = n^* D/u_\infty$, is reported in Fig. 4b. Since the vortex-shedding frequency is constant in time cells, but the instantaneous velocity monotonic increase, a decrease in the Strouhal number is

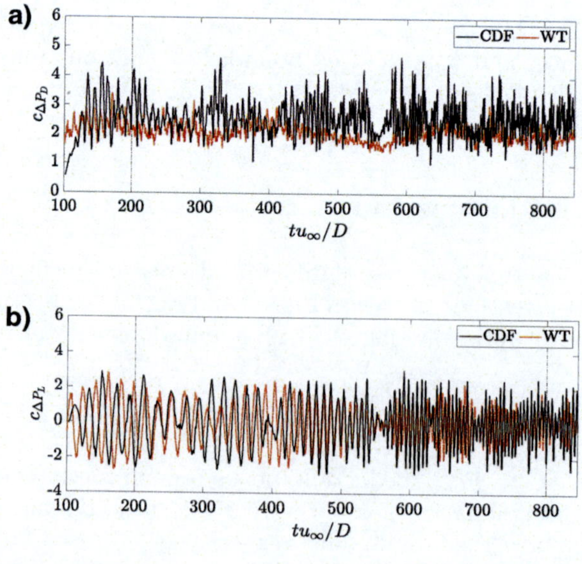

Fig. 3. a. Time history of the in-flow pressure coefficient (experiment from [3] in red). b. Time history of the cross-flow pressure coefficient (experiment from [3] in red) (Color figure online).

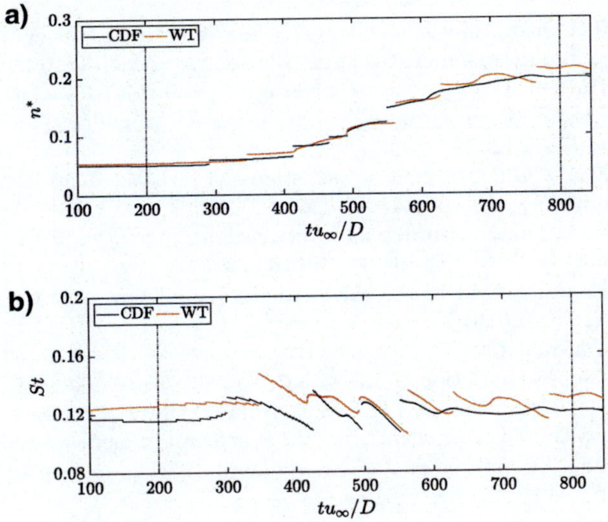

Fig. 4. a. Vortex-shedding frequency (experiment from [3] in red) b. Vortex-shedding Strouhal number (experiment from [3] in red) (Color figure online)

found within the time cells, followed by a sudden increase between the cells. The vortex-shedding Strouhal number is always inside the range $0.10 \leq St \leq 0.14$ and the agreement between LES and experiments ia again satisfactory.

4 Conclusions

The effect of the flow acceleration around a square cylinder has been investigated through a Large-Eddy Simulation and a good agreement is found with the experiments in [3], in terms of both the streamwise and the crossflow coefficient. Discrete increases of the vortex-shedding frequency occur with the inlet velocity, confirming the presence of the constant-frequency time cells. The results validate the simulation set-up and will be used in a future work for a systematic characterization of the effect of different values of the inflow acceleration/deceleration.

Acknowledgments. Stefano Brusco and Giuseppe Piccardo have been financially supported by the European Research Council under the European Union's Horizon 2020 research and innovation program (grant agreement No. 741273, Principal Investigator Professor Giovanni Solari) for the project THUNDERR - Detection, simulation, modelling and loading of thunderstorm outflows to design wind safer and cost-efficient structures - through an Advanced Grant 2016.

References

1. Bosh, G., Rodi, W.: Simulation of vortex shedding past a square cylinder with different turbulence models. Int. J. Num. Meth. Fluids **28**, 601–616 (1998)

2. Bruno, L., Oberto, D.: Effects of cell quality in grid boundary layer on the simulated flow around a square cylinder. Comput. Fluids **238**, 105351 (2022)
3. Brusco, S., Buresti, G., Lo, Y.L., Piccardo, G.: Constant-frequency time cells in the vortex-shedding from a square cylinder in accelerating flows. J. Wind Eng. Ind. Aerod. **230**, 105182 (2022)
4. Brusco, S., Solari, G.: Transient aeroelasticity of structures subjected to thunderstorm outflows. Eng. Struct. **245**, 112801 (2021)
5. Guo, F., Wu, G., Du, X.: Numerical investigation of flow around a square cylinder in accelerated flow. Phys. Fluids **33**, 104105 (2021)
6. Ke, J.: RANS and hybrid LES/RANS simulations of flow over a square cylinder. Adv. Aerod. **1**, 10 (2019)
7. Lunghi, G., Pasqualetto, E., Rocchio, B., Mariotti, A., Salvetti, M.V.: Impact of the lateral mean recirculation characteristics on the near-wake and bulk quantities of the BARC configuration. Wind Struct. **34**(1), 115–125 (2022)
8. Mariotti, A., Siconolfi, L., Salvetti, M.V.: Stochastic sensitivity analysis of large-eddy simulation predictions of the flow around a 5:1 rectangular cylinder. Eur. J. Mech. B/Fluids **62**, 149–165 (2017)
9. Mariotti, A.: Axisymmetric bodies with fixed and free separation: base-pressure and near-wake fluctuations. J. Wind Eng. Ind. Aerod. **176**, 21–31 (2018)
10. Norberg, C.: Flow around rectangular cylinders: pressure forces and wake frequencies. J. Wind Eng. Ind. Aerod. **49**, 187–196 (1993)
11. Oka, S., Ishihara, T.: Numerical study of aerodynamic characteristics of a square prism in a uniform flow. J. Wind Eng. Ind. Aerod. **97**, 548–559 (2009)
12. Pasqualetto, E., Lunghi, G., Rocchio, B., Mariotti, A., Salvetti, M.V.: Experimental characterization of the lateral and near-wake flow for the BARC configuration. Wind Struct. **34**(1), 101–113 (2022)
13. Perna, R., et al.: Flow characterization of pulsating heat pipe through the wavelet analysis of pressure signals. Appl. Therm. Eng. **171**, 115128 (2020)
14. Rocchio, B., Mariotti, A., Salvetti, M.V.: Flow around a 5:1 rectangular cylinder: effects of upstream-edge rounding. J. Wind Eng. Ind. Aerod. **204**, 104237 (2020)
15. Rodi, W.: Comparison of LES and RANS calculations of the flow around bluff bodies. J. Wind Eng. Ind. Aerod. **69–71**, 55–75 (1997)
16. Rodi, W., Mansour, N.: Low Reynolds number k-ϵ modelling with the aid of direct simulation data. J. Fluid Mech. **250**, 509–529 (1993)
17. Sohankar, A.: Flow over a bluff body from moderate to high Reynolds numbers using large eddy simulations. Comput. Fluids **35**, 1154–1168 (2006)
18. Sohankar, A., Davidson, L., Norberg, C.: Large Eddy simulation of flow past a square cylinder: comparison of different subgrid scale models. J. Fluids Eng. **122**, 39–47 (2000)
19. Solari, G.: Thunderstorm downbursts and wind loading of structures: progress and prospect. Front. Built Environ. **6**, 63 (2020)
20. Trias, F.X., Gorobets, A., Oliva, A.: Turbulent flow around a square cylinder at Reynolds number 22,000: a DNS study. Comput. Fluids **123**, 87–98 (2015)
21. Yang, T., Mason, M.S.: Aerodynamic characteristics of rectangular cylinders in steady and accelerating wind flow. J. Fluids Struct. **90**, 246–262 (2019)

DNS and POD Analysis of Separated Flow in a Three-Dimensional Diffuser

A. Miró[1](✉), B. Eiximeno[1,2], I. Rodríguez[2], and O. Lehmkuhl[1]

[1] Large Scale Computational Fluid Dynamics group, Barcelona Supercomputing Center, Barcelona, Spain
{arnau.mirojane,benet.eiximeno,oriol.lehmkuhl}@bsc.es

[2] TUAREG - Turbulence and Aerodynamics Research Group, Universitat Politècnica de Catalunya, Barcelona, Spain
{benet.eiximeno,ivette.rodriguez}@upc.edu

Abstract. A DNS of a three dimensional diffuser has been performed with the in-house solver Alya. The resulting dataset presents a good agreement with existing numerical and experimental data, even without smoothing of the sharp corners. Discrepancies in the duct area have been found to be due to uncertainties, however, in the diffuser region differences are more noticeably between experimental and numerical data, where they all exhibit an oscillating behavior. The POD analysis of this flow has quantified a slow mixing mechanism inside the diffuser with a period of 4 flow-through times. This is due to the slow motions in the diffuser that are only represented 3 times over the 21 flow-through times in which statistics are gathered. This is considered to be the main reason of discrepancy between all datasets. While DNS of turbulent flow at $Re = 10,000$ are feasible nowadays, temporal dynamics of the Standford diffuser present a challenge as the integration time needed becomes larger. The current dataset is, nevertheless, valid for further analysis and activities and will be made available online.

Keywords: DNS · POD · Separation · High-Fidelity Dataset

1 Introduction

The three-dimensional (Stanford) diffuser is a well documented case with complex internal corner flow and 3D separation while having a relatively simple geometry.

The physics of this flow was first analyzed by Cherry *et al.* [1] through an experimental setup, while the sharp corners of the diffuser were smoothed with a fillet radius of 6.0h. Such dataset provided detailed experimental data comprising the three components of the mean velocity field, the streamwise Reynolds stresses and the pressure distribution along the bottom wall of the diffuser. This provided information on the mean flow configuration, however, little insight was given on the more complex time-motions of this flow.

C. Marchioli et al. (Eds.): DLES 2023, ERCO 31, pp. 29–34, 2024.
https://doi.org/10.1007/978-3-031-47028-8_5

The first direct numerical simulation (DNS) was performed by Ohlsson *et al.* [6] at a Reynolds number of $Re = U_b h/\nu = 10,000$ based on bulk velocity U_b and channel height h. The computational domain was designed to have a close agreement with the diffuser geometry, with the corners resulting from the diffuser expansion smoothly rounded with a radius of 6.0. Malm *et al.* [4] conducted a Proper Orthogonal Decomposition (POD) analysis of the flow. This POD analysis contained 196 snapshots spanning a time interval of $\Delta t U_b/h = 392$, thus with a temporal resolution of $\Delta t U_b/h = 2$. It was found that 195 modes are necessary to cover about 72% of the total u_{rms} peaks, owing to a large-scale oscillation on the later part of the diffuser. These oscillations were related to low frequencies that resulted to large time periods.

The present work aims to enhance the current database of flows in the Stanford diffuser configuration by providing another DNS along with a global assessment of the flow, and to provide recommendations for further simulation campaigns on this configuration.

2 Mathematical and Numerical Model

The present DNS data have been obtained by solving the incompressible Navier-Stokes equations with the in-house solver Alya. Alya is a parallel multi-physics/multi-scale finite-element simulation code developed to run efficiently on high-performance computing environments. The general code structure is described in Vazquez *et al.* [7]. Details about the numerical schemes for the incompressible flow solver are described in Lehmkuhl *et al.* [3].

Regarding the diffuser geometry, the following configuration is considered: the upper-wall expansion angle is 11.3° and the side-wall expansion angle is 2.56°. The flow in the inlet duct (height $h = 1$, width $B = 3.33$) corresponds to fully-developed turbulent rectangular duct flow. The origin of coordinates is set at the entrance of the diffuser. The $L = 15h$ long diffuser section is followed by a straight outlet part ($12.5h$ long). Downstream of this, the flow goes through a $10h$ long contraction followed by a $5h$ straight duct in order to minimize the effect of the outlet to the diffuser. A difference from previous works [1,6] is that the geometry considered does keep the sharp angles on the walls transitioning between diffuser and the straight duct parts. The computational domain also includes a long inlet duct of $65h$ length in order to allow the flow in the inlet duct to fully develop. Before this, there is a section of $5h$ length with a small chevron placed $2h$ from the inlet in order to trigger the turbulent transition in the rectangular duct. This method is preferred over using a precursor calculation of rectangular duct flow with streamwise periodicity conditions. At the outlet ($x = 47.5h$) standard Dirichlet condition for the pressure is prescribed. An inflow Reynolds number 10,000, matching with the previous DNS and experimental data, is considered. The walls of the duct and the diffuser are set to no slip.

The computational grid resulted in about 250 million elements. With a stretched grid, the maximum grid resolution in the duct centre is $\Delta z^+ = 11.6$, $\Delta y^+ = 13.2$, $\Delta x^+ = 19.5$. At the wall, the resolution is $z^+ = 0.074$, $y^+ = 0.37$ in the spanwise and normal directions, respectively. A measure of mesh resolution can be obtained by comparing the grid resolution Δ with the Kolmogorov length scale η as $\Delta/\eta = \left(1/n_{grid} \int_V 1/\eta^3 dV\right)^{1/3}$, where n_{grid} is the number of grid points. For the present dataset, $\Delta/\eta \approx 2.7$ is obtained. Since $\Delta/\eta < 5$, it can be stated that the resolution achieved by the present grid is at DNS level.

For the temporal integration, a third order explicit Runge Kutta scheme was used with a dynamic time stepping that ensured a CFL below 0.9. The flow was computed for $tU_b/L = 13$ flow-through times based on the diffuser length before gathering statistics for $tU_b/L = 21$ flow-through times. This setup was deemed sufficient to compute the flow in the diffuser and is based on the previous work of Ohlsson *et al.* [6].

3 Validation and Results

3.1 Comparison with Experimental and DNS Data

The results of the present simulation campaign are compared with the DNS [6] and experimental [1] results. Figure 1 shows the mean streamwise velocity and its RMS values along characteristic lines of the diffuser compared to reference data. Moreover, Fig. 2 provides contours of the streamwise mean velocity and the RMS fluctuations at selected cross sections of the diffuser respectively.

Good agreement between the present DNS and the DNS by Ohlsson *et al.* [6] is generally observed. In terms of streamwise velocity, the current DNS agrees very well with the previously reported data. Deviations from the experiment are probably due to measurement uncertainties in the test rig. In terms of contour planes, the streamwise velocity is also very similar among the three datasets, with little discrepancies found in the near-wall regions. In terms of streamwise fluctuations, a good match is obtained in the duct area in both profile- and contour-wise. Nevertheless, in the diffuser region deviations of both DNS from experiments by Cherry *et al.* [1] are observed. Also, the contours of the streamwise velocity fluctuations are somewhat different between both DNS. In this case, the experimental data has considerable uncertainties due to the measuring technique. Nonetheless, it is observed that the present results compare better with the experimental ones.

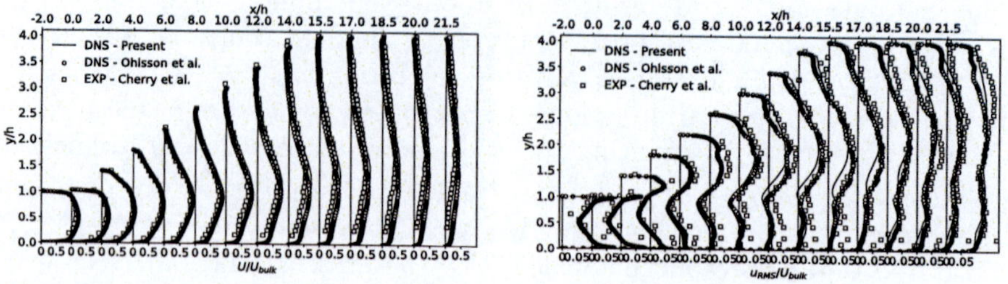

Fig. 1. Streamwise velocity (left) and average streamwise velocity fluctuations (right), present DNS vs Cherry *et al.* [1] and Ohlsson *et al.* [6].

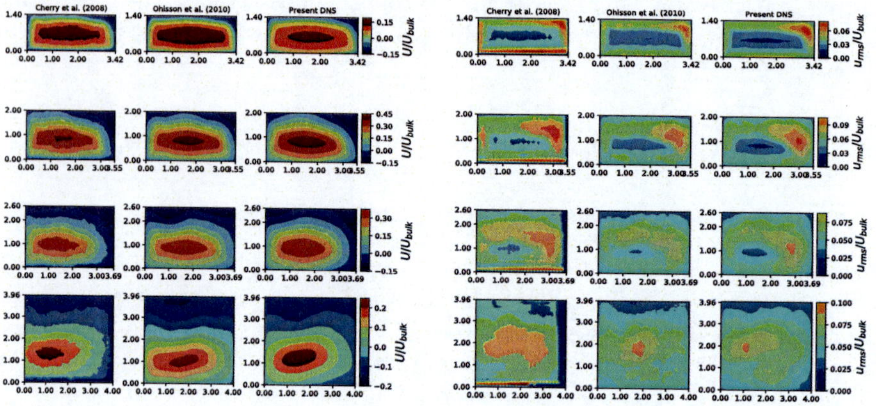

Fig. 2. Average streamwise velocity contours (left) and Average streamwise velocity fluctuation contours (right), present DNS vs Cherry *et al.* [1] and Ohlsson *et al.* [6]. The positions of the contour slices are, from top to bottom, x/h = 2, 5, 8 and 15.

3.2 POD Analysis

The POD analysis consisted of 626 snapshots over a time interval of $\Delta t U_b/h = 808$, leading to a temporal resolution of $\Delta t U_b/h = 1.29$. Hence, the present POD has 1.55 times more temporal resolution and spans about 2 times more than the work of Malm *et al.* [4]. The computation of the POD modes, temporal spectra and POD energy has been performed through an in-house parallel toolbox as described in Eiximeno *et al.* [2].

Information on the flow structures is contained on the first 200 POD modes, containing 60% of the total energy, while the rest could be considered small scale motions. Illustratively, Fig. 3 shows the contours of one of the highest energy modes (the 5th mode) in equispaced planes through the diffuser. Regarding the inlet duct, POD modes have yielded structures that are very similar to these of turbulent channel flows (see, for example, Nikolaidis *et al.* [5]). In the diffuser region, POD modes have shown the evidence of a large scale structure that is responsible of the mixing at the back region. This can be seen as the interaction of positive (orange) and negative (blue) motions of the mode, where

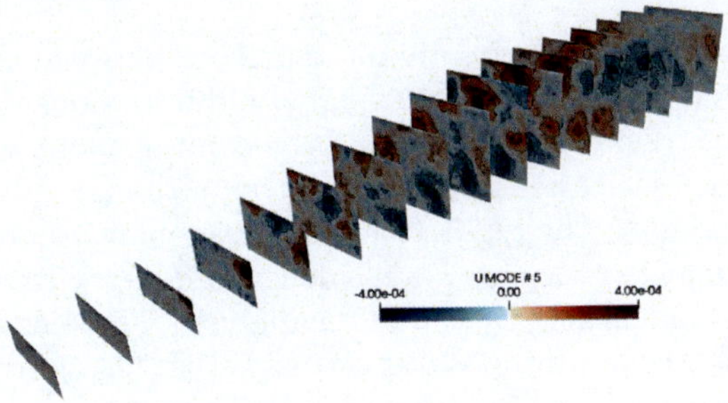

Fig. 3. Colour contours of the 5th POD mode. Planes are equispaced at $\Delta x/h = 2$ starting at $x/h = -3$ up to $x/h = 29$.

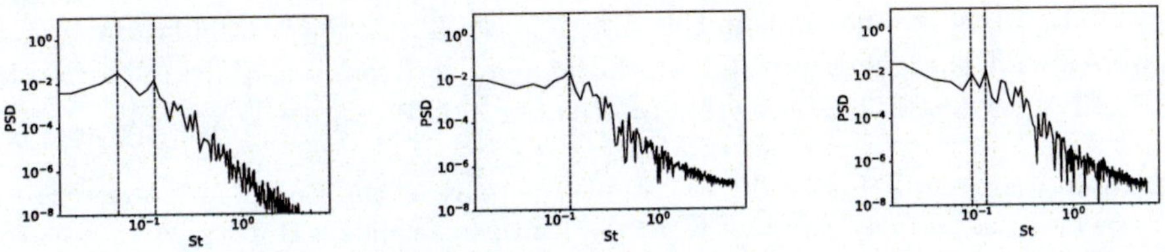

Fig. 4. POD temporal coefficient in frequency domain for mode 1 (left), 2 (center) and 3 (right). Dashed vertical lines denote the low frequencies of the mode.

they interact and twist exchanging energy. This is the consequence of the mixing of the corner vortices coming from the inlet duct that expand and detach due to the asymmetry of the diffuser. This creates a laminar recirculation bubble that is associated with slow motions inside the diffuser. Then, the resulting structures finally interact and merge at the beginning of the recovery section at $x/h = 15$.

Associated to these motions, low frequencies appear in the temporal POD coefficients as shown in Fig. 4, corresponding to the slow dynamics occurring at the back of the diffuser. Particularly, these frequencies correspond to periods of $tU_b/L \approx 7$, hence in the present 21 flow-through times are only represented 3 times. This might be the main reason of discrepancies between both DNS and the experimental data discussed in Sect. 3.1. This slow motion at the end of the diffuser section was also reported in the POD of Malm *et al.* [4]. This poses a serious question on the number of flow-through times necessary to correctly resolve this flow.

4 Conclusions

DNS of a separated flow in a three-dimensional diffuser have been performed with the in-house CFD code Alya. In the inlet duct area, good agreement is obtained between both DNS and the experiment. The spanwise velocity matches for the three datasets. Some small differences can be observed in terms of the spanwise fluctuations, mostly due to experimental and numerical uncertainties.

In the diffuser region, discrepancies are more evident between the three datasets for both the spanwise velocity and the spanwise fluctuations. These differences are seen in both the streamwise profiles and the contour plots, where the present data compares better with the experimental one, although all datasets present very oscillatory behaviors. The reason for these discrepancies are not only due to uncertainties, but to a slow mechanism inside the diffuser with a period of about 7 flow-through times, identified by the POD analysis. These flow features pose a challenge in the convergence of the statistics inside the (Stanford) diffuser and are found to be the reason of the main differences at that region between all datasets. While a DNS at $Re = 10,000$ is feasible nowadays, separated flow present more complex features that also need to be resolved. Subsequent simulation campaigns should increase the integration time at least 10 times more to be able to obtain reliable statistics at the end of the diffuser. Still, this high fidelity dataset is of interest and the averaged and temporal data will be available online in the ERCOFTAC wiki (https://kbwiki.ercoftac.org/w/index.php/DNS_1-3).

Acknowledgements. This work was funded as part of the European Project HiFi-TURB which has received funding from the European Union's Horizon 2020 research and innovation programme under grant agreement No 814837.

References

1. Cherry, E.M., Elkins, C.J., Eaton, J.K.: Geometric sensitivity of three-dimensional separated flows. Int. J. Heat Fluid Flow **29**, 803–811 (2008)
2. Eiximeno, B., Miró, A., Cajas, J.C., Lehmkuhl, O., Rodriguez, I.: On the wake dynamics of an oscillating cylinder via proper orthogonal decomposition. Fluids **7**, 292 (2022). https://doi.org/10.3390/fluids7090292
3. Lehmkuhl, O., Houzeaux, G., Owen, H., Chrysokentis, G., Rodriguez, I.: A low-dissipation finite element scheme for scale resolving simulations of turbulent flows. J. Comput. Phys. **390**, 51–65 (2019)
4. Malm, J., Schlatter, P., Henningson, D.S.: Coherent structures and dominant frequencies in a turbulent three-dimensional diffuser. J. Fluid Mech. **699**, 320–351 (2017). https://doi.org/10.1017/jfm.2012.107
5. Nikolaidis, M.-A., Farrell, B.F., Ioannou, P.J., Gayme, D.F., Lozano-Durán, A., Jiménez, J.: A POD-based analysis of turbulence in the reduced nonlinear dynamics system. J. Phys. Conf. Ser. **708**, 012002 (2016). https://doi.org/10.1088/1742-6596/708/1/012002
6. Ohlsson, J., Schlatter, P., Fischer, P.F., Henningson, D.S.: Direct numerical simulation of separated flow in a three-dimensional diffuser. J. Fluid Mech. **650**, 307–318 (2010)
7. Vazquez, A.M., et al.: Alya: multiphysics engineering simulation towards exascale. J. Comput. Sci. **14**, 15–27 (2016)

Large-Eddy Simulations of the Flow Around Rectangular Cylinders of Different Chord-to-Depth Ratios: Impact of Upstream-Edge Sharpness

M. V. Salvetti, G. Lunghi, M. Morello, and A. Mariotti(✉)

Dipartimento di Ingegneria Civile e Industriale, University of Pisa, Pisa, Italy
{maria.vittoria.salvetti,alessandro.mariotti}@unipi.it,
{gianmarco.lunghi,mario.morello}@phd.unipi.it

1 Introduction

Square and rectangular cylinders can be considered simplified models of structures of interest in civil engineering, such as, e.g., tall buildings or bridge sections. The chord-to-depth ratio, B/D, has a strong impact on the flow dynamics around the cylinder. In all cases, the flow separates at the upstream edges. The separated shear layers roll up forming vortical structures (Kelvin-Helmholtz instability), which are convected downstream and eventually interact by pairing with other vortical structures formed from the shear layers themselves. For large chord-to-depth ratios, the mean flow reattaches on the lateral side of the cylinder, and, thus, it is characterized by a closed mean recirculation region. Further downstream, the shear layers separate again from the trailing edge, and in the near wake the flow is characterized by the classical Von Karman vortex shedding. On the contrary, shear-layer vortex pairing and mean flow reattachment do not occur for low chord-to-depth ratios as, e.g., for the square cylinder. Many studies have identified the range $2.5 \leq B/D \leq 3$ as the cut-off between separated and attached mean flows on the cylinder lateral surface [13,14,17,18].

As for the elongated bodies, the flow around the rectangular cylinder having a 5:1 chord-to-depth ratio has been the most studied since it is the object of the international benchmark BARC [4]. In this context, LES simulations are characterized by a significant dispersion in the flow topology and in the mean and fluctuating pressure distributions on the cylinder side [4]. The sensitivity to grid resolution and subgrid-scale modeling [3,9] is also significant. This has been observed also for 4:1 rectangular cylinders [1]. Rocchio et al. [11] investigated whether this sensitivity could be a spurious effect of the perfect sharpness of the upstream corners in numerical simulations, while in reality the corners have a small degree of roundness. In particular, the effects of the degree of roundness of the upstream edges (for $r/D \leq 0.05$) have been investigated in [11], showing that even small values of the curvature radius significantly impact the results providing the lengthening of the mean recirculation region on the cylinder side. These roundings, in turn, modify the near-wake and bulk quantities [8] and

C. Marchioli et al. (Eds.): DLES 2023, ERCO 31, pp. 35–40, 2024.
https://doi.org/10.1007/978-3-031-47028-8_6

improve the agreement with experiments [10]. Indeed, sharp edges introduce significant velocity fluctuations in the shear layer at separation that, when not artificially damped by numerical or SGS dissipation, cause an upstream roll-up of the shear layers and, hence, a short mean recirculation region. This effect seems to be a numerical issue related to the flow resolution typical of LES simulations since the impact of upstream-edge rounding was found to be much smaller in DNS simulations in [5] at a lower Reynolds number. On the other hand, among the bodies having low chord-to-depth ratios, the square cylinder has been widely investigated in the literature (e.g. LES by [12] and DNS by [15]). Even if some dispersion is still present among the results of the numerical simulations, the flow field around the square cylinder is less sensitive to grid quality [2], whereas the effect of small corner roundings has not been systematically addressed. A few experiments are limited to large values of the corner roundings that provide significant changes of the cross-sectional geometries ($r/D = 0.10 - 0.30$ in [7,16] for 1:1 and 2:1 cylinders).

Here we want to investigate whether a similar effect of small upstream-corner roundings as in [11] is present also in LES of rectangular cylinders of different aspect ratios. To this aim, we considered also a square cross-section and a rectangular one having an aspect ratio of 3:1, which are characterized by different flow topologies. In particular, for the square cylinder, the mean flow does not reattach on the cylinder side, while for the 3:1 case it is about to reattach along the lateral side of the cylinder. We compare the case $r/D = 0$ (sharp-upstream edges) with $r/D = 0.0037$ for 1:1, 3:1, and 5:1 cylinders to highlight whether in LES even a small value of the curvature radius significantly changes the flow features.

2 Numerical Set-Up and Methodology

We consider the incompressible flow around rectangular cylinders having chord-to-depth ratios equal to 1:1, 3:1, and 5:1 and zero angles of attack. The upstream edges are sharp ($r/D = 0$) or rounded with a curvature radius of $r/D = 0.0037$, D being the depth of the cylinder. The cylinder center is located at $x/D = y/D = 0$ and the computational domain spans the following dimensions: $-75 \leq x/D \leq 125$, $-75 \leq y/D \leq 75$ and $0 \leq z/D \leq 5$. A uniform velocity with no turbulence is imposed at the inlet, no slip at the body surface, and traction-free boundary conditions for both the outflow and the upper and lower boundaries of the domain. Finally, periodicity is imposed in the spanwise direction. The Reynolds number, based on the free-stream velocity and on the cylinder depth, is $Re = 4 \times 10^4$.

The numerical simulations are performed through Nek5000, an open-source code based on a high-order spectral element method. Each spectral element is rectangular or a suitable coordinate mapping of a rectangle. The basis functions inside the elements are Legendre polynomials of order N for velocity and $N - 2$ for pressure in each direction; $N = 6$ has been used in this work, as in [8,9,11]. A third-order backward finite-difference scheme based on the high-order splitting method is employed for time advancing. A quadratic filter is applied to the three highest spectral modes. The effect of the filter can be considered as a subgrid-scale dissipation (see, e.g., [9,11]). The spectral element size and distribution are the same used in [11]. In particular, the element size in the streamwise and lateral directions is $x/D = y/D = 0.125$ near the cylinder, while, in the spanwise direction, the element size is uniform $z/D = 0.558$ as done in [8,11]. For rounded edges, the computational grid is the same as in the sharp-edge case, except that the rounded part is handled by curvilinear elements.

3 Results and Discussion

The flow field of the instantaneous vortex indicator λ_2 [6] and the mean-flow streamlines are shown in Figs. 1, 2, and 3 for 1:1, 3:1, and 5:1 cylinders, respectively. We compare the results for sharp-edged cylinders ($r/D = 0$) with the ones for cylinders having a rounding equal to $r/D = 0.0037$. Both square cylinders exhibit flow separation from the upstream corners and the separated shear layers directly form the von Karman vortex street. This implies that the corresponding

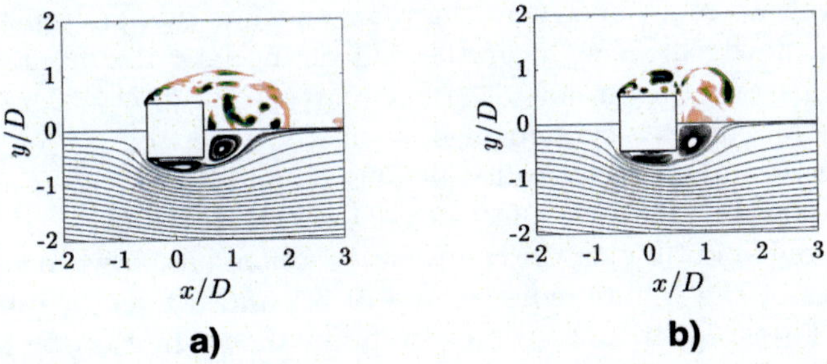

Fig. 1. a. Vortex indicator λ_2 (top) and mean flow streamlines (bottom) for the square cylinder with $r/D = 0$. b. Vortex indicator λ_2 (top) and mean flow streamlines (bottom) for the square cylinder with $r/D = 0.0037$.

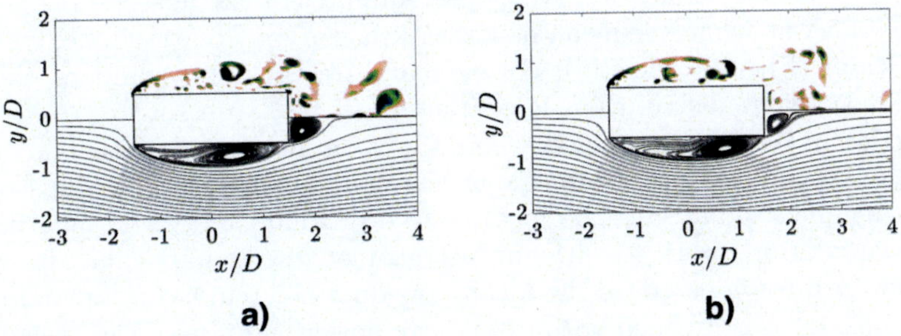

Fig. 2. a. Vortex indicator λ_2 (top) and mean flow streamlines (bottom) for the 3:1 cylinder with $r/D = 0$. b. Vortex indicator λ_2 (top) and mean flow streamlines (bottom) for the 3:1 cylinder with $r/D = 0.0037$.

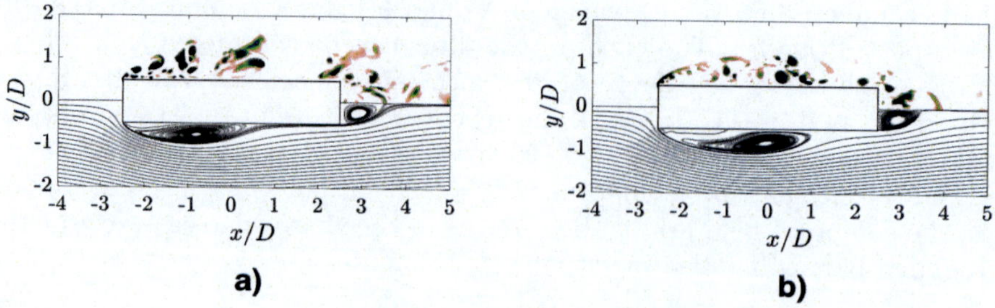

Fig. 3. a. Vortex indicator λ_2 (top) and mean flow streamlines (bottom) for the 5:1 cylinder with $r/D = 0$ [11]. b. Vortex indicator λ_2 (top) and mean flow streamlines (bottom) for the 5:1 cylinder with $r/D = 0.0037$ [11].

mean flow does not reattach on the cylinder's lateral side. For this cylinder shape, the degree of sharpness of the upstream edges has a negligible effect (compare Figs. 1a with 1b). This is not the case with the 3:1 geometry. Indeed, as can be seen by comparing Fig. 2a with Fig. 2b, a small curvature radius of the upstream edges changes the flow topology. For the cylinder with sharp edges, the separated shear layers roll up more upstream than the ones for rounded upstream edges, and the mean flow reattaches near the end of the lateral side at $x/D = 1.38$ (Fig. 2a), whereas the mean flow does not reattach for $r/D = 0.0037$ (Fig. 2b). Thus, the 3:1 rectangular cylinder has a sensitivity to r/D as for the 5:1 case already discussed in [11] and recalled in Figs. 3a and 3b.

Figures 4a and 4b show the effect of r/D on the distribution of the pressure coefficient averaged in time, in the spanwise direction and between the upper and the lower sides of the square cylinder and of 3:1 and 5:1 rectangular cylinders, respectively. The standard deviation of Cp are shown in Figs. 5a and 5b. The results of the 5:1 rectangular cylinder are extracted from [11]. The local abscissa s/D is evaluated on the cylinder surface from the front stagnation point to the rear one. For the square cylinder, the mean pressure distribution and the related standard deviation are not sensitive to the value of r/D (Figs. 4a and 5a). On the contrary, as previously observed, the roundness of the upstream edges plays an important role when the mean flow reattaches or may reattach, i.e. for the 5:1 and 3:1 cases (Figs. 4b and 5b). For elongated rectangular cylinders, the mean pressure coefficient distribution is characterized by a pressure recovery in the rear portion of the lateral surface, where the mean flow reattaches. The pressure fluctuations exhibit a maximum along the cylinder side, slightly upstream of the reattachment point [8]. The introduction of rounded corners lengthens the mean recirculations both for 3:1 and 5:1 geometries. In particular, for the 3:1 geometry with rounded edges the mean flow does not reattaches anymore along the lateral side, as confirmed both from mean flow streamlines (Fig. 2b) and from the increase of the maximum value of the pressure-coefficient fluctuations from $\sigma(C_p) = 0.15$ to roughly $\sigma(C_p) = 0.35$ (Fig. 5b).

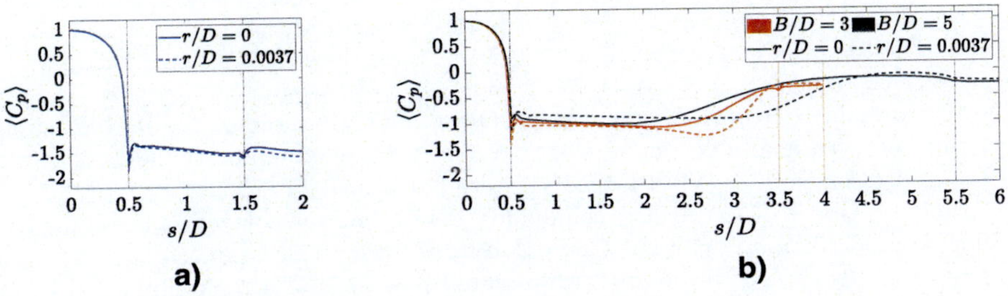

Fig. 4. a. Mean pressure coefficient distributions for the square cylinders. b. Mean pressure coefficient distribution for 3:1 and 5:1 rectangular cylinders. The results for the 5:1 cylinders from [11] are reported in black.

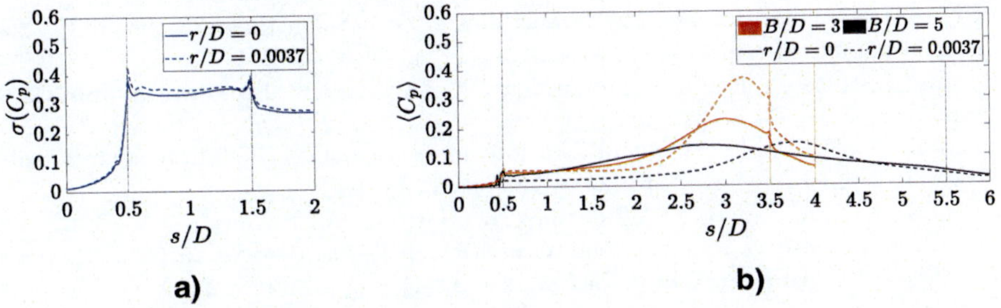

Fig. 5. a. Standard deviation of the pressure coefficient for the square cylinders. b. Standard deviation pressure coefficient for 3:1 and 5:1 rectangular cylinders. The results for the 5:1 cylinders from [11] are reported in black.

For the 3:1 geometry it is found that sharp edges introduce significant velocity fluctuations in the shear layer at separation, which cause an upstream roll-up of the shear layers and a short mean recirculation region, confirming the findings in [11] for the 5:1 rectangular cylinder. For the square cylinder, a similar increase of the velocity fluctuations is found for sharp edges compared with rounded ones. However, this has not a significant impact of the lateral and near-wake flow field.

4 Conclusions

For the square cross-section, characterized by a separated mean flow, a negligible effect of the upstream-corner sharpness is found. As found in [11] for the 5:1 rectangular cylinder, upstream-corner sharpness is important for the 3:1 rectangular cylinder, for which the mean flow is about to reattach along the lateral side of the cylinder. Even a small value of the curvature radius significantly changes the flow features. Sharp edges lead to premature onset of Kelvin-Helmholtz instability in the shear layers detaching from the upstream corners and this moves upstream the reattachment of the mean flow on the cylinder side for the 3:1 and 5:1 cases.

References

1. Alvarez, A.J., Nieto, F., Nguyen, D.T., Owen, J.S., Hernandez, S.: 3D LES simulations of a static and vertically free-to-oscillate 4:1 rectangular cylinder: effects of the grid resolution. J. Wind Eng. Ind. Aerod. **192**, 31–44 (2019)
2. Bruno, L., Oberto, D.: Effect of cell quality in grid boundary layer on the simulated flow around a square cylinder. Comput. Fluids **238**, 105351 (2022)
3. Bruno, L., Coste, N., Fransos, D.: Simulated flow around a rectangular 5:1 cylinder: spanwise discretization effects and emerging flow features. J. Wind. Eng. Ind. Aerod. **104–106**, 203–215 (2012)
4. Bruno, L., Salvetti, M.V., Ricciardelli, F.: Benchmark of the aerodynamic of the rectangular 5:1 cylinder: an overview after the first four years of activity. J. Wind Eng. Ind. Aerod. **126**, 87–106 (2014)
5. Chiarini, A., Quadrio, M.: The importance of corner sharpness in the BARC test case: a numerical study. Wind Struct. **34**(1), 43–58 (2022)
6. Jeong, J., Hussain, F.: On the identification of a vortex. J. Fluid Mech. **285**, 69–94 (1995)
7. Kawai, H.: Effect of corner modifications on aeroelastic instabilities of tall buildings. J. Wind Eng. Ind. Aerod. **74–76**, 719–729 (1998)
8. Lunghi, G., Pasqualetto, E., Rocchio, B., Mariotti, A., Salvetti, M.V.: Impact of the lateral mean recirculation characteristics on the near-wake an bulk quantities of the BARC configuration. Wind Struct. **34**(1), 115–125 (2022)
9. Mariotti, A., Siconolfi, L., Salvetti, M.V.: Stochastic sensitivity analysis of large-eddy simulation predictions of the flow around a 5:1 rectangular cylinder. Eur. J. Mech. B/Fluids **62**, 149–165 (2017)
10. Pasqualetto, E., Lunghi, G., Rocchio, B., Mariotti, A., Salvetti, M.V.: Experimental characterization of the lateral and near-wake flow for the BARC configuration. Wind Struct. **34**(1), 101–113 (2022)
11. Rocchio, B., Mariotti, A., Salvetti, M.V.: Flow around a 5:1 rectangular cylinder: effects of upstream-edge rounding. J. Wind Eng. Ind. Aerod. **204**, 104237 (2020)
12. Rodi, W.: Comparion of LES and RANS calculations of the flow around bluff bodies. J. Wind Eng. Ind. Aerod. **69–71**, 55–75 (1997)
13. Shimada, K., Ishihara, T.: Application of a modified k-ε model to the prediction of aerodynamic characteristics of rectangular cross-section cylinders. J. Fluids Struct. **16**(4), 465–485 (2002)
14. Sohankar, A.: Large eddy simulation on flow past rectangular-section cylinders: side ratio effects. J. Wind Eng. Ind. Aerod. **96**, 640–655 (2008)
15. Trias, F.X., Gorobets, A., Oliva, A.: Turbulent flow around a square cylinder at Reynolds number 22,000: a DNS study. Comput. Fluids **123**, 87–98 (2015)
16. van Hinsberg, N.P., Schewe, G., Jacobs, M.: Experiments on the aerodynamic behaviour of the square cylinders with rounded corners at Reynolds number up to 12 million. J. Fluids Struct. **74**, 241–233 (2017)
17. Yu, D., Kareem, A.: Parametric study of flow around rectangular prisms using LES. J. Wind Eng. Ind. Aerod. **77&78**, 653–662 (1998)
18. Yu, D., Butler, K., Kareem, A., Glimm, J., Sun, J.: Simulation of the influence of aspect ration on the aerodynamic of rectangular prisms. J. Eng. Mech. **139**(4), 429–438 (2013)

Reynolds-Number Effects of Separating Flow over a Bump in Spanwise Rotating Channels

B. Savino, D. Patel, and W. Wu(✉)

Department of Mechanical Engineering, University of Mississippi, Oxford, USA
wu@olemiss.edu

Abstract. Direct numerical simulations of turbulent channel flow rotating about the spanwise axis are performed at three Reynolds numbers. A parabolic bump is placed either on the anti-cyclonic or cyclonic wall of the channel and generates pressure-induced flow separation on its lee side. Reduced mean separation region is observed when the flow over the bump is unstabilized by the Coriolis force and/or at higher *Re*. The total drag decreases with rotation regardless of the rotation direction for this configuration. The form drag produced by the separated flow does not vary significantly with rotation or *Re*, indicating an inviscid-dominant drag production mechanism and the disadvantage of utilizing the extent of the separation region as an indicator of the magnitude of the form drag. The separated shear layer deviates from the canonical one due to the interactions of various types of vortices and the energy redistribution between Reynolds stresses.

Keywords: Direct numerical simulation · rotating channel · flow separation

1 Introduction

Rotation characterizes many turbulent flows, both in nature (*e.g.*, geophysical flows) and in engineering applications (for instance, turbines, pumps, cyclone separators, the flow in cooling systems in radars, and so on). The literature on the subject of rotating flow is quite large. For a channel that is rotating about its spanwise axis (z), the Coriolis force appears as terms $2\Omega v$ and $-2\Omega u$ in the streamwise (x) and wall-normal (y) momentum equations, respectively. It destabilizes the flow when the rotation has the same sign as the mean shear vorticity and stabilizes the turbulence when the two have opposite signs (Fig. 1). A flow phenomenon that often occurs and has a dramatic effect on the performance of a rotating device is massive flow separation. It is induced by either an abrupt geometrical expansion and/or an adverse pressure gradient in the flow. The stabilization/destabilization influence of the Coriolis force may promote or delay flow separation and reattachment. The effects of spanwise rotation on separation have been investigated in backward-facing steps, sudden expansion channel, rib-roughened channel and diffuser [1,2]. A common finding is that the separation bubble decreases with increasing anti-cyclonic rotation and increases with increase of cyclonic rotation rate. Since separation is enforced by the geometrical singularities in these studies, the separation point is fixed regardless how the reattachment is affected. Compared with the geometry-induced separation, limited investigation is available on the dynamics of

C. Marchioli et al. (Eds.): DLES 2023, ERCO 31, pp. 41–46, 2024.
https://doi.org/10.1007/978-3-031-47028-8_7

the adverse pressure gradient(APG)-induced flow separation in which the onset of separation may be changed by the rotation.

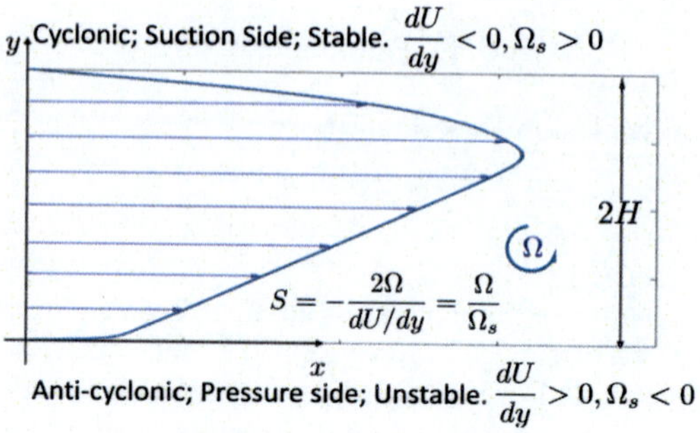

Fig. 1. Schematic of a spanwise rotating channel flow.

In this study, turbulent channel flows rotating in the spanwise direction at Reynolds number $Re_b = U_bH/\nu = 2{,}500$, 5,000 and 7,500 ($H$ is the channel half-height and U_b the bulk velocity) are simulated by DNS. The friction Reynolds number in the region where the flow is fully recovered from the wake of the bump ranges from 160 to 420. A calculation domain of $39H \times 2H \times 6H$ in the streamwise (x), wall-normal (y), and spanwise (z) directions is employed. A two-dimensional bump is placed on the bottom wall of the channel. Depending on the sign of the rotation rate, this side is either anti-cyclonic (*i.e.*, $Ro \equiv 2\Omega H/U_b > 0$) or cyclonic ($Ro < 0$). The height of the bump is $0.25H$. Nine simulations are performed at three rotation numbers $Ro = 0$ and $\pm\, 0.42$. The parameters of the cases are listed in Table 1.

Table 1. Simulation parameters and separation region measurements. $Re_{\tau,c}$ denotes the friction Reynolds number at the bump crest.

Cases	Re_b	Ro	Re_τ	$Re_{\tau,c}$	$N_i \times N_j \times N_k$	x_{sep}	L_{sep}
Ro0L	2,500	0	162	203	$1196 \times 192 \times 184$	4.55	1.79
P42L	2,500	0.42	154	209	$1196 \times 192 \times 184$	4.74	0.90
N42L	2,500	−0.42	155	174	$1196 \times 192 \times 184$	4.48	3.77
Ro0I	5,000	0	296	345	$2496 \times 386 \times 384$	4.49	1.40
P42I	5,000	0.42	254	346	$2496 \times 386 \times 384$	4.66	0.85
N42I	5,000	−0.42	256	275	$2496 \times 386 \times 384$	4.38	2.98
Ro0H	7,500	0	421	471	$3744 \times 580 \times 576$	4.46	1.25
P42H	7,500	0.42	346	470	$3744 \times 580 \times 576$	4.62	0.80
N42H	7,500	−0.42	342	360	$3744 \times 580 \times 576$	4.32	2.65

The equations of motion are solved using a well-validated finite difference code that solves the incompressible Navier-Stokes equations on a staggered grid [3,4]. Grid convergence study is performed to justify that the first and second-order statistics are grid independent.

2 Results and Discussion

The mean separation region enclosed by the streamline passing the reattachment point ($C_f = 0$, $dC_f/dx > 0$) is shown in Fig. 2 (a) for comparison. The mean separation point and separation region length are listed in Table 1. Regardless of the Reynolds number, separation is reduced (or enhanced) when the bump is on the unstable (stable) side, which agrees with previous studies of other rotating flow configurations. The mild geometrical variation in the current configuration allows the separation point to change with rotation: separation is delayed under anti-cyclonic rotation and promoted otherwise. An increase in Reynolds number, surprisingly, leads to early separation. In all the cases, a reduction of the recirculation region is correlated with the early onset of separation.

The change of the separation region size has been widely contributed to the augmented (or damped) turbulence and its ability to enhance (reduce) the mixing of momentum to compensate (facilitate) the deceleration and flow detachment. It should be noted that such an explanation assumes the incoming mean flow remains the same and the turbulent fluctuations vary between cases. It is not true when the channel is under rotation. The mean velocity profile at the crest of the bump is plotted in Fig. 2 (b). It can be seen that the non-rotating cases have the smallest momentum deficit (*i.e.*, highest velocity in the near-wall region) while cyclonic rotation leads to the largest. The anti-cyclonic rotation creates a slightly larger momentum deficit near the bump surface than the non-rotating case, which should have favored flow separation. The delayed separation observed in the anti-cyclonic cases, therefore, is because the enhanced mixing by the unstabilized turbulence overcomes the pre-existing momentum deficit. The early separation in the cyclonic rotation cases, on the other hand, is due to both a large momentum deficit and damped fluctuations.

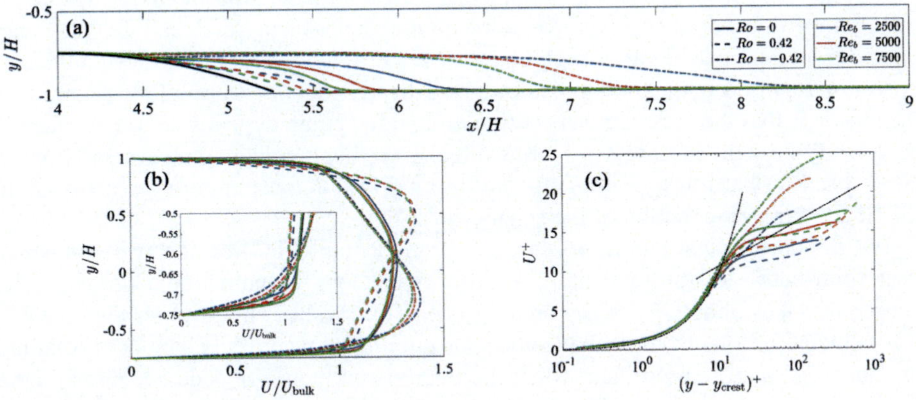

Fig. 2. Mean separating streamline (a); mean streamwise velocity at the bump crest in outer units (b) and in wall units (c).

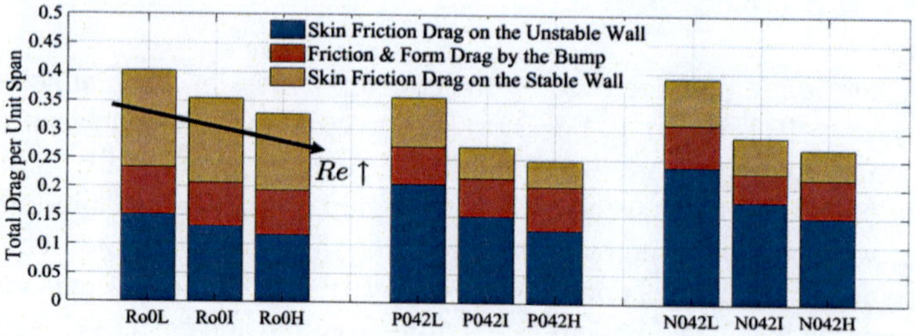

Fig. 3. Budget of the total drag. For the non-rotating case, the blue bar represents the bottom surface where the bump is placed and the yellow bar is the upper wall. (Color figure online)

The velocity profiles in wall units at the bump crest (Fig. 2 (c)) show that the flow is remarkably different from a canonical turbulent boundary layer. The contraction by the bump and the resultant favorable pressure gradient alters the inertia layer. In the non-rotating and anti-cyclonic cases, U^+ barely changes with y^+ in the buffer layer and lower log layer, leading to a logarithmic region shifting downward from the canonical one. As *Re* increases, this log region shifts upward due to the increase of the extent of the linear viscous layer. In the cyclonic rotating cases, the velocity profiles exhibit a laminar behavior augmented by the increase of *Re*.

The size of the recirculation region has been widely used as an indicator of the additional form drag due to flow separation. However, our results show that it is not necessarily true. The total drag produced by the bump and the walls of the channel is compared in Fig. 3. Overall, the total drag is decreased by rotation for the current configurations. It is also lowered as *Re* increases. Decomposing the total drag by the contributions of the walls and the bump, it can be seen that the drag created by the bump (mainly form drag) does not change much with the rotation and *Re*. Recall that the mean separation region indeed varies significantly between the cases, it indicates the size of separation region is not a good indicator of the drag production during flow separation. The reason is that the pressure distribution around the bump depends not on the sign of the velocity near the surface but on the velocity gradient; changing a reversed flow to a low-speed forward flow is not sufficient to modify the inviscid pressure distribution despite the evident reduction of the separation region.

For the current configuration, the skin friction drag (SFG) on the walls makes a great contribution to the total drag because of the long computational domain. The bump only takes about 7% of the area on one of the walls. The SFG on the unstable wall is increased due to the enhancement of turbulence when the channel is rotating. This increase is more prominent when the bump is on the stable side because i) there is no separation region on the unstable wall to negatively contribute to the SFG; ii) the region opposite to the bump gets higher SFG due to the acceleration by the blockage of the bump. The SFG on the stable wall is reduced due to the relaminarization. When the bump is on this side, the reduction is less because the fluctuations in the separating shear layer increase the near-wall velocity.

The separating shear layer determines the reattachment point by the roller vortices generated by the Kelvin-Helmholtz instability. Depending on which side the separation occurs in a rotating system, the separating shear layer may be unstabilized or stabilized by the Coriolis force. Moreover, the roller vortices may encounter non-canonical coherent structures in a boundary layer: streamwise-elongated Taylor vortices whose diameter is comparable to the boundary layer thickness have been observed near the unstable side in the literature of rotating channel flows. TS wave and associated λ-shape vortices may show up near the stable side at a high rotation rate. The interaction between these structures appears as changes in the Reynolds stresses and turbulent kinetic energy (TKE).

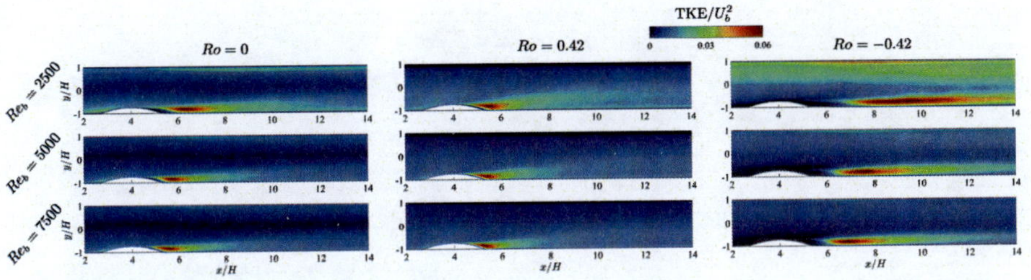

Fig. 4. Contours of the turbulent kinetic energy.

Figure 4 shows the contours of the TKE for the nine cases. In all cases, the region with intense TKE is associated with the forward-flow side of the separating shear layer. Anti-cyclonic rotation cases have shorter high-TKE regions compared with the non-rotating cases, and cyclonic rotation cases show very long high-TKE regions in the bump wake. Recall that the separating flow is quasi-laminar in the cyclonic cases, the high TKE in those cases represents the breakdown of the 2D roller vortices in the separating shear layer. There are no Taylor vortices near the stable side of the channel, thus the shear layer and associated fluctuations form straight bands along the stable wall. The increase of *Re* leads to a shortened high-TKE region in these cases, indicating a faster three-denationalization of the laminar separating shear layer. In contrast, the high-TKE regions in the anti-cyclonic cases represent the rapid dissipation of the turbulent separating shear layer. Since Taylor vortices occur on the unstable side, they cause vertical transport of the fluctuations in the shear layer. Compared with the non-rotating case, it is evident that the high TKE region is inclined towards the opposite wall in the bump wake.

The Coriolis term does not appear in the TKE budgets but plays an important role in the budgets of Reynolds stresses. The production terms for $\overline{u'u'}$ and $\overline{v'v'}$ are

$$\mathscr{P}_{11} + R_{11} = -2\overline{u'v'}\frac{\partial U}{\partial y} + 4\Omega\overline{u'v'}, \text{and } R_{22} = -4\Omega\overline{u'v'} \tag{1}$$

respectively. $\mathscr{P}$ is the production due to mean shear, and R denotes the effects of the Coriolis force. For $\Omega > 0$ and $dU/dy > 0$ (*i.e.*, flow near the bottom wall under counter-clockwise rotation), $\overline{u'v'}$ is negative and energy is taken from $\overline{u'u'}$ to $\overline{v'v'}$. Term $4\Omega\overline{u'v'}$ is compared in Fig. 5 for the rotating cases. When flow separation occurs on the unstable side, energy is redistributed by rotation from $\overline{u'u'}$ to $\overline{v'v'}$. Because the Taylor vortices are associated with $\overline{v'v'}$ (and $\overline{w'w'}$), such redistribution may augment them. It is probably why the upward-drifting of the high-TKE region is most prominent when this term is the most significant at the low Reynolds number. As *Re* increases, such a process is more confined to the downstream vicinity of the bump as the separation region decreases. The peak magnitude does not change much with *Re* though. When flow separation occurs on the stable side, rotation extracts energy from $\overline{v'v'}$ to $\overline{u'u'}$. The peak magnitude of this term increases with *Re*.

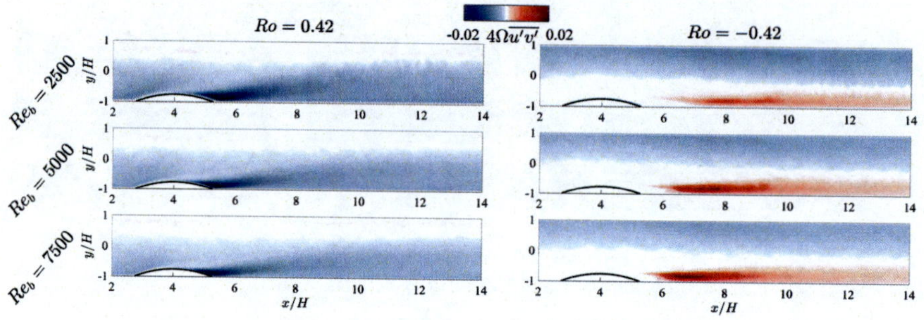

Fig. 5. Contours of the redistribution term $4\Omega\overline{u'v'}$ in the $\overline{u'u'}$ and $\overline{v'v'}$ budgets.

In summary, early separation and reduced recirculation region are observed when the separation occurs near the anti-cyclonic side of the channel and/or when *Re* increases. It does not alter the inviscid pressure distribution around the bump, thus the form drag due to separation barely changes with rotation or Reynolds number. It indicates the size of separation region is not a good indicator of hydro/aerodynamic performance. The turbulent generation mechanics in the separating shear layer exhibit rich diversity and dynamical details associated with the Coriolis effects.

References

1. Barri, M., Andersson, H.I.: Turbulent flow over a backward-facing step. Part 1. Effects of anti-cyclonic system rotation. J. Fluid Mech. **665**, 382–417 (2010)
2. Visscher, J., Andersson, H.I.: Particle image velocimetry measurements of massively separated turbulent flows with rotation. Phys. Fluids **23**, 075108 (2011)
3. Wu, W., Piomelli, U.: Effects of surface roughness on a separating turbulent boundary layer. J. Fluid Mech. **841**, 552–580 (2018)
4. Wu, W., Piomelli, U., Yuan, J.: Turbulence statistics in rotating channel flows with rough walls. Int. J. Heat Fluid Flow **80**, 108467 (2019)

Convection and Heat/MassTransfer

Effect of Variable Density on Subgrid Scales

A. Abbà[1(✉)], M.H. Aliyoldashi[1], A. Cimarelli[2], and M. Germano[3]

[1] Dipartimento di Scienze e Tecnologie Aerospaziali, Politecnico di Milano, Milan, Italy
antonella.abba@polimi.it, mohammad.aliyoldashi@mail.polimi.it

[2] Department of Engineering 'Enzo Ferrari', University of Modena and Reggio Emilia, Modena, Italy
andrea.cimarelli@unimore.it

[3] Department Civil and Environmental Engineering, Duke University, Durham, USA
mg234@duke.edu

1 Introduction

Variable density flows are very common in nature, in technology and in industry. The numerical simulation of complex flows of applicative interest and with relevant variable density flows requires the truncation of turbulent scales of the resolved velocity field, in order to reduce the required computational resources. By this way the averaged field is computed using the Reynolds averaged Navier-Stokes (RANS) equations, or the filtered in space equations if the Large Eddy Simulation (LES) approach is applied for which a closure model is required to represent the non resolved scales. The scales truncation applied to the mass conservation equation for compressible flows should require a closure model also for the mass flux. To avoid this, the Favre average [4] was introduced for the ensemble averaged quantities in RANS, and it has been extended to filter in space operator in LES [10]. The use of the Favre filter approach is prevalently preferred in LES of variable density flows. The works of [2, 8, 11], where a subgrid scale (SGS) eddy viscosity model is added to the density equation, and the paper of [7], where an equation for the unresolved density fluxes is solved, constitute exceptions. The filtered velocity $\overline{\mathbf{u}}$ and the Favre filtered velocity $\widetilde{\mathbf{u}}$ represent two different physical variables with different dynamics and evolution. The difference between the two velocities can be usually reasonably ignored also in compressible flows, but this is not justified in presence of strong density variation. Sidharth and Candler [7] observed, in decaying turbulence, that the effects of small scale density gradient can affect the dynamics of large scale velocity and vorticity. Particularly noteworthy is the influence of Favre filtering in the SGS modeling. Eddy viscosity type models are the most diffused models in LES and the eddy viscosity hypothesis is usually straightforward extended to model Favre filtered subgrid fluxes. This assumption is unjustified for compressible flows in presence of strong density gradient, as discussed in [5] where some additional terms are added to correctly compute the subgrid stress tensor in function of the Favre filtered velocity maintaining the eddy viscosity hypothesis.

C. Marchioli et al. (Eds.): DLES 2023, ERCO 31, pp. 49–54, 2024.
https://doi.org/10.1007/978-3-031-47028-8_8

An a priori test, finalised to analyse the effect of variable density flows in a multi-scale approach to turbulence, has been made in the present research and a correction to sgs models which take this into account is proposed.

2 The Subgrid Fluxes

The filtered Navier-Stokes equations for variable density flows are commonly solved for the Favre filtered velocity $\widetilde{\mathbf{u}} = \overline{\rho \mathbf{u}}/\overline{\rho}$ rather then the grid filter velocity $\overline{\mathbf{u}}$ to avoid subgrid mass flux in the density equation. By this way, the sgs Favre stresses τ_{ij}^F

$$\tau_{ij}^F = \overline{\rho u_j u_i} - \overline{\rho}\, \widetilde{u}_j\, \widetilde{u}_i = \overline{\rho}\,(\widetilde{u_j u_i} - \widetilde{u}_j\, \widetilde{u}_i) \tag{1}$$

derive from filtering the advection term of the momentum equation. The following decomposition

$$\tau_{ij}^F = \frac{1}{2}\left(\tau(\rho u_i, u_j) + \tau(\rho u_j, u_i)\right) - \frac{1}{2}\left(\widetilde{u}_i \tau(\rho, u_j) + \widetilde{u}_j \tau(\rho, u_i)\right) \tag{2}$$

is here considered where the subgrid momentum flux

$$\tau(\rho u_i, u_j) = \overline{\rho u_i u_j} - \overline{\rho u_i}\, \overline{u}_j \tag{3}$$

and the subgrid mass flux

$$\tau(\rho, u_i) = \overline{\rho u_i} - \overline{\rho}\, \overline{u}_i \tag{4}$$

highlight the dependence of the sgs stress tensor on the density. The tensorial eddy viscosity model presented in [3]

$$\tau(\theta, u_j) = -\nu_{kj} \partial_k \overline{\theta} \tag{5}$$

is here considered to model all these terms. This tensorial eddy viscosity model has been directly extended to compressible flows in [1]. Although the results presented in [1] exhibited a growth in accuracy with respect to current models, the model did not take into account specific corrections for intense density variations. The following proposal will overcome this lack. In Eq. (5) $\tau(\theta, u_j)$ represents the subgrid flux of a generic scalar θ advected by the velocity $\mathbf{u}$ and ν_{kj} are the components of a tensorial eddy viscosity expressed by the relation

$$\nu_{kj} = \mathscr{I}_{kh} \partial_h \overline{u}_j .$$

Here

$$\mathscr{I}_{kh} = -\frac{1}{2\Omega} \int_{\Omega} (\xi_k - x_k)(\xi_h - x_h) d\xi$$

is related to the inertial tensor of the computational grid elements. Since the computational grid is usually stretched to adapt to the flow properties, the eddy viscosity naturally adapts to the turbulence structure anisotropy.

We apply tensorial eddy viscosity procedure to Eqs. (3) and (4)

$$\tau(\rho u_i, u_j) = -\frac{1}{2}\nu_{kj}\partial_k(\overline{\rho u_i}) = -\frac{1}{2}\nu_{kj}\partial_k(\overline{\rho}\widetilde{u}_i), \qquad \tau(\rho, u_j) = -\frac{1}{2}\nu_{kj}\partial_k\overline{\rho}. \tag{6}$$

In the eddy viscosity expressed as

$$\nu_{kj} = \mathscr{I}_{kh}\partial_h\overline{u}_j = \mathscr{I}_{kh}\left[\partial_h\widetilde{u}_j - \partial_h\left(\frac{\tau(\rho, u_j)}{\overline{\rho}}\right)\right] \tag{7}$$

the dependence on the variable density is contained in the sgs mass flux. The Eqs. (6) and (7) contain a recursive relation which must be numerically approximated in actual LES simulations.

3 A Priori Analysis

An a priori analysis on the solution of a direct numerical simulation (DNS) of a turbulent compressible mixing layer has been performed in order to investigate the effect of variable density on small scale modeling. The DNS has been conducted by Trane [9] reproducing the flow simulated by Pantano and Sarkar [6] but in a larger computational domain of size $0 \leq x/\delta_\theta(0) \leq 344$, $-258 \leq y/\delta_\theta(0) \leq 258$, $0 \leq z/\delta_\theta(0) \leq 172$, where $\delta_\theta(0)$ is the initial momentum thickness. A density ratio $s = \rho_2/\rho_1 = 2$ has been imposed. Here the indeces 1 and 2 indicate quantities in the upper and lower stream respectively. The mean density $\rho_0 = (\rho_1 + \rho_2)/2$, the velocity difference $\Delta U = U_1 - U_2$, the temperature $T_0 = 2T_1T_2/(T_1 + T_2)$ and $\delta_\theta(0)$ are used to obtain nondimensional quantities. In the DNS the Navier-Stokes equations for a compressible flow are solved together with the pressure state equation of ideal gas. The nondimensional numbers Mach $Ma = \Delta U/(c_1 + c_2) = 0.7$, Reynolds $Re = \rho_0\Delta U\delta_\theta(0)/\mu = 160$ and Prandtl $Pr = 0.7$ have been applied. For the details about boundary and initial conditions see Pantano and Sarkar [6]. Grid filtered and Favre filtered quantities are computed applying box filter in space to the field obtained by the DNS at the non dimensional time $t\Delta U/\delta_\theta(0) = 800$. Three different filter size, as reported in Table 1 and named as in Table 2, have been considered. The profiles of the mean and rms values of density filtered from the DNS field are represented in Fig. 1. The density fluctuations are higher in the central region of the mixing layer, as expected, and the intensity decreases for the larger filter size. The profiles of the mean filtered velocity are shown in Fig. 2. Any relevant differences between the grid filtered and the Favre filtered velocity are not evident for the streamline component since they are few order of magnitude smaller respect to the range value. The effect of the variable density is instead visible for the normal component. Indeed the difference between the Favre and the grid filter in the central part of the mixing layer, where the density fluctuations are intense, are comparable with

Table 1. Grid resolution for the mixing layer DNS and for the filtered field.

	$\Delta_x/\delta_\theta(0)$	$\Delta_{y_{min}}/\delta_\theta(0)$	$\Delta_{y_{max}}/\delta_\theta(0)$	$\Delta_z/\delta_\theta(0)$
DNS	0.67	0.34	1.36	0.67
small filter	2.68	1.36	5.44	2.68
medium filter	5.36	2.72	10.88	5.36
large filter	10.72	5.44	21.76	10.72

Table 2. Terminology identifying the applied filters.

filter type	acronym
small grid filter	SGF
small Favre filter	SFF
medium grid filter	MGF
medium Favre filter	MFF
large grid filter	LGF
large Favre filter	LFF

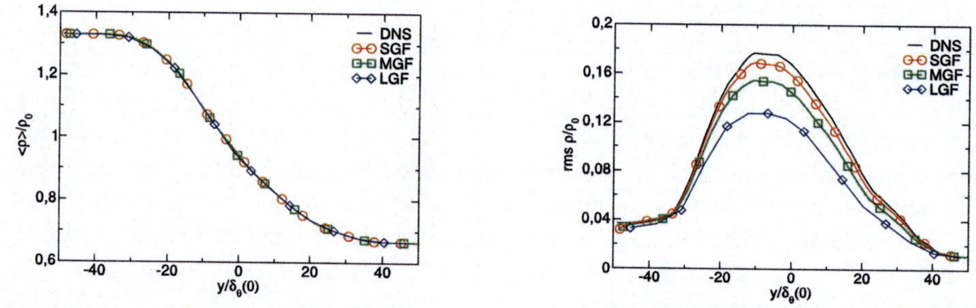

Fig. 1. Profiles of the mean density (left) and rms (right): comparison between the DNS and the filtered results.

the mean value of $\overline{u}_y$ and increases with the filter size. In order to highlight the correction for variable density, the modeled stresses for Favre filtering is decomposed as $\tau^F_{ij} = \tau^C_{ij} + \tau^{NC}_{ij}$ where

$$\tau^{NC}_{ij} = \frac{1}{4}\overline{\rho}\left(\nu^{NC}_{kj}\partial_k\widetilde{u}_i + \nu^{NC}_{ki}\partial_k\widetilde{u}_j\right), \qquad \nu^{NC}_{kj} = \mathscr{I}_{kh}\partial_h\widetilde{u}_j, \tag{8}$$

$$\tau^C_{ij} = \frac{1}{4}\overline{\rho}\left(\nu^C_{kj}\partial_k\widetilde{u}_i + \nu^C_{ki}\partial_k\widetilde{u}_j\right), \qquad \nu^C_{kj} = \mathscr{I}_{kh}\partial_h\left(\frac{\tau(\rho,u_j)}{\overline{\rho}}\right). \tag{9}$$

Figure 3 shows good agreement between the modeled and the exact sgs stresses computed using the DNS results, except for the large filter which is not located inside the inertial range of the turbulent spectrum but is very close to energetic scales. No relevant differences are visible in the mean stress profiles if correction for variable density is not taken into account, but the correction terms could locally assume important values. The pdf of the ratio $\tau^C_{ij}/(\tau^C_{ij} + \tau^{NC}_{ij})$, evaluated in the region of higher density fluctuations $-20 \leq y/\delta_\theta(0) \leq 20$ and for the medium filter, is depicted in Fig. 4 in order to investigate this aspect. The pdf for the normal components demonstrates that the corresponding correction terms are small, and the larger values are for the τ^{corr}_{yy} of the order 5% in the 75% of the considered volume. The pdf of the shear modeled stresses present different behaviour. The τ^C_{xy} and the τ^C_{xz} reach the 45% and 130% respectively of the value of the corresponding modeled τ^F_{ij} in the 95% of the considered volume.

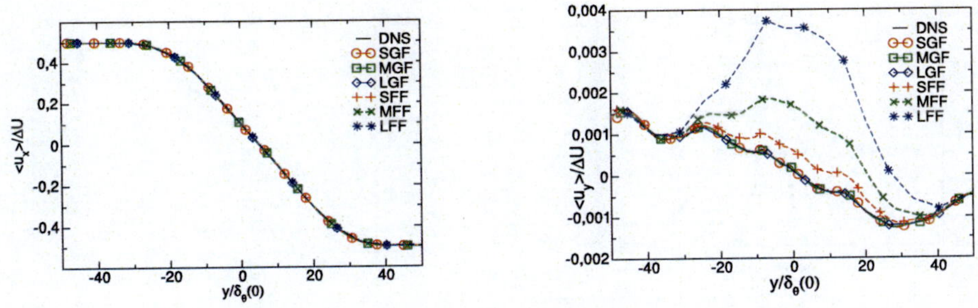

Fig. 2. Profiles of the mean filtered velocity. Left: streamline component; right:normal component.

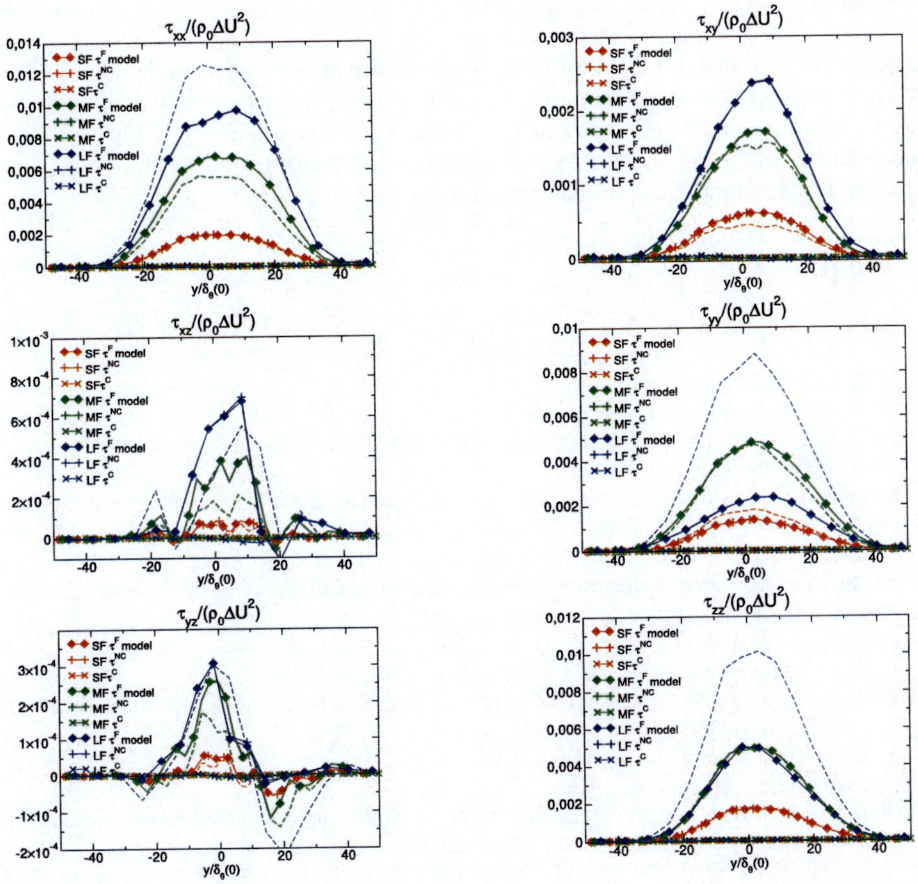

Fig. 3. Profiles of the mean sgs stresses. The modeled stresses are compared to the exact ones obtained filtering the DNS data, represented with dashed lines with colors corresponding to different filters.

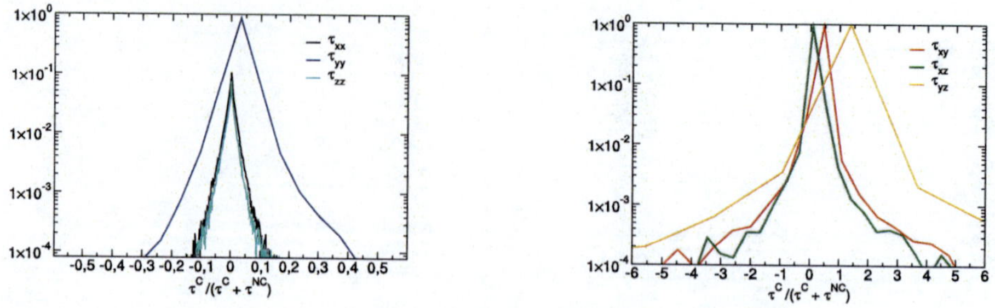

Fig. 4. pdf of the ratio between the correction term and the modelled stress, for the medium filter, for the normal components (left) and for the shear ones (right).

4 Conclusion

The analysis demonstrates that relevant differences are present between the filtered and the Favre filtered velocity. These differences locally affect the sgs turbulent stresses. The a priori test for the variation of a previously proposed tensorial eddy viscosity model, suitable for variable density flows, show good performances of the model and demonstrates the importance of the correction terms.

References

1. Abbà, A., Cimarelli, A., Germano, M.: Dynamic tensorial eddy viscosity model: effects of compressibility and of complex geometry. Phys. Fluids, **34**(2) (2022). https://doi.org/10.1063/5.0076341
2. Boersma, B.J., Lele, S.K.: Large eddy simulation of compressible turbulent jets. Center for Turbulence Research: Annual Research Briefs (Center for Turbulence Research, Stanford, CA), 365 (1999)
3. Cimarelli, A., Abbà, A., Germano, M.: General formalism for a reduced description and modelling of momentum and energy transfer in turbulence. J. Fluid Mech. **866**, 865–896 (2019)
4. Favre, A.: Turbulence: space-time statistical properties and behavior in supersonic flows. Phys. Fluids **26**, 2851 (1983)
5. Germano, M., Abbà, A., Arina, R., Bonaventura, L.: On the extension of the eddy viscosity model to compressible flows. Phys. Fluids **26**(4), 041702 (2014)
6. Pantano, C., Sarkar, S.: A study of compressibility effects in the high-speed turbulent shear layer using direct simulation. J. Fluid Mech. **451**, 329–371 (2002)
7. Sidharth, G.S., Candler, G.V.: Subgrid-scale effects in compressible variable-density decaying turbulence. J. Fluid Mech. **846**, 428–459 (2018)
8. Sun, B., Lu, X.Y.: A large eddy simulation approach of compressible turbulent flow without density weighting. Phys. Fluids **18**, 118101 (2006)
9. Trane, D.: Study of the Mach and density ratio effects in mixing layers. Master Degree Thesis, Università degli studi di Modena e Reggio Emilia (2021)
10. Yoshizawa, A.: Statistical theory for compressible turbulent shear flows, with the application to subgrid modeling. Phys. Fluids A **29**, 2152 (1986)
11. Zhang, C., Yuan, Z., Wang, Y., Zhang, R.: Density-unweighted subgrid-scale models for large-eddy simulations of compressible turbulence. Phys. Fluids **34**, 065137 (2022)

DNS of Mass Transfer in Bi-dispersed Bubbly Flows in a Vertical Pipe

N. Balcazar-Arciniega(✉), J. Rigola, and A. Oliva

Heat and Mass Transfer Technological Center (CTTC),
Universitat Politècnica de Catalunya - BarcelonaTech (UPC), Barcelona, Spain
nestorbalcazar@yahoo.es,
{nestor.balcazar,joaquim.rigola,asensio.oliva}@upc.edu

1 Introduction

Mass transfer in bubbly flows is ubiquitous in nature and industry. For instance, bubble chemical reactors and mass-transfer operations of chemical engineering (e.g., distillation columns) employ bubbly flows to produce chemical products or to promote separation processes. Correlations have been reported to estimate mass transfer rates in bubbles [14], however the interplay between fluid mechanics and mass transfer in bubble swarms still needs to be better understood [9–11,16,17]. This work investigates the mass transfer in bi-dispersed gravity-driven turbulent bubbly flows, by Direct Numerical Simulation (DNS) of the Navier-Stokes equations and the multiple-marker unstructured conservative level-set (UCLS) method for bubble swarms, introduced by Balcazar et al.(2019) [2,5,6,9,11].

2 Mathematical Formulation and Numerical Methods

The Navier-Stokes equations for incompressible two-phase flows are presented in the framework of the so-called one-fluid formulation:

$$\frac{\partial}{\partial t}(\rho \mathbf{v}) + \nabla \cdot (\rho \mathbf{v}\mathbf{v}) = -\nabla p + \nabla \cdot \mu (\nabla \mathbf{v}) + \nabla \cdot \mu (\nabla \mathbf{v})^T + (\rho - \rho_0)\mathbf{g} + \mathbf{f}_\sigma, \quad (1)$$

$$\nabla \cdot \mathbf{v} = 0, \quad (2)$$

where $\mathbf{v}$ is the velocity, p is the pressure, $\mathbf{g}$ is the gravitational acceleration, μ is the dynamic viscosity, ρ is the fluid density, $\mathbf{f}_\sigma$ is the surface tension force per unit volume concentrated on the interface (Γ), subscripts d and c represents the continuous phase and dispersed phase respectively. Physical properties are constant with a jump discontinuity at Γ: $\rho = \rho_d H_d + \rho_c H_c$, $\mu = \mu_d H_d + \mu_c H_c$. Here H_c denotes the Heaviside step function, which is one in Ω_c (continuous phase) and zero in Ω_d (dispersed phase), whereas $H_d = 1 - H_c$. Since periodic boundary conditions are employed, the force $-\rho_0 \mathbf{g}$, with $\rho_0 = V_\Omega^{-1} \int_\Omega (\rho_d H_d + \rho_c H_c)\, dV$, avoids the acceleration of the flow field in the downward vertical direction [2,6,9].

C. Marchioli et al. (Eds.): DLES 2023, ERCO 31, pp. 55–61, 2024.
https://doi.org/10.1007/978-3-031-47028-8_9

A multiple marker approach [2,15], as introduced by [9] in the framework of the UCLS method [1,9,12] for mass transfer in bubble swarms is used in this research. Indeed, the ith marker is represented by the UCLS function $\phi_i = 0.5\,(\tanh{(d_i(\mathbf{x},t)/(2\varepsilon))}+1)$ [1], where d_i is a signed distance function. Given the incompressible constraint (Eq.(2)), the ith interface transport equation can be written in conservative form. Moreover, a re-initialization equation is solved up to the steady state to keep a sharp and constant level-set profile [1]:

$$\frac{\partial \phi_i}{\partial t} + \nabla \cdot \phi_i \mathbf{v} = 0,\ \frac{\partial \phi_i}{\partial \tau} + \nabla \cdot \phi_i(1-\phi_i)\mathbf{n}_i^0 = \nabla \cdot \varepsilon \nabla \phi_i,\ i = \{1, ..., N_m\} \tag{3}$$

where N_m is the number of UCLS markers, here equivalent to the number of fluid particles $N_m = n_d$. At Ω_P, $\varepsilon_P = 0.5h_P^{0.9}$, h_P is the grid-size [1,5,9]. As proposed by Balcazar et al. (2014) [1], the diffusion term of the re-initialization equation is computed as $\nabla \cdot \varepsilon \nabla \phi_i$, to deal with unstructured meshes. Indeed ε_f is linearly interpolated at the cell-faces (f) [1,4,9,12]. Normal vectors ($\mathbf{n}_i$) and curvatures (κ_i) are computed as follows: $\mathbf{n}_i(\phi_i) = \nabla\phi_i \|\nabla\phi_i\|^{-1}$ and $\kappa_i = -\nabla \cdot \mathbf{n}_i$. The Continuous Surface Force model computes surface tension forces [13], extended to the multiple markers UCLS method by [2,5,9] as follows: $\mathbf{f}_\sigma = \sum_{i=1}^{n_d} \sigma \kappa_i \nabla \phi_i$, where σ is the surface tension coefficient. Fluid properties (ρ and μ) are regularized by using a global level-set function, i.e., $H_d^s = \phi$, defined as $\phi = \min\{\phi_1, ..., \phi_{n_d}\}$ [2,5,9]. Advantages of the multiple-marker UCLS method include mass conservation of the fluid phases, accurate computation of surface tension forces, and efficient parallelization [1–3,5,6,9,10]. Numerical stability at high Reynolds numbers and high physical properties ratios is achieved by using unstructured flux-limiter schemes proposed by Balcazar et al. [1,9], which avoids numerical oscillations around discontinuities and minimize the so-called numerical diffusion.

This research focuses in bubbly flows with external mass transfer. The transport equation for the chemical-species concentration (C) is defined as:

$$\frac{\partial C}{\partial t} + \nabla \cdot (\mathbf{v}C) = \nabla \cdot (\mathscr{D} \nabla C) + \dot{r}(C), \text{ in } \Omega_c, \tag{4}$$

where $\mathscr{D} = \mathscr{D}_c$ is the diffusion coefficient in Ω_c, $\dot{r} = -k_1 C$ is the chemical reaction rate, k_1 is the reaction rate constant. Thermodynamic equilibrium is assumed, therefore $C_{\Gamma,c} = HC_{\Gamma,d}$, where H is the so-called Henry constant. Here, C is constant inside bubbles [16,17], whereas its value on the bubble interface is $C_{\Gamma,c}$. An interpolation method proposed by Balcazar et al.(2019) [9] is used to calculate the concentration on the interface cells. Further technical details on the finite-volume discretization of the multiple marker UCLS method can be found in [9].

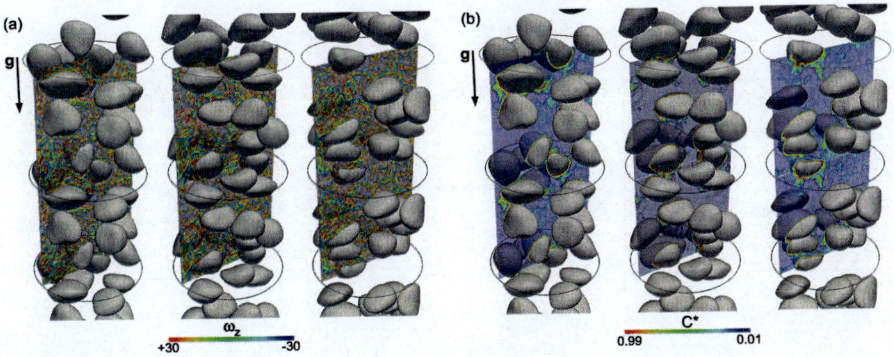

Fig. 1. Mass transfer in bi-dispersed bubbly flow in a vertical pipe. Ω discretized by 10.44×10^6 triangular-prisms, 960 CPU-cores. Case 3 (Table 1). Images at $t^* = t\,(\|\mathbf{g}\|/d)^{1/2} = \{7.5, 8.75, 10\}$. (a) Vorticity ($\omega_z = \mathbf{e}_z \cdot (\nabla \times \mathbf{v})$). (b) Concentration ($C^* = C/C_{\Gamma,c}$) on the plane $x - z$.

Table 1. Bi-dispersed bubble swarms: Eo $= 3.125$, Mo $= 10^{-11}$, Sc $= 1$, Da $= 400$, $\eta_\rho = 100$, $\eta_\mu = 100$, $C_R = 0.25$, $d^*/d = 1.2$.

Case	$\alpha = \alpha_{d^*} + \alpha_d$	α_{d^*}	α_d	V_R
1	7.20%	7.20%	0.00%	∞
2	11.36%	7.20%	4.17%	1.728
3	15.53%	7.20%	8.33%	0.864

3 Numerical Experiments and Results

Systematic validations and verifications of the UCLS method [1,2,5,6,9,11] include: gravity-driven motion of single bubbles [1,3,4], binary droplet collision [2], collision of a droplet against a fluid interface [2], mono-dispersed bubbly flow in a vertical channel [6,9,11], falling droplets [7], two-phase flow with variable surface tension [5,8], and liquid-vapor phase change [12]. Furthermore, validations for external mass transfer, with and without chemical reaction, have been reported in our previous works [9–11]. Indeed, this research is a further step in mass transfer in bi-dispersed bubbly flows in a vertical pipe for $Re \sim O(1000)$.

Bi-dispersed bubbly flow in a vertical pipe is characterized by the Morton number Mo $= \|\mathbf{g}\|\mu_c^4(\rho_c - \rho_d)\rho_c^{-2}\sigma^{-3}$, Eötvös number Eo $= \|\mathbf{g}\|\, d^2(\rho_c - \rho_d)\sigma^{-1}$, density ratio $\eta_\rho = \rho_c/\rho_d$, viscosity ratio $\eta_\mu = \mu_c/\mu_d$, bubble volume-fraction $\alpha = V_{\Omega_d}/V_\Omega$, confinement ratio $\mathrm{C}_R = d/D_\Omega$, volume ratio $\mathrm{V}_R = V_{\Omega_{d,d^*}}/V_{\Omega_{d,d}}$, diameter ratio $\mathrm{D}_R = d_b^* \,/\, d_b$, and Reynolds number Re $= \rho_c\, U_T\, d \,/\, \mu_c$. Here, U_T is the time averaged vertical component (aligned with $\mathbf{g}$) of the drift velocity, defined as the difference between the bubble velocity and whole domain space-average velocity, d is the smaller bubble diameter, d^* denotes bigger bubble diameter, $V_{\Omega_{d,d^*}}$ is the volume of bubbles with diameter d^*, and $V_{\Omega_{d,d}}$ is the volume of bubbles with diameter d. Reactive mass transfer is characterized by

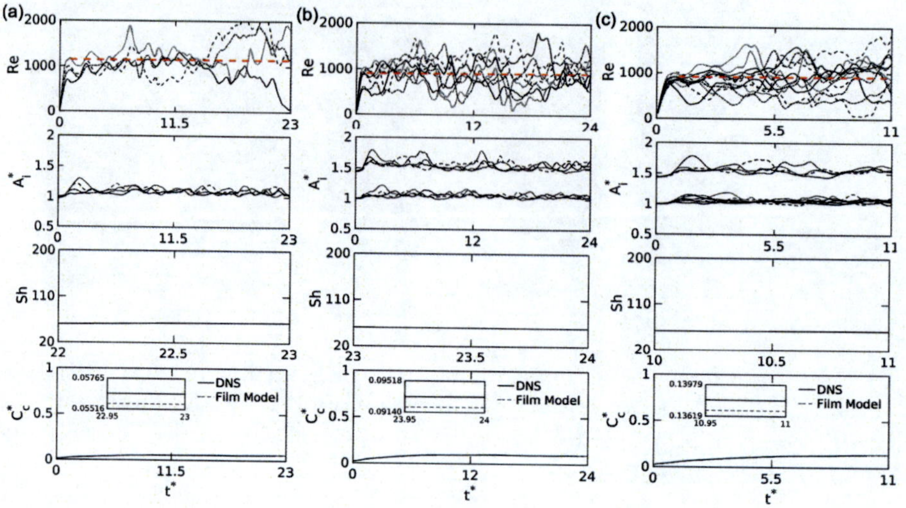

Fig. 2. Mass transfer from bubble swarms: Red discontinuous line for time-averaged Re. Technical details on the film model in [9]. (**a**) Case 1 (Table 1). (**b**) Case 2 (Table 1). (**c**) Case 3 (Table 1).

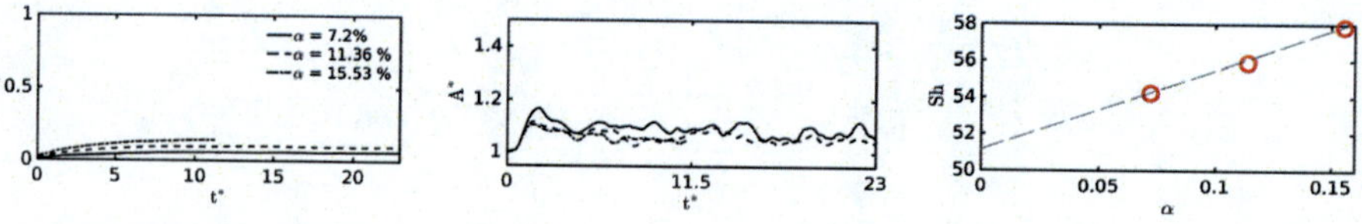

Fig. 3. Mass transfer from bubble swarms: Summary of numerical results for cases detailed in Table 1.

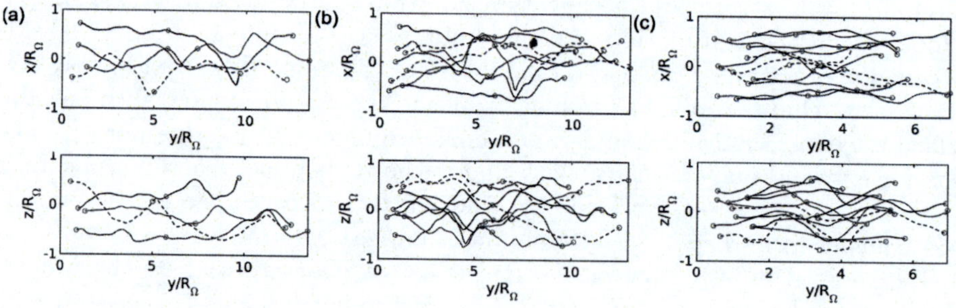

Fig. 4. Trajectories of bubbles: (**a**) Case 1 (Table 1). (**b**) Case 2 (Table 1). (**c**) Case 3 (Table 1).

the Damköler (Da) number ($\text{Da} = k_1 d^2/\mathscr{D}_c$), Schmidt number ($\text{Sc} = \mu_c/(\rho_c \mathscr{D}_c)$) or Péclet number ($\text{Pe} = \text{Re}\,\text{Sc}$), and Sherwood number ($\text{Sh} = k_c d/\mathscr{D}_c$), where k_c is the mass transfer coefficient at Ω_c.

The domain (Ω) is a cylinder of height $L_\Omega = 4d$ and radius $R_\Omega = 4d$, discretized by 10.44×10^6 triangular prisms, distributed on 960 CPU-cores. Bubbles of diameter d have a grid resolution of 40 control volumes per bubble diameter, as suggested in [9–11]. Ω is periodic along the y-axis (parallel to $\mathbf{g}$). The mesh is refined close to the wall to capture the thin fluid films that arise in bubble-wall interactions. Bubbles are initially distributed in Ω following a random pattern, whereas both fluid phases are quiescent. Dimensionless numbers for the simulated cases are summarized in Table 1. Non slip boundary condition (BC) is applied to $\mathbf{v}$ on the wall, and Neumann BC ($\mathbf{e}_w \cdot (\nabla C)_w = 0$) for the concentration on the wall (subindex w).

Figure 1a shows instantaneous images of vorticity ($\omega_z = \mathbf{e}_z \cdot (\nabla \times \mathbf{v})$) generated by bubble-bubble and bubble-wall interactions on the plane x-y. Figure 1b depicts the dimensionless concentration ($C^* = C/C_{\Gamma,c}$) on the plane $x - z$, and the mixing of chemical species promoted by the bubble induced turbulence. Here $C_{\Gamma,c}$ is the concentration on the interface from the continuous fluid side. Images correspond to case 3 of Table 1, at the dimensionless time $t^* = t\,(\|\mathbf{g}\|/d)^{1/2} = \{7.5, 8.75, 10\}$.

Figure 2 shows the time evolution of Re, dimensionless area $A_i^* = A_i(t)/A_0$ of each bubble, $A_0 = \pi d^2$, Sh and dimensionless concentration $C_c^* = C_c/C_{\Gamma,c}$. Here $A_i(t) = \int_\Omega ||\nabla \phi_i|| dV$, $C_c = V_c^{-1} \int_{\Omega_c} C dV$, Ω_c denotes the continuous phase domain, V_c is the volume of Ω_c. On the other hand, a good agreement of C_c computed by DNS with predictions of the so-called Film-model for external mass transfer with first-order chemical reaction [9] is observed. Figure 3 shows a linear relation for Sh = Sh(α), concerning the cases of Table 1, where the bubble volume fraction of bigger bubbles (α_{d^*}) is kept constant. Finally, Fig. 4 depicts the bubble trajectories.

4 Conclusions

DNS of mass transfer in gravity-driven bi-dispersed bubbly flow (Re $\sim O(1000)$) has been performed using a parallel multiple marker conservative level-set method [2,6,9]. A repulsion effect arises from horizontal bubble-bubble and wall-bubble interactions. The so-called drafting-kissing-tumbling mechanism is followed by initially vertically aligned bubbles. Turbulence induced by bubble interactions promotes the mixing of chemical species in Ω_c. Numerical results for the volume-averaged concentration (C_c) are in close agreement with the film model [9]. Numerical results show a linear increment of Sh as the bubble volume fraction increases.

Acknowledgements. The main author N. Balcazar-Arciniega, as a Serra Húnter Fellow (UPC-LE8027), acknowledges the Catalan Government for the financial support trough this program. The computer resources at MareNostrum IV and the technical support provided by BSC, RES (IM-2022-2-0009, IM-2021-3-0013, IM-2021-2-0020) are acknowledged. The financial support of MINECO - Spain (PID2020-115837RB-100), and FusionCAT (001-P-001722) co-financed by the European Union Regional Development Fund within the framework of the ERDF Operational Program of Catalonia 2014-2020 with a grant of 50% of total cost eligible, is acknowledged.

References

1. Balcázar, N., Jofre, L., Lehmkhul, O., Castro, J., Rigola, J.: A finite-volume/level-set method for simulating two-phase flows on unstructured grids. Int. J. Multiph. Flow **64**, 55–72 (2014)
2. Balcázar, N., Lehmkhul, O., Rigola, J., Oliva, A.: A multiple marker level-set method for simulation of deformable fluid particles. Int. J. Multiph. Flow **74**, 125–142 (2015)
3. Balcázar, N., Lemhkuhl, O., Jofre, L., Oliva, A.: Level-set simulations of buoyancy-driven motion of single and multiple bubbles. Int. J. Heat Fluid Flow **56**, 91–107 (2015)
4. Balcázar, N., Lehmkhul, O., Jofre, L., Rigola, J., Oliva, A.: A coupled volume-of-fluid/level-set method for simulation of two-phase flows on unstructured meshes. Comput. Fluids **124**, 12–29 (2016)
5. Balcázar, N., Rigola, J., Castro, J., Oliva, A.: A level-set model for thermocapillary motion of deformable fluid particles. Int. J. Heat Fluid Flow **62**, 324–343 (2016)
6. Balcázar, N., Castro, J., Rigola, J., Oliva, A.: DNS of the wall effect on the motion of bubble swarms. Procedia Comput. Sci. **108**, 2008–2017 (2017)
7. Balcázar, N., Castro, J., Chiva, J., Oliva, A.: DNS of falling droplets in a vertical channel. Int. J. Comput. Methods Exp. Meas. **6**, 398–410 (2018)
8. Balcazar-Arciniega, N., Rigola, J., Oliva, A.: A level-set model for two-phase flow with variable surface tension: thermocapillary and surfactants. In: ECCOMAS2022 (2022). https://www.scipedia.com/public/Balcazar_Arciniega_et_al_2022a, https://doi.org/10.23967/eccomas.2022.011
9. Balcázar, N., Antepara, O., Rigola, J., Oliva, A.: A level-set model for mass transfer in bubbly flows. Int. J. Heat Mass Transf. **138**, 335–356 (2019)
10. Balcázar-Arciniega, N., Rigola, J., Oliva, A.: DNS of mass transfer from bubbles rising in a vertical channel. In: Rodrigues, J.M.F., et al. (eds.) ICCS 2019. LNCS, vol. 11539, pp. 596–610. Springer, Cham (2019). https://doi.org/10.1007/978-3-030-22747-0_45
11. Balcázar, N., Antepara, O., Rigola, J., Oliva, A.: DNS of drag-force and reactive mass transfer in gravity-driven bubbly flows. In: García-Villalba, M., Kuerten, H., Salvetti, M.V. (eds.) DLES 2019. ES, vol. 27, pp. 119–125. Springer, Cham (2020). https://doi.org/10.1007/978-3-030-42822-8_16
12. Balcazar, N., Rigola, J., Oliva, A.: Unstructured level-set method for saturated liquid-vapor phase change. In: WCCM-ECCOMAS2020 (2021). https://www.scipedia.com/public/Balcazar_et_al_2021a. https://doi.org/10.23967/wccm-eccomas.2020.352
13. Brackbill, J.U., Kothe, D.B., Zemach, C.: A continuum method for modeling surface tension. J. Comput. Phys. **100**, 335–354 (1992)
14. Clift, R., Grace, J.R., Weber, M.E.: Bubbles, Drops and Particles. Academin Press, New York (1978)
15. Coyajee, E., Boersma, B.J.: Numerical simulation of drop impact on a liquid-liquid interface with a multiple marker front-capturing method. J. Comput. Phys. **228**, 4444–4467 (2009)

16. Roghair, I., Van Sint Annaland, M., Kuipers, J.A.M. : An improved front-tracking technique for the simulation of mass transfer in dense bubbly flows. Chem. Eng. Sci. **152**, 351–369 (2016)
17. Aboulhasanzadeh, B., Thomas, S., Taeibi-Rahni, M., Tryggvason, G.: Multiscale computations of mass transfer from buoyant bubbles. Chem. Eng. Sci. **75**, 456–467 (2012)

Application of LES to the Thermal–Hydraulics of Target Station 2 of the ISIS Muon and Neutron Source

G. Cartland-Glover[1(✉)], S. Rolfo[1], D. R. Emerson[1], D. Wilcox[2], D. Blanco-Lopez[3], L. G. Jones[3], D. M. Jenkins[3], and S. Jago[3]

[1] Scientific Computing Department, Daresbury Laboratory, Science and Technology Facilities Council, UK Research and Innovation, Warrington, Cheshire, UK
{greg.glover,stefano.rolfo,david.emerson}@stfc.ac.uk

[2] Technology, Rutherford Appleton Laboratory, Science and Technology Facilities Council, UK Research and Innovation, Didcot, Oxfordshire, UK
dan.wilcox@stfc.ac.uk

[3] ISIS Neutron and Muon Source, Rutherford Appleton Laboratory, Science and Technology Facilities Council, UK Research and Innovation, Didcot, Oxfordshire, UK
{daniel.blanco-lopez,leslie.jones,david.jenkins,
stephen.jago}@stfc.ac.uk

1 Introduction

The ISIS Muon and Neutron source uses the spallation of protons accelerated to 84 % of the speed of light to generate neutrons at two tungsten targets clad in tantalum and cooled by recirculating water. The neutrons are then moderated and directed to a number of instruments that can measure the diffraction and scattering of the neutrons with applications in materials science and electronics. The first target station has an operational lifetime of 5 years with spallation products remaining in the target geometry, while there is some egress of the spallation products after 18 months of the use of target station 2. As the targets are encased in radiologically shield zones, we rely on numerical simulations to understand the impact of various factors that could contribute to the shorter operational lifetime of the second target. The second target is a single piece of tungsten clad in several pieces of tantalum with an overall radius of 0.034 m and length ~0.3 m (see Fig. 1). Two coolant channels situated in the cladding feed cooling water around the nose of the target. Each channel is 2 mm thick with an arched span of around 40 mm, and a flow length of about 600 mm with two long sections on either side of the target and a u-bend at the nose of the target. Note that the span of the channel closer to the head of the tungsten target tapers in the region of the head resulting in a narrower channel with sloping sides.

Studies have been performed that have examined residual and beam induced stresses in the target and shell [8–10], radiation damage [6,7,10], water hammer [10], flow induced erosion of tantalum [2], hydrogen generation and accumulation in the coolant via radiolysis, cavitation, and how they affect fatigue.

Here we describe preliminary calculations that examine the potential for thermal striping in the target.

C. Marchioli et al. (Eds.): DLES 2023, ERCO 31, pp. 62–68, 2024.
https://doi.org/10.1007/978-3-031-47028-8_10

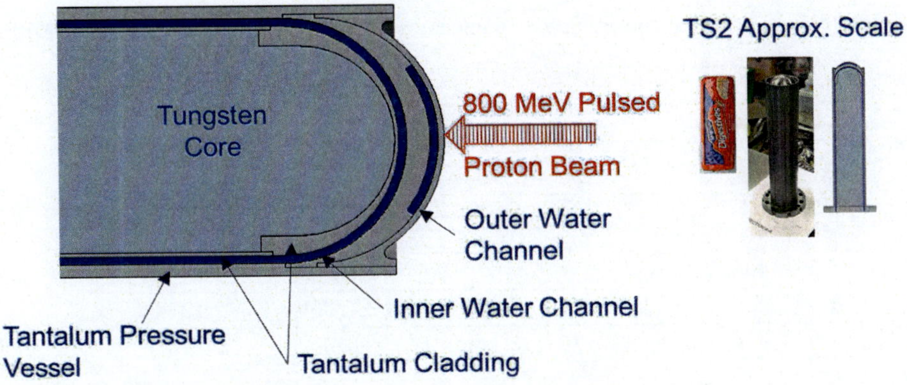

Fig. 1. Configuration of target station two of the ISIS Muon and Neutron Source

2 Numerical Models

LES of the fluid dynamics and conjugate heat transfer occurring in Target Station 2 of the ISIS neutron source are modelled using versions 6 and 7 of *Code_Saturne* [4]. Water is fed to each channel at a temperature of 33.3 °C, a pressure of 6 bar and a flow rate of 45 l min^{-1}, which results in a channel Reynolds number of around 23000 and a Prandtl number of ∼5.0 at the inlet temperature. The pressure drop over the target is about 2.7 bar with temperatures up to 75 °C in the fluid boundary layer at the nose of the target and about 3 °C higher than the inlet temperature at the outlet of both channels. Tables 1–3 give an indication of the variation of the physical properties in both channels (Table 1), the tantalum cladding (Table 2) and tungsten target (Table 3). Note that the viscosity is the key parameter contributing to the variation of the Prandtl number with temperature and pressure in the channel. The heat arising in the target by the spallation process and the subsequent transport of neutrons was obtained from a cylindrical model of the second target station modelled by FLUKA [3,8]. Polynomial functions were used to approximate the heat distribution in the target and the shell that could approximate the steady state heat flux applied to the target by the pulsed beam. We assume incompressible flow and that the flow is turbulent. LES was applied using the Smagorinsky sub-grid scale with a constant coefficient of the rate of strain tensor. The constant was set to 0.065. A laminar inlet profile was obtained from a RANS calculation of a short 3-D section of just the region near to the inlet of both channels. Two geometries were considered, an untapered inner channel with a heat flux obtained from the inner wall of the inner channel of a RANS conjugate heat transfer calculation of the target (inner channel heated wall flow case or ICHWF case) and geometry that comprised the target, the cladding and both cooling channels up to 148.5 mm from the nose of the target (conjugate heat transfer case or CHT case).

Table 1. Variation of the coolant Prandtl number, *Pr* (-), with temperature and pressure [5]

Parameter	Channel Inlet	Nose of target
T (K)	306.45	348.15
p (1×10^6kgm^{-1}s^{-2})	0.566325	0.566325
Pr (–)	5.02	2.38
δPr (–) for $\delta T \pm 1$ K	± 0.116	± 0.034
δPr (–) for $\delta T \pm 4$ K	± 0.493	± 0.140
δPr (–) for $\delta T \pm 10$ K	± 1.389	± 0.379
δPr (1×10^3) for $\delta p \pm 0.135 \times 10^6$kgm^{-1}s^{-2}	± 0.98	± 0.20

Table 2. Variation of the thermal diffusivity, α (-), of the tantalum cladding with temperature [1]

Parameter	Channel Inlet	Nose of target
T (K)	306.45	348.15
α (1×10^5 m^{2}s^{-1})	2.405	2.399
$\delta\alpha$ (1×10^8 m^{2}s^{-1}) for $\delta T \pm 1$ K	± 0.143	± 0.143
$\delta\alpha$ (1×10^8 m^{2}s^{-1}) for $\delta T \pm 4$ K	± 0.572	± 0.572
$\delta\alpha$ (1×10^8 m^{2}s^{-1}) for $\delta T \pm 10$ K	± 1.430	± 1.430

Table 3. Variation of the thermal diffusivity of the tungsten target with temperature [1]

Parameter	Channel Inlet	Nose of target
T (K)	306.45	521.15
α (1×10^5 m^{2}s^{-1})	4.732	3.206
$\delta\alpha$ (1×10^7 m^{2}s^{-1}) for $\delta T \pm 1$ K	± 0.847	± 0.576
$\delta\alpha$ (1×10^7 m^{2}s^{-1}) for $\delta T \pm 4$ K	± 3.394	± 2.310
$\delta\alpha$ (1×10^7 m^{2}s^{-1}) for $\delta T \pm 10$ K	± 8.523	± 5.812

ICHWF case: The mesh comprised of about 10 million hexahedral cells with a resolution of 128 by 256 nodes in the cross-section and ~316 cells along the length of the channel. The first node at the wall is 0.5 μm from the wall. A mapped inlet velocity profile is applied to the inlet without using the synthetic eddy method perturbing the flow.

CHT case: The mesh comprised of about 214 million hybrid cells constructed from three parts with four regions modelled focussing on the 148.5 mm closest to the nose of the target, where most of the heat is produced. Two hexahedral meshes were used to model both the inner and outer fluid channels plus about 0.5 mm of the cladding surrounding the channels (16 cells over the wall depth). The first node from either side of the solid-fluid interface at the wall was 1.0 μm with a resolution of 64 by 512 nodes in the cross-section of the fluid. There were 1640 cells along the length (~0.3 m) of the inner channel and 1768 cells along the length (~0.32 m) of the outer channel. Thus, there are 54 million and 58 million hexahedral cells in the respective channels. The maximum values of $y^+ = 11.0$ and the $T^+ = 9.3$ were obtained with the respective mean values of 0.18 and 0.45 for the LES. The third part comprised two regions corresponding to the tungsten target with 17.5 million hexahedral cells and the tantalum cladding with 85 million hybrid cells. The hybrid cells were used due to the overlapping of the two channels. A mapped inlet velocity profile is applied to the inlet of the inner channel without the application of the synthetic eddy method to perturb the flow, while a mean velocity is applied to the inlet of the outer channel.

3 Results and Conclusions

Profiles of the instantaneous, mean and the standard deviation of the temperature of the inner wall of the ICHWF are plotted in Figs. 2 and 3. The LES contours show high temperature regions that correspond to the location of the proton beam, but hotter regions are also observed at the entrance and exit of the bend section, as well as stripes that form near to the side walls of the channels. The stripes of hot and cold fluid at the walls persist after ~5 s of averaging, noting that the flow-through time for the section modelled is ~0.03 s and the frequency of the proton beam pulse is 0.1 s. In Fig. 4, a frequency analysis of the temperature profiles obtained at the nose of the target indicate that there are low frequency vortices (40 Hz–50 Hz), with some higher frequency events at around 3500 Hz.

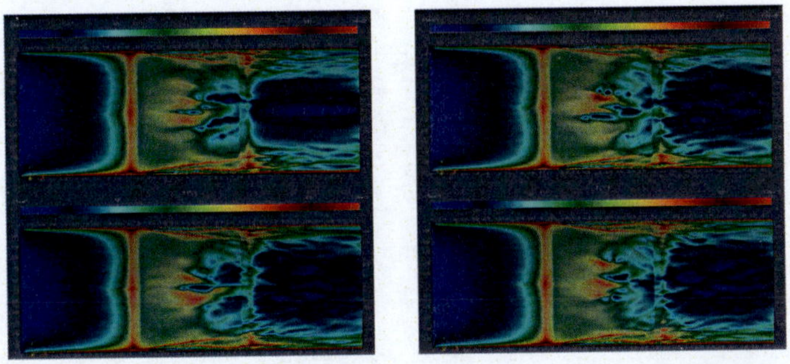

Fig. 2. Contours of the instantaneous temperature on the inner wall of the inner channel for the ICHWF case

The simulation for the CHT case were initialized with a solution obtained from a lower resolution case, where an initial flow field had been allowed to develop for a couple of flow-throughs, then the only the temperature field was resolved until a steady state was achieved. The lower resolution case took 25 s to reach a steady-state temperature. After the solution of the lower resolution case was interpolated onto the 214M cell mesh, the full flow equations and temperature field were resolved for a further 0.024 s.

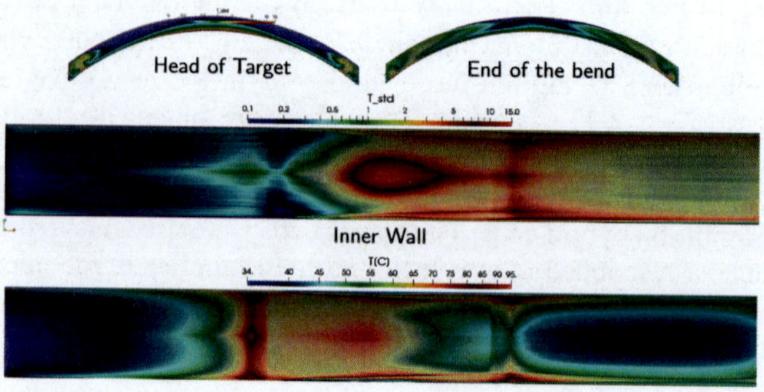

Fig. 3. Contours of the time averaged temperature and the standard deviation of the temperature for the ICHWF case

The calculations for the 214M cell case are on-going, while the CHT case converges to new thermal steady state. Therefore, we only present instantaneous contour fields of the temperature and the heat flux in the shell of the target to show that thermal striping may occur around the bend of the inner channel (Fig. 5). Figure 6 shows the regions in the cladding where the standard deviation of the temperature is greater than 1 K, with the plot on the left showing the RMS temperature and the plot on the right the instantaneous temperatures.

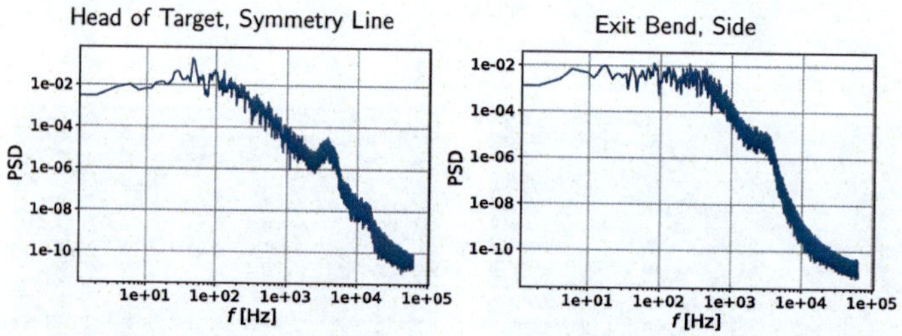

Fig. 4. Frequency analysis of profiles obtained from the nose and exit of the bend of the ICHWF case

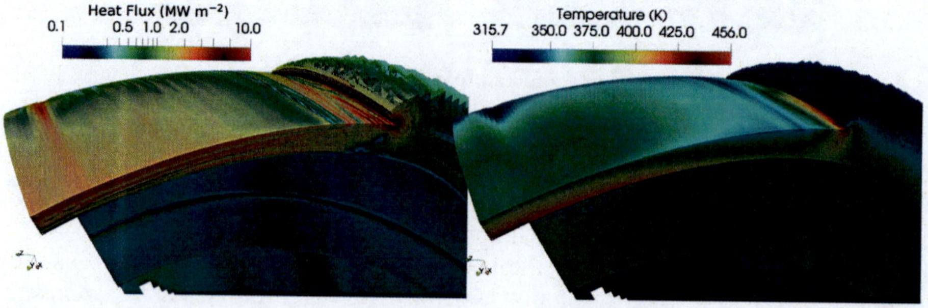

Fig. 5. Contours of heat flux and temperature on a quarter section of the solid tungsten cladding, which is in contact with both the inner channel and the target from the CHT case

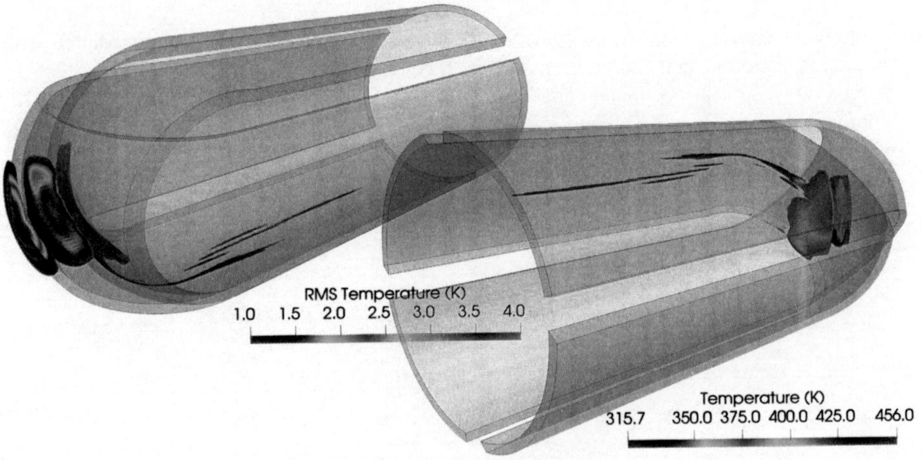

Fig. 6. Profiles of the instantaneous temperature on the fluid side of the inner wall of the inner channel at the nose of the target from the CHT case

Studies are on going to increase the resolution of the ICHWF and CHT cases LES to further elucidate the features we observe in these simulations and to have a better resolution of the very thin thermal boundary layer. Our concern is that low frequency temperature fluctuations may have caused thermal fatigue, alongside the stresses applied by the beam pulses, while the higher frequency oscillations may exert other stresses on the cladding and the target.

Acknowledgements. The authors are grateful to the UK Science and Technology Facilities Council (STFC) for the funding under the grant ST/W002388/1. The authors are also grateful to the UK Turbulence Consortium (UKTC) for the computational time on the Archer2 HPC system and to the UK Engineering and Physical Sciences Research Council (EPSRC) under grants EP/L000261/1, EP/N016602/1, EP/K038427/1 and EP/N030028/1.

References

1. Bejan, A., Kraus, A.D.: Heat Transfer Handbook. John Wiley & Sons, Hoboken (2003). Tables 2.8z and 2.8ac. https://app.knovel.com/hotlink/toc/id:kpHTH00003/heat-transfer-handbook/heat-transfer-handbook. Cited 14 Nov 2022
2. Caretta, O., Davenne, T., Densham, C.: Water erosion tests on a tantalum sample: a short communication. J. Nucl. Mater. **492**, 52–55 (2017)
3. Ferrari, A., Sala, P.R., Fassò, A., Ranft, J.: FLUKA: a multi-particle transport code. Technical report CERN-2005-10, INFN/TC_05/11, SLAC-R-773, CERN (2005)
4. Fournier, Y., Bonelle, J., Moulinec, C., Shang, Z., Sunderland, A., Uribe, J.: Optimizing code saturne computations on petascale systems. Comput. Fluids **45**, 103–108 (2011)
5. Gómez Romera, J.J.: IAPWS Python Module – jjgomera/iapws: (v1.5.2). Zenodo. https://doi.org/10.5281/zenodo.4744318. Cited 11 Nov 2022
6. Habainy, J., Dai, Y., Lee, Y., Iyengar, S.: Thermal diffusivity of tungsten irradiated with protons up to 5.8 dpa J. Nucl. Mater. **509**, 152–157 (2018)

7. Habainy, J., Dai, Y., Lee, Y., Iyengar, S.: Mechanical properties of tungsten irradiated with high-energy protons and spallation neutrons. J. Nucl. Mater. **514**, 189–195 (2019)
8. Wilcox, D.: Thermo-mechanical analysis of ISIS TS2 target. Technical report, STFC (2013)
9. Wilcox, D., Loveridge, P., Davenne, T., Jones, L., Jenkins, D.: Stress levels and failure modes of tantalum-clad tungsten targets at ISIS. J. Nucl. Mater. **506**, 76–82 (2018)
10. Wilcox, D.: ISIS Target modeling. (2021) https://conference.sns.gov/event/267/contributions/485/. Cited 27 Feb 2022

Modulation of Turbulence Flux Budgets by Varying Fluid Properties in Heated High Prandtl Number Flow

C. Irrenfried and H. Steiner(✉)

Institute of Fluid Mechanics and Heat Transfer, Graz University of Technology, Graz, Austria
helfried.steiner@tugraz.at

1 Introduction

High Prandtl number liquids are most commonly used for lubrication and cooling of various technical devices. The large disparity of the smallest turbulent thermal and dynamic scales strongly challenges the numerical modelling of the transport of heat and momentum. The inherently steeper temperature gradients due to the high thermal resistance generally leads to a marked variation of the fluid properties, whose often assumed neglect further adds to predictive uncertainty. The present Direct Numerical Simulation (DNS) study specially addresses this issue considering a typical heat transfer oil with a realistic variation in material properties in the heated near wall region. The effect of this variation on the turbulence budgets is highlighted by comparison against corresponding results with constant fluid properties, as it has been mostly assumed in existing DNS or LES studies on high Prandtl number flow in literature [1,2].

2 Governing Equations and Numerical Setup

The present DNS solved the conservation equations of mass, momentum, and energy, generally written in non-dimensional representation as

$$\frac{\partial \rho^*}{\partial t} + \nabla \cdot (\rho^* \mathbf{U}^+) = 0 \tag{1}$$

$$\frac{\partial \rho^* \mathbf{U}^+}{\partial t} + \nabla \cdot (\rho^* \mathbf{U}^+ \, \mathbf{U}^+) = -\nabla P^+ + \nabla \cdot \left[\frac{\mu^*}{Re_\tau} \left(2\underline{S} - \frac{2}{3} tr(\underline{S}) \underline{\mathbf{I}} \right) \right] + \mathbf{f}_w \tag{2}$$

$$\frac{\partial \chi^+}{\partial t} + \nabla \cdot (\mathbf{U}^+ \, \chi^+) - \chi^+ \nabla \cdot \mathbf{U}^+ = \frac{1}{\rho^* Re_\tau Pr_w} \nabla \cdot \left(\frac{\lambda^*}{c_p^*} \nabla \chi^+ \right) + f_\chi, \tag{3}$$

respectively, using the pipe diameter D, the wall friction velocity $w_\tau = (\tau_w/\rho_w)^{1/2}$, and friction enthalpy $h_\tau = \langle q_w \rangle / \rho_w w_\tau$ as reference scales. $\langle q_w \rangle$ represents the imposed constant average wall heat flux. $\mathbf{U}^+ = \mathbf{U}/w_\tau = (u^+, v^+, w^+)^T$ and $\chi^+ = (h_w - h)/h_\tau$ represent the non-dimensional velocity vector in cylindrical coordinates and enthalpy difference to the wall value, respectively. The asterisked material properties always refer to the corresponding wall values. $Re_\tau = \rho_w w_\tau D/\mu_w$ and $Pr_w = \mu_w c_{p,w}/\lambda_w$ denote the Reynolds and Prandtl number based on wall conditions, respectively. The terms $\mathbf{f}_w = 4\,\mathbf{e}_z$ and $f_\chi = w^+/\dot{m}^*$ with $\dot{m}^* = \dot{m}/D^2 \pi \rho_w w_\tau$ arise from axial periodicity.

C. Marchioli et al. (Eds.): DLES 2023, ERCO 31, pp. 69–74, 2024.
https://doi.org/10.1007/978-3-031-47028-8_11

Figure 1 shows the computational domain. No-slip boundary conditions are prescribed for the velocity, $\mathbf{U}^+ = 0$, and an uniform time averaged wall heat flux with with $\chi^+ = 0$ is imposed at the wall. The computational mesh consists of $256 \times 1024 \times 2048$ cells, with the corresponding near wall resolution of $\Delta r^+ = 0.05$, $R^+\Delta\varphi = 1.17$, and $\Delta z^+ = 0.90$ wall units in the radial, azimuthal, and axial direction, respectively.

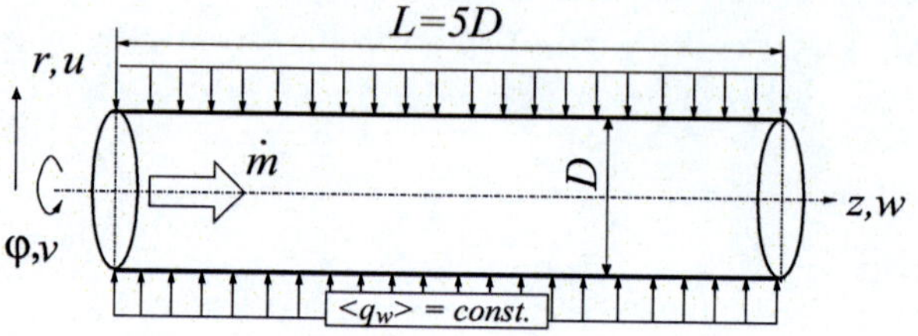

Fig. 1. Computational domain

3 Results

The investigated fully developed pipe flow is specified by the Reynolds number $Re_\tau = 380$ and the molecular Prandtl number $Pr_w = 56$, based on the average wall temperature $\langle T_w \rangle = 386.0K$. The uniform wall heat flux is $\langle q_w \rangle = 1.6 \times 10^4 W/m^2$. The results for the variable and constant fluid properties are always referred to as "VFP" and "CFP", respectively.

The variation of the material properties of the considered oil relative to the wall values is shown in Fig. 2. The dynamic viscosity quantitatively varies strongest, while the changes in density, thermal conductivity, and specific heat capacity are negligibly small. The viscosity induced decrease of the local Reynolds number consistently leads

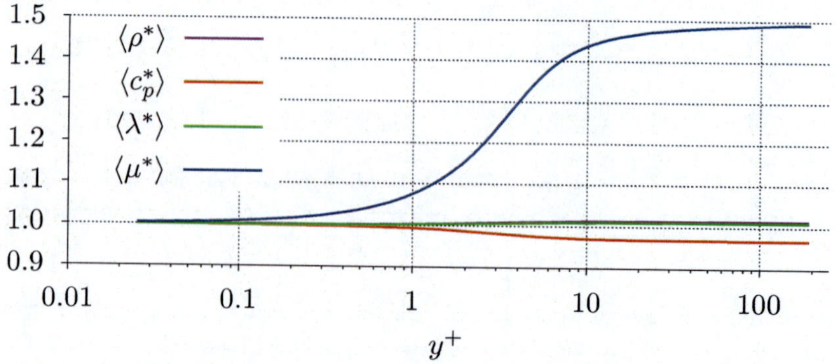

Fig. 2. Average fluid properties vs. wall distance y^+

to strong dampening of the turbulent fluctuations of the radial and azimuthal velocity components, while the peak of the axial component is rather increased and shifted towards the center, as shown in Fig. 3. The thermal fluctuations $\langle \chi'^{+2} \rangle^{1/2}$ are strongly enhanced due to the viscosity induced increase in the local molecular Prandtl number.

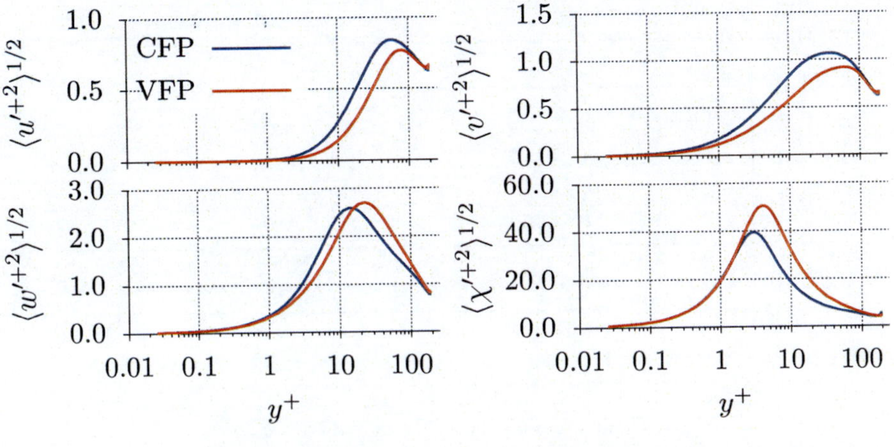

Fig. 3. Fluctuations of velocity and enthalpy vs. y^+

Despite the observed opposite tendencies of the turbulent velocity and thermal fluctuations, the increased viscous dampening of the radial velocity fluctuation still dominates the turbulent components of the flux budgets for moment and heat, written as

$$\tau_{tot}^{+} = \frac{\tau_{tot}}{\tau_w} = \tau_{lam}^{+} + \tau_{turb}^{+} = \mu^* \frac{\partial \langle w^+ \rangle}{\partial y^+} + \rho^* \langle u'^+ w'^+ \rangle \tag{4}$$

$$q_{tot}^{+} = \frac{q_{tot}}{\langle q_w \rangle} = q_{lam}^{+} + q_{turb}^{+} = \frac{1}{Pr_w} \frac{\lambda^*}{c_p^*} \frac{\partial \langle \chi^+ \rangle}{\partial y^+} + \rho^* \langle u'^+ \chi'^+ \rangle, \tag{5}$$

respectively. As shown in Fig. 4, the turbulent contributions are always decreased near the wall, while in turn the laminar component increase. The increased laminar momentum flux seen for VFP is however largely covered by the also notably increasing molecular viscosity, which effectively keeps the average axial velocity for VFP very close to the profile for CFP near the wall as shown in Fig. 5(a). The resulting wall friction coefficient only marginally changes from $C_{f,CFP} = 0.0093$ to $C_{f,VFP} = 0.0091$.

Such compensatory effect of a locally changing transfer coefficient is not equivalently seen for the heat transfer.

The almost uniform thermal conductivity and specific heat capacity do not cover the increase in the laminar thermal flux. This inherently leads to larger enthalpy gradients near the wall, which finally produces a significantly higher bulk-to-wall difference of

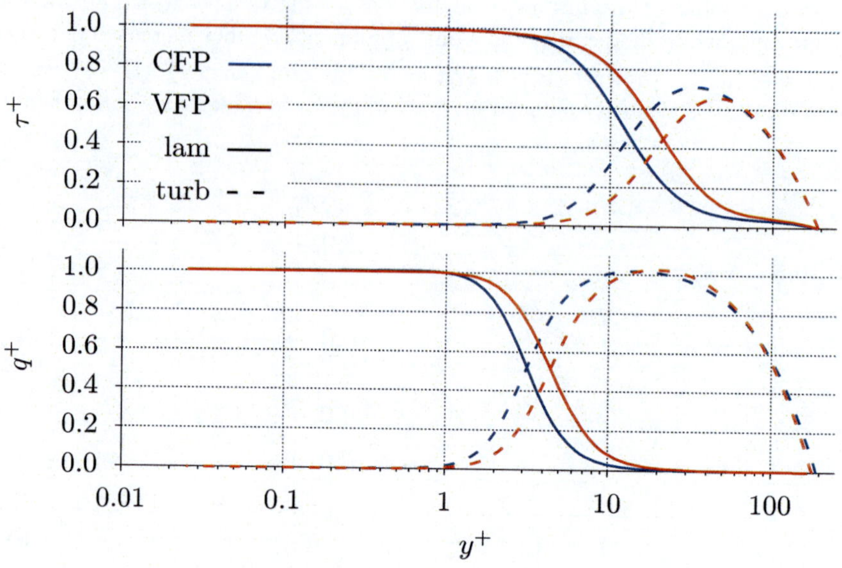

Fig. 4. Shear stress and heat flux vs. wall distance y^+

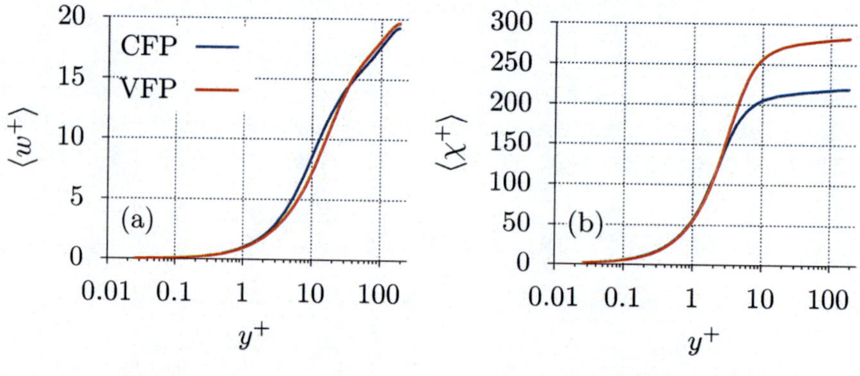

Fig. 5. (a) Axial velocity and (b) heat flux vs. wall distance y^+

the enthalpy for VFP, as seen from Fig. 5(b), indicating a markedly decreased global heat transfer coefficient. Accordingly, the Nusselt number drops from $Nu_{CFP} = 98.4$ to $Nu_{VFP} = 76.5$.

The different trends observed for VFP in the velocity fluctuations are reflected by the turbulence budgets for the axial and the radial velocity in Fig. 6. Due to the reduced turbulent flux component $\langle u'^+ w'^+ \rangle$ and shear rate close to the wall, the production term of the axial velocity, written as

$$P_{zz} = \langle u'^+ w'^+ \rangle \frac{\partial \langle w^+ \rangle}{\partial y^+} \tag{6}$$

remains consistently lower near the wall for VFP, before it rises up to a peak. Further inside beyond $y^+ \approx 20$, it even exceeds the values for CFP, which explains the occurrence of the peak of the axial fluctuation $\langle w'^{+2} \rangle^{1/2}$ seen for VFP at this position in Fig. 3. The turbulence energy produced by P_{zz} is primarily redistributed by the velocity-pressure gradient interaction term, written as

$$\Pi_{rr} = \langle 2u'^+ \frac{\partial p'^+}{\partial y^+} \rangle. \tag{7}$$

Accordingly, the notably lower Π_{rr} up to $y^+ \approx 60$ seen for VFP in Fig. 6(b) clearly indicates an attenuated redistribution of turbulence energy, effectively leading to more anisotropic fluctuating motion in the turbulent inner region. The opposite tendencies exhibited by $\langle u'^{+2} \rangle^{1/2}$ and $\langle \chi'^{+2} \rangle^{1/2}$ for VFP are also clearly reflected by the variation of the production terms of the budgets of $\langle \chi'^{+2} \rangle$ and the turbulent heat flux $\langle u'^+ \chi'^+ \rangle$, defined as

$$P_{\chi\chi} = \langle u'^+ \chi'^+ \rangle \frac{\partial \langle \chi^+ \rangle}{\partial y^+}, \tag{8}$$

$$P_{u\chi} = \langle u'^+ u'^+ \rangle \frac{\partial \langle \chi^+ \rangle}{\partial y^+}, \tag{9}$$

respectively. As shown in Fig. 7, the shift and increase in the thermal fluctuations χ'^+ seen for VFP in Fig. 3 essentially follows from the radial shift of the peak for the production term $P_{\chi\chi}$, with a rather marginal change in magnitude. On the other hand, the production term for the turbulent heat flux $P_{u\chi}$, consistently remains well below the values for CFP.

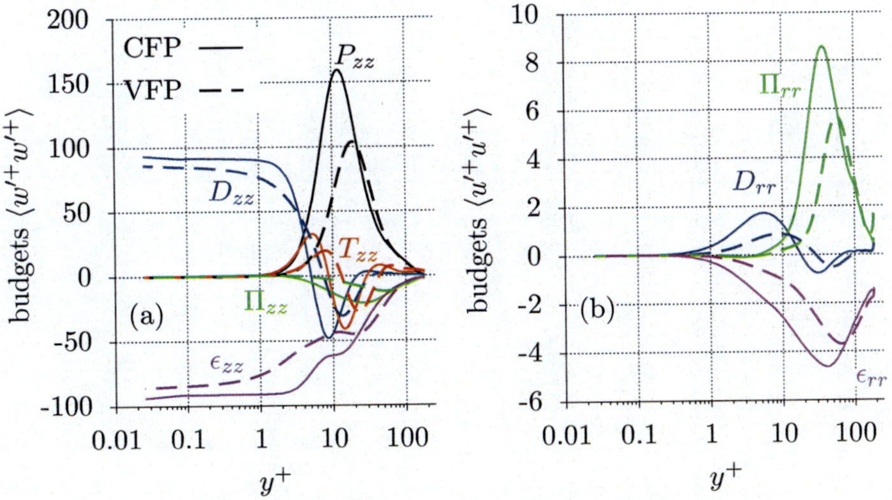

Fig. 6. Budgets of $\langle w'^+ w'^+ \rangle$ and $\langle u'^+ u'^+ \rangle$

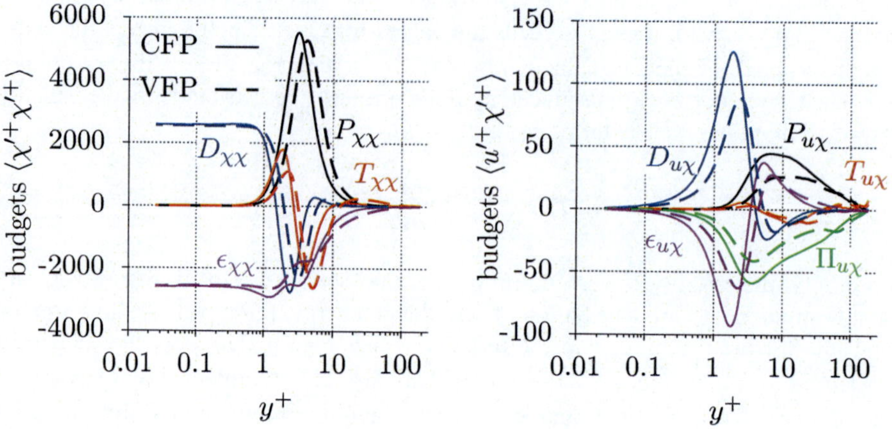

Fig. 7. Budgets of $\langle\chi'^+\chi'^+\rangle$ and $\langle u'^+\chi'^+\rangle$

4 Conclusions

The different trends in the local Reynolds and Prandtl numbers with distance to the heated wall give rise to different tendencies in the turbulent fluctuations. The strong dampening of the radial velocity essentially determines the significant reduction in heat transfer rate seen for the variable property case. The increase in thermal fluctuations appears as not sufficient to compensate for the deficit in turbulent convective mixing. The observed differences in the production and redistribution of the turbulence budgets counteract the general tendency towards a more isotropic fluctuating motion away from the wall.

References

1. Kawamura, H., Ohsaka, K., Abe, H., Yamamoto, K.: DNS of turbulent heat transfer in channel flow with low to medium-high Prandtl number fluid. Int. J. Heat Mass Trans. **19**, 482–491 (1998)
2. Schwertfirm, F., Manhart, M.: DNS of passive scalar transport in turbulent channel flow at high Schmidt numbers. Int. J. Heat Fluid Flow **28**, 1204–1214 (2007)

A Posteriori LES of Forced Convection Along Heated and Cooled Walls with Temperature-Dependent Fluid Properties

L. Sufrà and H. Steiner(✉)

Institute of Fluid Mechanics and Heat Transfer, Graz University of Technology,
Inffeldgasse 25F, 8010 Graz, Austria
helfried.steiner@tugraz.at

1 Introduction

The forced convective transfer of momentum and heat is frequently subject to a strong variation in local material properties, especially for working fluids with a high molecular Prandtl number, typically involving steep near wall gradients of temperature. The inherent disparity of smallest thermal and dynamic length scales does not only seriously challenge the spatial resolution, it also questions the commonly assumed Reynolds analogy, relating the turbulent heat flux directly to the turbulent flux of momentum. The present LES study particularly assesses three popular subgrid-scale models [1–3] in capturing adequately the feedback of the variation of the material properties on the turbulent motion near heated or cooled walls. Data from highly resolved Direct Numerical Simulations (DNS) [4] are used for validation. Fully developed turbulent pipe flow is considered as generic test configuration, assuming a real-life coolant, a $50/50Vol\%$ mixture of ethylene glycol/water, with a molecular Prandtl number $Pr_w = 10$, based on wall conditions. The strongly varying dynamic viscosity significantly affects here the local molecular Prandtl number and Reynolds number inducing either a tendency to relaminarization or enhancement of turbulent mixing. Thus far, rather few DNS and LES studies addressed this issue, examining mainly the effect of a temperature-dependent viscosity on the turbulent flow field, while giving no or comparatively little attention to the turbulent heat transfer [5,6].

2 Mathematical Formulation and Tested Sgs-Models

The present LES solved the Favre-filtered conservation equations of mass, momentum, and energy, generally written in non-dimensional representation as

$$\frac{\partial \widetilde{\rho}^*}{\partial t} + \nabla \cdot \left(\widetilde{\rho}^* \overline{\mathbf{U}}^+ \right) = 0 \tag{1}$$

$$\frac{\partial \widetilde{\rho}^* \overline{\mathbf{U}}^+}{\partial t} + \nabla \cdot (\widetilde{\rho}^* \overline{\mathbf{U}}^+ \, \overline{\mathbf{U}}^+) = -\nabla \widetilde{P}^+$$
$$+ \nabla \cdot \left[\frac{\widetilde{\mu}^*}{Re_\tau} \left(2 \underline{\widetilde{S}} - \frac{2}{3} tr(\underline{\widetilde{S}}) \underline{\mathbf{I}} \right) \right] - \nabla \cdot \tau_{sgs} + \widetilde{\mathbf{f}}_w \tag{2}$$

C. Marchioli et al. (Eds.): DLES 2023, ERCO 31, pp. 75–80, 2024.
https://doi.org/10.1007/978-3-031-47028-8_12

$$\frac{\partial \overline{\chi}^+}{\partial t} + \nabla \cdot (\overline{\mathbf{U}}^+ \, \overline{\chi}^+) - \overline{\chi}^+ \nabla \cdot \overline{\mathbf{U}}^+ =$$

$$\frac{1}{\widetilde{\rho}^* Re_\tau Pr_w} \nabla \cdot \left(\frac{\widetilde{\lambda}^*}{\widetilde{c}_p^*} \nabla \widetilde{\chi}^+ \right) - \frac{1}{\widetilde{\rho}^*} \nabla \cdot \mathbf{q}_{sgs} + \widetilde{f}_\chi , \tag{3}$$

respectively, using the pipe diameter D, the wall friction velocity $w_\tau = (\tau_w/\rho_w)^{1/2}$, and friction enthalpy $h_\tau = \langle q_w \rangle / \rho_w w_\tau$ as reference scales. $\langle q_w \rangle$ represents the imposed constant average wall heat flux, which is positive/negative in the heated/cooled case. $\overline{\mathbf{U}}^+ = \overline{\mathbf{U}}/w_\tau = (\bar{u}^+, \bar{v}^+, \bar{w}^+)^T$ and $\overline{\chi}^+ = (h_w - \bar{h})/h_\tau$ represent the filtered non-dimensional velocity vector in cylindrical coordinates and filtered enthalpy difference to the wall value, respectively. The temperature-dependent asterisked material properties always refer to the corresponding wall values. $Re_\tau = \rho_w w_\tau D/\mu_w$ and $Pr_w = \mu_w c_{p,w}/\lambda_w$ denote the Reynolds and Prandtl number based on wall conditions, respectively. The terms $\widetilde{\mathbf{f}}_w = 4\mathbf{e}_z$ and $\widetilde{f}_\chi = \overline{w}^+/\dot{m}^*$ with $\dot{m}^* = \dot{m}/D^2 \pi \rho_w w_\tau$ enforce axial periodicity.

The computational domain is sketched in Fig. 1. At the wall, no-slip boundary conditions are assumed for the velocity, $\overline{\mathbf{U}}^+ = 0$, and isothermal boundary conditions are imposed for the instantaneous enthalpy difference, $\overline{\chi}^+ = 0$, The near wall resolution of the LES grid is $\Delta r^+ = 0.29$, $R^+ \Delta\varphi = 8.8$, and $\Delta z^+ = 28.1$ wall units in the radial, azimuthal, and axial direction, respectively.

The unresolved subgrid-scale fluxes are modelled based on the Boussinesq eddy viscosity/diffusivity concept

$$\tau_{sgs}^d = -2\widetilde{\rho}^* \frac{\nu_{sgs}^*}{Re_\tau} \widetilde{S}^d, \quad \mathbf{q}_{sgs} = -\frac{1}{Re_\tau \widetilde{c}_p^*} \frac{\nu_{sgs}^*}{Pr_{sgs}} \nabla \widetilde{\chi}^+ \tag{4}$$

with constant $Pr_{sgs} = 0.5$. Three popular models for the subgrid-scale eddy viscosity $\nu_{sgs}^* = \nu_{sgs}/\nu_w$ are tested: the Smagorinsky (SMAG) [1], the Wall-Adapting Local Eddy Viscosity (WALE) [3], and the Coherent Structure Model (CSM) [2]. The Smagorinsky model (SMAG) enforces vanishing eddy viscosity near the wall applying a Van Driest-type damping function. CSM and WALE inherently provide this important feature in their model formulation for ν_{sgs}^*.

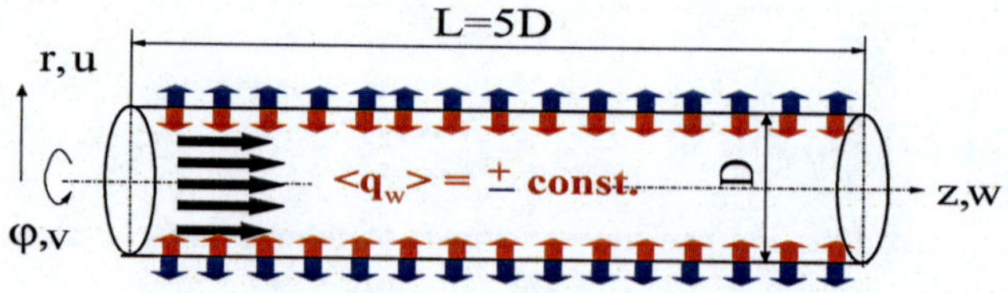

Fig. 1. Computational domain

3 Results

We considered fully developed pipe flow at Reynolds number $Re_\tau = 360$ and molecular Prandtl number $Pr_w = 10$, based on wall conditions, specified by the average wall tem-

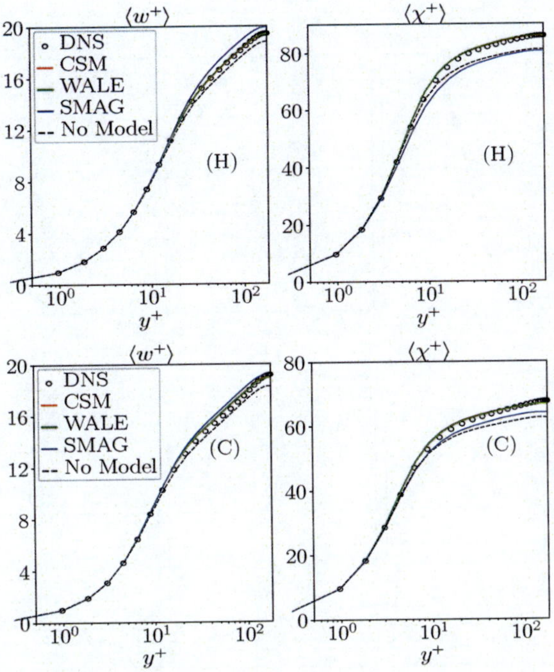

Fig. 2. Mean axial velocity and enthalpy difference vs. wall distance y^+, (H) heated case, (C) cooled case, $Re_\tau = 360$

perature being always $\langle T_w \rangle = 344.8K$. The wall is uniformly heated/cooled by a wall heat flux $\langle q_w \rangle = \pm 2 \cdot 10^4 W/m^2$, respectively. Figure 2 compares the predicted first order statistics against DNS, including also results from LES without a sgs-model. CSM and WALE produce evidently very good agreement with DNS over the whole y^+-range, while the LES with SMAG generally overestimates the velocity and underestimates the enthalpy difference. This discrepancy can be explained by the budgets for the total shear stress and heat flux, rewritten as

$$\tau_{tot}^+ = -\langle \mu^* \rangle \frac{\partial \langle \overline{w}^+ \rangle}{\partial y^+} - \langle \tilde{\rho}^* \overline{u}'^+ \overline{w}'^+ \rangle - \tau_{rz,sgs} \tag{5}$$

$$q_{tot}^+ = -\frac{\langle \lambda^* \rangle}{Pr_w \langle c_p^* \rangle} \frac{\partial \langle \overline{\chi}^+ \rangle}{\partial y^+} - \langle \tilde{\rho}^* \overline{u}'^+ \overline{\chi}'^+ \rangle - q_{r,sgs} \tag{6}$$

respectively. SMAG notably overestimates the turbulent heat flux component near the wall, while the turbulent shear stress is predicted somewhat too low, as shown in Fig. 3 for the heated case. The accordingly lower/higher laminar components translate into a faster/slower increase in the axial mean velocity $\langle w^+ \rangle$ and mean enthalpy difference $\langle \overline{\chi}^+ \rangle$, respectively, always exhibited by SMAG. The underestimated $\langle \overline{\chi}^+ \rangle$ also implies higher near wall temperatures leading to lower predicted molecular viscosities, as seen in Fig. 4. This effectively yields a higher velocity gradient in the laminar shear

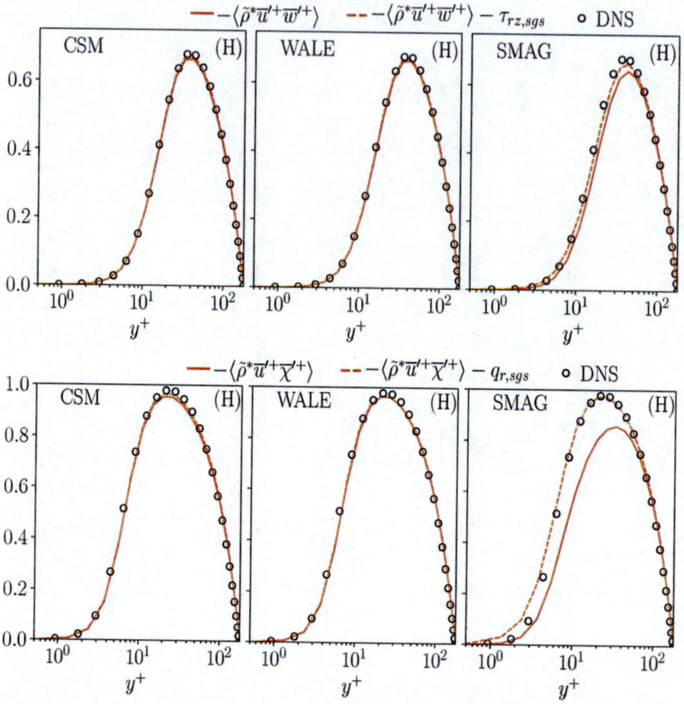

Fig. 3. Resolved and modelled turbulent fluxes of momentum and heat, heated case, $Re_\tau = 360$

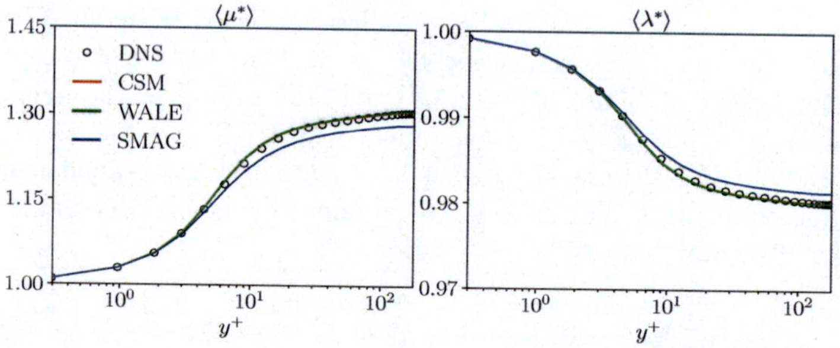

Fig. 4. Material properties relative to wall value vs. wall distance y^+: molecular viscosity and thermal conductivity, heated case, $Re_\tau = 360$

stress component and, hence, further increases the overestimate of $\langle w^+\rangle$. Similarly, the underestimated thermal conductivity additionally increases the deviation of $\langle \chi^+\rangle$. For the cooled case, sgs-model contributions of SMAG basically exhibit the same, albeit less pronounced, deficiencies. The deviations in the predicted first order results are still smaller, as indicated by the cooled case results in Fig. 2. The observed better agreement of SMAG can be partly attributed to the fact, that inaccuracies in the predicted material properties rather have here a beneficial effect, unlike in the heated case.

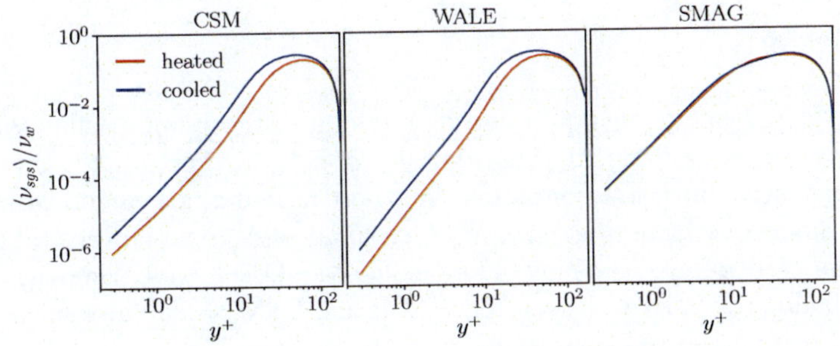

Fig. 5. Subgrid-scale eddy viscosity $\nu^*_{sgs} = \nu_{sgs}/\nu_w$ vs. wall distance y^+, $Re_\tau = 360$

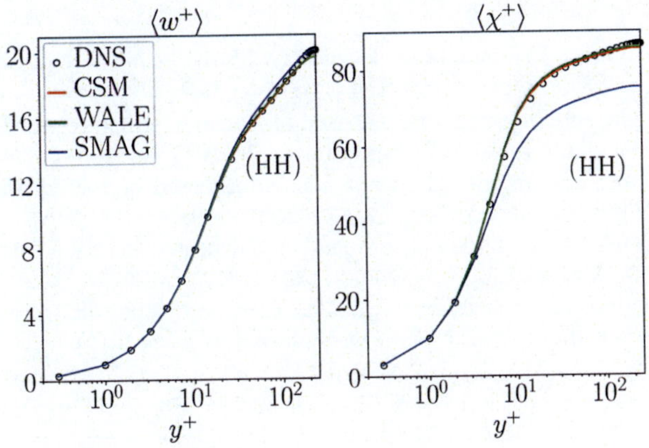

Fig. 6. Mean axial velocity and enthalpy difference vs. wall distance y^+, $Re_\tau = 500$ heated case with $\langle q_w\rangle = 4 \cdot 10^4 W/m^2$

Overall, CSM and WALE apparently provide a small, but still essential subgrid-scale model contribution, as indicated by the notable discrepancy of the LES without sgs-model in Fig. 2. The SMAG predicted subgrid-scale eddy viscosity shown in Fig. 5 appears as fairly insensitive to the decrease/increase of the local Reynolds number, as the molecular viscosity decreases/increases with distance to the heated/cooled wall, while CSM and WALE reasonably reflect this trend.

The validity of the present observations was also confirmed for a more challenging case, assuming a higher Reynolds number $Re_\tau = 500$ and a doubled wall heat flux $\langle q_w\rangle = 4 \cdot 10^4 W/m^2$. For all sgs models, the observed trends are consistently similar to the lower Reynolds number case, as exemplarily indicated by the average profiles in Fig. 6. As a general remark on the influence of the Reynolds number, it is noted, that the enhanced turbulent transport at higher Re_τ inherently reduces the wall-to-bulk temperature difference, for given wall heat flux. This trend towards more thermal uniformity with increasing Reynolds number basically mitigates the issue of temperature-dependent varying fluid properties in the sgs modelling, unless the imposed wall heat flux becomes extremely high.

4 Conclusions

For the considered operating liquid and flow conditions, the approaches WALE and CSM are proven as well-suited subgrid-scale models for capturing adequately the dampened/enhanced turbulent convective transport near the heated/cooled pipe wall. The SMAG model overly dissipates the resolved small-scale motion instead. This deficit is not appropriately compensated by the modelled Van Driest-style dampened subgrid-scale contribution, as otherwise seen for WALE and CSM, which also need to model considerably less unresolved small-scale content.

References

1. Smagorinsky, J.: General circulation experiments with the primitive equations: I. The basic experiment. Mon. Weather Rev. **91**, 99–164 (1963)
2. Kobayashi, H.: The subgrid-scale models based on coherent structures for rotating homogeneous turbulence and turbulent channel flow. Phys. Fluids **17**, 45–104 (2005)
3. Nicoud, F., Ducros, F.: Subgrid-scale stress modelling based on the square of the velocity gradient tensor. Flow Turbul. Combust. **62**, 183–200 (1999)
4. Sufrà, L., Steiner, H.: Forced turbulent convection along heated and cooled walls with variable fluid properties. ASME. J. Fluids Eng. **144**, 011401 (2022)
5. Zonta, F., Marchioli, C., Soldati, A.: Modulation of turbulence in forced convection by temperature-dependent viscosity. J. Fluid Mech. **697**, 150–174 (2012)
6. Lee, J., Jung, S.Y., Sung, H.J., Zaki, T.A.: Effect of wall heating on turbulent boundary layers with temperature-dependent viscosity. J. Fluid Mech. **726**, 196–225 (2013)

Industrial and Environmental Applications

Wind Gust-Induced Flutter of an Elastically Mounted Airfoil: A Fluid-Structure Interaction Study Based on LES

K. Boulbrachene and M. Breuer(✉)

Professur für Strömungsmechanik, Helmut-Schmidt-Universität Hamburg, Hamburg, Germany
{boulbrak,breuer}@hsu-hh.de

1 Introduction

Classical investigations on the aeroelasticity of wings used for the analysis of flutter stability are often carried out based on simplifying model configurations given by elastically mounted rigid wings. For this purpose, either one degree of freedom (1-DOF) considering solely the pitch motion or two degrees of freedom (2-DOF: pitch and heave) are taken into account [5,6]. Concerning the Reynolds number, flows in the transitional regime are of specific interest due to two reasons. First, unmanned air vehicles including Micro Air Vehicles operate in this range. Second, specific flow phenomena such as laminar separation bubbles and transition render this a challenging case. Aeroelastic effects like laminar separation flutter that are different compared to high-Re investigations typically carried out for airplanes are expected. In a precursor study [9] experimental investigations on the fluid-structure interaction (FSI) of a 2-DOF elastically mounted wing were carried out in a wind tunnel. Complementary numerical predictions of the identical coupled FSI setup were performed [4] relying on large-eddy simulations (LES). The results show that coupled high-fidelity solvers allow to predict the FSI problem with an acceptable accuracy leading to either small/large-amplitude oscillations or flutter depending on the case and Re number considered. Note that no wind gusts were taken into account in this preceding study and thus the airfoil had to be initially excited by mechanical perturbations. The objective of the present contribution is to study the aeroelastic response of the same 2-DOF NACA 0012 airfoil in the transitional Reynolds number regime subjected to vertical wind gusts encountered in the atmosphere or the wake of a preceding aircraft. The effect of different gust parameters (gust lengths and amplitudes) on the dynamics of the airfoil is examined. Especially the question of interest is whether certain gusts can induce limit-cycle oscillations or even flutter.

2 Applied Methodology

The numerical methodology applied consists of a partitioned coupled solver combining large-eddy simulations on the fluid side with a solver for the governing equations of the translation and rotation of the rigid wing. The finite-volume Navier-Stokes solver works with curvilinear, block-structured body-fitted grids in the Arbitrary Lagrangian-Eulerian formulation and is second-order accurate in space and time [1]. The standard

C. Marchioli et al. (Eds.): DLES 2023, ERCO 31, pp. 83–88, 2024.
https://doi.org/10.1007/978-3-031-47028-8_13

Smagorinsky model ($c_s = 0.1$) with van-Driest damping near solid walls is applied. The rigid-body solver receives the fluid forces and moments, and predicts the translational and rotational motions based on Newton's second law using the standard Newmark method of second-order accuracy. The predicted structural displacements of the surface of the airfoil are the input parameters for the adaption of the inner grid. The hybrid grid adaption algorithm (IDW-TFI) developed by Sen et al. [7] is applied. Since in the present application the added-mass effect is negligible, a loose coupling algorithm is preferred. To inject vertical wind gusts into the computational domain, the split velocity method [8] was found to be superior [2] since it is capable of capturing the full interaction between the gust and the structure. Furthermore, the gust can be injected directly in front of the structure, which avoids the strong numerical dissipation associated with regions of coarse grids.

3 Test Case and Setup

As mentioned before, the FSI setup is inspired by precursor investigations [4,9]. The airfoil exposed to a constant free-stream velocity and vertical gravity undergoes heave and pitch motions modeled by a mass-spring-damper system for both DOFs. These are described by six structural parameters: The mass and the mass moment of inertia of the airfoil as well as one material stiffness and one material damping coefficient for each direction of movement; for details refer to [3,4]. In the configuration considered the center of gravity nearly coincides with the center of rotation. In [4,9] this setup was investigated for a wide range of transitional Re numbers. The most interesting case was found to be Re = 33,900, which is employed in the present study. Initiated by external perturbations, a series of sustained large bounded amplitude oscillations was detected in [4,9]. Here, the effect of gust-induced perturbations and their effects on the stability of the airfoil will be examined. For this purpose, vertical gusts which vary solely in the streamwise direction are introduced upstream of the airfoil and convected downstream with the free-stream velocity. Note that a deterministic gust model is applied using an Extreme Coherent Gust (ECG) often denoted "1-cosine" shape defined in the IEC-Standard [10]. In the present study gusts of different wavelengths ($L_g^{x_1}/c$ = 2, 4 and 6) and amplitudes (A_g/u_∞ = 1, 2 and 3) are investigated. Here c denotes the chord length and u_∞ the free-stream velocity.

A C-type grid with about 1.9 million control volumes (CVs) is used and 60 CVs are distributed in the spanwise direction assuming periodic boundary conditions. The grid points are clustered near the surface such that the viscous sublayer can be resolved and no-slip conditions can be applied. At the inlet and the bottom boundary the free-stream velocity is assumed. At the outlet and the top boundary a convective outflow boundary condition is prescribed to assure that the generated vortical structure can leave the computational domain without perceptible disturbances.

4 Results

The aerodynamic forces exerted on the airfoil are compared for the fixed and the elastically mounted airfoil. Figure 1 depicts the lift and pitching moment coefficients of the airfoil encountering a 1-cosine vertical gust with $L_g^{x_1}/c = 2$ and $A_g/u_\infty = 2$. Although identical curves are obtained during the airfoil-gust interaction (i.e., until $t^* = t\,u_\infty/c = 3$), the curves of the FSI case continue to oscillate around zero while their CFD counterparts recover to the steady-state values. Oscillating aerodynamic forces contribute to dampening as well as to amplifying the motion as found in [3]. The high-frequency components in the lift and moment coefficients are due to the laminar flow separation taking place at the lower and upper surface of the airfoil.

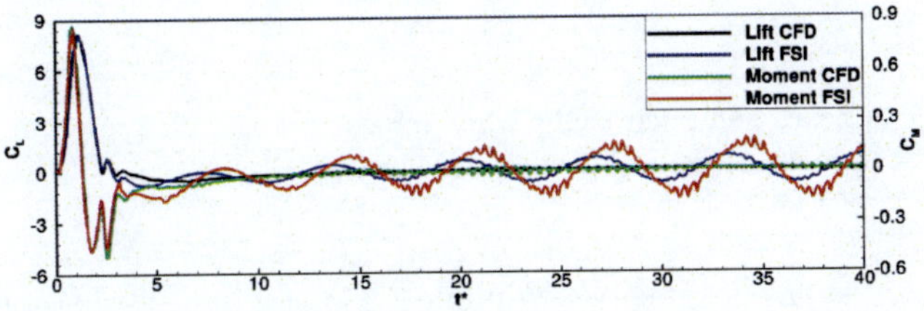

Fig. 1. Instantaneous lift and pitching moment coefficients of the fixed and elastically mounted airfoils subjected to a vertical gust with $L_g^{x_1}/c = 2$ and $A_g/u_\infty = 2$.

Figure 2 depicts four instants in time showing the development and transport of flow structures during the airfoil-gust encounter of the elastically mounted wing. Leading-edge vortices are formed and convected downstream when the airfoil undergoes heave and pitch motions as a result of the imposed aerodynamic loads.

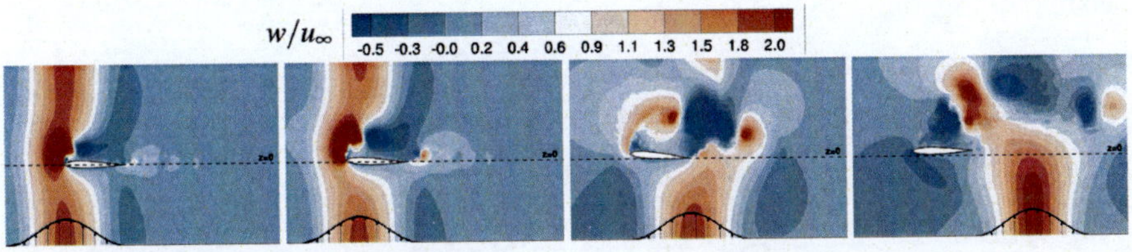

Fig. 2. Snapshots of the total vertical velocity for the elastically mounted airfoil. From left to right: Gust center at leading edge, gust center when lift is maximum, gust center at trailing edge, gust completely left the airfoil.

The aeroelastic response is found to be strongly dependent on the parameters of the gust. Figure 3 shows the instantaneous heave and pitch motion of an airfoil experiencing interaction with a vertical gust of length $L_g^{x_1}/c = 4$. For $A_g/u_\infty = 1$ and 2 the amplitudes of both DOFs initially increase upon the gust encounter followed by a steady decrease related to an overall positive damping effect of the aerodynamic forces. Both DOFs are found to eventually settle at a stage in which their amplitudes are sustained and bounded, i.e., a limit-cycle oscillation (LCO) appears. Similar patterns are observed for gusts with $\left(L_g^{x_1}/c = 2, A_g/u_\infty = 1, 2, 3\right)$ and $\left(L_g^{x_1}/c = 6, A_g/u_\infty = 1, 2\right)$. In contrast, for

$A_g/u_\infty = 3$ gusts with wavelength of $L_g^{x_1}/c \geq 4$ induces significantly larger excitations which trigger a pitch governed diverging motion of the airfoil known as stall flutter (see blue curve in Fig. 3).

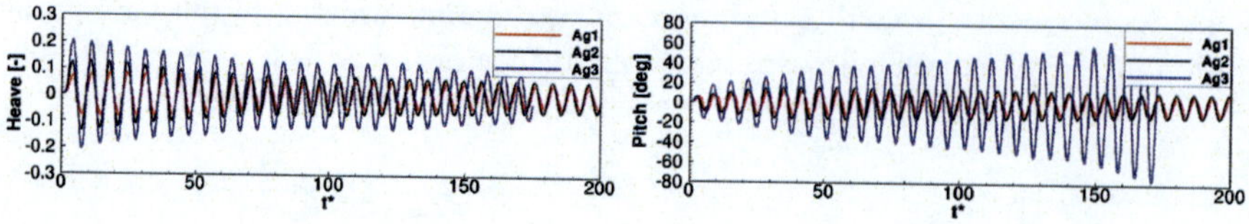

Fig. 3. Instantaneous heave and pitch DOFs when imposing a vertical gust with $L_g^{x_1}/c = 4$ and three different gust strength $A_g/u_\infty = 1, 2$ and 3. (Color figure online)

To investigate the physics involved within a cycle of a limit-cycle oscillation, snapshots of the spanwise y-vorticity ω_y, the skin-friction coefficient C_f, and the location of flow separation are depicted in Fig. 4. When the airfoil starts to pitch up reaching its maximum angle of attack (see Figs. 4a to 4d), the separation starting in the vicinity of the trailing edge moves upstream on the suction side. During the phase when the angle of attack is already decreasing again, vortices are shed from the suction side (see Figs. 4f and 4g). These vortices disappear when the angle of attack tends towards the other extremum (see Fig. 4h) and thus a favorable pressure gradient occurs on the corresponding side of the airfoil. At that stage the same process begins on the lower surface. The laminar separation taking place is mostly two-dimensional and transition to turbulence is detected further downstream justifying the spanwise varying separation line visible in Fig. 4h.

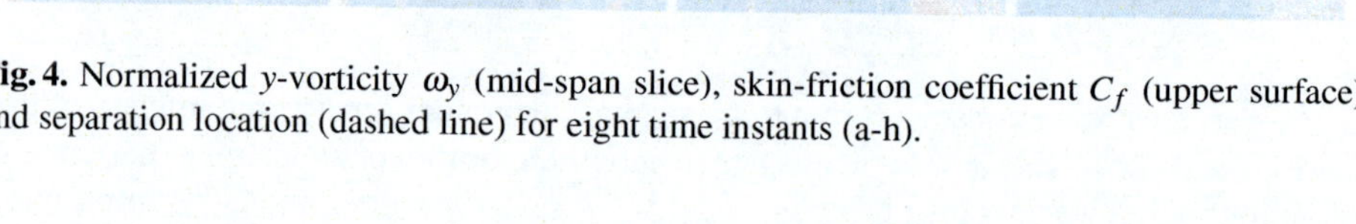

Fig. 4. Normalized y-vorticity ω_y (mid-span slice), skin-friction coefficient C_f (upper surface), and separation location (dashed line) for eight time instants (a-h).

Figure 5 depicts the aerodynamic hysteresis curves of the heave and pitch DOFs upon the interaction with two gusts either with ($L_g^{x_1}/c = 2$, $A_g/u_\infty = 2$) or ($L_g^{x_1}/c = 4$, $A_g/u_\infty = 3$). The former is a LCO case, while the latter is a flutter case. The arrows indicate the direction of advancing time. The black and red circles are used to identify the phases in which energy is being extracted from the flow or delivered to the flow. The direction in which the ellipse is formed indicates the sign of the work done by the aerodynamic coefficients (clockwise positive and counterclockwise negative) [5]. The loops of the two cases can be clearly distinguished. For the first case the clockwise rotation in Fig. 5a signifies positive work of the heave and Fig. 5b points to the overall damping effect of the pitch motion. The total energies shown in Fig. 5c prove that the system undergoes a LCO as the energies of the heave and pitch DOFs cancel out leading to a vanishing total energy of the system. Note that the heave motion is found to be responsible for the energy extraction in the LCO case which is consistent with the outcome of the hysteresis curves.

In contrast, the diverging motion case experiences an increase in the total energy as visible in Fig. 5f. The main part of the energy is extracted by the pitch motion and that continues until the numerical solver can no longer handle the strong deformation of the grid and finally crashes. The hysteresis curves in Figs. 5d and 5e do not have a single turning direction but have a clockwise middle loop and counterclockwise end loops instead. This points to the alternating role that the flow plays in supplying and dampening the energy of the system within a single cycle.

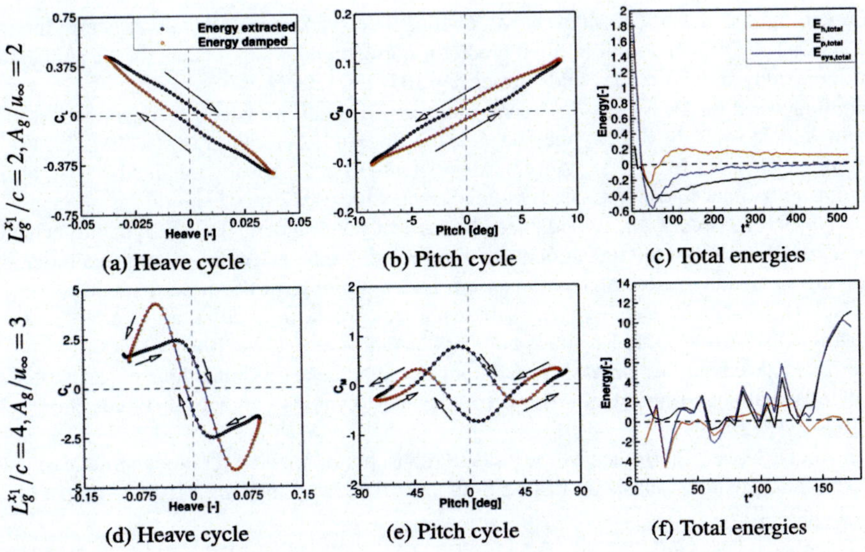

Fig. 5. (a), (b), (d), (e) Aerodynamic hysteresis curves within a single cycle. (c), (f) Instantaneous dimensionless energies of the system.

5 Conclusions

The split velocity method is integrated into a partitioned FSI solver to study the gust-induced oscillations of an elastically mounted 2-DOFs NACA 0012 airfoil. Within a transitional Reynolds number regime, the effect of gust length and strength on the aeroelastic response of the airfoil is investigated. For the same gust width, the initial heave excitation is found to be strongly dependent on the gust strength while that of the pitch showed no dependency. For the same gust strength, wider gusts lead to higher amplitudes in the aerodynamic coefficients and the degrees of freedom. Seven out of nine investigated cases went into a limit-cycle oscillation driven by the laminar separation taking place on the upper and lower surfaces of the airfoil. Two cases excited the airfoil to levels beyond its critical angle of attack and led to a diverging pitch motion. An energy analysis of the coupled system showed that in the LCO case, the heave DOF extracts energy from the flow sustaining the bounded motion of the system. In contrast, in the flutter scenario the energy is accumulatively fed from the flow into the system by the pitch DOF. For further results refer to [3].

Acknowledgements. The project is financially supported by the *German Research Foundation (Deutsche Forschungsgemeinschaft)* under the contract number BR 1847/17-1.

References

1. Breuer, M., De Nayer, G., Münsch, M., Gallinger, T., Wüchner, R.: Fluid-structure interaction using a partitioned semi-implicit predictor-corrector coupling scheme for the application of large-eddy simulation. J. Fluids Struct. **29**, 107–130 (2012)
2. Boulbrachene, K., De Nayer, G., Breuer, M.: Assessment of two wind gust injection methods: field velocity vs. split velocity method. J. Wind Eng. Ind. Aerodyn. **218**, 104790 (2021)
3. Boulbrachene, K., Breuer, M.: Aeroelastic response of an elastically mounted 2-DOF airfoil and its gust-induced oscillations. J. Fluids Struct. **117**, 103828 (2023)
4. De Nayer, G., Breuer, M., Wood, J.N.: Numerical investigations on the dynamic behavior of a 2-DOF airfoil in the transitional Re number regime based on fully coupled simulations relying on an eddy-resolving technique. Int. J. Heat Fluid Flow **85**, 108631 (2020)
5. Poirel, D., Harris, Y., Benaissa, A.: Self-sustained aeroelastic oscillations of a NACA 0012 airfoil at low-to-moderate Reynolds numbers. J. Fluids Struct. **24**, 700–719 (2008)
6. Poirel, D., Métivier, V., Dumas, G.: Computational aeroelastic simulations of self-sustained pitch oscillations of a NACA 0012 at transitional Reynolds numbers. J. Fluids Struct. **27**, 1262–1277 (2011)
7. Sen, S., De Nayer, G., Breuer, M.: A fast and robust hybrid method for block-structured mesh deformation with emphasis on FSI-LES applications. Int. J. Numer. Methods Eng. **111**(3), 273–300 (2017)
8. Wales, C., Jones, D., Gaitonde, A.: Prescribed velocity method for simulation of aerofoil gust responses. J. Aircraft **52**(1), 64–76 (2014)
9. Wood, J.N., Breuer, M., De Nayer, G.: Experimental investigations on the dynamic behavior of a 2-DOF airfoil in the transitional Re number regime based on digital-image correlation measurements. J. Fluids Struct. **96**, 103052 (2020)
10. DIN EN 61400-1:2019, Wind energy generation systems – Part 1: Design requirements. Technical report, IEC (2019)

Euler-Lagrange LES Predictions of a Powder Disperser Including a Multiscale Wall-Impact Breakage Model

Ali Khalifa and Michael Breuer(✉)

Professur für Strömungsmechanik, Helmut-Schmidt-Universität Hamburg, Hamburg, Germany
{khalifa,breuer}@hsu-hh.de

1 Introduction

The evolution of the particle size in turbulent particle-laden flows is a complex phenomenon, which plays an important role in various environmental (e.g., air pollution) and industrial applications (e.g., dry powder inhalers). The main challenge for investigating such flows is the extensive range of length and time scales to be dealt with, especially in the case of turbulent flows. Thus, efficient and reliable methods for numerically simulating particle-laden flows are of interest. Namely, multiscale strategies, in which models derived at different scales are combined to efficiently describe the flow system, offer a practical alternative to expensive direct methods. The present study demonstrates the viability of a multiscale Euler–Lagrange technique for predicting the flow in a lab-scale powder disperser while effectively describing the development of the particle size. Of specific interest are breakup processes taking place in turbulent flows due to fluid forces and wall impact. For the latter models based on artificial neural networks were developed.

2 Applied Methodology

The continuous phase is predicted in the Eulerian frame of reference relying on the large-eddy simulation (LES) technique. The code LESOCC applies the finite-volume method for block-structured curvilinear grids [1]. Furthermore, an efficient Lagrangian tracking scheme is employed, which deterministically detects inter-particle collisions [2]. The collisions are handled based on an extended hard-sphere model taking possible agglomeration processes into account [3]. Moreover, the effect of subgrid-scale motions on the particles is accounted for and the feedback effect of the particles on the carrier fluid is considered. Hence, the method is four-way coupled [2].

To maintain manageable computational costs, the detailed structure of agglomerates is substituted by a single sphere possessing an effective diameter. In this context, models describing the breakage of agglomerates due to fluid-induced stresses [4,5,10] and wall impacts [6–10] have been developed and incorporated into the described simulation methodology.

C. Marchioli et al. (Eds.): DLES 2023, ERCO 31, pp. 89–94, 2024.
https://doi.org/10.1007/978-3-031-47028-8_14

3 ANN-Based Wall-Impact Breakage Model

In general, to fully describe a wall-impact breakage event a model needs to provide descriptions for (1) the number, (2) the size distribution, and (3) the post-impact velocities of the arising fragments. The proposed model is data-driven and is derived as follows. First, an extensive number of discrete element simulations (DEM) of single agglomerates impacting a wall in a vacuum environment are carried out (see Fig. 1). Wide ranges of different impact conditions are taken into account: Three silica powders distinguished by the size of the primary particle (0.97, 2.47 and 5.08 μm), eight agglomerate size classes ranging between 5 and 10^3, nine impact angles between a flat and a normal impact, and a range of velocities able to reproduce the full spectrum of breakage events from a rebound of an intact agglomerate to a full disintegration of the agglomerate structure.

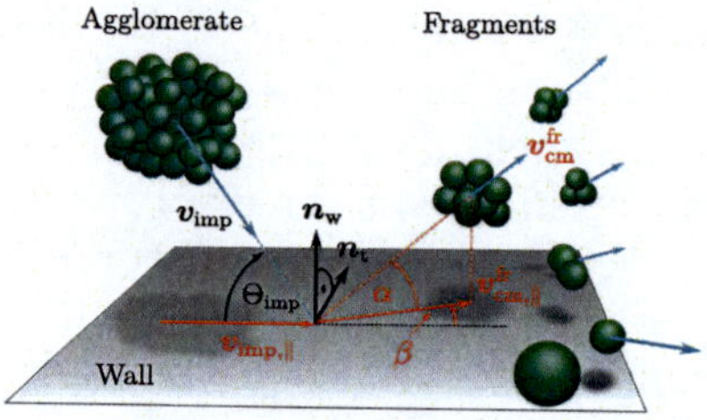

Fig. 1. DEM wall-impact events used for generating a database [7,9] for training artificial neural networks [9].

Second, the results of the DEM simulations are quantified based on useful parameters allowing a further modeling. Specifically, the number and the size distribution of the arising fragments are described in terms of the fragmentation ratio (FR) and the fragment size parameters (ζ_i), for details refer to [6]. Furthermore, the collective post-breakage motion of the fragments is expressed relying on the Weibull probability density functions (PDFs) of the reflection angle α, the spreading angle β and the ratio of the velocity of the fragment with respect to the impact velocity of the agglomerate v_{ratio} (see Fig. 1). Additionally, the ratio of the kinetic energy ER_{kin} of the fragments to the incident kinetic energy of the agglomerate is considered [6,7].

Third, the relationships between the impact conditions and the breakage parameters are approximated using two feed-forward artificial neural networks. The first network is trained to predict the number of fragments and the particle size distribution, whereas the second network is devoted for learning the shape (k) and the scale (λ) parameters of the Weibull PDFs as well as the energy ratio ER_{kin}. The training is performed in MATLAB applying the Bayesian regularization on the basis of the backpropagation algorithm [9]. After a reasonable training performance is achieved, the details of the trained networks are extracted and incorporated into the Euler–Lagrange code to be used during the coupled Euler–Lagrange simulations.

4 Results for the Particle-Laden Flow in a Dry Powder Disperser

The design of the dry powder disperser and the flow settings are inspired by the experimental study of Weiler [11]. The configuration consists of a duct (blue) with a square cross-section (hydraulic diameter d_h) and a funnel (orange) protruding into the duct as depicted in Fig. 2. The geometry is discretized by a block-structured grid. The resolution of the grid in the near-wall regions allows a wall-resolved LES.

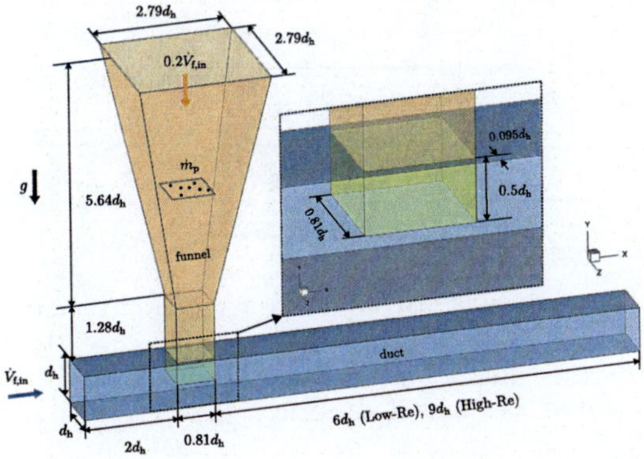

Fig. 2. Schematic sketch of the dry powder disperser [5].

At the inlet of the main duct, inflow data generated by an auxiliary simulation of a single periodic duct are applied. The flow rate of the fluid is equal to 25 norm-liter/min leading to a Reynolds number of $\mathrm{Re} = U_b\, d_h/\nu = 8698$. At the top of the funnel agglomerates each consisting of 100 silica particles are discharged at a mass flow rate of 10 mg/s. To examine the influence of cohesion on breakage, the diameter of the single primary particle d_{pp} is varied between 0.97, 2.47, and 5.08 μm defining three powders (**A**, **B**, and **C**), which are separately investigated. In earlier studies [4,5] the described Euler–Lagrange simulation methodology was applied to investigate the deagglomeration solely due to fluid-induced stresses in this disperser. However, breakage due to wall impact was not considered since a corresponding model was missing at that time.

Figure 3a depicts the typical flow field in the device by contours of the instantaneous streamwise velocity in the symmetry plane at an arbitrarily chosen snapshot. As visible, the funnel protrudes into the duct acting as an obstacle and thus leading to a strong contraction of the cross-section. Consequently, the flow is significantly accelerated below the funnel generating strong shear layers. Furthermore, a recirculation region is observed behind the obstacle. As a results, agglomerates reach the outlet of the funnel under the effect of gravity and emerge

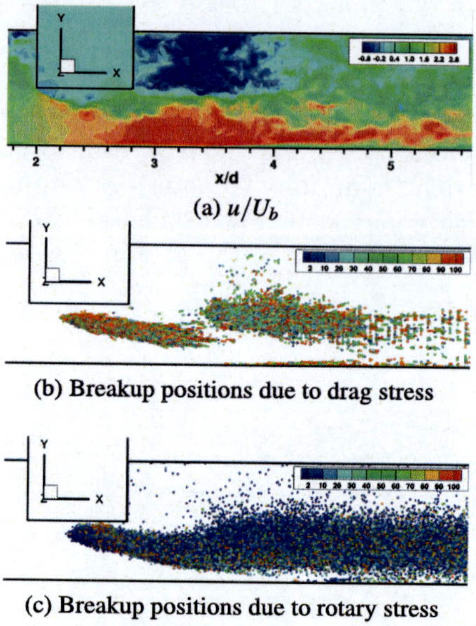

Fig. 3. Results: (a) Instantaneous streamwise velocity u/U_b in the symmetry plane, (b) Breakup positions by drag stress and (c) Breakup positions by rotary stress. Note that (b) and (c) refer to the simulations of powder **A** and that the color levels describe the number of primary particles included in the agglomerate at the time of breakage [4,10].

with negligible streamwise velocities into the main duct where they abruptly accelerate and pass through regions of high shear. That gives rise to breakup by mechanisms such as the drag and the rotary stress as depicted in Figs. 3b and 3c depicting the breakup locations for both mechanisms, respectively.

In the present investigation the computations are repeated while taking additionally the wall-impact breakage into account. The motivation is to better explore the role of different breakage phenomena and to attempt enhancing the agreement between the numerically predicted and the experimentally measured dispersion efficiency. The latter is defined in the reference study of Weiler [11] as the ratio of the median volumetric diameter $d_{50,3}^{\mathrm{CD}}$ of a high-end commercial disperser (CD) to the median volumetric diameter $d_{50,3}^{\mathrm{exp}}$ obtained in the considered lab-scale funnel-duct disperser.

Figure 4 summarizes the main results, which are obtained over a dimensionless time period of $\Delta T^* = \Delta T\, U_b/d_h = 18$. In Fig. 4a the percentage contributions of the different breakage mechanisms are compared. Obviously, the drag stress is the dominant mechanism for all powders followed by the rotary stress. The wall-impact breakage is responsible for about 10% of the total number of breakage events for all powders. Such a small contribution is expected since the majority of the agglomerates breaks due to the strong drag or due to the rotary stress induced by either the fluid torque or the inter-particle collisions before arriving at the walls.

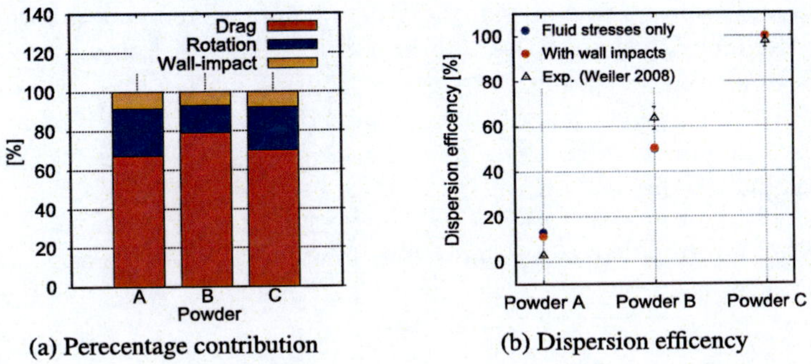

Fig. 4. Comparison between results achieved for all powders: (a) Percentage contribution of the different deagglomeration mechanisms and (b) Dispersion efficiency.

Furthermore, Fig. 4b depicts the dispersion efficiency obtained at the outlet of the disperser. The red and blue symbols refer to the numerically predicted results with and without taking the wall-impact breakage into account, respectively, whereas the experimentally measured values [11] are represented by triangles. In general, a reasonable agreement between the numerical results and the measurements is found. Astonishingly, the dispersion efficiency slightly decreases for powder **A**, when the wall-impact breakage is taken into account. A possible explanation is as follows. In the model of the wall-impact breakage the simplifying assumption of a binary breakage (i.e., split into two fragments) considered in the models of the fluid-induced stresses (drag, rotary, turbulence) is avoided, since the number of arising fragments is directly predicted by the neural network. Hence, the probability that more fragments are generated in the vicinity of each other increases which in turn encourages agglomeration. Since this behavior is more realistic than the binary disintegration mode, the dispersion efficiency tend towards the experimentally measured values. However, the inclusion of the wall-impact breakage improves the agreement only slightly. This is again explained by the fact that the wall-impact breakage does not play a significant role in the considered funnel-duct flow configuration.

5 Conclusions

Euler–Lagrange LES simulations have been carried out to investigate the agglomerate-laden flow in a funnel-duct disperser taking three powders distinguished by the size of the primary particles into account. The efficient simulation methodology is equipped with models for the agglomeration and the breakup of agglomerates due to turbulence, drag and rotation. In addition, a recently developed machine-learning assisted data-driven model for the wall-impact breakage is applied. The results show that agglomerates mainly break by the drag stress and the rotary stress in the foreseeable regions, where agglomerates are abruptly accelerated and exposed to a strong shear flow before arriving at the walls. Consequently, the contribution of the wall-impact breakage is found to be less significant. In addition, analyzing the size distribution of the population reaching the outlet of the disperser shows that by taking the wall-impact breakage into account the agreement with the experimental data concerning the dispersion efficiency slightly improves.

Acknowledgements. Financial supported by Deutsche Forschungsgemeinschaft: BR 1847/13-2.

References

1. Breuer, M.: A challenging test case for large-eddy simulation: high Reynolds number circular cylinder flow. Int. J. Heat Fluid Flow **21**(5), 648–654 (2000)
2. Breuer, M., Alletto, M.: Efficient simulation of particle-laden turbulent flows with high mass loadings using LES. Int. J. Heat Fluid Flow **35**, 2–12 (2012)
3. Breuer, M., Almohammed, N.: Modeling and simulation of particle agglomeration in turbulent flows using a hard-sphere model with deterministic collision detection and enhanced structure models. Int. J. Multiph. Flow **73**, 171–206 (2015)
4. Breuer, M., Khalifa, A.: Revisiting and improving models for the breakup of compact dry powder agglomerates in turbulent flows within Eulerian-Lagrangian simulations. Powder Technol. **348**, 105–125 (2019)
5. Breuer, M., Khalifa, A.: Refinement of breakup models for compact powder agglomerates exposed to turbulent flows considering relevant time scales. Comput. Fluids **194**, 104315 (2019)
6. Khalifa, A., Breuer, M.: Data-driven model for the breakage of dry monodisperse agglomerates by wall impact applicable for multiphase flow simulations. Powder Technol. **376**, 241–253 (2020)
7. Khalifa, A., Breuer, M.: An efficient model for the breakage of agglomerates by wall impact applied to Euler-Lagrange LES predictions. Int. J. Multiph. Flow **142**, 103625 (2021)
8. Khalifa, A., Gollwitzer, J., Breuer, M.: LES of particle-laden flow in sharp pipe bends with data-driven predictions of agglomerate breakage by wall impacts. Fluids **6**(12), 424 (2021)
9. Khalifa, A., Breuer, M., Gollwitzer, J.: Neural-network based approach for modeling wall-impact breakage of agglomerates in particle-laden flows applied in Euler-Lagrange LES. Int. J. Heat Fluid Flow **94**, 108897 (2022)
10. Khalifa, A.: Modeling and simulation of the breakage of cohesive particle agglomerates in turbulent wall-bounded flows. Ph.D. thesis, Helmut-Schmidt-Univ. Hamburg, Germany (2022)
11. Weiler, C.: Generierung leicht dispergierbarer Inhalationspulver mittels Sprühtrocknung. Ph.D. thesis, Univ. Mainz, Germany (2008)

A Four Mixture Fraction FPV-LES for the Co-firing of Coal and Ammonia

D. Meller(✉), L. Engelmann, S.-J. Baik, and A. M. Kempf

Fluid Dynamics, Institute of Combustion and Gasdynamics, University of Duisburg-Essen, Duisburg, Germany
dominik.meller@uni-due.de

1 Introduction

The rising energy demand of emerging countries will continue to cause coal to play an essential role in the supply of electricity in the coming years. However, coal is considered an environmentally harmful fuel that releases large amounts of greenhouse gases and pollutants during firing. Measures must therefore be taken to improve Pulverized Coal Combustion (PCC) and reduce emissions. Numerical simulations have become increasingly important for the detailed understanding of PCC in recent years. Combustion models for gaseous combustion were extended to describe PCC. Among these models, the flamelet model (Peters, 1984) has proven to be a precise model that can represent detailed chemical mechanisms with low computational effort. The flamelet/progress variable (FPV) model by Pierce and Moin (Pierce and Moin, 2004) is an extension of this model, in which a progress variable is used as trajectory variable. This model has been widely applied to PCC simulations. In order to describe the important processes in coal gasification, devolatilisation and char burn-up, flamelet models with two mixture fractions Z were developed. However, it is difficult to stabilise the pulverized coal flame itself, which is why an additional pilot gas stream is usually used. This additional fuel stream makes strong assumptions necessary, i.e. assuming that char reactions do not take place or assuming the pilot gas to be modelled as volatile gas. A complex extension of the 2Z-flamelet model was the 3Z-flamelet model by Wen et al. (Wen et al., 2019), which offers the possibility to describe all three fuel streams with moderate computational effort. New measures to reduce i.e. CO_2 emissions, require the admixture of additional fuels. A promising measure of continuing to operate coal-fired power plants in the future and drastically reducing CO_2 emissions is the co-firing of coal and ammonia. This relatively new method has not yet been widely investigated, and there are hardly any numerical studies on it. Difficulties that arise in the numerical simulation of co-firing are caused by the additional fuel stream ammonia. Four mixture fractions would now have to be considered as fuel streams (volatiles, char-off gases, pilot gas, ammonia) in order to describe the co-firing with a stabilizing pilot stream in detail. In a previous study (Meller et al., 2022), the co-firing was already investigated numerically and the first LES with a detailed chemical description was delivered. In this study, char-off gases were neglected and a 3Z-flamelet model was used. In the present study, a flamelet model based on a flamelet/progress variable (FPV) approach will be presented, which represents four mixture fractions and thus considers all necessary fuel streams for

C. Marchioli et al. (Eds.): DLES 2023, ERCO 31, pp. 95–100, 2024.
https://doi.org/10.1007/978-3-031-47028-8_15

an accurate description of co-firing of coal and ammonia with stabilizing pilot stream - volatiles, char-off gases, pilot gas, and ammonia.

2 Flamelet Modeling

For piloted pulverized coal and ammonia co-firing, four fuel streams exist, resulting from 1) the pilot stream, 2) the NH_3 and air stream, and 3) and 4) the volatile matter and the char off-gases, from the coal particles. To characterize the mixing between the four fuel sources and oxidizer (ox) (ambient air), four mixture fractions $Z_j = Y_j/(Y_{ox} + Y_{vol} + Y_{char} + Y_{ammo} + Y_{pil})$ are needed, indicated by $j \in \{\text{pil, ammo, vol, char}\}$ for the pilot stream, the ammonia stream, the volatile stream, and the char off-gases. Following the approach of Wen et al. (Wen et al., 2019), a coordinate transformation is used to avoid numerical problems. The newly introduced parameters (Z, A, B, C) are as follows:

$$Z = Z_{pil} + Z_{vol} + Z_{ammo} + Z_{char} \tag{1}$$

$$A = \frac{Z_{pil}}{Z_{pil} + Z_{vol} + \varepsilon} \tag{2}$$

$$B = \frac{Z_{pil} + Z_{vol}}{Z_{pil} + Z_{vol} + Z_{ammo} + \varepsilon} \tag{3}$$

$$C = \frac{Z_{pil} + Z_{vol} + Z_{ammo}}{Z_{pil} + Z_{vol} + Z_{ammo} + Z_{char} + \varepsilon}, \tag{4}$$

where ε denotes a small positive number. The new parameters (ranging from 0 to 1) describe all states of mixing. C was limited to a lower value of 0.2 (not zero), as the flame did not ignite at $C = 0.0$. Thus, only flamelet solutions for $C > 0.2$ are considered. To represent these states in the flamelet table, one-dimensional steady non-premixed flamelet equations are solved using FlameMaster (Pitsch, 1998) based on the assumption of unity Lewis number for varying values of A, B and C. To represent the interphase heat transfer between the particle and gas phase, different input stream temperatures (300 K, 600 K, 800 K) are set to vary the enthalpy levels. Scalar dissipation rates χ are varied when solving the non-premixed flamelet equations in mixture fraction space to account for different strain rates. Finally, the thermochemical quantities from the flamelet solutions are parameterized by the control variables. In addition to the four mixture fraction parameters Z, A, B and C, a progress variable is used, which results from the linear combination of different key species in the exhaust gas, following $Y_{PV} = Y_{CO_2} + Y_{CO} + Y_{H_2O} + 10Y_{NO}$, and the total enthalpy. The final parameterization follows $\Psi = F(A, B, C, Z, Y_{PV,norm}, H_{norm})$ for the thermochemical quantities Ψ, where $Y_{PV,norm}$ and H_{norm} are the normalized quantities for the progress variable and the enthalpy. The table dimensions are $6 \times 6 \times 5 \times 132 \times 51 \times 6$ in $A \times B \times C \times Z \times Y_{PV,norm} \times H_{norm}$, reading a total of 7.3 Mio points and requiring approximately 3.2 GBytes of total storage in uncompressed ASCI format. A reduced CRECK mechanism consisting of 129 species and 1644 elementary reactions is used, which was previously designed to model co-firing of coal and ammonia (Meller et al., 2022).

3 Flame and Numerical Setup

Figure 1 shows the geometry of the investigated burner. It consists of an inner tube and a coaxial pilot channel, using hydrogen to stabilize the flame. The burner has three operation modes which can be investigated, case C) a pure coal combustion mode, where only coal particles and air are injected in the main tube, case A) a pure ammonia combustion mode, where ammonia and air are injected, and case CA), a co-firing mode, where ammonia, air, and coal particles are injected. Experimental data is taken from (Tainaka et al., 2018, Tainaka et al., 2019, Stein and Watanabe, 2019). In this work, cases C and CA are investigated numerically, as these cases include the newly added mixture fraction for char off-gases.

The massively parallel LES tool PsiPhi is used in this study. A low-Mach Finite Volume Method (FVM) with conjugated gradient solver and Jacobi preconditioner is applied. The solver uses an equidistant, orthogonal Cartesian grid. For diffusive fluxes and the convective flux of momentum, a second-order accurate Central Differencing Scheme (CDS) is applied. Convective scalar fluxes are discretized through a Total Variation Diminishing (TVD) scheme with a CHARM limiter. Time integration is done through an explicit third-order low-storage Runge-Kutta scheme. Radiation effects are neglected in this study, as previous studies showed negligible influence. An Euler-Lagrange framework is used to couple particle-gas phases. Settings have been taken from our previous work on this burner and are applied here (Meller et al., 2021).

Coal conversion is described by the two main processes devolatilization and char combustion. As devolatilization model the single first order reaction (SFOR) model by Badzioch & Hawksley is used (Badzioch and Hawksley, 1970). The parameters were kept consistent to our previous works (Meller et al., 2021, Meller et al., 2022) and are $k = 875.06\text{E}5\,\text{s}^{-1}$ and $E = 688.69\text{E}5\,\text{J}/(\text{kmolK})$ for the pre-exponential factor and the activation energy, respectively. Char conversion is modeled using Smith's intrinsic model (Smith, 1982). It is modeled such that it starts after 90 % of the volatiles were released.

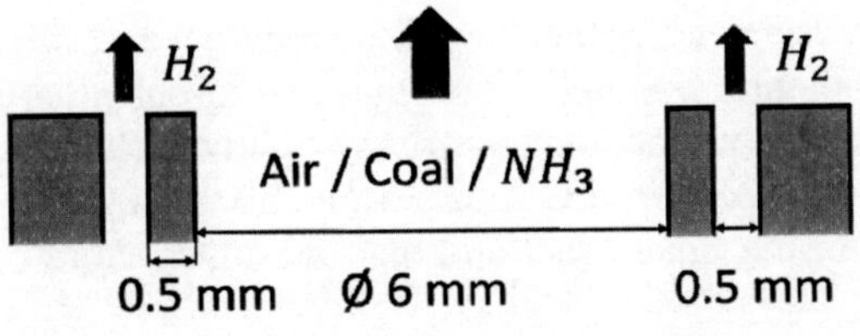

Fig. 1. Setup of the investigated coaxial burner.

4 Results

Figure 2 shows the mean (left) and instantaneous (right) temperature (left side) and O_2 mole fraction fields (right side) for both cases CA and C.

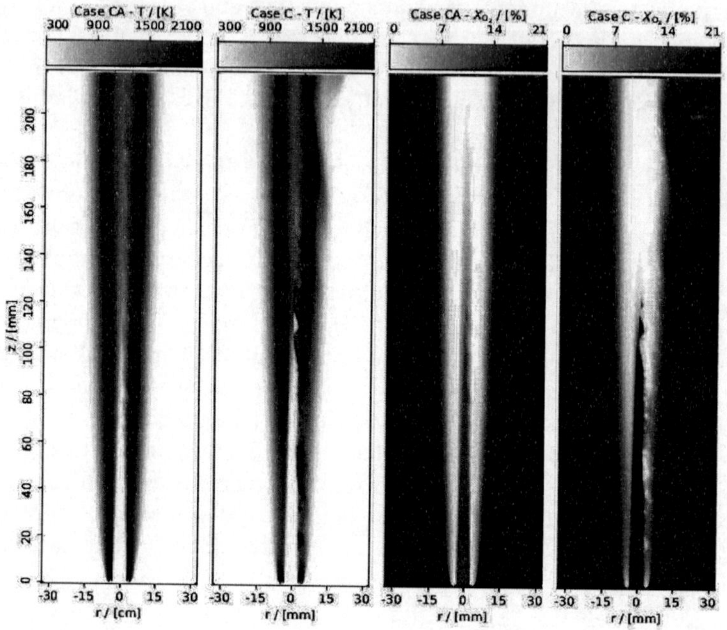

Fig. 2. Mean (left) and instantaneous (right) temperature (left side) and O_2 mole fraction fields (right side) of cases CA and C.

While case C shows an earlier breakup of the main jet and stronger turbulent structures, case CA shows much smoother fields and less turbulent structures. In case CA, the temperature in the outer reaction zone rises slightly earlier due to the combustion of ammonia and ambient air. Case C shows a much higher oxygen consumption, the volatile combustion in the inner reaction zone consumes more oxygen compared to case CA.

Figures 3 and 4 show the respective mixture fraction fields for volatiles, hydrogen pilot, char off-gases and ammonia of each case. In both cases, the hydrogen pilot stream is burned immediately after injection. The hot hydrogen flame leads to the heating of incoming particles and the release of volatiles. In case CA, the hydrogen flame also ignites the ammonia/air mixture and an ammonia flame is formed. In both cases, a volatile accumulation zone forms further downstream. The volatile accumulation zone for case C is thicker and starts earlier, as oxygen is consumed faster due to the stronger turbulent structures, and thus, less oxygen for volatile combustion is available. Char combustion starts after 90 % of the volatiles are released and consumes oxygen to form carbon oxide. As the oxygen level in the inner jet decreases strongly further downstream, the char off-gas mixture fraction rises at the outer edge areas of the jet, where oxygen from the ambient air is available for char combustion. In both cases C and CA, the char off-gas concentration is relatively low; case C shows a maximum mean value of around 4.5 %, while the maximum mean value of case CA is below 1 %. Since the influence of char combustion in the investigated domain in both cases C and CA is very small, the results are in line with the results from our previous study (Meller et al., 2022). As the focus of this work is on the development of the four mixture fraction flamelet strategy, no comparison of the two works with and without consideration of char combustion is given at this point. This will be part of a future study.

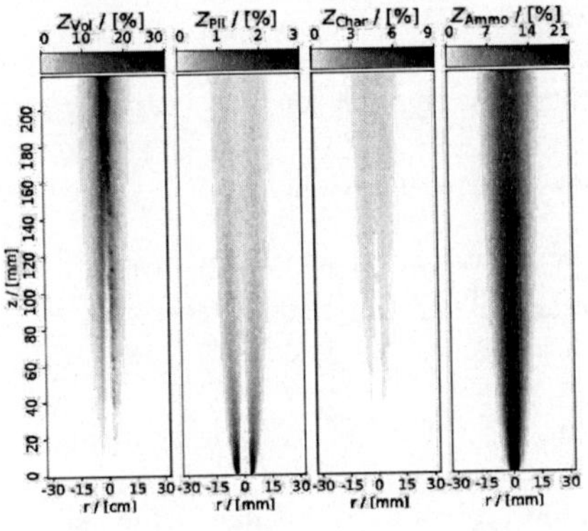

Fig. 3. Mixture fractions of case CA

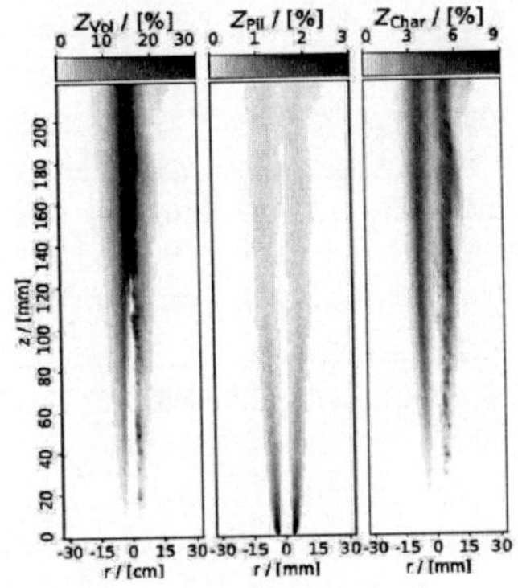

Fig. 4. Mixture fractions of case C

5 Conclusion

The co-firing of different fuels such as coal and ammonia makes it necessary to consider four mixture fractions to accurately describe its combustion. In this work, a simple flamelet model based on a flamelet/progress variable approach to account for four mixture fractions is proposed. The results are in line with the results of the three mixture fraction approach neglecting char off-gases and show that the influence of the newly added mixture fraction of char off-gases is negligibly small for both investigated cases C and CA (Meller et al., 2022). Future work will apply this model to more complex, highly-turbulent semi-industrial combustors.

Acknowledgements. The authors gratefully acknowledge the financial support through the German Research Foundation (DFG, project 238057103) and the Gauss Centre for Supercomputing e.V. (www.gauss-centre.eu) (grant DUEHAWK2021) for providing computing time on the GCS Supercomputer HAWK at Höchst-leistungsrechenzentrum Stuttgart (www.hlrs.de).

References

Badzioch, S., Hawksley, P.G.: Kinetics of thermal decomposition of pulverized coal particles. Ind. Eng. Chem. Process Des. Dev. **9**, 521–530 (1970)

Meller, D., et al.: Numerical analysis of a turbulent pulverized coal flame using a flamelet/progress variable approach and modeling experimental artifacts. Energy Fuels **35**, 7133–7143 (2021)

Meller, D., et al.: Evaluation of ammonia co-firing in the CRIEPI coal jet flame using a three mixture fraction FPV-LES. Proc. Combust. Inst. **39**, 3615–3624 (2022)

Peters, N.: Laminar diffusion flamelet models in non-premixed turbulent combustion. Prog. Energy Combust. Sci. **10**, 319–339 (1984)

Pierce, C.D., Moin, P.: Progress-variable approach for large-eddy simulation of non-premixed turbulent combustion. J. Fluid Mech. **504**, 73–97 (2004)

Pitsch, H.: FlameMaster, a computer code for homogeneous combustion and one-dimensional laminar flame calculations. Institute for Technical Mechanic, RWTH Aachen (1998). https://www.itv.rwth-aachen.de/downloads/flamemaster/

Smith, I.W.: The combustion rates of coal chars: a review, Symp. (Int.) Combust. **19**, 1045–1065 (1982)

Stein, O.T., Watanabe, H.: CRIEPI coal jet flame. Proc. CBC Workshop **3**, 69–110 (2019)

Tainaka, K., Yamamoto, A., Kimoto, M., Ozawa, Y., Hara, S.: Investigation of co-combustion characteristics of pulverized coal and ammonia in a jet flame. Symp. (Japan.) Combust. **56** (2018)

Tainaka, K., Sawada, S., Okada, D., Akamatsu, F., Run, L.: Effects of co-firing ammonia on structure of a pulverized coal jet flame. Symp. (Japan.) Combust. **57** (2019)

Wen, X., Luo, Y., Wang, H., Luo, K., Jin, H., Fan, J.: A three mixture fraction flamelet model for multi-stream laminar pulverized coal combustion. Proc. Combust. Inst. **37**, 2901–2910 (2019)

Large Eddy Simulation of a Low Pressure Turbine Cascade with Turbulent End Wall Boundary Layers

C. Morsbach[1(✉)], M. Bergmann[1], A. Tosun[1], E. Kügeler[1], and M. Franke[2]

[1] Institute of Propulsion Technology, German Aerospace Center (DLR), Cologne, Germany
christian.morsbach@dlr.de

[2] MTU Aero Engines AG, Munich, Germany

Abstract. We present first results of an implicit large eddy simulation of the MTU T161 low pressure turbine at a Reynolds number of 90,000 and Mach number of 0.6, both based on isentropic exit conditions, using a high order discontinuous Galerkin method. The aim is to validate the numerical setup with respect to available experimental data. We discuss the steps taken to create realistic inflow boundary conditions in terms of end wall boundary layer thickness and free stream turbulence intensity. This is achieved by tailoring the input distribution of Reynolds stresses and turbulent length scale to a Fourier series based synthetic turbulence generator. Both blade loading and total pressure losses at midspan show excellent agreement with the measurements. Following a short discussion of the secondary flow structures emerging due to the interaction of the incoming boundary layer and the turbine blade, we show that this simulation is also able to reproduce loss distribution behind the blade over the whole channel height.

1 Introduction

Many large eddy simulation (LES) studies of turbomachinery flows focus on the low-pressure turbine (LPT) due to its low Reynolds number regime of 10^5. To further reduce the simulation costs, the assumption of a statistically 2D flow at the midspan of a turbine blade is often made by applying periodic boundary conditions in the spanwise direction. However, a significant amount of the aerodynamic losses is generated in the secondary flow regions influenced by the interaction of end wall and blade boundary layers [1]. Hence, the next logical step to evaluate the performance of the cascade is to conduct 3D simulations, including the endwall boundary layers, e.g. [2].

The MTU T161, considered in this work, is representative of high lift low-pressure turbine airfoils used in modern jet engines [3]. The blades with a chord length of $C = 0.069935\,\mathrm{m}$ and an average aspect ratio of 2.65 are staggered at an angle of 61.72°. The cascade is arranged with pitch to chord ratio of $l_{\mathrm{pitch}}/C = 0.956$. It features diverging end walls at an angle of 12°, such that the flow cannot be studied using a simple spanwise periodic setup. Its geometry and boundary conditions are in the public domain and it has been the subject of both experimental [4] and numerical [5–7] investigations. The numerical investigations have focused on operating points with a Mach number

C. Marchioli et al. (Eds.): DLES 2023, ERCO 31, pp. 101–107, 2024.
https://doi.org/10.1007/978-3-031-47028-8_16

of 0.6 and Reynolds numbers of 90,000 and 200,000 based on isentropic exit conditions. Müller-Schindewolffs et al. [5] performed a direct numerical simulation of a section of the profile in which the effect of the diverging end walls was modelled using inviscid walls (termed quasi 3D, Q3D). Recently, results obtained with a second order FV code were presented, which included the end wall boundary layers but the analysis was focussed on the flow physics in the mid-section [8]. Various computations of the full 3D configuration were conducted using high-order codes during the EU project TILDA [6, 7]. However, due to the specification of laminar end wall boundary layers and no freestream turbulence at the inflow, no satisfactory results could be obtained. With a full 3D LES including appropriate turbulent end wall boundary layers and freestream turbulence obtained with a high order Discontinuous Galerkin (DG) method, we aim to provide a high-quality reference dataset for this configuration.

2 Numerical Method

We use DLR's flow solver for turbomachinery applications, TRACE, to perform an implicit LES with a kinetic-energy-preserving DG scheme for the spatial discretisation of the implicitly filtered Navier-Stokes equations [9]. The scheme is based on the collocated nodal Discontinuous Galerkin Spectral Element Method (DGSEM) on Legendre-Gauss-Lobatto nodes with a polynomial order of 5. The anti-aliasing is performed by the split-formulation of Kennedy and Gruber [10], cf. [11]. Due to the non-uniqueness of the solution at the element interfaces, Roe's approximate Riemann solver is applied for the advective part and the viscous terms are discretised by the Bassi-Rebay 1 scheme [12]. To advance in time, a third-order explicit Runge-Kutta scheme of [13] has been used. Resolved turbulent scales are injected at the inflow boundary using a synthetic turbulence generation (STG) method based on randomized Fourier modes [14]. The fluctuating velocity is introduced using a Riemann boundary condition [15] and, in contrast to the inner faces, no numerical flux function is used at the inflow faces.

3 Inflow Boundary Conditions

The operating conditions are specified by centerline inlet flow angle of $\alpha_1 = 41°$, total pressure of $p_{t,1} = 11636\,\text{Pa}$ and total temperature of $T_{t,1} = 303.25\,\text{K}$ as well as outlet Mach number $\text{Ma}_{2,\text{th}} = 0.6$ and Reynolds number $\text{Re}_{2,\text{th}} = 90{,}000$ based on isentropic relations computed with the centerline outflow pressure. Furthermore it is required to match the momentum thickness of the incoming end wall boundary layers as well as the decay of freestream turbulence. The boundary conditions are then set by a spanwise varying profiles of $p_{t,1}(z)$ and $T_{t,1}(z)$ with a constant α_1 as well as spanwise varying Reynolds stress tensor $\overline{u_i'u_j'}(z)$ and turbulent length scale $L_T(z)$. All these profiles will be obtained by scaling a boundary layer profile in wall units at $\text{Re}_\theta = 670$ (https://www.mech.kth.se/~pschlatt/DATA/, [16]) and combining it with freestream turbulence values. We perform a precursor channel flow simulation to determine the distance from the inlet required for the boundary layer to develop and meet the specifications.

Starting from the above centerline inflow conditions, we use the isentropic relations

$$\frac{T_t(z)}{T_1} = \left(\frac{p_t(z)}{p_1}\right)^{\frac{\gamma-1}{\gamma}} = \left(1+\frac{\gamma-1}{2}\mathrm{Ma}(z)^2\right) \tag{1}$$

to obtain a spanwise variation. Here, we assume a constant static temperature T_1 and pressure p_1 throughout the boundary layer. These two quantities can be computed from (1) by introducing a centerline Mach number $\mathrm{Ma}_{1,\mathrm{center}}$, which essentially controls the value of the T_t and p_t at the end wall. It has to be chosen such that the total pressure is always greater than the static pressure resulting from the simulation. Hence, an iteration might be necessary. It has to be noted, though, that small adaptations of $\mathrm{Ma}_{1,\mathrm{center}}$ mainly influence the total pressure profile close to the wall and do not have a significant impact on the quantities of interest in the simulation. For this concrete case, we chose $\mathrm{Ma}_{1,\mathrm{center}} = 0.362$.

The discrete DNS velocity profile $u^+_{i,\mathrm{DNS}}$ and the corresponding distances to the wall $z^+_{i,\mathrm{DNS}}$ are now transformed to obtain a Mach number profile $\mathrm{Ma}(z)$ simply using the definitions of u^+ and z^+ via

$$\mathrm{Ma}(z_i) = u^+_{i,\mathrm{DNS}} \cdot \frac{u_\tau}{a_1}, \quad z_i = z^+_{i,\mathrm{DNS}} \cdot \frac{\nu_1}{u_\tau} \quad \text{with} \quad u_\tau = a_1 \cdot \mathrm{Ma}_{1,\mathrm{center}}/u^+_{\infty,\mathrm{DNS}} \tag{2}$$

and the DNS freestream velocity $u^+_{\infty,\mathrm{DNS}}$. The speed of sound a_1 and viscosity ν_1 can be computed from T_1 and p_1 using ideal gas and Sutherland laws, respectively. This allows to describe the spanwise variation of T_t and p_t using (1). The non-dimensional turbulent stress tensor is scaled analogously via $\overline{u'_j u'_k}(z_i) = \overline{u'_j u'_k}^+|_{i,\mathrm{DNS}} \cdot u_\tau^2$. Finally, the integral turbulent length scale in the boundary layer is estimated from the turbulent kinetic energy and dissipation rate as $L_T^+ = (k^+)^{3/2}/\varepsilon^+$ and dimensionalised as the z-coordinate in (2).

Optimally, the freestream Reynolds stresses and turbulent length scale should be chosen such that the measured turbulent decay is matched. However, in this case, this would result in a length scale in the order of cascade pitch. This could, in principle, be accommodated for by simulating more than one blade at the respective expense of computational effort. Instead, we chose to decrease the turbulent length scale and scale the non-zero components of the Reynolds stress tensor at the inflow of the domain, to reproduce the turbulence intensity in the blade leading edge plane according to the experiment. This leads to a stronger decay of turbulent structures and has to be considered when assessing the quality of the results.

The boundary layer profile and freestream values of the turbulence quantities are combined where they intersect at the edge of the boundary layer. Because of the large freestream turbulent length scale compared to the smaller length scale in the boundary layer, we use $L_T = \max(L_{T,\mathrm{BL}}, L_{T,\mathrm{freestream}})$ for distances to the wall greater than δ_{99}. Finally, the Reynolds stress tensor is rotated from the streamline-aligned to a Cartesian coordinate system. Figure 1 (*left*) shows both the development of momentum thickness Reynolds number Re_θ and freestream turbulence intensity Tu in comparison to fits to the measured data. As discussed, the turbulent decay is steeper in our setup but the turbulence intensity at the blade leading edge (dashed vertical line) is well captured.

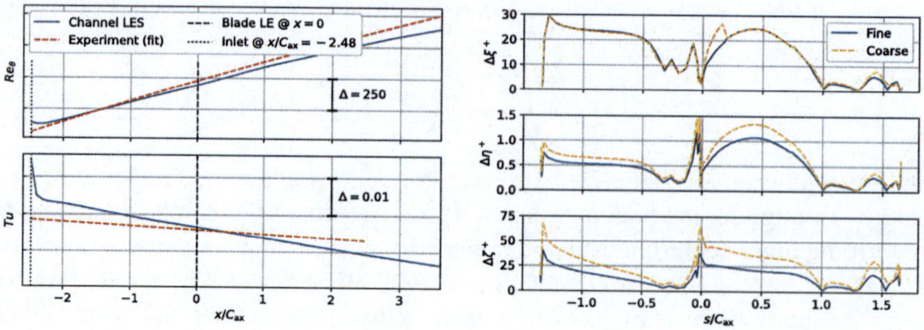

Fig. 1. Momentum thickness Reynolds number and turbulence intensity over axial distance from inlet plane (*left*) and average solution point distances in streamwise, wall normal and spanwise direction on the blade at midspan (*right*).

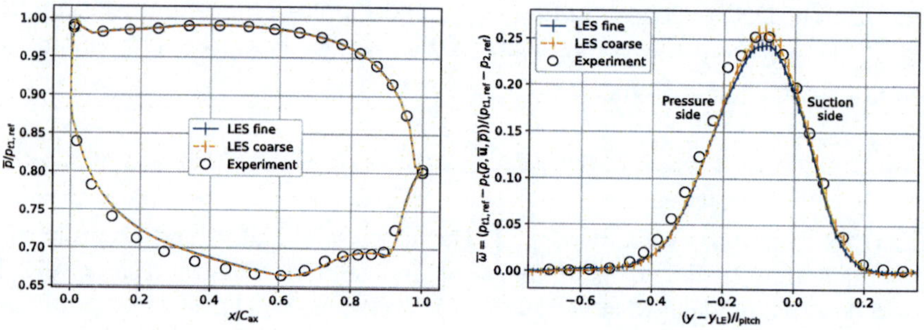

Fig. 2. Midspan blade pressure distribution (*left*) and wake total pressure loss coefficient at $x/C_{\mathrm{ax}} = 1.4$ (*right*). Errorbars indicate 68% confidence intervals.

4 Validation and Analysis of Flow Structures

The mesh consisting of 876,960 hexahedral elements (geometry polynomial order $q = 4$, solution polynomial order $p = 5$) with 189.4M degrees of freedom (DoF) for the final computation was designed to meet widely accepted criteria for wall resolution required for LES. This is demonstrated in Fig. 1 (*right*), which shows the streamwise, wall normal and spanwise solution point distances to be below 30, 1 and 30, respectively. The resolution in the free stream was ensured by the ratio of solution point distance and estimated Kolmogorov scale below 6 along a mid-passage streamline. To be able to assess mesh dependence, results from a preliminary study with a significantly coarser mesh in spanwise direction with 312,424 elements ($q = 4$, $p = 5$) and 67.9M DoF are also included.

After clearing the initial transient, the statistics on the fine mesh were sampled for 100 convective time units based on chord length and outflow velocity. Figure 2 shows a comparison of the midspan blade surface pressure distribution and wake total pressure losses with the experiment. Both simulations agree very well within the 68% confidence intervals [17] for these first order statistics. While LES and experiment agree very well on the pressure side, a difference in surface pressure can be found on the

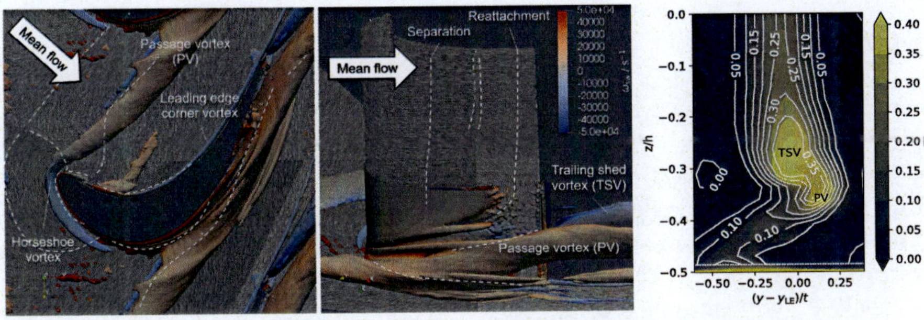

Fig. 3. Mean secondary flow structures viewed towards end wall (*left*) and towards suction side (*middle*). Pressure loss coefficient at $x/C_{ax} = 1.4$ (*right*) - LES contour colours vs. experiment contour lines.

suction side between $x/C_{ax} = 0.1$ and 0.6. Similar offsets have been found in previous studies of this configuration, e.g. [5]. The laminar separation bubble and subsequent turbulent reattachment indicated by the pressure plateau between 0.7 and 0.9 and recovery of the base pressure, on the other hand, is captured very well. The total pressure loss coefficient $\overline{\varpi}$ is computed from time-averaged primitive variables in the wake plane at $x/C_{ax} = 1.4$ and the upstream reference plane at $x/C_{ax} = -0.099$. The pitch coordinate y is given with respect to the point y_{LE} on the blade at its minimum axial coordinate $x/C_{ax} = 0$. The simulation on the coarse mesh shows larger confidence intervals in the highly turbulent region of the wake due to its shorter averaging time of only 31 convective time units. Both simulations and the experimental data agree very well within the statistical confidence intervals.

With the simulation setup validated against the experiment, we can now discuss the system of secondary flows developing at the intersection of blade and end wall. It is shown in Fig. 3 (*left* and *middle*) in terms of time-averaged surface streaklines visualised with line integral convolution and vortices visualised using an isosurface of $Q = 10^7 \mathrm{s}^{-2}$ coloured with streamwise vorticity ω_{sw} to indicate the sense of rotation. Limiting surface lines have been added by hand to ease the interpretation. The horseshoe vortex develops due to a roll-up of the incoming boundary layer. While its suction side leg (blue) dissipates well before the suction peak, its pressure side leg follows the passage cross flow towards the suction side of the next blade lifting off the end wall and becoming the passage vortex (PV). Behind the blade, a second large structure can be identified as the trailing shed vortex (TSV), which rotates in the opposite direction. Both vortices persist well downstream of the blade and become visible as the two regions of strong loss in Fig. 3 (*right*). The loss distribution in the wake plane also agrees very well with the measured data. Another small but very distinct vortex developing along the pressure side of the blade is the leading edge corner vortex. The pressure side features a short separation bubble close to the leading edge. Due to the diverging end walls, the backflow within the bubble is driven towards the endwalls where it rolls up, lifts off and mixes with the newly developing boundary layer in the passage between the passage vortex and the pressure side of the blade.

5 Conclusions

We have presented a new well-resolved dataset of the MTU T161 at $Ma_{2,th} = 0.6$, $Re_{2,th} = 90,000$ and $\alpha_1 = 41^\circ$ with focus on appropriate reproduction of inflow turbulent boundary layers and freestream turbulence. Average blade loading and loss distribution due to secondary flow features agree well with the experiment underlining the validity of the presented approach. After the brief discussion of the secondary flow structures in this paper, more in depth analysis of the unsteady data set is ongoing. Future analyses will introduce a numerical experiment featuring unsteady wakes at engine-relevant Strouhal numbers.

References

1. Denton, J.D.: The 1993 IGTI scholar lecture: loss mechanisms in turbomachines. J. Turbomach. **115**(4), 621–656 (1993)
2. Pichler, R., et al.: Large-eddy simulation and RANS analysis of the end-wall flow in a linear low-pressure turbine cascade, Part I: flow and secondary vorticity fields under varying inlet condition. J. Turbomach. **141**(12), 121005 (2019)
3. Gier, J., et al.: Preparation of aero technology for new generation aircraft engine LP turbines. In: 1st CEAS European Air and Space Conference, Berlin, Germany (2007)
4. Martinstetter, M., Niehuis, R., Franke, M.: Passive boundary layer control on a highly loaded low pressure turbine cascade. Turbo Expo: Power Land Sea Air **7**, 1315–1326 (2010)
5. Müller-Schindewolffs, C., Baier, R.-D., Seume, J.R., Herbst, F.: Direct numerical simulation based analysis of RANS predictions of a low-pressure turbine cascade. J. Turbomach. **139**(8), 081006 (2017)
6. Rasquin, M., et al.: Computational campaign on the MTU T161 cascade. In: Hirsch, C., et al. (eds.) TILDA: Towards Industrial LES/DNS in Aeronautics. NNFMMD, vol. 148, pp. 479–518. Springer, Cham (2021). https://doi.org/10.1007/978-3-030-62048-6_18
7. Iyer, A., et al.: High-order accurate direct numerical simulation of flow over a MTU-T161 low pressure turbine blade. Comput. Fluids **226**, 104989 (2021)
8. Afshar, N.F., Deutsch, J., Henninger, S., Kožulović, D., Bechlars, P.: Turbulence anisotropy analysis at the middle section of a highly loaded 3D linear turbine cascade using large eddy simulation. In: Global Power & Propulsion Technical Conference, Chania, Greece (2022)
9. Bergmann, M., Morsbach, C., Ashcroft, G.: Assessment of split form nodal discontinuous Galerkin schemes for the LES of a low pressure turbine profile. In: García-Villalba, M., Kuerten, H., Salvetti, M.V. (eds.) DLES 2019. ES, vol. 27, pp. 365–371. Springer, Cham (2020). https://doi.org/10.1007/978-3-030-42822-8_48
10. Kennedy, C.A., Gruber, A.: Reduced aliasing formulations of the convective terms within the Navier-Stokes equations for a compressible fluid. J. Comput. Phys. **227**(3), 1676–1700 (2008)
11. Gassner, G.J., Winters, A.R., Kopriva, D.A.: Split form nodal discontinuous Galerkin schemes with summation-by-parts property for the compressible Euler equations. J. Comput. Phys. **327**, 39–66 (2016)
12. Bassi, F., Rebay, S.: A high-order accurate discontinuous finite element method for the numerical solution of the compressible Navier-Stokes equations. J. Comput. Phys. **131**(2), 267–279 (1997)
13. Shu, C.-W., Osher, S.: Efficient implementation of essentially non-oscillatory shock-capturing schemes. J. Comput. Phys. **77**(2), 439–471 (1988)

14. Shur, M.L., Spalart, P.R., Strelets, M.K., Travin, A.K.: Synthetic turbulence generators for RANS-LES interfaces in zonal simulations of aerodynamic and aeroacoustic problems. Flow Turbul. Combust. **93**(1), 63–92 (2014). https://doi.org/10.1007/s10494-014-9534-8
15. Leyh, S., Morsbach, C.: The coupling of a synthetic turbulence generator with turbomachinery boundary conditions. In: García-Villalba, M., Kuerten, H., Salvetti, M.V. (eds.) DLES 2019. ES, vol. 27, pp. 349–355. Springer, Cham (2020). https://doi.org/10.1007/978-3-030-42822-8_46
16. Schlatter, P., Örlü, R.: Assessment of direct numerical simulation data of turbulent boundary layers. J. Fluid Mech. **659**, 116–126 (2010)
17. Bergmann, M., Morsbach, C., Ashcroft, G., Kügeler, E.: Statistical error estimation methods for engineering-relevant quantities from scale-resolving simulations. J. Turbomach. **144**(3), 031005 (2021)

Modelling SGS-Turbulent Transport of Fine Particles with Application to Cyclone Separator Performance

M. Sommerfeld(✉), M. A. Taborda, and O. Sgrott

Multiphase Flow Systems, Institute for Process Engineering, Otto-von-Guericke-University, Hoher Weg 7B, 06120 Halle (Saale), Germany
Martin.sommerfeld@ovgu.de

Abstract. Unsteady, technically relevant particle-laden flows are nowadays efficiently simulated with an LES-Euler/Lagrange approach. The present contribution is related to fine particle separation in a gas cyclone which involves an unsteady central vortex being responsible for particle separation. Particle motion is largely controlled by near-wall transport processes including wall collisions as well as near-wall turbulence. Naturally, in a LES this region is not very well resolved so that a remarkable contribution to the local turbulence is included in the modelled sub-grid-scale (SGS) turbulence. In this paper the importance of SGS turbulence on particle transport is analysed. It is shown that neglecting the SGS contribution to the particle transport yields a wrong prediction of the particle size-dependent separation of a cyclone.

Keywords: Euler/Lagrange approach · gas cyclone · particle separation · SGS turbulence · particle dispersion

1 Introduction

Dispersed particle-laden flows of industrial relevance are very often numerically computed by using the point-mass Euler/Lagrange approach as described in Sommerfeld [1]. For unsteady flow fields with strongly anisotropic turbulence, for example observed in a cyclone separator, large eddy simulation (LES) is most suitable, still requiring acceptable computational resources. During the last two decades numerous research work on LES applications to dispersed particle-laden has been publishes [2, 3]. Since large-scale turbulence and vortical structures are resolved by this method, only sub-grid-scale turbulence (SGS) needs to be modelled using either the eddy-viscosity concept or solving for an additional transport equation for the turbulent kinetic energy (see e.g. Lee and Cant [4]). Based on the assumption that the SGS turbulence is more or less isotropic, such turbulence models are however quite simple. Regarding the modelling of the influence of SGS turbulence onto the particles, several approaches may be used (Huilier [5]). According to the review of Marchioli [6] one may distinguish between structural and stochastic dispersion models.

C. Marchioli et al. (Eds.): DLES 2023, ERCO 31, pp. 108–117, 2024.
https://doi.org/10.1007/978-3-031-47028-8_17

In the present article, LES-Euler/Lagrange simulations are applied to one of the most important equipment in powder handling and environment, namely a gas cyclone separator, which was experimentally analysed by Hoekstra [7] and also considered in the numerical studies of Sgrott and Sommerfeld [8] modelling fine particle agglomeration and analysing this effect on separation efficiency. In the present studies the importance of SGS turbulence on fine particle motion and separation in a small-scale gas cyclone is analysed using a stochastic single-step Langevin model.

2 Summary of Numerical Approach

The numerical calculations of a high efficiency Stairmand cyclone based on the LES-Euler/Lagrange approach were conducted using the customised open-source code Open-FOAM®. The gas phase transport equations were solved by Large Eddy Simulations (LES) combined with the dynamic Smagorinsky sub-grid-scale model as proposed by Germano et al. [9] and extended by Lilly [10]. The PIMPLE (merged PISO-SIMPLE) algorithm is applied to couple velocity and pressure fields. The conservation of mass and the filtered Navier Stokes equations for an incompressible and Newtonian fluid may be written as:

$$\frac{\partial u_i}{\partial x_i} = 0 \tag{1}$$

$$\frac{\partial \overline{u}_i}{\partial t} + \frac{\partial \overline{u}_i \overline{u}_j}{\partial x_j} = -\frac{1}{\rho}\frac{\partial \overline{p}}{\partial x_i} - \frac{1}{\rho}\frac{\partial \boldsymbol{\tau_{ij}}}{\partial x_j} + \frac{\partial}{\partial x_j}\left[\nu\left(\frac{\partial \overline{u}_i}{\partial x_j} + \frac{\partial \overline{u}_j}{\partial x_i}\right)\right] + \overline{S}_{ui,p} \tag{2}$$

where the overbars denote the filtered quantity, $\overline{p}$ is the resolved static pressure and ν is the kinematic gas viscosity. The SGS stress tensor $\boldsymbol{\tau_{ij}}$ may be calculated based on the turbulent viscosity ν_T and the strain rate tensor based on the resolved scales $\overline{S}_{ij}$, as follow:

$$\boldsymbol{\tau_{ij}} = -\frac{1}{3}\tau_{kk}\delta_{ij} = 2\nu_T \overline{S}_{ij}; \quad \overline{S}_{ij} = \frac{1}{2}\left(\frac{\partial \overline{u}_i}{\partial x_j} + \frac{\partial \overline{u}_j}{\partial x_i}\right); \quad \nu_T = (C_s\Delta)^2\sqrt{2\overline{S}_{ij}\overline{S}_{ij}} \tag{3}$$

Here the SGS length scale $\Delta = (V_{\mathrm{CV}})^{1/3}$ is based on the volume of the computational cell and C_s has been determined dynamically. The influence of the particles onto the flow field (i.e. two-way coupling) is presently not considered, so that $\overline{S}_{ui,p} = 0$.

The dispersed phase is treated in the Lagrangian framework, in which all injected particles are simultaneously and time-dependent tracked through the flow domain obtained at the previous flow time step. The particles are treated as point-masses and their shape is assumed to be spherical. Particle transport is calculated by considering all relevant forces, which include, drag, gravity/buoyancy, slip-shear force F_{ls} and slip-rotational lift force F_{lr}, summarised in Sommerfeld et al. [11]. The change of the angular velocity along the particle trajectory due to the viscous interaction with the fluid (i.e. torque T_i) requires the solution of an additional partial differential equation. Hence, the complete equations of motion for the particles are given by:

$$\frac{dx_{pi}}{dt} = u_{Pi} \tag{4}$$

$$m_P \frac{du_{Pi}}{dt} = \frac{3}{4}\frac{\rho}{D_p} m_p c_D (u_i - u_{Pi}) \left| \vec{u} - \vec{u}_P \right| + F_{ls} + F_{lr} + m_P g_i \left(1 - \frac{\rho}{\rho_P}\right) \tag{5}$$

$$I_P \frac{d\omega_{pi}}{dt} = T_i \tag{6}$$

where u_{Pi} and ω_{pi} are, respectively, the linear and angular velocity components of the particle, x_{pi} are the particle position components, D_p is the particle diameter, m_P is the particle mass, ρ and ρ_P are, respectively, the fluid and particle densities, $u_i = U_i + u_i'$ is the instantaneous velocity of the gas phase (U_i is the time-dependent resolved velocity, Eq. (2) and u_i' is the instantaneous SGS fluctuating velocity), c_D is the drag coefficient, g_i is the gravity and I_P is the moment of inertia. More details about the forces acting on the particles and the respective coefficients allowing the extension of the equations of motion to higher particle Reynolds numbers are presented by Sommerfeld et al. [11]. For solving the partial differential equations, a dynamic time step for the particle tracking is adapted, which is automatically adjusted according to the local restricting time scales such as particle response time and integral time scale of SGS turbulence. Even though in this work very fine particles are considered (i.e. below 20 μm) the transverse lift forces play a large role due to the existing strong shear gradients in the cyclone and the frequent wall collisions of the particles inducing extremely high rotational velocities.

In confined flows, the modelling of particle–wall collisions is of utmost importance. The change of the linear and angular particle velocity during a wall collision process is calculated based on the solution of the impulse equations coupled with Coulomb's law of friction including the wall friction coefficient as presented by Sommerfeld and Huber [12]. Since the experimentally studied cyclone was made of a smooth plastic material (i.e. PMMA) no wall roughness effects are considered here.

The sub-grid turbulent particle dispersion is calculated using the isotropic temporally correlated Langevin model described by Sommerfeld et al. [11] and extended for LES simulations by Lipowsky and Sommerfeld [13]. In this modelling approach the new fluctuating fluid velocity seen by the particle at the next location is determined using a correlated (first term) and a random part (second term), both depending on Lagrangian and Eulerian correlation functions:

$$u_i^{\prime n+1} = R_{P,i}(\Delta t_{\mathrm{L}}, \Delta r) u_i^{\prime n} + \sigma_{\mathrm{F}} \sqrt{1 - R_{P,i}^2(\Delta t_{\mathrm{L}}, \Delta r)}\, \xi_i \tag{7}$$

where the superscripts denote the time step and the subscripts the spatial component. Δt_{L} is the Lagrangian time step and Δr is the spatial separation between the virtual fluid element and the particle during the time Δt_{L}. The SGS turbulence may be considered to be isotropic so that σ_{F} represents the rms value of the fluid velocity fluctuation and ξ_i denote Gaussian distributions with zero mean and unit variance. The correlation functions $R_{P,i}(\Delta t_{\mathrm{L}}, \Delta r)$ has Lagrangian and Eulerian components, depending on the SGS turbulence time and length scale. The integral time scale $T_{\mathrm{L}} = c_T \sigma_F^2 / \varepsilon_{\mathrm{SGS}}$ and the turbulent length scale of turbulence $L_{\mathrm{E}} = c_L \sigma_{\mathrm{F}} T_{\mathrm{L}}$ are estimated with $\sigma_{\mathrm{F}} = \sqrt{2/3 k_{\mathrm{SGS}}}$, where the constants $c_T = 0.24$ and $c_L = 3.0$ were used [11]. From the LES results, it is possible to estimate the unresolved turbulent kinetic energy and the dissipation rate, reconstructed base on the approximations introduced by Lilly [14], using the dissipation constant selected as $C_\varepsilon = 0.7$:

$$k_{\mathrm{SGS}} = \frac{\mu_{t,c}^2}{\rho_{\mathrm{F}}^2 (0.094\Delta)^2}; \qquad \varepsilon_{\mathrm{SGS}} = \frac{C_\varepsilon k_{\mathrm{SGS}}^{3/2}}{\Delta} \tag{8}$$

3 Cyclone Discretisation and Boundary Conditions

In order to investigate the effect of particle transport by unresolved turbulence on the performance of a cyclone, a high efficiency Stairmand cyclone was considered where also experimental data on flow field and particle separation are available [7]. The geometrical dimensions as well the mesh design used in the simulations are shown in Fig. 1. The cyclone consists of a cylindrical-conical main body with a rectangular tangential inlet and a dust collection bin. The cyclone diameter is $D = 290\,mm$ and all dimensions provided in Fig. 1 left) are normalised by this value. The inlet is a tangential rectangular channel with dimensions $0.5D \times 0.2D$. The exit pipe has a diameter of $0.5D$ and is immersed into the main body with the same distance. The flow velocity measurements were conducted at an inlet gas velocity of $U_0 = 16.1$ m/s and are used here for validation of the LES. In the present study however, inflow velocities of $U_0 = 5$, 10 and 20 m/s were considered since measurements of the separation efficiency are available for these conditions. The continuous phase was air with constant density of 1.1147 kg/m^3 and a dynamic viscosity of 18.587 10^{-6} N s/m^2. With the hydraulic diameter of the inlet channel (i.e. $D_{e,h} = 82.86\,mm$), Reynolds numbers of 24,846, 49,693 and 99,386 are obtained. At the inlet a plug flow was assumed for simplicity and the majority of simulations were conducted with 10 m/s.

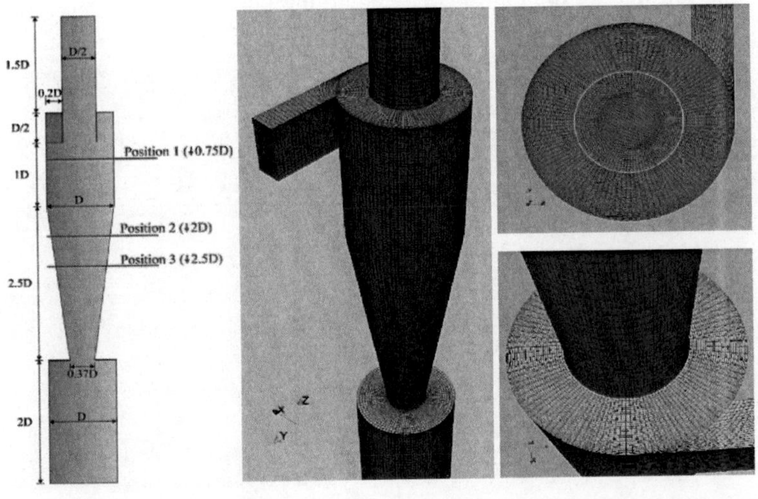

Fig. 1. Geometry of the high efficiency Stairmand cyclone with body diameter D = 0.29 m, a rectangular tangential slit inlet (0.5 D × 0.2 D), dust collection bin and an exit pipe with 0.5 D (left) and several graphs showing the numerical grid with 1 million elements (right).

The numerical grid was constructed as a structured hexahedral mesh with four different resolutions, namely, about 0.5, 1.0, 2.0 and 4.0 million grid cells. In Fig. 1b) some views of the 1.0 million mesh are shown. All the simulations for the flow field are performed with a fixed Eulerian time-step of $\Delta t_E = 5 \times 10^{-5}$ s, ensuring a Courant number less than one. For determining a pseudo-steady-state flow field the simulations were run first for 1.0 s and thereafter continuously particles were randomly injected at the local fluid velocity until 3.0 s. Since the Lagrangian time step for the considered fine particles was normally much smaller than the Eulerian time step, multiple Lagrangian time steps were conducted using a dynamic time step adaptation [1]. For each control volume, the mean velocities as well as their standard deviations (rms-values for the resolved flow) were obtained and additionally the kinetic energy of the SGS turbulence (see Eq. 8).

The particles were spherical beads with a density of 2,740 kg/m^3 and a log-normal-shaped size distribution between 0.5 to 20 μm (i.e. selected according to Hoekstra [7]), discretised by 12 classes. The particle number mean diameter for the selected distribution was about 3.3 μm. In total 500,000 particles were injected at the inlet plane with a size randomly selected from the prescribed size distribution. A mass loading of 0.1 is considered, however, two-way coupling was not considered in this study.

4 Results Cyclone Separator

For illustrating the flow structure and particle separation due to centrifugal effects, Fig. 2 shows the simulation results for a grid of 1 million cells at a gas inlet velocity on 10 m/s. Typical for a cyclone flow, a strong vortex develops in the cyclone with a solid body rotation in the core and a free vortex in the outer regions visualised by the gas phase velocity magnitude (Fig. 2a)). Like a tornado the vortex extends from the bottom of the collection bin throughout the entire cyclone and the outlet pipe. The highest resolved

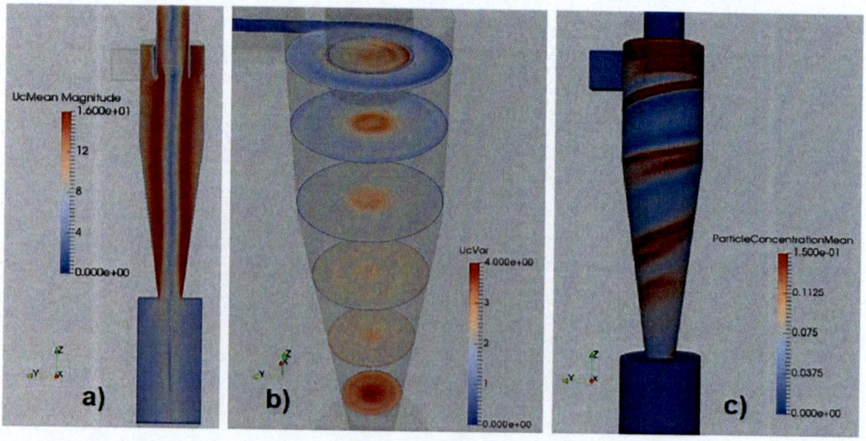

Fig. 2. Visualisation of the particle-laden flow through the Stairmand cyclone a) magnitude of gas phase velocity in m/s; b) magnitude of the resolved gas phase velocity fluctuations in m/s; c) instantaneous particle wall concentration in kg/m^3 (dynamic Smagorinsky model, 1 million cells, $U_0 = 10$ m/s, averaged values between 1 and 3 s).

gas phase fluctuations are found in the core of the vortex (Fig. 2b). The injection of the particles through the tangential slit-type inlet causes the formation of a spirally downwards moving dust rope, which eventually enters the dust collection bin (Fig. 2c)).

For validating the numerical results, here obtained with an inlet velocity of $U_0 =$ 10 m/s, the velocity measurements conducted by Hoekstra [7] at $U_0 = 16.1$ m/s are considered. In Fig. 3, the normalised velocity profiles are compared for a measurement location of z = −0.2175 m (position 1) downstream of the cyclone roof (i.e. profile 1 in Fig. 1a)). The general trends of all velocity profiles are predicted very well, concerning mean and resolved rms velocities. Only the measured low axial velocities in the core of the cyclone (i.e. close to zero) are not entirely captured by the simulations. With increasing the mesh resolution, the tangential mean velocity profile becomes steeper in the core and in the free vortex higher tangential velocities are reached, which is of course essential for particle separation. As a consequence of that, the resolved fluctuation velocities for both axial and tangential component also increase with refining the numerical grid. The agreement with the experiments is reasonably good and the largest differences are observed in the region outside the exit pipe cylinder extension. This however was also observed in the simulations of Sgrott and Sommerfeld [8] with the exact inlet velocity.

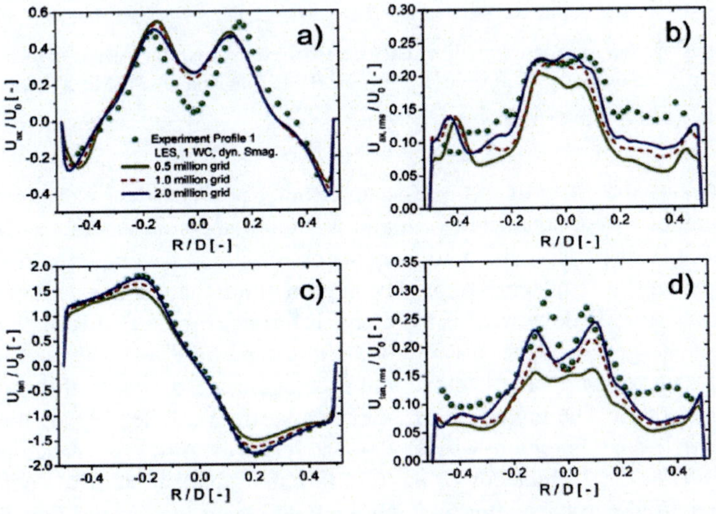

Fig. 3. Normalised velocity profiles for Profile 1 at z = −0.2175 m (see Fig. 1) computed with different grid resolution at $U_0 = 10.0$ m/s and comparison with experiments obtained at 16.5 m/s; a) axial mean velocity; b) axial rms value; c) tangential mean velocity; d) tangential rms velocity.

As expected from a LES, when refining the numerical grid, the kinetic energy of resolved turbulence will increase (see Fig. 3) and the unresolved SGS turbulence should of course decrease accordingly as shown on Fig. 4 for the profiles 1 and 2. Here the SGS turbulence (Eq. 8) is normalised by the total turbulent kinetic energy. The highest values are expectedly found near the wall, although a grid refinement towards the wall was applied (see Fig. 1). In the first profile the near-wall values are between 3 to 6% of the total kinetic energy without considering the very coarse grid. In the second profile the downward flow in the cyclone has more developed and the near-wall values are between 3 to 25% of the total kinetic energy for the grids with 4 to 1 million cells, respectively. Also, in the core region expectedly the SGS turbulent kinetic energy decreases when refining the mesh. For a reasonable dimensioned grid with about 1 million cells, the SGS turbulence in this region is only about 2% of the total kinetic energy. Since particle motion in a cyclone mainly occurs in the near-wall region (see Fig. 2c) where LES cannot resolve turbulence so well, the modelling of unresolved turbulence by the proposed correlated Langevin equation model may have a drastic effect on cyclone performance.

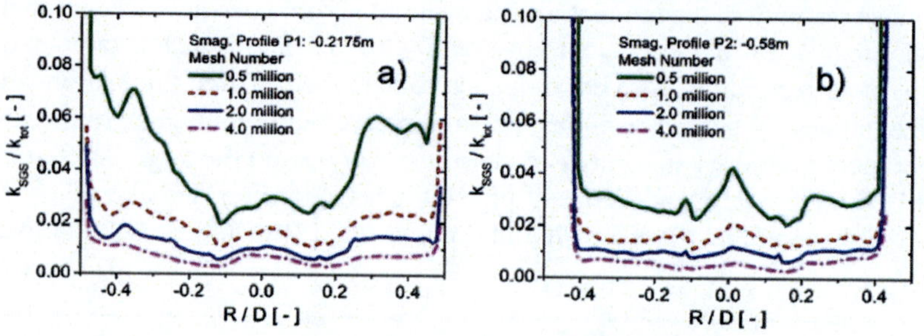

Fig. 4. Profiles of SGS turbulent kinetic energy normalised by the total fluctuating kinetic energy for different grid resolutions; a) Profile 1 at $z = -0.2175$ m and b) Profile 2 at $z = -0.58$ m (dynamic Smagorinsky model; $U_0 = 10.0$ m/s).

For analysing the effect of sub-grid-scale turbulence on particle transport and separation, simulations were conducted with and without dispersion model. For illustrating the effect of SGS dispersion, the particle concentration field developing throughout the cyclone is shown (Fig. 5). Here it should be kept in mind, that the dust entering through the inlet moves spirally downward is form of a dust rope (Fig. 2c)), driven by the downward airflow and gravity. After a wall collision, the particles are reflected backwards towards the core region of the cyclone and then again are driven to the wall through the centrifugal force. The high SGS turbulence near the wall (see Fig. 4) helps to disperse the particles in this region, whereby a relatively wide dust layer develops near the cyclone wall (Fig. 5a)). Neglecting this SGS turbulent dispersion of the particles, the larger more inertial particles, after wall rebound, will enter the upward flow in the core of the cyclone and escape. It should be noted that larger particles contribute more to the local concentration. The smaller particles with less inertia are less strongly rebound and then are carried along the wall with the rotating flow. Therefore, the dust layer near the wall has much lower concentration values without dispersion compared to the case with turbulent dispersion (compare Fig. 5a) and b)).

These arguments about the importance of near-wall SGS turbulence are supported by the particle-wall collision number, which naturally also depends on the grid resolution. With consideration of SGS turbulent dispersion and with 1 million grids about 260 wall collisions per injected parcel are counted. Without modelling SGS particle dispersion the wall collision number is an order of magnitude larger, namely, 2640 wall collisions per injected parcel.

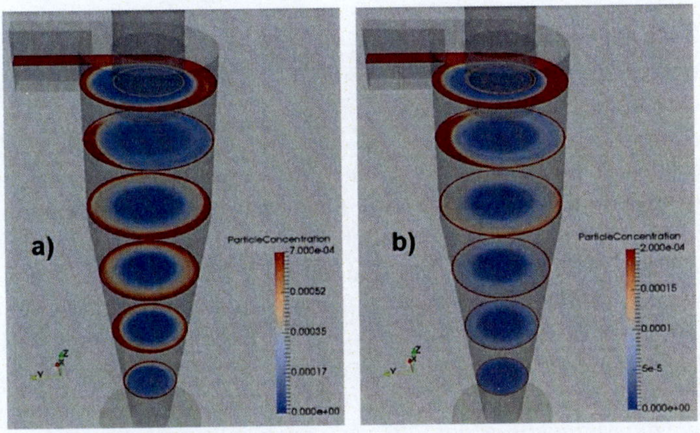

Fig. 5. Particle concentration [kg/m^3] distribution in 6 cross-sections through the cyclone measured from the roof at vertical locations z = −0.08, −0.22, −0.40, −0.60, −0.80 and − 1.00 m; a) with particle SGS dispersion b) without particle SGS dispersion (U_0 = 10.0 m/s).

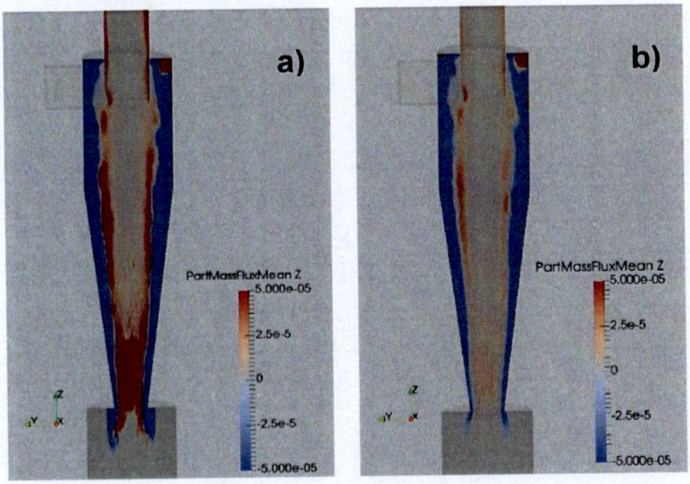

Fig. 6. Vertical-central plains of particle mass flux [kg/(s m^2)] distribution in the z-direction; a) with particle SGS dispersion b) without particle SGS dispersion (dynamic Smagorinsky model; U_0 = 10.0 m/s).

Another region where turbulent dispersion may remarkably alter the performance of a cyclone separator is the dust bin. Here turbulence is responsible for a re-entrainment of fine particles into the upward flow of the core vortex; this mainly occurs near the entrance to the dust bin (Fig. 6). Thereby the particles are transported upwards, however, due to the rotational flow again are separated and transported downwards along the walls to the dust bin (i.e., in the angular blue region), at least the larger particles. This is revealed by coloured vertical panes through the cyclone centre for the z-component of the particle mass flux. In the case with dispersion there is a considerably larger amount of particles transported out of the bin (Fig. 6a) compared to the case without dispersion modelling (Fig. 6b). Such a circulating particle transport will of course remarkably alter the grade efficiency distribution of a cyclone.

The separation efficiency (also called grade efficiency) of the considered Stairmand cyclone is presented in Fig. 7, again for simulations with and without particle dispersion by SGS turbulence. These values constitute the amount of separated particles in a size class divided by the injected amount of particles in the same size class. In addition to that, the effect of grid resolution on the computed separation is highlighted, showing clearly that a mesh with 0.5 million cells is not sufficient. The curves for the other finer grids provide more or less the same result. The points are the measurements of Hoekstra [7] and the red dash-dot line is the reference result of Sgrott and Sommerfeld [8] obtained however with two-way coupling. It is clear, that the steepness of the separation curves compared to the measurements is not fully captured by all simulations. Interestingly, the results obtained without dispersion (Fig. 7b) show efficiencies less than one for particles larger than about 4μm. The location of the minimum in large particle separation (i.e. the worse separation) depend on the grid resolution. This phenomenon was described above (i.e., discussion of Fig. 5); larger particles rebounding from the wall, are not captured by the high near-wall SGS turbulence and consequently bounce back to the cyclone core, where they are captured by the upward outlet flow. Hence, without considering fine particle SGS turbulent dispersion the grade efficiency of a cyclone cannot be predicted correctly.

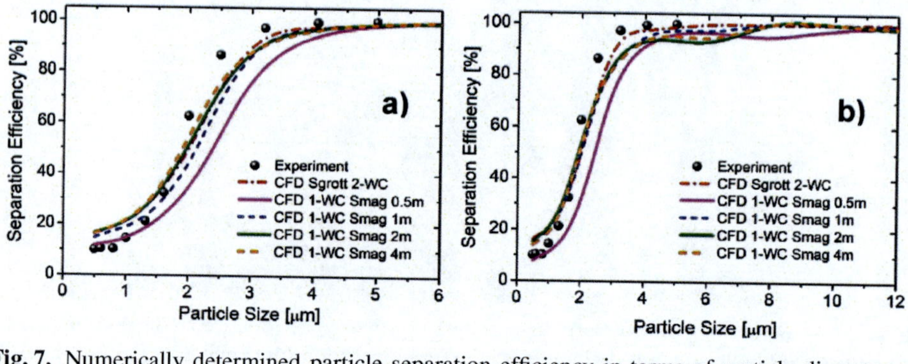

Fig. 7. Numerically determined particle separation efficiency in terms of particle diameter; a) with particle SGS dispersion b) without particle SGS dispersion (dynamic Smagorinsky model; $U_0 = 10.0$ m/s, please note the different scales for the horizontal axis).

5 Conclusions

Fine particle separation in a small-dimension Stairmand cyclone was numerically simulated based on the LES-Euler/Lagrange approach by neglecting two-way coupling. The study revealed the importance of particle turbulent transport by the SGS turbulence, modelled through a stochastic approach. With increasing mesh resolution (i.e., from 0.5

to 4.0 million cells) the resolved turbulent kinetic energy increases associated with a decrease of the energy in the SGS. Both effects are considered when modelling particle dispersion and the separation efficiency is not strongly modified when the mesh is fine enough. However, due to the limited grid resolution in the near-wall region, the SGS turbulent energy in this region is very high. Hence, when neglecting SGS dispersion one important transport mechanism for the particles is neglected and as a result dominated by wall collisions. Consequently, larger particles rebounding from the walls are not re-entrained by SGS turbulence and may escape through the outlet. Thereby, when neglecting fine particle transport by SGS turbulence in a cyclone, the separation efficiency is reduced remarkably for particles being larger than about 2.5 μm.

References

1. Sommerfeld, M.: Numerical methods for dispersed multiphase flows. In: Bodnár, T., Galdi, G.P., Nečasová, Š (eds.) Particles in Flows. AMFM, pp. 327–396. Springer, Cham (2017). https://doi.org/10.1007/978-3-319-60282-0_6
2. Apte, S.V., Mahesh, K., Moin, P., Oefelein, J.C.: Large-eddy simulation of swirling particle-laden flows in a coaxial-jet combustor. Int. J. Multiphase Flow **29**, 1311–1331 (2003)
3. Atzori, M., Chibbaro, S., Duwig, C., Brandt, L.: LES and RANS calculations of particle dispersion behind a wall-mounted cubic obstacle. Int. J. Multiphase Flow **151**, 104037 (2022)
4. Lee, C.Y., Cant, S.: Assessment of LES subgrid-scale models and investigation of hydrodynamic behaviour for an axisymmetrical bluff body flow. Flow Turbul. Combust. **98**, 155–176 (2017)
5. Huilier, D.G.F.: An overview of the Lagrangian dispersion modeling of heavy particles in homogeneous isotropic turbulence and considerations on related LES simulations. Fluids **6**, 145 (2021)
6. Marchioli, C.: Large-eddy simulation of turbulent dispersed flows: a review of modelling approaches. Acta Mech. **228**(3), 741–771 (2017). https://doi.org/10.1007/s00707-017-1803-x
7. Hoekstra, A.J.: Gas flow field and collection efficiency of cyclone separators. The Netherlands: Ph.D. thesis - Delft University of Technology (2000)
8. Sgrott Junior, O.L., Sommerfeld, M.: Influence of inter-particle collisions and agglomeration on cyclone performance and collection efficiency. Can. J. Chem. Eng. **97**, 511–522 (2019)
9. Germano, M., Piomelli, U., Moin, P., Cabot, W.H.: A dynamic subgrid-scale eddy viscosity model. Am. Inst. Phys. **7**, 1760–1795 (1990)
10. Lilly, D.K.: A proposed modification of the Germano subgrid-scale closure method. Phys. Fluids **4**, 633–635 (1992)
11. Sommerfeld, M., van Wachem, B., Oliemans, R.: Best practice guidelines for computational fluid dynamics of dispersed multiphase flows. ERCOFTAC (2008)
12. Sommerfeld, M., Huber, N.: Experimental analysis and modelling of particle-wall collisions. Int. J. Multiphase Flow **25**, 1457–1489 (1999)
13. Lipowsky, J., Sommerfeld, M.: LES-simulation of the formation of particle strands in swirling flows using an unsteady Euler-Lagrange approach. In: Proceedings of the 6th International Conference on Multiphase Flow, ICMF2007, Leipzig Germany, Paper No. S3_Thu_C_54 (2007)
14. Lilly, D.K.: The representation of small scale turbulence in numerical simulation experiments. In: Proceedings of the IBM Scientific Computing Symposium on Environmental Sciences (1967)

Direct and Large–Eddy Simulation of Turbulent Oscillatory Flow Through a Hexagonal Sphere Pack

L. Unglehrt and M. Manhart(✉)

Professorship of Hydromechanics, Technical University of Munich, Munich, Germany
michael.manhart@tum.de

1 Introduction

In coastal engineering, the description of wave transmission and attenuation through rubble-mound breakwaters requires a model for the wave-induced flow inside the pore space of the breakwater [10,16]. Extensive experimental investigations of oscillatory porous media flow were performed [3,7,15] in order to determine the coefficients of the unsteady Forchheimer equation [5] which relates the superficial velocity and the pressure gradient. These coefficients do not solely depend on the geometry, but they vary with time or frequency [2,3,15], raising doubts about the validity of this equation.

In order to gain insight into the flow physics and to obtain a high-fidelity data base for modelling, we simulated the flow through a hexagonal close-packed arrangement of spheres with uniform diameter d driven by a sinusoidal pressure gradient $f_x \sin(\Omega t)\,\mathbf{e}_x$. The flow problem is governed by two independent parameters: the Hagen number $Hg = f_x d^3/(\rho \nu^2)$ and the Womersley number $Wo = \sqrt{\Omega d^2/\nu}$. We define the Reynolds number as $Re = \max \langle u \rangle_\mathrm{s}\, d/\nu$ with the superficial velocity

$$\langle u \rangle_\mathrm{s} = \frac{1}{L_x L_y L_z} \int_{V_\mathrm{f}} u \,\mathrm{d}V \tag{1}$$

where L_x, L_y, L_z are the domain sizes, $V_\mathrm{f} = \varepsilon\, L_x L_y L_z$ is the fluid volume and $\varepsilon = 0.259$ is the porosity. We performed a direct numerical simulation (DNS) at $Hg = 10^8$ and $Wo = 100$ resulting in $Re = 1086$ and a large–eddy simulation (LES) at $Hg = 10^9$ and $Wo = 100$ resulting in $Re = 3580$. Both simulations lie within the range of the experiments of [3].

In this contribution, we investigate the temporal behaviour of the superficial velocity and relate it to characteristic events in the instantaneous flow fields. We present instantaneous velocity fields which, despite the strong confinement of the flow, exhibit features that are predominantly found in external bluff body flow. For instance, the flow has a boundary layer structure and shows massive flow separations. The high frequency of the forcing leads to a strongly varying turbulence intensity over the course of the cycle. Based on symmetry- and phase-averaged statistics, we decompose the volume-averaged kinetic energy as well as the volume-averaged dissipation rate into their respective mean flow and turbulence contributions.

C. Marchioli et al. (Eds.): DLES 2023, ERCO 31, pp. 118–123, 2024.
https://doi.org/10.1007/978-3-031-47028-8_18

2 Methodology

2.1 Numerical Method

Our in-house code MGLET [9] solves the incompressible Navier-Stokes equations on a staggered Cartesian grid using a finite volume discretisation with second-order central differencing. A third-order explicit Runge-Kutta scheme is employed for the time integration. The no-slip boundary condition on the spheres is enforced using an immersed boundary method [12]. For the LES the sub-grid stresses are formulated with the WALE viscosity model [11] with a coefficient $C_w^2 = 0.1$.

2.2 Domain and Grid

The domain size was chosen as $L_x = 2d$, $L_y = \sqrt{3}d$ and $L_z = 2\sqrt{6}/3\,d$ with triple periodic boundary conditions. The domain contains two spheres in every lattice direction and every pore is repeated eight times. For their DNS of stationary turbulent flow, He et al. [4] found that a domain of equal volume "showed little variation in statistics compared to a larger domain". At $Wo = 100$, the flow belongs to the high frequency regime and features very thin boundary layers and high velocities near the contact points of the spheres. Thus, a higher resolution than in stationary flow is required. In order to assess the effect of the grid resolution, the DNS at $Re = 1086$ was simulated at resolutions of 192, 384 and 768 cells per diameter (cpd). We observed good agreement between the simulations at 384 and 768 cpd. The LES at $Re = 3580$ was performed using a resolution of 384 cpd. Table 1 summarises the simulation parameters. In the following, the results at 384 cpd will be analysed.

2.3 Turbulence Statistics

The turbulence statistics are estimated by a phase average of instantaneous velocity fields. As we are only concerned with single-point statistics, the symmetries of the sphere pack are used to generate additional realisations. When the flow is driven by a body force along the x-direction, the velocity field has four spatial symmetries; this leads to an increase of the sample size by a factor of 16. The sampling is started after one cycle, when the transient of the superficial velocity has sufficiently decayed.

Table 1. Summary of computational cases

Re	Hg	Wo	Cells per diameter	Number of cells	Simulated cycles	CPU-h
1073^a	10^8	100	192	$42 \cdot 10^6$	10	20000
1088^a	10^8	100	384	$199 \cdot 10^6$	10	138000
1086^a	10^8	100	768	$1.07 \cdot 10^9$	3.9	661000
3580^b	10^9	100	384	$199 \cdot 10^6$	10	669000

^{a} DNS, ^{b} LES

3 Results

3.1 Superficial Velocity

The time series of the superficial velocity is plotted in Fig. 1. For both Reynolds numbers, the amplitude of $\langle u \rangle_s$ lies within 2% of the steady state value obtained from the correlations of [8]. This is in line with the observation of [3] that "the Forchheimer unsteady-stationary flow law described the oscillatory measurements well when velocities and energy losses were maximum". Based on the time of the zero crossings, the superficial velocity lags behind the sinusoidal forcing by a phase angle of 0.2π at $Re = 1086$ and of 0.1π at $Re = 3580$. At $Re = 3580$ the superficial velocity shows a strong acceleration followed by a low acceleration plateau. The instantaneous velocity fields indicate that this change is caused by flow separations at the contact points and an increase in turbulence intensity.

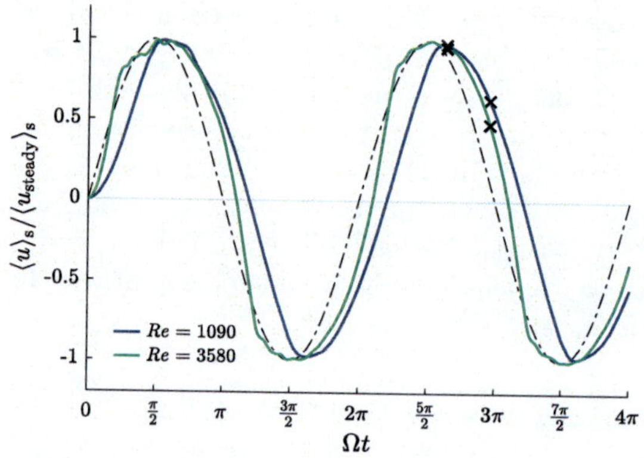

Fig. 1. Time series of the superficial velocity at $Re = 1086$ (blue) and $Re = 3580$ (green). The crosses mark the time of the instantaneous fields presented in Figs. 2 and 3. The reference velocity was obtained from the steady state correlations of [8]. The black dash-dotted curve denotes the sinusoidal body force.

3.2 Instantaneous Velocity Fields

The spatial distribution of the velocity magnitude at the maximum superficial velocity ($\Omega t = 8.3$) and during the deceleration phase ($\Omega t = 9.3$) is displayed in Figs. 2 and 3 for $Re = 1086$ and $Re = 3580$, respectively. At $Re = 1086$, the most prominent feature of the flow is the separation at the contact points between the large pores. Due to the confined pore space, the flow around the separation bubble is concentrated into two jets that are diverted towards the wake by the spheres bounding the pore. The flow topology appears similar to the results of [14] for stationary turbulent flow. Inside the separation bubble we can observe some complicated vortex structures which disintegrate into turbulence in the course of the cycle. Whereas the velocity field is quite smooth at the end of the acceleration phase, a lot of small vortical structures appear during the deceleration. This behaviour is common in oscillatory flow [1]. At $Re = 3580$, the flow topology is similar to the lower Reynolds number, but the flow at the maximum superficial velocity is already turbulent; this is consistent with the formation of the plateau in the superficial velocity.

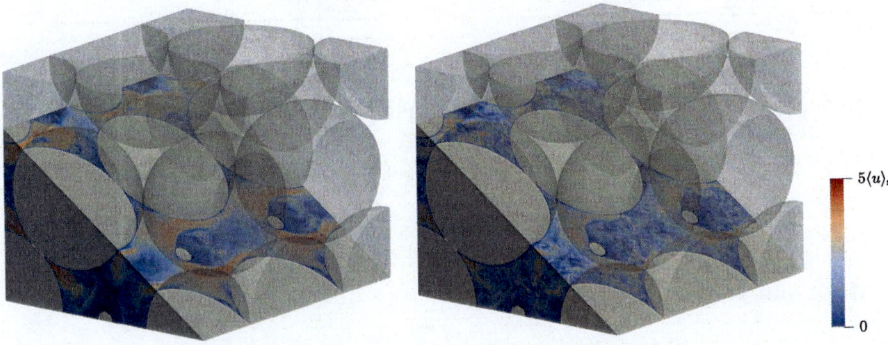

Fig. 2. Velocity magnitude for $Re = 1086$ at the maximum superficial velocity (left) and during deceleration (right). The colourmap is based on the maximum superficial velocity $\langle u\rangle_s$.

Fig. 3. Velocity magnitude for $Re = 3580$ at the maximum superficial velocity (left) and during deceleration (right). The colourmap is based on the maximum superficial velocity $\langle u\rangle_s$.

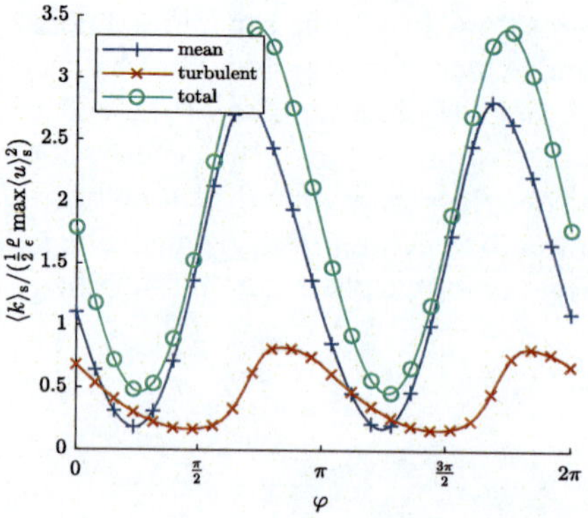

Fig. 4. Volume-averaged MKE, TKE, and sum of MKE and TKE over the cycle for $Re = 1086$.

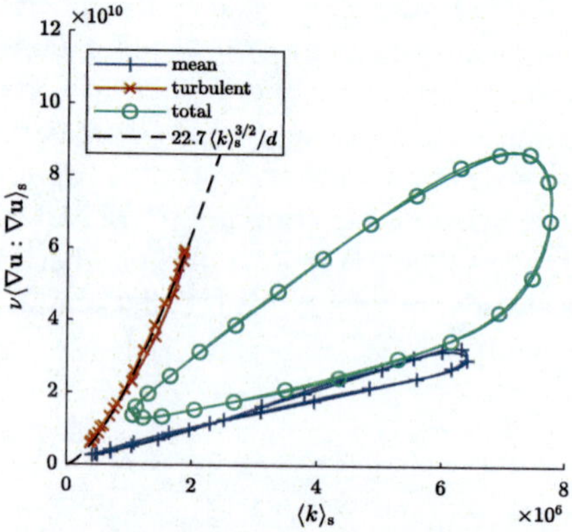

Fig. 5. Volume-averaged dissipation over kinetic energy for $Re = 1086$. The dashed line indicates a fit to the $\mathrm{TKE}^{3/2}$ power law.

3.3 Mean and Turbulent Kinetic Energy

In order to investigate the evolution of the turbulence over the course of a cycle, we analyse the volume-averaged mean and turbulent kinetic energy (MKE and TKE) for $Re = 1086$. It can be seen in Fig. 4 that TKE is present throughout the entire cycle, indicating that no complete relaminarisation occurs. Moreover, a phase shift can be observed between the MKE and the TKE: During the acceleration phase (indicated by rising MKE) the TKE has only a small contribution, whereas during the deceleration phase the TKE attains its maximum of approximately 25% of the total kinetic energy amplitude.

3.4 Relation Between Kinetic Energy and Dissipation Rate

A phase lag between the volume-averaged kinetic energy and the dissipation rate can be observed at $Re = 1086$ which results in the green hysteresis loop in plotted in Fig. 5. When the Reynolds decomposition is introduced, the hysteresis disappears and a nearly one-to-one relation between the TKE and the turbulent dissipation rate emerges (red curve). This matches well to a $\mathrm{TKE}^{3/2}$ power law similar to the ones used by Prandtl [13] and Lilly [6] to model the dissipation rate. This suggests that the length scale of the turbulence production remains relatively constant which could be explained by the dominant flow separation pattern.

The LES at $Re = 3580$ follows approximately the same curve in the acceleration phase, whereas it follows another $\mathrm{TKE}^{3/2}$ power law with a smaller prefactor in the deceleration phase. It will be further investigated whether this discrepancy between DNS and LES is influenced by the WALE model coefficient C_w.

4 Conclusion

We performed DNS and LES of a turbulent oscillatory flow through a hexagonal sphere pack at $Re = 1086$ and 3580 and $Wo = 100$. A separation occurs at the contact points of the spheres that leads to a significant increase in the drag. At $Re = 1086$ turbulence can be observed mainly during the deceleration phase whereas at $Re = 3580$ turbulence can be observed over a longer part of the cycle. The turbulent dissipation rate was shown to depend on the TKE with the well known $\mathrm{TKE}^{3/2}$ relation.

Acknowledgements. We gratefully acknowledge the financial support of the Deutsche Forschungsgemeinschaft (DFG, German Research Foundation) under grant no. MA2062/13-1. Computing time was granted by the Leibniz Supercomputing Centre on its Linux-Cluster.

References

1. Akhavan, R., Kamm, R., Shapiro, A.: An investigation of transition to turbulence in bounded oscillatory stokes flows part 1. Experiments. J. Fluid Mech. **225**, 395–422 (1991)
2. Burcharth, H., Andersen, O.: On the one-dimensional steady and unsteady porous flow equations. Coast. Eng. **24**, 233–257 (1995)
3. Hall, K., Smith, G., Turcke, D.: Comparison of oscillatory and stationary flow through porous media. Coast. Eng. **24**, 217–232 (1995)
4. He, X., Apte, S., Finn, J., Wood, B.: Characteristics of turbulence in a face-centred cubic porous unit cell. J. Fluid Mech. **873**, 608–645 (2019)
5. Polubarinova-Kochina, P., De Wiest, R.: Theory of Ground Water Movement (1962)
6. Lilly, D.: The representation of small-scale turbulence in numerical simulation experiments. In: Proceedings Of IBM Scientific Computing Symposium On Environmental Sciences, Yorktown Heights, New York, pp. 195–210 (1967)
7. Losada, I., Losada, M., Martín, F.: Experimental study of wave-induced flow in a porous structure. Coast. Eng. **26**, 77–98 (1995)
8. Macdonald, I., El-Sayed, M., Mow, K., Dullien, F.: Flow through porous media - the Ergun equation revisited. Ind. Eng. Chem. Fundam. **18**, 199–208 (1979)
9. Manhart, M., Tremblay, F., Friedrich, R.: MGLET: a parallel code for efficient DNS and LES of complex geometries. Parallel Comput. Fluid Dyn. **2000**, 449–456 (2001)
10. Muttray, M.: Wellenbewegung an und in einem geschütteten Wellenbrecher - Laborexperimente im Großmaßstab und theoretische Untersuchungen. Technische Universität Braunschweig (2000)
11. Nicoud, F., Ducros, F.: Subgrid-scale stress modelling based on the square of the velocity gradient tensor. Flow Turbul. Combust. **62**, 183–200 (1999)
12. Peller, N.: Numerische simulation turbulenter Strömungen mit immersed boundaries. Technische Universität München (2010)
13. Prandtl, L.: Über ein neues Formelsystem für die ausgebildete Turbulenz. Nachrichten Akademie Wissenschaften zu Göttingen Mathematisch-Physikalische Klasse **2**, 6–19 (1945)
14. Sakai, Y., Manhart, M.: Consistent flow structure evolution in accelerating flow through hexagonal sphere pack. Flow Turbul. Combust. **105**(2), 581–606 (2020). https://doi.org/10.1007/s10494-020-00168-4
15. van Gent, M.: Stationary and oscillatory flow through coarse porous media. Commun. Hydraul. Geotech. Eng. **9** (1993)
16. van Gent, M.: The modelling of wave action on and in coastal structures. Coast. Eng. **22**, 311–339 (1994)

Check for updates

High-Fidelity Large-Eddy Simulation of a Pulsed Jet Actuator

Özgür Yalçın[1(✉)], Xavier Gloerfelt[1], Georges Saliba[2], Ahmad Batikh[3], and Lucien Baldas[2]

[1] Laboratoire DynFluid, Arts et Métiers Institute of Technology, Paris, France
{ozgur.yalcin,xavier.gloerfelt}@ensam.eu

[2] Institut Clément Ader (ICA), Univ. Toulouse, CNRS, INSA, ISAE-SUPAERO, Mines Albi, UPS, Toulouse, France
{saliba,baldas}@insa-toulouse.fr

[3] Institut Clément Ader (ICA), Univ. Toulouse, CNRS, INSA, ISAE-SUPAERO, Mines Albi, UPS, Icam Toulouse, Toulouse, France
ahmad.batikh@icam.fr

1 Introduction

Pulsed jet actuators (PJAs) are a type of fluidic oscillators that can be served as an active flow control device on lifting bodies. PJAs are able to provide an unsteady periodic blowing in a wide frequency range, which adds momentum to the boundary layer to prevent separation, and thereby enhancing the aerodynamic performances [2]. This kind of fluidic oscillator has no moving parts and induces the oscillation using only its internal unsteady flow dynamics, which makes it attractive for applications given its robust design and installation [9].

The geometry of the PJA examined in this study is given in Fig. 1. A jet through a nozzle that is supplied by a pressurized flow in a reservoir attaches on the side wall of one of the branches due to the Coanda effect [5]. While the flow partially exits from the actuator outlet located on the same side, the bi-stable attached flow starts to fill the feedback loop with a compression wave. When this wave reaches the control nozzle, it triggers a detachment of the attached jet, which is followed by a recirculation bubble growing between the side wall and the jet. Meanwhile, an expansion wave reflected from the control nozzle travels back to the branch exit. This back-and-forth traveling of the pressure wave inside the loop causes pressure differences between the two branch exits as well as the two control nozzles [10]. The combined effect of these pressure differences and the recirculation zone growth make the jet switch to the other side, and the same procedure is repeated. The switching of the jet provides an oscillatory blowing at the outlets.

The oscillation period, which is one of the critical design parameters in active flow control, is simply a combination of the travel time of the pressure waves inside the feedback loops and the switching time of the main jet. As the feedback loop length increases, the switching time becomes negligible. For this scenario, some simple algebraic formulas suggested in literature can provide reasonable prediction of the oscillation frequencies. However, the switching time still remains a mystery. There have

C. Marchioli et al. (Eds.): DLES 2023, ERCO 31, pp. 124–129, 2024.
https://doi.org/10.1007/978-3-031-47028-8_19

been very few studies focused on understanding the switching period, which are mainly restricted to subsonic nozzle jets [7]. Besides, most of the numerical studies on this topic are based on unsteady Reynolds-Averaged Navier-Stokes (URANS) simulations. Although URANS approaches could predict the frequency of the oscillations having negligible switching time, they failed to predict the magnitude and profile of the exit velocity [1, 10], which is another critical design parameter. In order to obtain internal and external flow fields accurately, and to understand the unsteady and highly compressible flow dynamics of the switching phenomenon, a high-fidelity numerical analysis is required [6]. For this purpose, a high-order Large-Eddy Simulation (LES) inside of the PJA studied in [8] has been performed and analyzed in the present study. This PJA, also shown in Fig. 1, has relatively short feedback loops, yielding a non negligible switching time. It thus constitutes a good test case to investigate the mechanisms behind the switching time.

This paper includes the solver description, the computational setup, and the numerical results with discussions. The result section is mainly focused on the comparisons with measurements and URANS results, which are described in [8].

2 Flow Solver and Computational Setup

The numerical simulation is performed using a research code, called MUSICAA (MUltiscale Solver In Computational Aeroacoustics and Aerodynamics). The 3D filtered compressible Navier-Stokes equations are solved by a wall-resolved LES (WRLES) strategy that models the subgrid motions numerically with a Regularization Term approach, i.e. without using an explicit subgrid scale model. The inviscid fluxes are discretized by means of 10th-order standard centred differences whereas 4th-order is used for viscous fluxes. In order to introduce a minimal amount of numerical dissipation while ensuring computational robustness for compressible flow simulations, the centered scheme is supplemented by a high-order nonlinear filtering term, combining 2nd- and 10th-order derivatives approximated by standard central differences. The activation of the low-order term rests on a modified Jameson's pressure-based shock sensor. A four-stage Runge-Kutta algorithm is used for time integration and a 4th-order implicit residual smoothing method (IRS4) is implemented to enlarge the stability and allow the use of larger timesteps. The curvilinear grid is taken into account by a coordinate transformation. More details can be found in [4].

The computational domain covers both the internal and external fields of the PJA, from the inlet reservoir to the farfield. Figure 1 shows the multiblock domain with structured grids. A subsonic inlet boundary condition based on Riemann invariants is enforced at the inlet by specifying the ratios of the inlet total pressure to the freestream static one ($p_{t,in}/p_{fs}$). At the farfield boundaries, non-reflective boundary conditions based on asymptotic solutions of linearized Euler equations are applied. All walls are treated by no-slip conditions. The 2D domain given in Fig. 1 has the same internal geometry as the experimented PJA, where the nozzle width is 0.2 mm. The domain is extruded uniformly along the third direction where periodicity is enforced. The reason is that in the experiment, the depth-to-width ratio is 35, which is assumed to be large enough to neglect the depthwise wall effects. The depth of the computational domain is chosen as 2.5 times of the nozzle exit width, considering the jet size which is limited by the distance between the nozzle exit and the splitter edge as well as the branch inlet widths. The mesh generation for each block is started by taking the non dimensional first cell height as 1 wall unit. The maximum streamwise grid spacing is 30 wall units, whereas the spanwise one is 15. In the whole domain, the grid stretching ratios are no more than 1.05. In total, the grid consists of approximately 1 million nodes in 2D, and has 30 nodes in the third direction.

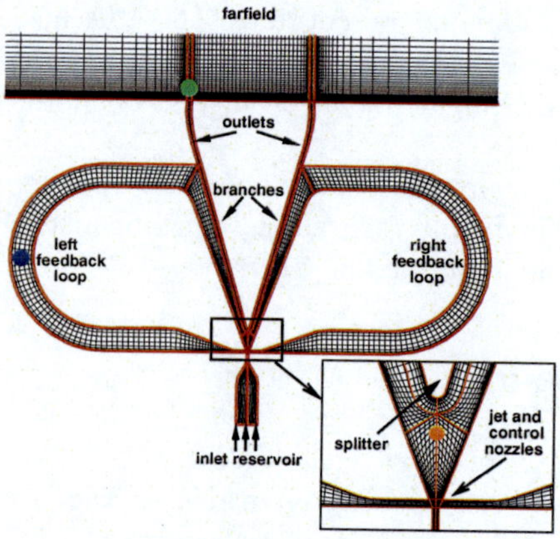

Fig. 1. Multiblock structured grid of the PJA. Every four cells are shown. The subframe zooms the switching area.

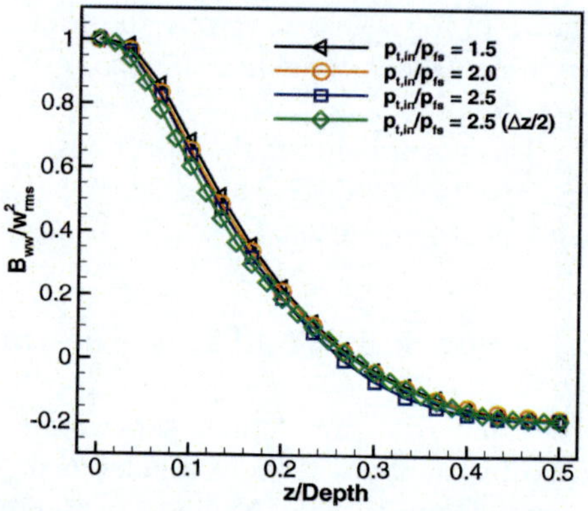

Fig. 2. Nondimensional two-point spanwise correlation at the center of the jet switching region.

3 Results

The simulations are performed for three different $p_{t,in}/p_{fs}$ values: 1.5, 2.0, and 2.5. For each simulation, approximately 1000 CPU processors are used, each of which has a single domain with $30 \times 30 \times 30$ grid nodes. The time steps are 5.0×10^{-9} s, 3.5×10^{-9} s, and 1.5×10^{-9} s for the lowest to the highest pressure ratios, respectively. Note that thanks to the IRS4 method, these time steps are around 8 times those limited by the explicit time scheme.

The simulations are carried out for more than 15 oscillation periods after the initial transients have left the computational domain. Before collecting data, a grid dependency study has been done to make sure that the grid resolution is sufficient. In this respect, two-point correlations are investigated, as proposed in [3]. Figure 2 shows the nondimensional two-point spanwise correlation variations, B_{ww}/w^2_{rms}, which are computed using time signals of third velocity component, w, for all pressure ratios at a point located at the center of the jet switching region (orange point in Fig. 1). The figure also includes the results for the highest pressure ratio when the spanwise grid spacing (Δz) is divided by two. It is observed that the correlations go to zero within 10 grid points when Δz is used. It means the largest eddy scale is discretized by 10 nodes, which is sufficient [3]. Besides, using $\Delta z/2$ does not change the trend. Hence, the present grid spacing is kept for the rest of the study.

The oscillation frequency variations as a function of the pressure ratios are compared in Fig. 3 with the measured data as well as URANS simulations based on two different modeling approaches, k-γ SST and k-ω SST with $\gamma - Re_\theta$ transition model. It is worth noting that the jet is choked beyond $p_{t,in}/p_{fs}$ of 1.7 such that the case with $p_{t,in}/p_{fs}$ of 1.5 is the only one where the jet is subsonic at the throat. It is observed from the measurements that after the jet is choked, the oscillation frequency decreases as the pressure ratio increases up to a certain point, and then remains almost constant. In the region of interest, the frequencies of the current study, indicated as WRLES in Fig. 3, are in much more better agreement with the measurements than the URANS ones, even if WRLES predictions show a slight overestimation.

Instantaneous Mach contours are presented in Fig. 4 for the case with $p_{t,in}/p_{fs}$ of 2.0, at an instant corresponding to the jet switching from left to right. The unsteadiness of the switching process, which includes the evolution of the recirculation and separation bubbles, can be observed by looking at the streamlines in the right subframe of the figure. Moreover, the flow is choked at the nozzle throat at this pressure ratio, and during the switching shocklets appear in the vicinity of the impingement onto the splitter. In addition, this region involves not only the most energetic turbulent structures but also the smallest ones in the domain. This is why the grid dependency study was conducted at the center of this region. On the other hand, this is an indication of the necessity of proper eddy resolution during the jet switching, explaining the failure of URANS when switching time is non negligible. One can also observe from the left subframe of Fig. 4 that the jet is unsteady throughout the branch after it is detached from the corresponding side wall.

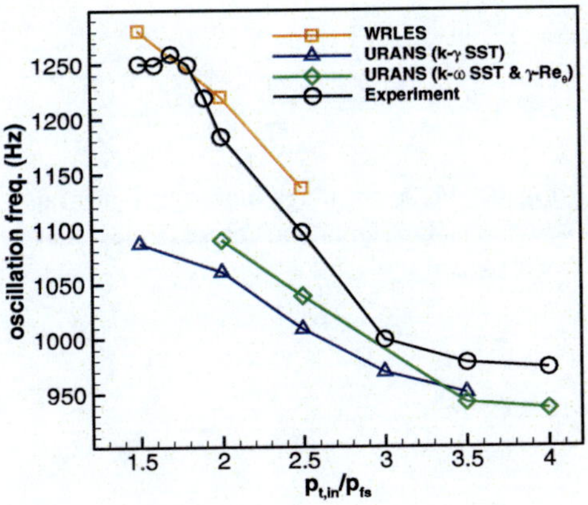

Fig. 3. Comparison of the oscillation frequencies changing with $p_{t,in}/p_{fs}$.

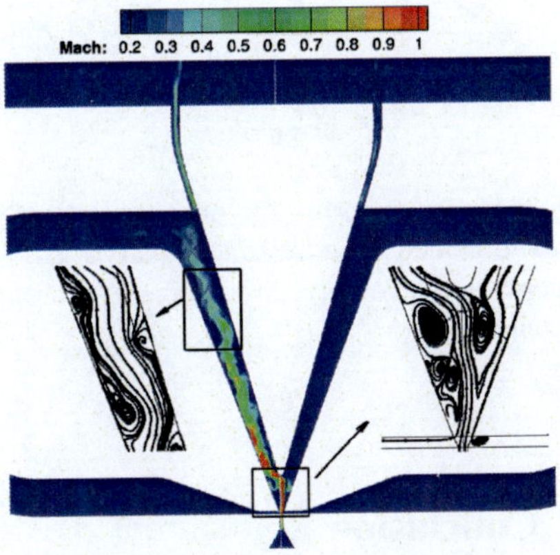

Fig. 4. Instantaneous Mach contours with streamlines during the jet switching ($p_{t,in}/p_{fs} = 2.0$).

Figure 5 shows a comparison of the pressure norms (Z-score) for one oscillation period at the location of the left loop center (blue point in Fig. 1). The results are given for the case with $p_{t,in}/p_{fs}$ of 2.5. The WRLES prediction is in fair agreement with the measurements although some discrepancies are observed in the high frequency signal components, possibly due to the diameter size of the subminiature pressure transducer being of the same order as the loop width (see [8] for details about experimental apparatus).

Lastly, the exit velocity magnitudes are compared with the experimental data. The signal point (corresponding to the hot wire location in the experiment) is located at a distance less than a half of the outlet width from the left outlet exit, and aligned with its center (green point in Fig. 1). Figure 6 displays the comparison of velocity magnitudes in a spectral domain for the $p_{t,in}/p_{fs}$ of 1.5 case. The magnitudes of the first peak which corresponds to the oscillation frequency, and of the next two peaks are in perfect agreement with the experimented ones. The differences in low and high frequency contents are caused by discrepancies due to signal durations and due to the sampling frequencies, respectively. For the rest, the spectrum, thus the pulsed jet dynamics, is well predicted by WRLES.

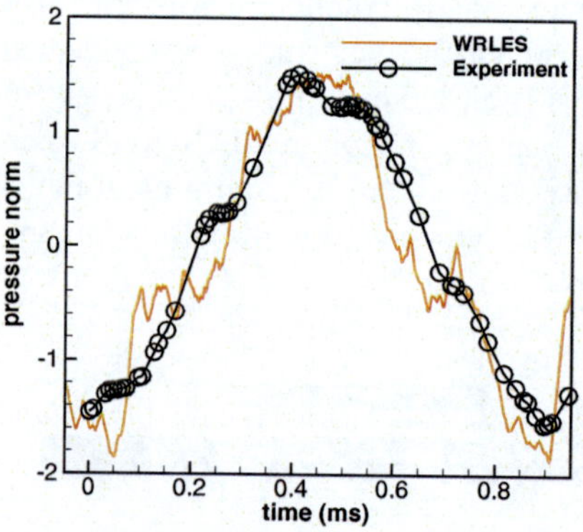

Fig. 5. Pressure norms for one oscillation period obtained from the time signals at the center of the left loop center using Z-score normalization ($p_{t,in}/p_{fs} = 2.5$).

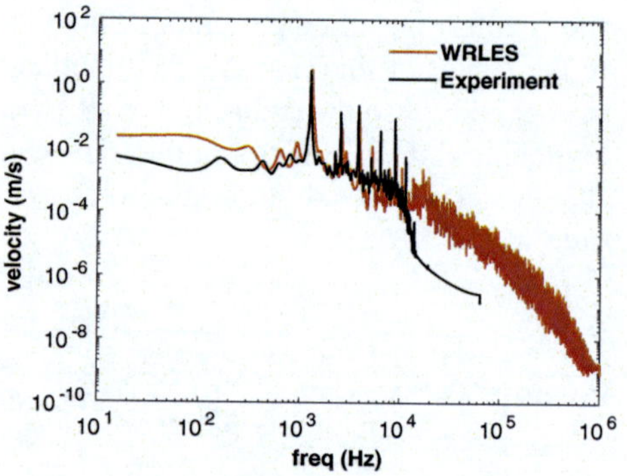

Fig. 6. Velocity magnitudes in spectral domain at the signal point close to the outlet exit ($p_{t,in}/p_{fs} = 1.5$).

4 Conclusion

In this study, PJAs having different inlet pressure ratios are simulated by a high-fidelity WRLES. Oscillation frequencies as well as outlet velocities, which are the critical design parameters, are in good agreement with measurements. This study shows the importance of high-fidelity eddy resolution inside the region where recirculation and separation zones are growing during the jet switching, and where the compressible flow is highly unsteady. The next step of this study will be to analyze the time signals at different points to get further insights into the switching time as the pressure ratio changes.

Acknowledgements. This study has received funding from the Clean Sky 2 Joint Undertaking (JU) under grant agreement No 887010. The JU receives support from the European Union's Horizon 2020 research and innovation program and the CleanSky 2 JU members other than the Union. This work was granted access to the HPC resources of IDRIS and TGCC under the allocation A0092A01736 made by GENCI (Grand Equipement National de Calcul Intensif).

References

1. Aram, S., Lee, Y.T., Shan, H., Vargas, A.: Computational fluid dynamic analysis of fluidic actuator for active flow control applications. AIAA J. **56**(1), 111–120 (2018)
2. Cattafesta, L.N., III., Sheplak, M.: Actuators for active flow control. Annu. Rev. Fluid Mech. **43**, 247–272 (2011)
3. Davidson, L.: Large eddy simulations: how to evaluate resolution. Int. J. Heat Fluid Flow **30**(5), 1016–1025 (2009)
4. Gloerfelt, X., Cinnella, P.: Large eddy simulation requirements for the flow over periodic hills. Flow Turbul. Combust. **103**(1), 55–91 (2019)
5. Gregory, J., Tomac, M.N.: review of fluidic oscillator development and application for flow control. In: 43rd AIAA Fluid Dynamics Conference, p. 2474 (2013)
6. Kim, J., Moin, P., Seifert, A.: Large-eddy simulation-based characterization of suction and oscillatory blowing fluidic actuator. AIAA J. **55**(8), 2566–2579 (2017)
7. Muller, H.R.: A study of the dynamic features of a wall-reattachment fluid amplifier. J. Basic Eng. **86**, 819–826 (1964)
8. Saliba, G.: Study and development of fluidic micro-oscillators for cooling electronic systems. Ph.D. thesis, Université Paul Sabatier-Toulouse III (2022)
9. Schlösser, P., Bauer, M.: Design of a pulsed jet actuator for separation control. CEAS Aeronaut. J. **11**(4), 805–812 (2020)
10. Wang, S., et al.: On the modelling of the switching mechanisms of a Coanda fluidic oscillator. Sens. Actuat. A **299**, 111618 (2019)

Internal and External Aerodynamics

Transition and Acoustic Excitation of Stenotic Pipe Flows at Different Reynolds Numbers

A. Abdelsamie[1(✉)], S.-R. Koh[2], G. Janiga[1], and D. Thévenin[1]

[1] Laboratory of Fluid Dynamics and Technical Flows (LSS/ISUT), University of Magdeburg "Otto von Guericke", Universitätsplatz 2, 39106 Magdeburg, Germany
{abouelmagd.abdelsamie,janiga,thevenin}@ovgu.de

[2] Forschungszentrum Jülich GmbH, Wilhelm-Johnen-Straße, 52425 Jülich, Germany
s.koh@fz-juelich.de

1 Introduction

A human respiratory system consists of phonation components that are coupled in a complex manner in order to ensure various vital functions, in particular voice generation. The interdisciplinary nature of the processes controlling sound generation complicates the analysis. Analytical studies are limited and can only be used to characterize the main acoustic sources in connection to various types of fluid motion. Numerical investigations of sound generation require an accurate simulation of the flow field with a proper representation of the respiratory pathways and process conditions in order to get the acoustic source terms. Furthermore, voice formation is closely related to the resonance of acoustic modes in and around the mouth cavity; in order to be able to model this properly, it is essential to identify first the sound sources excited within the vocal tract.

To better elucidate this complex process, the fundamental mechanisms in a biofluid flow mimicking such configurations have been first investigated based on a simplified stenotic pipe using high-resolution discretization methods [1,2]. The variation in cross-section was regarded as critical to trigger flow transition. The geometry used in a previous experimental study [3] was applied to a numerical analysis in the Reynolds number range of Re = 400–800 based on the pipe diameter [4]. Results from direct numerical simulation (DNS) reveal the development of flow instabilities as a function of the selected Reynolds number. Later on, still based on DNS, a global linear stability analysis [5] showed that a symmetric flow becomes linearly unstable to an eigenmode when increasing Reynolds number. Over a critical value of Re, the instability triggers non-axisymmetric flow oscillations. The transition occurs in connection with the break-up of streamwise hairpin vortices generated by the jet downstream of the stenotic throat. Exchange processes are most intense at a turbulent spot where the sound generation is dominated by the dipole contribution on the surface [6]. On the other hand, the contribution of viscous shear stress is rapidly dissipated due to relaminarization further downstream [2]. Though already very interesting, these findings must be refined further to better understand the contribution of dipole and higher-order sources to acoustics in stenotic pipe flows. In the present work, a wider range of Reynolds number will be examined (Re = 400, 500, 750, 1000, 1800), for axisymmetric and eccentric pipes as

C. Marchioli et al. (Eds.): DLES 2023, ERCO 31, pp. 133–138, 2024.
https://doi.org/10.1007/978-3-031-47028-8_20

well. The flow regimes (laminar, transition, and turbulence) are first examined using the spectral entropy indicator. Then, two different solvers are examined, one based on the incompressible Navier-Stokes equation and the other based on the Lattice-Boltzmann method, to check if both can give similar information about the acoustic (noise) sources.

2 Numerical Methods

The present work focuses on DNS using either a finite-difference or a lattice-Boltzmann formulation to solve for stenotic pipe flows at relevant Reynolds numbers in the range Re = 500–1800. The two in-house solvers are (1) DINO, providing a finite-difference solution of the incompressible Navier-Stokes equations (NSE), and (2) a lattice Boltzmann (LB) solver, providing a weakly-compressible solution. Proper Orthogonal Decomposition (POD) and spectral entropy S_d [7] are applied to analyze transition. The noise sources are obtained by calculating the source terms of the Acoustic Perturbation Equations (APE) [8], i.e., the contribution of hydrodynamic pressure fluctuations for the incompressible flow fields determined by DINO and the Lamb vector fluctuations obtained from the LB solutions.

For the present study, the reference results are those obtained by the well-established DNS code DINO, first described in [9] and used since then for many DNS studies for a variety of configurations. DINO relies on a sixth-order central finite-difference scheme for spatial discretization and a fourth-order Runge-Kutta method for time integration. The code has been optimized for massively parallel High-Performance Computing (HPC) leading to excellent scalability on various HPC systems [10]. Any complex geometry can be handled in DINO using a novel high-order directional ghost-cell immersed boundary method [11].

As an alternative, the LB method is used to solve the stenotic pipe flows in the weakly compressible regime. The single-relaxation-time (SRT) method has been selected. The Boltzmann equation with a Bhatnagar-Gross-Krook (BGK) approximation of the collision term is solved to get the fluid particle probability distribution functions (PPDFs) in a discretized form. The D3Q27 discretization scheme has been selected as a set of lattice vector weights [12], with 27 directions in the three-dimensional flow field. The macroscopic variables are obtained from the moments of the PPDFs. The numerical approach has been validated in generic flow configurations [13] and later applied to bio-fluid mechanical problems such as the circulation of a brain fluid [14].

3 Flow Configuration

Steady flows (i.e., non-pulsatile inflow) through both axisymmetric and eccentric stenosis models have been computed and analyzed. The baseline geometry is given in [2] and defined by a cosine function dependent on the axial coordinate x. The cross-section coordinates y and z are determined by the function of the axial coordinate x such that the profile realizes 25% diameter reduction at the stenosis throat. The eccentric case offsets the stenosis axis by $0.05D$ at $x/D = 0$. The details of the stenosis geometry are defined in [2]. The inflow condition corresponds to a parabolic profile of the streamwise velocity component u. Based on the cross-sectional average velocity and the pipe diameter D

= $2R$ at the inlet ($x = -4D$) the Reynolds number Re based on the pipe diameter is 400, 500, 750, 1000, or 1800, for five different flow simulations. The computational domain extends up to the outflow boundary at $x = 18D$. For DINO simulations, the number of grid points is varied with the Reynolds number; where the range of the required mesh is 33.6 to 279 million grid points corresponding to Re=400 to 1800, respectively. With this number of grid points, the value of y^+ is kept smaller than 1 for all considered cases.

4 Results

The impact of the Reynolds number on the flow can be presented by examining the streamwise velocity at the center plane for the axisymmetric and eccentric cases as it is shown in Fig. 1. As can be observed from this figure, for the axisymmetric case, the jet break is observed at Re = 750, whereas for the eccentric pipe the jet break starts at a smaller Reynolds number (Re = 500). It has been observed that by increasing the Reynolds number, the jet break (and therefore, the transition) starts earlier.

In order to detect the transition more precisely, spectral entropy S_d is used. Depending on the value of the S_d at a specific region, the flow can be classified as laminar, transition, or turbulent. In Abdelsamie et al. (2017), the flow regions that have $S_d \leq 0.67$ are classified as laminar, whereas the regions with $S_d \geq 1.1$ are classified as turbulent flow region; the regions with $0.67 < S_d < 1.1$ can be considered as transitional flow. Figure 2 shows the S_d versus the tested cases (Reynolds number) at three different planes that are normal on the streamwise direction at distance $1D$, $2D$, $3D$ from the origin (axial center of the stenosis region). As is clear from the figure, in all cases at an axial distance $1D$, the flow is laminar ($S_d < 0.67$), except the eccentric pipe with Re = 1800, where the transitional behavior is observed ($0.67 < S_d < 1.1$). At a longer distance $2.5D$, the flow in the eccentric pipe becomes turbulent at cases with Re ≥ 700, whereas in the axisymmetric pipe, the flow first becomes turbulent at Re ≥ 1000. At an axial distance of $4D$, the flow in axisymmetric pipe shows a transitional behavior for Re > 500, whereas, in the case of eccentric pipes, the flow becomes transitional at a lower Reynolds number, Re > 400.

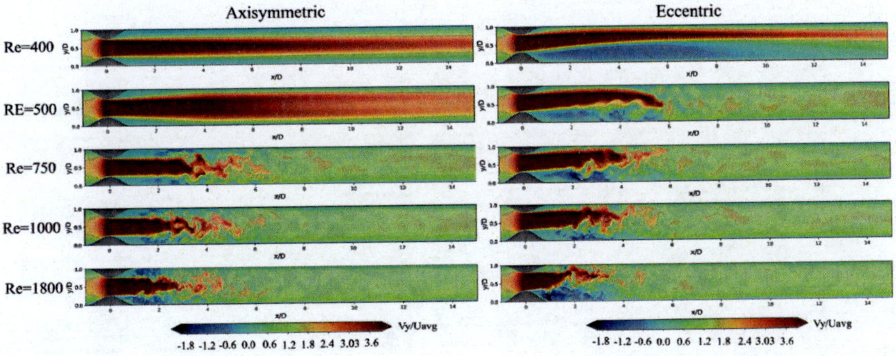

Fig. 1. Temporal-averaged values of streamwise velocity along centerline plane for different Reynolds numbers.

This transition scenario dominates the level of the sound noise in the flow. This noise level can be predicted by investigating either the temporal/total derivative of hydrodynamic pressure in incompressible flow solver or pressure and Lamb vector from the Lattice-Boltzmann flow solvers as it will be clarified from Figs. 3 and 4.

Using the incompressible flow solver DINO, the temporal derivative of the pressure shows that for axisymmetric and eccentric pipes, the acoustic (noise) sources are maximal downstream of the stenosis and increase rapidly with Re. For eccentric stenosis, the sources come close to the pipe wall, which could induce an acoustic dipole due to the modified distribution of near-wall shear stresses, which can be confirmed by the lattice-Boltzmann flow solver as shown in Fig. 4. In this figure, the LB simulations were used to determine the flow fields of the axisymmetric and the eccentric case at Reynolds number 1000. The contours of pressure fluctuations in Fig. 4(left) show that the eccentric case leads to a higher magnitude close to the wall, as it was also concluded from Fig. 3.

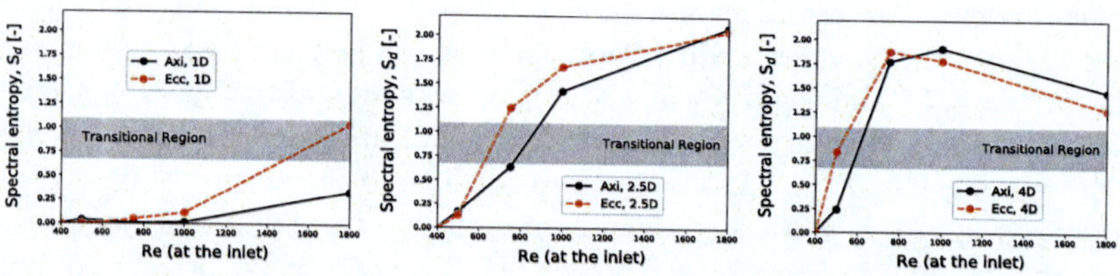

Fig. 2. Spectral entropy at different Reynolds numbers along cross-section planes at a distance of $1D$, $2.5D$, and $4D$

The acoustic excitation occurs at a position further upstream than for the axisymmetric stenosis setup. The acoustic source is then calculated to plug in the Acoustic Perturbation Equations (APE) developed for a compressible flow [8]. The acoustic generation in the current isothermal condition is dominated by the momentum source term, i.e., fluctuations of Lamb vector components. The root-mean-square value of the momentum source is illustrated in Fig. 4(right).

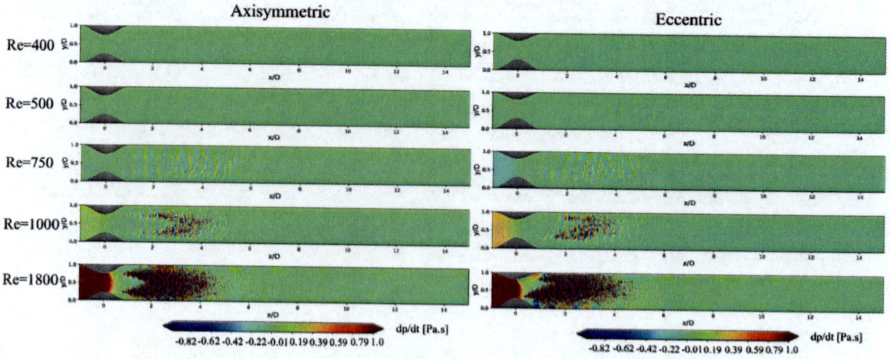

Fig. 3. Temporal-averaged values of $\frac{dp}{dt}$ along centerline plane for different Reynolds numbers.

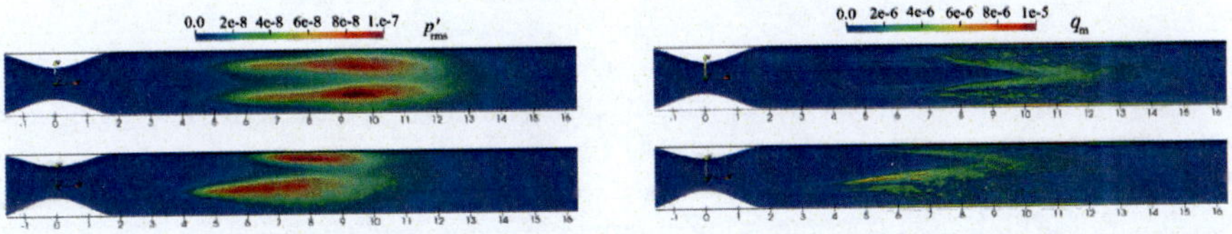

(a) Pressure (b) Lamb vector

Fig. 4. Contours of root-mean-square values determined by LB simulations for an axisymmetric (top) and an eccentric (bottom) case at Re = 1000, (a) pressure fluctuations p'_{rms}, (b) magnitude of Lamb vector fluctuations $(\omega \times \mathbf{u})'_{rms}$.

5 Conclusion

In this work, DNS was performed to simulate flows of a straight pipe with an axisymmetric and an eccentric stenotic geometry. A detailed analysis based on POD/spectral entropy has been carried out at different Reynolds numbers: Re = 400, 500, 750, 1000, and 1800. It is found that eccentricity considerably speeds-up transition, starting at Re = 500 (from the theory by Samuelsson et al., (2015); Re_{crit} = 525) instead of Re = 750 for the axisymmetric case (theory by Samuelsson et al. (2015): Re_{crit} = 713). Increasing Reynolds number also leads to an earlier transition, as expected. A first acoustic analysis has been started by calculating acoustic source terms. This procedure will now be continued.

Acknowledgements. The authors gratefully acknowledge the Gauss Centre for Supercomputing e.V. (www.gauss-centre.eu) for funding this project by providing computing time through the John von Neumann Institute for Computing (NIC) on the GCS Supercomputer JUWELS [15] at Jülich Supercomputing Centre (JSC).

References

1. Mallinger, F., Drikakis, D.: Instability in three-dimensional, unsteady, stenotic flows. Int. J. Heat Fluid Flow **23**(5), 657–663 (2002)
2. Varghese, S., Frankel, S., Fischer, P.: Direct numerical simulation of stenotic flows. Part 1. Steady flow. J. Fluid Mech. **582**, 253–280 (2007)
3. Ahmed, S.A., Giddens, D.P.: Pulsatile poststenotic flow studies with laser doppler anemometry. J. Biomech. **17**(9), 695–705 (1984)
4. Sherwin, S.J., Blackburn, H.M.: Three-dimensional instabilities and transition of steady and pulsatile axisymmetric stenotic flows. J. Fluid Mech. **533**, 297–327 (2005)
5. Samuelsson, J., Tammisola, O., Juniper, M.P.: Breaking axi-symmetry in stenotic flow lowers the critical transition Reynolds number. Phys. Fluids **27**, 104103 (2015)
6. Hu, Z., Morfey, C., Sandham, N.: Sound radiation in turbulent channel flows. J. Fluid Mech. **475**, 269–302 (2003)
7. Abdelsamie, A., Janiga, G., Thévenin, D.: Spectral entropy as a flow state indicator. Int. J. Heat Fluid Flow **68**, 102–113 (2017)
8. Ewert, R., Schröder, W.: Acoustic perturbation equations based on flow decomposition via source filtering. J. Comput. Phys. **188**, 365–398 (2003)

9. Abdelsamie, A., Fru, G., Dietzsch, F., Janiga, G., Thévenin, D.: Towards direct numerical simulations of low-Mach number turbulent reacting and two-phase flows using immersed boundaries. Comput. Fluids **131**(5), 123–141 (2016)
10. Abdelsamie, A., Lartigue, G., Frouzakis, C.E., Thévenin, D.: The Taylor-Green vortex as a benchmark for high-fidelity combustion simulations using low-Mach solvers. Comput. Fluids **223**, 104935 (2021)
11. Chi, C., Abdelsamie, A., Thévenin, D.: A directional ghost-cell immersed boundary method for incompressible flows. J. Comput. Phys. **404**, 109122–109142 (2020)
12. Qian, Y.H., D'Humières, D., Lallemand, P.: Lattice BGK models for Navier-Stokes equation. Europhys. Let. **17**(6), 479–484 (1992)
13. Freitas, R., Henze, A., Meinke, M., Schröder, W.: Analysis of lattice-Boltzmann methods for internal flows. Comput. Fluids **47**(1), 115–121 (2011)
14. Koh, S.R., Kim, J.-H., Lintermann, A.: Numerical analysis of oscillatory flows in the human brain by a lattice-Boltzmann method. In: Proceedings of the 14th WCCM-ECCOMAS (2021)
15. Jülich Supercomputing Centre: JUWELS cluster and booster: exascale pathfinder with modular supercomputing architecture at Juelich supercomputing centre. J. Large-Scale Res. Facil. **7**(A183) (2021). http://dx.doi.org/10.17815/jlsrf-7-183

Leading-Edge Effects in Freestream Turbulence Induced Transition in a Dense Gas Flow

A. Bienner[1(✉)], X. Gloerfelt[1], and P. Cinnella[2]

[1] DynFluid Laboratory, Arts et Metiers Institute of Technology, 75013 Paris, France
{aurelien.bienner,xavier.gloerfelt}@ensam.eu

[2] Jean Le Rond D'Alembert Institute, Sorbonne University, 75005 Paris, France
paola.cinnella@sorbonne-universite.fr

1 Introduction

Organic Rankine cycle (ORC) is widely used for waste-heat recovery and eco-friendly power generation. Similar to Rankine cycles using water steam, they make use of low boiling molecurlaly complex organic fluid to optimize cycle efficiency when working with low temperature sources. Most ORC rely on a turbine to expand the heated organic vapor and generate the power output. Such turbines generally use a few highly loaded stages, working in the high subsonic or supersonic regimes. To gain insights into loss mechanisms in ORC turbines, the CLOWT facility has been built at University of Muenster. It consists in a continuously running pressurized closed-loop wind tunnel using Novec^{TM}649 in the high subsonic speed range. An experimental and numerical Franco-German collaboration was then undertaken, with the main goal of studying an idealized blade vane configuration (Fig. 1a).

As a first step in this project, a numerical study of a turbulent boundary layer (TBL) of Novec^{TM}649 has been realised [1]. Despite $M = 0.9$ in the freestream, we found that turbulent profiles were very close to profiles in incompressible databases due to the high specific heat of Novec^{TM}649, which significantly reduces friction heating. Then, free-stream turbulence (FST) transition, the principal mechanism of laminar-turbulent transition for turbine blades, was investigated for a zero-pressure gradient (ZPG) flat plate (FP) [2], in particular for large-scale disturbances using LES. The results show that, for an inlet turbulence intensity $T_{u,in} \sim 4\%$, a high integral length scales L_f tend to delay transition and increase the spanwise scales of the Klebanoff streaks. To get further insight into FST induced transition and specifically into leading edge (LE) receptivity effects, we focus in this study on the LE of a blade configuration representative of ORC turbines.

2 Numerical Methods and Set-Up

The in-house finite-difference code MUSICAA is used to solve the compressible Navier-Stokes equations over a curvilinear domain using coordinate transformation. The inviscid fluxes are discretized by means of 10th-order standard centred

C. Marchioli et al. (Eds.): DLES 2023, ERCO 31, pp. 139–144, 2024.
https://doi.org/10.1007/978-3-031-47028-8_21

differences whereas 4th-order is used for viscous fluxes. The scheme is supplemented with a 10th-order selective filtering to eliminate grid-to-grid unresolved oscillations and provide a selective regularization of subgrid-scales. A four-stage Runge-Kutta algorithm is used for time integration and a fourth-order implicit residual smoothing method ([3,6]) is applied to enlarge the stability domain and enable larger timesteps. The fluid Novec^{TM}649 is modeled with the Peng-Robinson-Stryjek-Vera equation of state [8] and the Chung-Lee-Starling model [5] for transport properties.

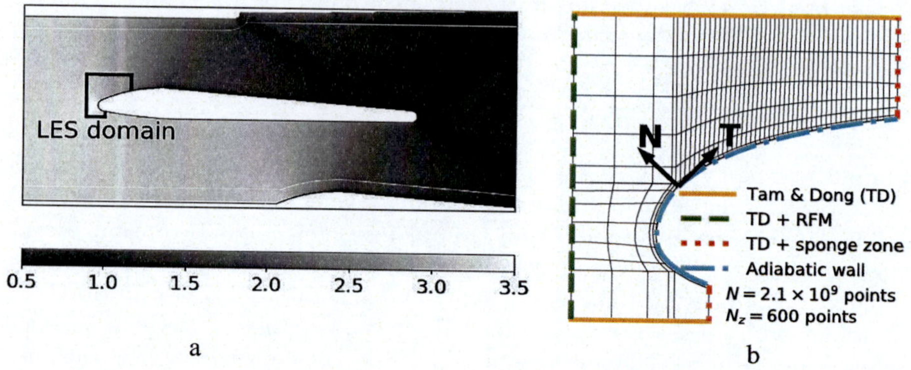

Fig. 1. (a) Pressure field over the blade of the precursor RANS simulation. Colorbar in bars. (b) Mesh (1 line out of 100).

The computational domain is a subregion of the turbine vane (see Fig. 1a). Concerning boundary conditions (see Fig. 1b), non-reflecting Tam & Dong's conditions are applied at the inlet, top and outflow boundaries, where mean flow quantities at inlet and top boundaries are extracted from a companion RANS simulation. A sponge zone combining grid stretching and a Laplacian filter is added at the oulet. At the inlet, synthetic freestream turbulence is generated based on Random Fourier Modes (RFM) [2]. The homogeneous isotropic turbulent field is generated with a finite number of independent RFM, whose amplitudes are based on a von Kármán spectrum with imposed $T_{u,in}$ and L_f values. Adiabatic no-slip conditions are applied at the wall and periodicity is enforced in the spanwise direction.

We select operating conditions representative of the future experimental campaign in CLOWT facility (see Table 1) corresponding to a flow of Novec^{TM}649 in relatively dilute conditions. The fundamental derivative of gas dynamics Γ has a value below 1 so that the flow occurs in the so called dense-gas thermodynamic region, as used in ORC applications. This leads to high Reynolds conditions, with $Re_{ch,in} = 9 \times 10^6$, where ch is the chord of the full blade. The flow accelerates from $M = 0.5$ up to $M = 1.1$ at the outlet. The conditions are also kept similar to the one used in the ZPGFP simulations ($P_\infty = 4$ bars, $T_\infty = 100$ °C, $M_\infty = 0.9$), to allow comparisons. The integral length scale and

Table 1. Thermodynamic and aerodynamic conditions at the inlet of LES domain.

M_{in}	c_{in} [m/s]	p_{in} [$bars$]	p_{in}/p_c	T_{in} [°C]	T_{in}/T_c	ρ_{in}	ρ_{in}/ρ_c [kg/m^3]	μ_{in} [Pa.s]	Γ_{in}
0.52	74	3.3	0.17	60	0.75	44.1	0.07	1.26 $\times 10^{-5}$	0.8

the corresponding Reynolds number $Re_{L_f} = 11 \times 10^3$ (Table 2) are of the order of magnitude of those used in the FP-high L_f cases, i.e. are representative of large-scale incoming disturbances. The inlet turbulence intensity, chosen to be representative of turbines, corresponds to $T_{u,in} = 3.9\%$ and decays to 3.3 % immediatly upstream of the LE.

Table 2. FST characteristics and computational grid properties for current LES configuration (Leading edge) and LES realised over a ZPGFP [2].

Case	$L_{f,in}$ [μm]	$Re_{L_{f,in}}$	$T_{u,in}$ (%)	$2\pi/k_{min}$	$1/k_{max}$	$\Delta x^+ \times \Delta y_w^+ \times \Delta z^+$
Leading edge (LE)	75	11000	3.9	$9.0L_{f,in}$	$L_{f,in}/9.9$	$13 \times 0.45 \times 11$
FP - Low L_f	6	1728	3.9	$9.7L_{f,in}$	$L_{f,in}/9.1$	$13 \times 1.0 \times 11$
FP - High L_f-high T_u	60	17280	4.0	$9.6L_{f,in}$	$L_{f,in}/9.9$	$28 \times 0.9 \times 11$
FP - High L_f-low T_u	60	17280	2.5	$9.6L_{f,in}$	$L_{f,in}/41$	$27 \times 0.9 \times 11$

The timestep is fixed to $\Delta t = 2.36 \times 10^{-9}$ s, which corresponds to a CFL of approximately 5. The total number of points of the mesh is 2.1 billion points, with 600 points in the spanwise direction. The resolution along the suction side, given in Table 2, is typical of wall-resolved LES. In the freestream region, the resolution is kept sufficiently fine to fully capture the injected RFM modes. The LES strategy used in the present calculations has been validated in a previous work [1] against a reference DNS for turbulent ZPG boundary layer of Novec^{TM}649.

3 Results

To characterize the boundary layer (BL) evolution along the suction side, various flow quantities are plotted as a function of the curvilinear abscissa s. In Fig. 2 (top), we report the acceleration parameter $K = \frac{\nu}{U_e^2}\frac{dU_e}{ds}$ (with ν the kinematic viscosity and U_e the mean velocity at the edge of BL). The positions of minimal and maximal C_f are also reported to figure out the transition region. The flow exhibits a highly favourable pressure gradient (FPG) close to the LE, which decreases further downstream, tending toward a nearly constant value of about 1.5×10^{-7} (slightly favourable PG) in the fully turbulent region, where the TBL reaches equilibrium. This is also confirmed by the nearly linear growth of the boundary layer thickness δ_{99} (see Fig. 2 bottom). The shape factor $H = \delta^\star/\theta$ (with $\delta^\star$ the displacement thickness and θ the momentum thickness), a good indicator of the boundary layer state, is also depicted in Fig. 2 (bottom). As the

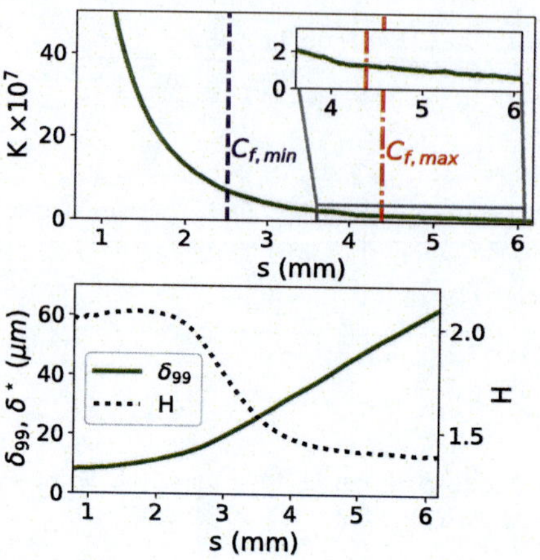

Fig. 2. Acceleration parameter K (top), BL thickness δ_{99} and shape factor H (bottom).

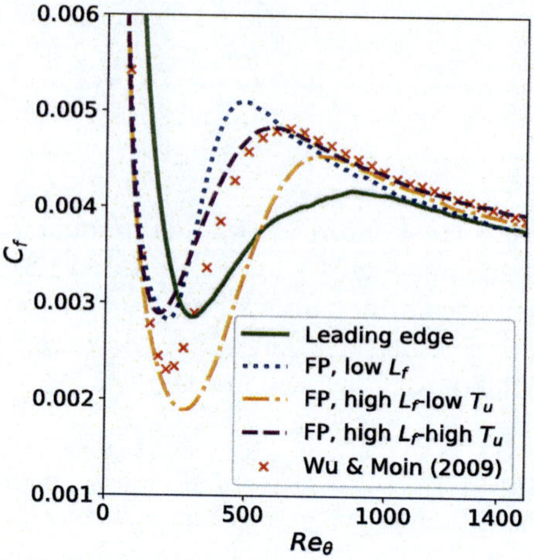

Fig. 3. Friction coefficient C_f evolution.

FPG weakens in the laminar region, H increases until the onset of transition where it drastically drops before reaching a plateau in the fully turbulent BL.

The distribution of C_f is plotted in Fig. 3 along with the ZPGFP results of [2] for comparison. Compared to FP cases with similar L_f, the strong FPG in the laminar region delays the onset of transition and the weaker FPG afterwards significantly affects the slope of the curve and the length of the transitional region. The mean tangential velocity $\overline{u_T}$ profiles at 4 stations in the TBL are plotted in Fig. 4 (left) in wall units. Compared with the ZPG reference of Wu & Moin [9], we observe a modification of the wake law due to the FPG. Moreover, a slight deviation of the log law from canonical ZPG is observed. The turbulent intensities, reported in Fig. 4 (right), present a self-similar behavior characteristic of an equilibrium TBL.

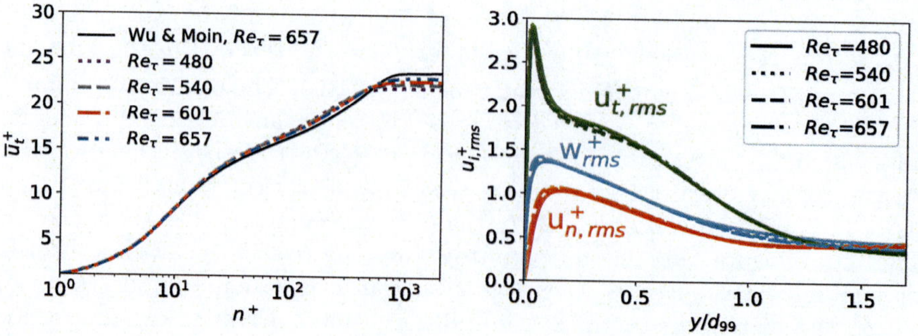

Fig. 4. Mean velocity $\overline{u}_T$ (left) and turbulent intensities (right) profiles in the TBL.

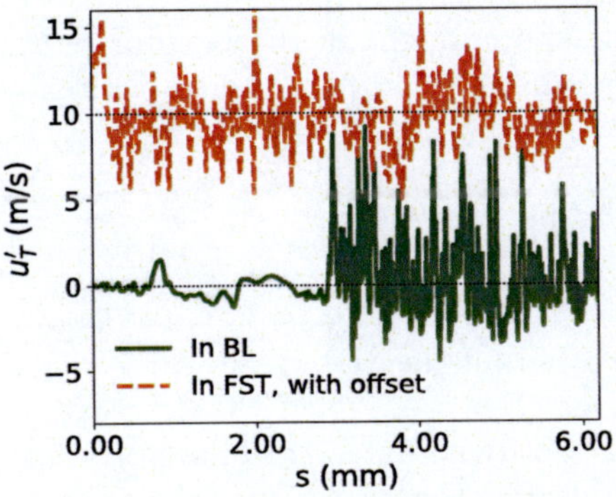

Fig. 5. u'_T evolution in the streamwise direction along a line inside and outside the BL.

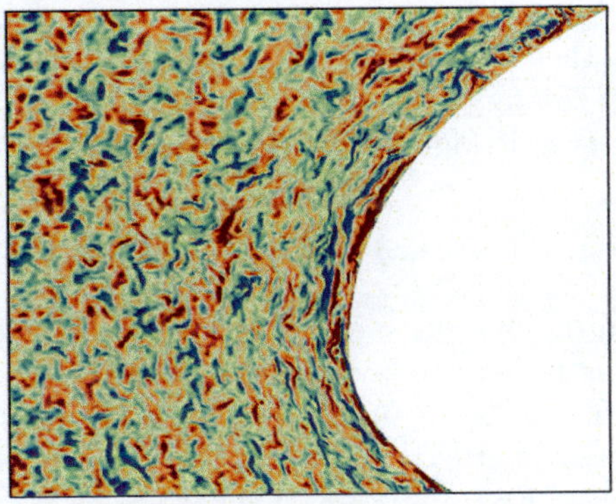

Fig. 6. Snapshot of w' at LE ($\pm 0.04 \times c_{in}$).

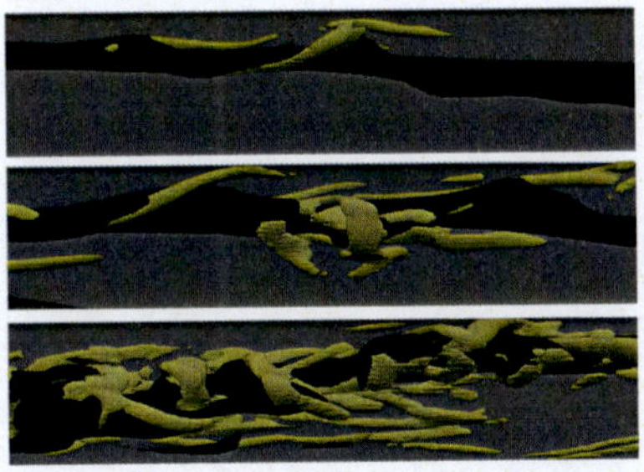

Fig. 7. Evolution of a low-speed streak instability. λ_2 criterion in yellow.

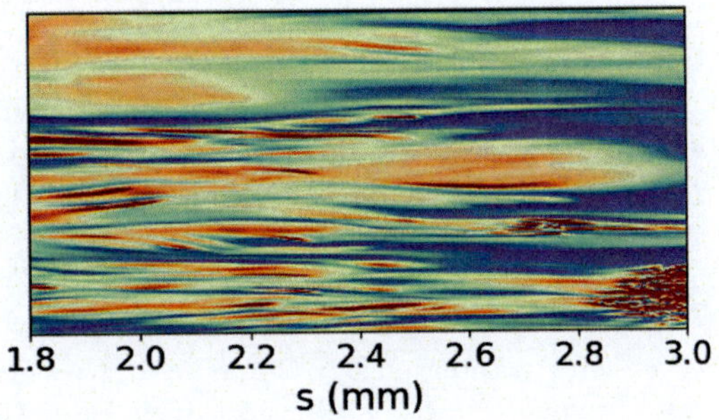

Fig. 8. Snapshot of u'_T ($\pm 0.12 \times c_{in}$) at $n^+ \sim 15$.

Figure 5 displays the tangential velocity fluctuations u'_T along mesh lines located inside and outside the BL. We observe that high frequencies present in the outer signal are damped by the laminar BL and are absent in the inner signal while the low-frequency content penetrates in the inner region. Such a phenomenon is known as shear sheltering [7]. A close-up view of the spanwise velocity fluctuations w' in the LE region is reported in Fig. 6. As the turbulent structures approach the LE, they wrap around it, leading to the amplification and the streching of the streamwise vortices. These elongated streamwise vortices then lead to the formation of streamwise streaks (Klebanoff modes) through the lift-up effect [4]. The destabilization of these streaks, illustrated in Fig. 7, conducts to the formation of turbulent spots in most of FST induced transition. However, streak instabilities do not seem to be the only spot precursors here. A snapshot of u'_T in a plane normal to the wall is shown in Fig. 8. Contrary to FP cases, the streaks in the LE case are not present everywhere in the BL and rather appear sporadically. For instance, the spot present in the bottom right of the figure does not seem to originated from a streak. In fact, the transitional region is highly perturbated by the external disturbances.

4 Conclusions

In the present study, we have investigated the FST induced transition of a dense-gas boundary layer past a realistic leading-edge configuration. Due to the high $Re_{ch,in}$, transition occurs early on the blade ($x/ch \sim 5\%$). The FPG delays and smooths the transitional region over a wider chordwise extent compared to flat plate FST induced transition with comparable FST characteristics. Moreover, laminar streak breakdown is disrupted by intense vortical structures generated at the leading edge. In future work, we will investigate the influence of the LE and the associated FPG over the transitional region in more details, before tackling the full blade vane configuration.

Acknowledgements. This research has been funded by the Agence Nationale de la Recherche through the ANR-20-CE92-0019-02 project Regal-ORC. This work was granted access to the HPC resources of IDRIS and TGCC under the allocation A0112A01736 made by GENCI (Grand Equipement National de Calcul Intensif).

References

1. Bienner, A., Gloerfelt, X., Cinnella, P.: Numerical study of boundary-layer transition in a high-subsonic organic vapor flow. In: 56th 3AF Conference, 28–30 March 2022, Toulouse, France, FP108-AERO2022 (2022)
2. Bienner, A., Gloerfelt, X., Cinnella, P., Hake, L., aus der Wiesche, S., Strehle, S.: Study of bypass transition in dense-gas boundary layers. In: 12th International Symposium on Turbulence and Shear Flow Phenomena Conference (TSFP12), 19–22 July 2022, Osaka, Japan (Online) (2022)
3. Bienner, A., Gloerfelt, X., Cinnella, P.: Assessment of a high-order implicit residual smoothing time scheme for multiblock curvilinear meshes. In: 11th International Conference on Computational Fluid Dynamics (ICCFD11), 11–15 July 2022, Maui, HI, USA (2022)
4. Brandt, L.: The lift-up effect: the linear mechanism behind transition and turbulence in shear flows. Eur. J. Mech. **47**, 80–96 (2014)
5. Chung, T.H., Ajlan, M., Lee, L.L., Starling, K.E.: Generalized multiparameter correlation for nonpolar and polar fluid transport properties. Ind. Eng. Chem. Res. **27**(4), 671–679 (1988)
6. Cinnella, P., Content, C.: High-order implicit residual smoothing time scheme for direct and large eddy simulations of compressible flows. J. Comput. Phys. **326**, 1–29 (2016)
7. Jacobs, R.G., Durbin, P.A.: Shear sheltering and the continuous spectrum of the Orr-Sommerfeld equation. Phys. Fluids **10**(8), 2006–2011 (1998)
8. Stryjek, R., Vera, J.H.: PRSV: an improved Peng-Robinson equation of state for pure compounds and mixtures. Can. J. Chem. Eng. **64**(2), 323–333 (1986)
9. Wu, X., Moin, P.: Direct numerical simulation of turbulence in a nominally zero-pressure-gradient flat-plate boundary layer. J. Fluid Mech. **360**, 5–41 (2009)

Transition Prediction on a Wind Turbine Blade at Re = 10^6 Under Varying Inflow Turbulence Based on Wall-Resolved LES

Michael Breuer[1(✉)], Brandon Arthur Lobo[2], and Alois Peter Schaffarczyk[2]

[1] Professur für Strömungsmechanik, Helmut-Schmidt Universität, Hamburg, Germany
breuer@hsu-hh.de

[2] Mechanical Engineering Department, Kiel University of Applied Sciences, Kiel, Germany
lobobrandon@gmail.com, alois.schaffarczyk@fh-kiel.de

1 Introduction

Wind energy plays a leading role as a renewable energy source, especially in countries with a low annual solar yield. In 2019 more than half of the renewable energy supply in Germany was generated by wind turbines. The blades of the rotor are the most important components for optimizing the performance. To construct special aerodynamic profiles which possess large areas of low-resistance, requires the understanding of the laminar-to-turbulent transition process taking place at such profiles. That is a non-trivial task since the rotor blades operate under atmospheric inflow turbulence and partially in the wake of preceding wind turbines. Furthermore, the Reynolds numbers are very high. In a first step to predict these kinds of flows based on wall-resolved LES with modeled atmospheric turbulence, a precursor study [5] at a rather low Re number of 10^5 was carried out for a 20% thickness airfoil corresponding to the experiment by Reichstein et al. [6]. Five different turbulence intensities (TI) were studied. A laminar separation bubble roughly between 50 and 75% of the chord disappeared for higher turbulence intensities. Depending on the TI level different transition modes (inflectional instability vs. transition mechanism influenced by the presence of streaks within the boundary layer) were found. In summary, the results showed that the applied methodology of wall-resolved LES with injected inflow turbulence works reliably and provides physically meaningful results. In the present study the Reynolds number is increased to $\mathrm{Re} = 10^6$ which is close to the real case. The objective is to better understand the transition scenario under varying atmospheric inflow conditions and realistic Re numbers.

2 Applied Methodology

Three main ingredients are required to investigate the effect of the inflow turbulence on the transition process taking place at wind turbine blades.

C. Marchioli et al. (Eds.): DLES 2023, ERCO 31, pp. 145–150, 2024.
https://doi.org/10.1007/978-3-031-47028-8_22

1. Simulation Methodology
A wall-resolved LES relying on a dynamic Smagorinsky model is used. The filtered Navier-Stokes equations are solved based on a finite-volume method on block-structured grids, which is second-order accurate in space and time [1,2]. This central scheme possesses low numerical dissipation, which is important for LES and especially the simulation of transitional flows. However, it is prone to numerical oscillations in case of high Re. In Lobo et al. [5] a blended scheme with a high percentage of the central difference scheme and a very low percentage of an upwind scheme was found to be a good compromise. Presently, the shares are 95% and 5%, respectively. The time marching within the predictor-corrector method uses a low-storage Runge-Kutta scheme. A C-type grid with about 52 million control volumes is used. This grid already includes the angle of attack of $\alpha = 4°$. It is refined near the wall with a wall-normal distance of the first cell center of $y^+_{1st} < 0.5$. The streamwise resolution is $\Delta x^+ \leq 30$ on the suction side, $\Delta x^+ \leq 60$ on the pressure side and $\Delta z^+ \leq 25$ for the spanwise direction satisfying the requirements for a wall-resolved LES. The spanwise extension of the airfoil is only 6% of the chord length c and resolved by 105 control volumes.

2. Synthetic Turbulence Inflow Generator (STIG)
The method proposed by Klein et al. [4] is applied. It relies on digital non-recursive filters which depend on statistical properties. These spatial and temporal correlations and the resulting length and time scales allow an adjustment to the particular flow case. Appropriate inflow data are generated by multiplying filter coefficients, which describe the two-point correlations and the autocorrelation of the inflow turbulence, with a series of random numbers. A required 3-D correlation between the filter coefficients is achieved by convolution of the three 1-D filter coefficients. Moreover, the cross-correlations between all velocity components are taken into account guaranteeing the representation of realistic inflow data.

The generated inflow turbulence is anisotropic in nature and follows the Kaimal spectrum. The input parameters of the STIG are based on those suggested by the IEC-61400-1 standard [8]. However, the corresponding length scales in longitudinal, lateral and vertical direction ($8.1\,\Lambda_1$, $2.7\,\Lambda_1$, $0.66\,\Lambda_1$ with $\Lambda_1 = 42\,\text{m}$ as the relative length scale) are far too large for the limited spanwise extension of the computational domain. Thus, they are scaled down to maintain the anisotropic nature such that the eddies have a similar form but not the same absolute size. The resulting maximal length scales in the three directions normalized by the chord length c are 0.211, 0.07 and 0.017, respectively. Note that the presently applied STIG only allows to define one length scale per direction. However, this disadvantage can be compensated by superimposing the solutions of different length scales given by the maximal length scale divided by the factor 2^n with n varying between 0 and 5.

Based on TI and the ratios of standard deviations of the fluctuations in the different directions according to the IEC-61400-1 standard [8] ($\sigma_1 = \overline{u'u'}$, $\sigma_2 = \overline{v'v'} = 0.8\,\sigma_1$, $\sigma_3 = \overline{w'w'} = 0.5\,\sigma_1$), the normal Reynolds stress components can be determined for the anisotropic case. Furthermore, a non-zero Reynolds shear stress component $\overline{u'w'}$ is included by an estimation based on the negative value of the square of the friction velocity u_τ, which is assumed to be $u_\tau/u_\infty \approx 0.05$.

3. Source Term Concept of the STIG
Instead of applying the inflow data at the inlet, where the grid resolution is coarse and thus leads to a strong damping of small flow structures, a recently developed source-term formulation [7] is applied as done in Breuer [3]. That allows to shift the artificial turbulence injection closer to the airfoil, where a higher resolution guarantees the correct development of the flow structures. The source terms are superimposed to the momentum equations in a predefined influence area which is located about one chord length upstream of the airfoil. Based on a Gaussian bell-shaped distribution the source terms are scaled within this influence area. The streamwise dimension of this region is defined by the integral length scale in main flow direction calculated with the help of the integral time scale and the Taylor hypothesis [3,7]. Thus, all required parameters are directly coupled to the STIG.

3 Results and Discussion

The profile considered corresponds to the section of the LM 43P blade of a REpower MM92 wind turbine [6] at a radius of 35 m and a relative thickness of 20%. Figures 1(a) and (b) show the distributions of the pressure coefficient c_p and the friction coefficient c_f for the averaged flow field, where the averaging is carried out over 8 dimensionless time units and additionally in spanwise direction. The results of four different turbulence intensities of the inflow ($TI = 0$, 0.6, 2.4 and 4.5%) are depicted. Obviously, a laminar separation bubble is observed in the time-averaged flow field in case of no inflow turbulence. In this case the pressure distribution on the suction side exhibits a plateau indicating the presence of a separated flow region between $x/c = 0.45$ and 0.55. Accordingly, c_f is negative in this range. The values of the resulting lift and drag coefficients are summarized in Table 1. Transition to turbulence is taking place between $x/c = 0.52$ and 0.59 as indicated by the sharp drop and subsequent steep rise in c_f. Further downstream the friction coefficient decreases again indicating a fully turbulent flow. Analyzing the Reynolds shear stress in the second cell in the wall-normal direction (Fig. 1c) the laminar, transitional and turbulent regimes can be clearly distinguished. Especially, the strong momentum exchange taking place during transition is obvious, e.g. for $TI = 0\%$ the Reynolds shear stress is zero along the suction side up to $x/c \approx 0.52$. Further downstream a strong increase of $|\overline{u'v'}|$ and a peak at $x/c \approx 0.55$ indicates the transition process.

For visualizing boundary layer streaks, snapshots of the instantaneous streamwise velocity disturbances $u' = u - \overline{u}$ are depicted in Fig. 2. At $TI = 0\%$ spanwise rolls of Kelvin-Helmholtz type are visible in Fig. 2a at $0.52 \leq x/c \leq 0.56$

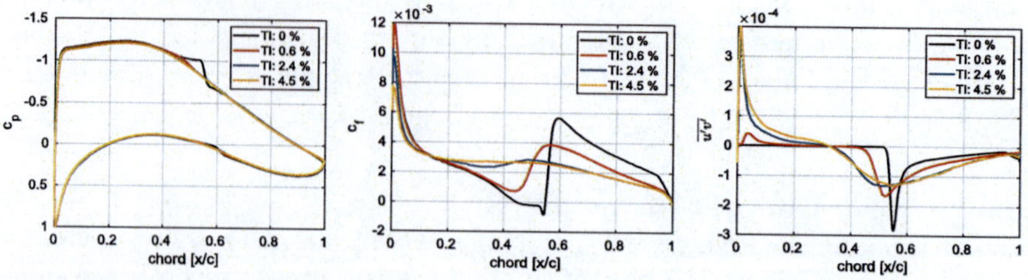

(a) c_p based on mean flow (b) c_f based on mean flow (c) $\overline{u'v'}$ close to wall

Fig. 1. Distribution of the pressure coefficient c_p, the friction coefficient c_f and the Reynolds shear stress $\overline{u'v'}$ on the suction side at Re $= 10^6$ and $\alpha = 4°$ and four different inflow turbulence intensities.

Table 1. Transition locations and force coefficients. Transition start determined from the location of the shape factor H_{12} maximum. Transition end determined from location of highest turbulent kinetic energy (PSD) at δ_1.

Inflow turbulence intensity TI [%]	Transition start: x/c	Transition end: x/c	Lift coefficient	Drag coefficient	Lift to drag ratio
0.0	0.52	0.59	0.8964	0.0111	80.67
0.6	0.43	0.53	0.8811	0.0110	79.36
2.4	0.03	0.50	0.8773	0.0154	56.95
4.5	0.03	0.53	0.8790	0.0151	58.36

discerning the laminar and turbulent regions. By mildly increasing the inflow turbulence intensity to 0.6%, no separation of the mean flow is observed anymore (see Fig. 1b). The dark blue region visible in Fig. 2b between $x/c \approx 0.48$ and 0.52 indicates a negative disturbance velocity u' (inflectional roll) which is not necessarily a hint for a temporary separation bubble since the averaged flow velocity is positive. Most likely it is due to an inflectional instability. Overall, it is obvious that transition to turbulence moves upstream due to the added inflow turbulence. Based on the friction coefficient depicted in Fig. 1b a clear distinction between the laminar, transitional and turbulent flow regime is still possible at $TI = 0.6\%$. The laminar region is evident up to $x/c \approx 0.43$. The flow is transitional up to $x/c \approx 0.56$ indicated by the rise in c_f. Already upstream of these phenomena boundary layer streaks are generated as a response of the boundary layer on account of the penetration of external disturbances. The formation of these boundary layer streaks is visible in Fig. 2b as elongated longitudinal regions of the flow, where blue zones represent regions that are slower than the mean flow and red regions those which are faster than the mean flow.

A further increase of TI to 2.4% and 4.5% only yields marginal changes in the c_p distributions between $x/c \approx 0.4$ and 0.5. A strong increase in c_f as observed in the two previous cases indicating transitional flow is not apparent in Fig. 1b. Thus, a distinction between laminar, transitional and turbulent flow is not as easy as before. No evidence of instantaneous separation is found probably

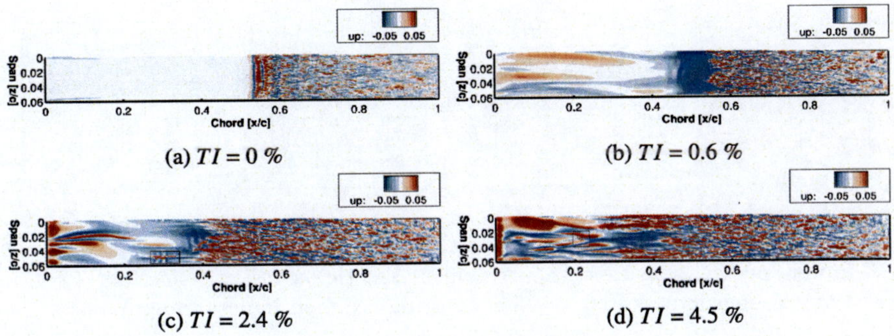

Fig. 2. Snapshots of the instantaneous streamwise velocity disturbances $u' = u - \overline{u}$ for the visualization of boundary layer streaks. Slices are taken at a wall-normal height corresponding to the displacement thickness at 0.5% chord.

due to the increased frequency of streak formation within the boundary layer as visible in Figs. 2c and 2d. The flow seems to be already transitional near the leading edge at approximately 3% chord. Secondary instabilities formed by the interaction of high- and low-speed streaks are observed and highlighted by rectangles in Figs. 2c and 2d.

For further investigations on the mode and process of transition and the influence of the inflow turbulence on the boundary layer receptivity, power spectral densities (PSD) are analyzed for different locations along the chord at every 10% chord length. Using a Hann windowing function on the data collected at a height corresponding to the boundary layer displacement thickness δ_1, PSD plots of the turbulent kinetic energy are depicted in Fig. 3.

For the case without inflow turbulence a clear increase in energy is seen at 40% chord with the separation bubble showing a sharp rise in energy at a frequency corresponding to 10 dimensionless units at 50% chord. The region where the flow first turns fully turbulent is expected to have the highest turbulent kinetic energy. That is found at about 60% chord, and a decrease is observed further downstream. Taking the inflow turbulence into account ($TI = 0.6\%$) the energy at 10% chord is approximately five orders of magnitude larger than without inflow turbulence. Between 40 and 50% chord a strong increase in energy is observed indicating transitional flow. Finally, at 50% chord the flow is turning turbulent as discussed earlier reaching the highest energy level with a decrease observed further downstream.

At the two highest inflow turbulence intensities ($TI = 2.4\%$ and 4.5%) it is more difficult to distinguish laminar and turbulent regimes. The energy content at 10% chord is obviously lower than at the other chord locations. Nevertheless, it is evidently higher than in the cases with $TI = 0$ and 0.6%. Furthermore, at $TI = 4.5\%$ the highest energy is found between 40 and 50% chord indicating that this is the location around which the flow turns fully turbulent. This is in agreement with the discussion above.

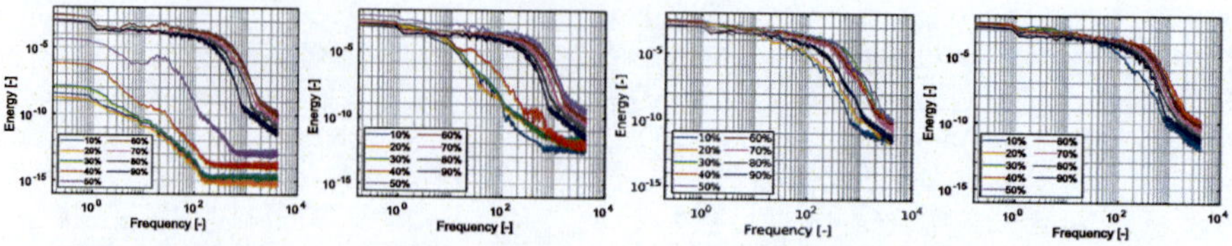

(a) $TI = 0$ % (b) $TI = 0.6$ % (c) $TI = 2.4$ % (d) $TI = 4.5$ %

Fig. 3. Power spectral densities of the turbulent kinetic energy at δ_1 for four TI. Data collected at a dimensionless freq. of 8300 covering 6 dimensionless time units.

4 Conclusions

To optimize the aerodynamic design of turbine blades, it is essential to better understand the transition process under varying inflow conditions in the free-atmosphere. Based on wall-resolved LES the present study analyzes transition on the LM 43P blade at $\mathrm{Re} = 10^6$ and four turbulence intensities. Without inflow turbulence or at low TI a differentiation into laminar, transitional and turbulent areas is straightforward, for example using c_f or $\overline{u'v'}$ close to the wall. Obviously transition is due to an inflectional instability. At the two highest TI investigated, however, the flow is transitional from close to the leading edge. Secondary instabilities formed by the interaction of high- and low-speed streaks are found. The transition location can be approximately determined based on the highest turbulent kinetic energy in the PSD plots. The increase in energy in the lower frequency range with rising inflow turbulence can be attributed to boundary layer streaks increasing in number and intensity.

References

1. Breuer, M.: Large-eddy simulation of the sub-critical flow past a circular cylinder: numerical and modeling aspects. Int. J. Numer. Meth. Fluids **28**(9), 1281–1302 (1998)
2. Breuer, M.: A challenging test case for large-eddy simulation: high Reynolds number circular cylinder flow. Int. J. Heat Fluid Flow **21**(5), 648–654 (2000)
3. Breuer, M.: Effect of inflow turbulence on an airfoil flow with laminar separation bubble: an LES study. J. Flow Turbul. Combust. **101**, 433–456 (2018)
4. Klein, M., Sadiki, A., Janicka, J.: A digital filter based generation of inflow data for spatially-developing direct numerical or large-eddy simulations. J. Comput. Phys. **186**, 652–665 (2003)
5. Lobo, B.A., Schaffarczyk, A.P., Breuer, M.: Investigation into boundary layer transition using wall-resolved LES and modeled inflow turbulence. Wind Energy Sci. **7**(3), 967–990 (2022)
6. Reichstein, T., et al.: Investigation of laminar-turbulent transition on a rotating wind-turbine blade of multi-megawatt class with thermography and microphone array. Energies **12**, 2102 (2019)
7. Schmidt, S., Breuer, M.: Source term based synthetic turbulence inflow generator for eddy-resolving predictions of an airfoil flow including a laminar separation bubble. Comput. Fluids **146**, 1–22 (2017)
8. DIN EN 61400-1, Wind turbines – Part 1: Design requirements. Technical report, IEC (2006)

Laminar-Turbulent Transition in Swept-Wing Flows with a Supercritical Forward-Facing Step

J. Casacuberta(✉), S. Hickel, and M. Kotsonis

Aerodynamics Group, Faculty of Aerospace Engineering,
Delft University of Technology, Kluyverweg 1, 2629HS Delft, The Netherlands
J.CasacubertaPuig@tudelft.nl

1 Introduction

Achieving and maintaining laminar flow on large swept lifting surfaces of subsonic aircraft poses a considerable challenge. Surface roughness, such as imperfect joints introducing sharp surface distortions, is a source of significant laminar flow deterioration that promotes laminar-turbulent transition.

The present work considers a laminar-turbulent transition route initiated by primary (stationary crossflow) eigenmode amplification, which is the prevalent scenario in low-disturbance environments as in free-flight [9]. This instability kind manifests in the developed flow field as stationary co-rotating vortices whose axes of rotation are practically aligned with the direction of the streamlines in the outer-flow region. Under Decomposition (1), the developed flow $\mathbf{q} = [u\ v\ w\ p]^T$ is conceived as the superposition of the laminar unperturbed base flow, $\mathbf{q}_\mathrm{B}$, with steady, $\mathbf{q}'$, and unsteady, $\mathbf{q}''$, perturbation fields:

$$\mathbf{q}(x, y, z, t) = \mathbf{q}_\mathrm{B}(x, y) + \mathbf{q}'(x, y, z) + \mathbf{q}''(x, y, z, t). \tag{1}$$

In stationary-crossflow-dominated flows, forward-facing steps significantly altering the laminar-turbulent transition path may be classified as critical or supercritical. Critical steps advance the transition front upstream, as compared to a reference clean (i.e. no step present) case. Supercritical steps trip the flow immediately and have been found to drastically move the transition front to the vicinity of the step. The main goal of the present work is to investigate the step-flow mechanisms responsible for inducing a supercritical transition scenario. To this end, we carry out Direct Numerical Simulations (DNS) and perform a modal analysis of the steady and unsteady perturbation fields.

2 Flow Problem and DNS Setup

The incompressible swept-wing flow is modeled as flat-plate flow with an imposed airfoil-like pressure gradient in the chordwise direction at the free-stream. The main coordinate system reads $\mathbf{x} = [x\ y\ z]^T$, where x, y, z indicate the chordwise (i.e. normal to the virtual leading edge), wall-normal, and spanwise (i.e.

C. Marchioli et al. (Eds.): DLES 2023, ERCO 31, pp. 151–156, 2024.
https://doi.org/10.1007/978-3-031-47028-8_23

parallel to the virtual leading edge) directions, respectively. The coordinate $x_{\mathrm{st}} = x - 177.62\,\delta_0$ is additionally introduced to express the relative chordwise distance to the step. The inlet free-stream velocity is decomposed into a chordwise component, u_∞, and a spanwise component, $w_\infty = -1.24u_\infty$, to model the effect of sweep angle. Pressure measurements from wind-tunnel experiments on a 45° swept wing [7] are used to guide the DNS setup. The inflow boundary layer thickness, $\delta_0 = 7.71 \times 10^{-4}$ m, and free-stream velocity, $u_\infty = 15.10$ m/s, are chosen as global characteristic quantities.

A stationary crossflow mode, computed as solution to a local linear Orr-Sommerfeld analysis on the base flow profile, is prescribed at the inflow. The step height, $h = 0.97\delta_0$, corresponds to approximately 50% of the undisturbed boundary layer thickness at the virtual step location. The spanwise domain length (i.e. fundamental spanwise wavelength) is set to $\lambda_z = 7.5$ mm. This wavelength yields the integrally most amplified perturbation at the end of the computational domain in reference no-step conditions.

In Sects. 3 and 4, we will analyze steady and unsteady perturbation effects. The steady perturbation field, $\mathbf{q}'$, is decomposed in spanwise Fourier modes, i.e.

$$\mathbf{q}'(x, y, z) = \sum_{j=-N}^{N} \tilde{\mathbf{q}}_j(x, y)\mathrm{e}^{\mathrm{i}j\beta_0 z}, \tag{2}$$

where N is the number of modes considered, $\tilde{\mathbf{q}}_j$ is the Fourier coefficients of mode j, $\beta_0 = 2\pi/\lambda_z$, and $\mathrm{i}^2 = -1$.

The DNS of the incompressible Navier-Stokes equations are performed with INCA, a conservative finite-volume solver. The Navier-Stokes equations are marched in time with a third-order Runge-Kutta method. A fifth-order upwind scheme is used to discretize the convective terms. The Selective Frequency Damping (SFD) technique [1] is applied to numerically compute the stationary isolated form of $\mathbf{q}'$ following a recently proposed methodology to compute the control parameters [3].

The computational domain encompasses $0 \leq x/\delta_0 \leq 517$ and $y/\delta_0 \leq 26$ and the grid contains $N_x = 6760$, $N_y = 1008$, and $N_z = 144$ points in the chordwise, wall-normal, and spanwise directions, respectively. These values yield $\Delta x^+ = 1.8$, $\Delta y^+ = 0.9$, and $\Delta z^+ = 9$ in the steady perturbed flow (i.e. $\mathbf{q}_{\mathrm{B}} + \mathbf{q}'$) near the step. Numerical computations of the unperturbed base flow, $\mathbf{q}_{\mathrm{B}}$, and the steady perturbed flow, $\mathbf{q}'$, are performed independently for a similar numerical setup. In the former, the equations are converged in time up to a threshold value of 10^{-8} based on an L_2-norm of the temporal derivatives. In the latter, unsteady perturbation content is triggered by a multi-modal harmonic forcing of the wall-normal velocity at the wall near the inflow ($20 \leq x/\delta_0 \leq 32$) as follows:

$$v(x, 0, z, t) = f_s(x) \sum_{k=1}^{M} A_k^{\mathrm{BS}} \cos(\beta_0 z + 2\pi k f_0 t + \phi_k), \tag{3}$$

with f_0 representing a fundamental temporal frequency, A_k^{BS} indicates the initial amplitude of a temporal component k, ϕ_k is a random phase, and f_s modulates smoothly the response in x. Unsteady forcing is applied ranging from 3 kHz to 14 kHz in intervals of 1 kHz. For the present setup, the DNS results reproduce the transitional-flow scenario observed in experiments in a low-disturbance wind tunnel [7].

3 Steady Perturbation Evolution

To elucidate on the modification of the transition path by the step, the evolution of the steady perturbation field ($\mathbf{q}'$) is first addressed. An overview of the subsequent unsteady flow evolution and the ultimate transition mechanisms will be discussed in the next section.

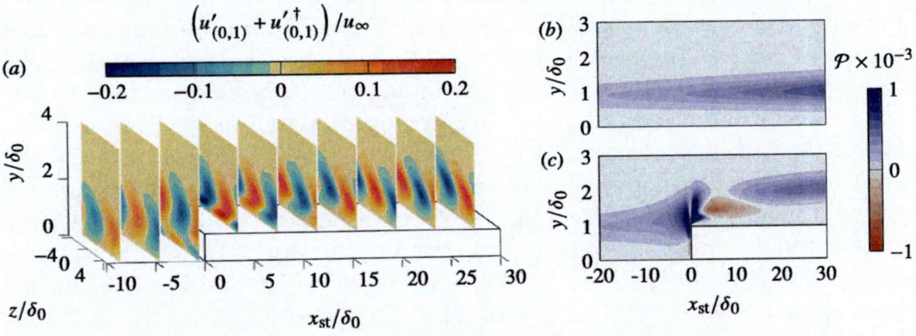

Fig. 1. Evolution of the fundamental perturbation Fourier mode at the step (a). Spatial evolution of linear energy production for $\beta = \beta_0$ in the no-step (b) and step (c) cases.

Around the step, two major flow regimes are identified. *Far* from the wall, the pre-existing crossflow perturbation lifts off and passes over the step. *Close* to the wall, chordwise-velocity (u') perturbation streaks of alternating sign in z are induced at the step apex (Fig. 1(a)). This scenario develops analogously for each Fourier mode, implying that a near-wall streak system contained in a particular Fourier mode *has adopted* the spanwise wavelength of the incoming crossflow component that induces it.

The origin and evolution of such new streaky structures induced at the step is scrutinised next by formulating energy-balance equations of the spanwise-harmonic perturbation modes [5]. They read

$$0 = \mathcal{P}_{n\beta_0} + \mathcal{T}_{n\beta_0} + \mathcal{D}_{n\beta_0} + \mathcal{W}_{n\beta_0} + \mathcal{N}_{n\beta_0}, \; n = 0, 1, \ldots, N, \tag{4}$$

where $\mathcal{P}_{n\beta_0}$ denotes linear energy production, $\mathcal{T}_{n\beta_0}$ is the advection of perturbation kinetic energy by the base flow, and $\mathcal{D}_{n\beta_0}$, $\mathcal{W}_{n\beta_0}$, $\mathcal{N}_{n\beta_0}$ express the work done by viscous forces, pressure forces, and non-linear perturbation interactions, respectively.

In essence, Eq. (4) expresses a balance between different perturbation mechanisms which remain in temporal equilibrium. The balance is harmonic, implying that it accounts for perturbation mechanisms acting on a particular perturbation Fourier space $n\beta$, solely. Equation (4) reduces to the well-known Reynolds-Orr equation under certain flow conditions.

When the budget analysis based on Eq. (4) is applied to the fundamental perturbation field, $\beta = \beta_0$, it is found that linear production ($\mathcal{P}_{\beta_0}$) is the dominant perturbation mechanism in both the near- and far-wall regimes. Similar results have been reported recently by use of stability tools applied on the step flow [4]. In particular, the spatial evolution of the dominant term $\mathcal{P}_{\beta_0}$ is portrayed in Fig. 1(*b*,*c*). The enhanced blue region with $\mathcal{P}_{\beta_0} > 0$ around the step is associated to the mechanism responsible for inducing streaks; essentially, it expresses that kinetic energy is transferred from the base flow to the perturbation field. At the same time, in the region further from the wall, an algebraic instability associated to the lift-up effect is proposed as a main mechanism responsible for the (linearly-dominated) alteration of the fundamental crossflow perturbation at the step. This is elaborated upon in an article in preparation [6].

When the initial amplitude of the pre-existing crossflow instability is sufficiently small, the near-wall streaks vanish rapidly downstream of the step (Fig. 1(*a*)). This is observed for a choice of amplitude of the pre-existing instability that yields linear behaviour until the virtual step location in reference conditions (i.e. without the step). By the increase of the amplitude of the incoming instability, it is observed that, eventually, the perturbation streaks originally induced at the step apex are sustained in space and amplified in x. By evaluating the relative contribution of the terms of Eq. (4), this is ascribed to the work of non-linear interactions between perturbation streaks among all harmonic spaces, i.e. for $\beta = \beta_0$ as well as for $\beta > \beta_0$. This (non-linear) growth of perturbation streaks close downstream of the step plays a main role in significantly deforming both, the shear layer close to the wall and the topology of the incoming stationary crossflow vortex.

4 Unsteady Perturbation Evolution

For the presently investigated step height, laminar-turbulent transition is captured shortly downstream of the step. This is line with observations of recent experimental investigations [7] under similar flow conditions and step geometry. The results of our DNS indicate that the shear layer significantly deformed by the step is prone to unsteady perturbation amplification. The explosive chordwise growth of unsteady perturbation structures downstream of the step initiates the breakdown of the laminar flow. Following the discussion in the previous section, the occurrence of supercritical transition appears to be exclusive to cases with a sufficiently large pre-existing stationary crossflow perturbation.

Figure 2 depicts an instantaneous isosurface of the Q-criterion for the unsteady developed flow. Close downstream of the step, wedges of unsteady contamination approximately aligned with the crossflow-vortex direction are the

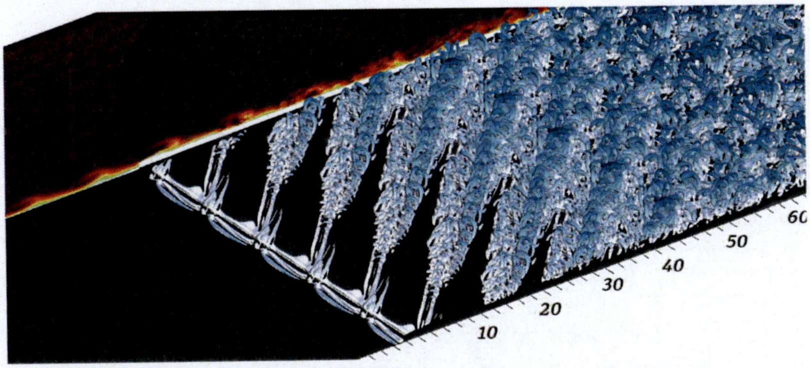

Fig. 2. Instantaneous Q-criterion isosurface colored by wall distance (white indicates close to the wall) and x-y plane of instantaneous chordwise-velocity. The x_{st}-axis is expressed in units of δ_0.

precursor of turbulence. Even though the unsteady fluctuations arise rather close to the wall, they manifest at significantly larger frequencies than those associated typically to the *type*-III secondary crossflow instability. The latter has been reported as well in corresponding experimental investigations [8] for a similar setdup. This finding suggests that a new transition mechanism is introduced by the step; it would effectively by-pass the growth of the secondary crossflow instabilities, which induce typically the breakdown of the stationary crossflow vortices in reference no-step conditions.

It is noted that the unsteady fluctuations initiating the laminar breakdown process develop in the shear layers embedding the stationary streaks induced at the step apex. In particular, two main families of large-scale hairpin vortices are captured in the DNS driving the unsteady wedges. The origin of both families of hairpin vortices is found at the upper part of regions of low-speed streaks. Accordingly, in future work it will be investigated whether the unsteady structures responsible for inducing supercritical transition by a forward-facing step are related to the so-called varicose and sinuous instabilities [2]; these develop typically in streaky boundary-layer flows subject to high levels of free-stream turbulence in scenarios of classic by-pass transition.

References

1. Åkervik, E., Brandt, L., Henningson, D.S., Hoepffner, J., Marxen, O., Schlatter, P.: Steady solutions of the Navier-Stokes equations by selective frequency damping. Phys. Fluids **18**, 068102 (2006)
2. Brandt, L.: Numerical studies of the instability and breakdown of a boundary-layer low-speed streak. Eur. J. Mech. B/Fluids **26**, 64–82 (2007)
3. Casacuberta, J., Groot, K.J., Tol, H.J., Hickel, S.: Effectivity and efficiency of selective frequency damping for the computation of unstable steady-state solutions. J. Comput. Phys. **375**, 481–497 (2018)

4. Casacuberta, J., Hickel, S., Westerbeek, S., Kotsonis, M.: Direct numerical simulation of interaction between a stationary crossflow instability and forward-facing steps. J. Fluid Mech. **943**, A46 (2022)
5. Casacuberta, J., Hickel, S., Kotsonis, M.: Laminar-turbulent transition by crossflow instability interacting with forward-facing step: direct numerical simulations. Manuscript in preparation
6. Casacuberta, J., Hickel, S., Kotsonis, M.: Passive stabilization of crossflow instabilities by a reverse lift-up effect. Manuscript in preparation
7. Rius-Vidales, A.F., Kotsonis, M.: Impact of a forward-facing step on the development of crossflow instability. J. Fluid Mech. **924**, A34 (2021)
8. Rius-Vidales, A.F., Kotsonis, M.: Unsteady interaction of crossflow instability with a forward-facing step. J. Fluid Mech. **939**, A19 (2022)
9. Saric, W.S., Reed, H.L., White, E.B.: Stability and transition of three-dimensional boundary layers. Annu. Rev. Fluid Mech. **35**, 413–440 (2003)

Analysis of a Transonic Cascade with Wall-Modeled LES Based on DGSEM

B. F. Klose[1(✉)], E.J. Munoz Lopez[2], A. Hergt[2], J. Klinner[2], M. Bergmann[2], and C. Morsbach[2]

[1] Institute of Test and Simulation for Gas Turbines, German Aerospace Center (DLR), Am Technologiezentrum 5, 86159 Augsburg, Germany
bjoern.klose@dlr.de

[2] Institute of Propulsion Technology, German Aerospace Center (DLR), Linder Höhe, Cologne 51147, Germany

1 Introduction

High-Reynolds number flows, which typically occur in turbomachinery applications, require sufficiently well-resolved boundary layer discretizations to accurately capture the small spatial and temporal scales of vortical flow within these turbulent regions. Large-eddy simulations (LES) have seen a growing popularity in modeling these complex flows and generating numerical references, despite being computationally expensive [1]. A remedy to the strict resolution requirements of *wall-resolved* LES (WRLES) is the use of wall functions to model the inner part of the turbulent boundary layer based on analytic or empirical wall laws. In this so-called *wall-modeled* LES (WMLES) approach, the boundary layer is no longer explicitly resolved and the wall-shear stress entering the viscous fluxes is computed from the velocity some distance away from the wall [2,3]. Because the grid sizing in WMLES is not governed by the $y^+ \leq 1$ condition, this family of schemes allows for significantly larger elements near the wall, resulting in a reduction of the number of degrees of freedom and larger time-step sizes for a given CFL number. The WMLES thereby enables the simulation of large-scale turbomachinery configurations or operating points that are otherwise computationally too expensive.

Being computationally efficient, geometrically flexible, and numerically robust [4,5], the discontinuous Galerkin spectral element method (DGSEM) is a popular choice among the family of high-order spectral methods for simulations of complex, compressible flows. While WMLES is already established in the finite volume community [3], applications based on spectral methods are scarce. Only recently, Frere [6] presented results of the wall-modeled LES approach with the discontinuous Galerkin method and elaborated on guidelines based on the canonical channel flow.

In this work, we assess a WMLES strategy for DGSEM adopted from Frere [6] and present results for a periodic channel flow at $Re_\tau = 2003$ and a transonic

C. Marchioli et al. (Eds.): DLES 2023, ERCO 31, pp. 157–163, 2024.
https://doi.org/10.1007/978-3-031-47028-8_24

compressor cascade at $Re_{\text{in}} = 1.4 \times 10^6$. For simplicity, the WMLES simulations are based on constant velocity sampling locations at the bottom of the second high-order element in accordance with [6]. The cascade is simulated using WRLES, under-resolved WRLES (URLES), WMLES and a mixed boundary condition (zones of low-Reynolds and wall function) to evaluate the effect of the wall-model on the transitional boundary layer characteristics.

2 Numerical Method

DLR's flow solver for turbomachinery applications TRACE is used for the discretization of the governing Navier-Stokes equations. A discontinuous Galerkin spectral element method (DGSEM) with nodal collocation, i.e. the solution and quadrature points are both taken to be the Legendre-Gauss-Lobatto nodes, is applied in this work and yields an efficient numerical scheme with diagonal mass matrix. Numerical errors arising from the non-linearity of the advective fluxes and the limited precision of integration are addressed by employing kinetic-energy or entropy conserving split-form approximations of the inviscid fluxes [4,7]. All simulations are conducted with the implicit (no-model) LES approach, i.e. the dissipation is added implicitly via the numerical dissipation of the Riemann solver. The Bassy-Rebay-1 (BR1) scheme [8] is applied for the viscous part and an explicit third-order Runge-Kutta scheme is used for the integration in time. For more details on the scheme, we refer the interested reader to Bergmann [5,9].

For transonic flow, the finite-volume (FV) subcell shock capturing approach by Hennemann [10] is applied to locally blend the high-order operator with a first-order FV scheme to reduce unphysical oscillations across the shock. We employ the feature-based indicator by Fernandez [11], which is a linear function of the product of a dilatation sensor and a Ducros vorticity sensor.

In the WMLES approach, the Dirichlet no-slip boundary condition (BC) at the wall is relaxed and the viscous flux at the wall-boundary is imposed in form of wall-shear stress τ_w calculated from an empirical wall function. We have implemented and tested the wall-model approach by Frere [6] and evaluate the velocity at the bottom of the second cell from the wall. The shear stress is computed from the friction velocity, which is based on the wall function by Reichardt [12] and obtained iteratively via a Newton-Raphson method at each time step.

3 Results

3.1 Turbulent Channel Flow

We validate the implementation of the wall function for WMLES with DGSEM on the periodic, turbulent channel flow at a Reynolds number of $\text{Re}_\tau = 2003$. The flow is simulated in a box of dimensions $[2\pi\delta,\ 2\delta,\ \pi\delta]$ consisting of $20 \times 10 \times 10$ elements with periodic boundaries in x- and z- direction. A constant volume source term in x direction compensates for the viscous losses. The elements are clustered towards the no-slip walls at $y = [0, 2\delta]$ for URLES, but evenly spaced in case of WMLES (according to the setup by [6]). The element size in case of the WMLES is therefore such that the velocity sampling location (bottom of the second element from the wall) is located at 0.2δ in accordance with the recommendation by Larsson [2]. Polynomial orders of $N = (4, 5, 7)$ are tested for URLES and $N = (4, 5)$ for WMLES.

Figure 1 shows that URLES are not able to capture the wall-shear stress accurately and the friction velocity profiles substantially deviate from the reference DNS. Applying the wall function (WMLES) improves the results and the log-region is captured accurately despite having fewer points in the boundary layer and none within the viscous sublayer.

3.2 Transonic Compressor Cascade

We simulate the flow over a transonic compressor cascade aiming to match the experiments by Klinner [14] for the off-design operating point with $Ma_{MP1} = 1.05$, $Re_{MP1} = 1.4 \times 10^6$ and $\alpha_{MP1} = 60.6°$, where MP1 refers to the 'measurement plane 1' located half a pitch upstream of the leading edge (LE). A non-reflecting inlet boundary [15] is located one pitch length upstream of the leading edge and a non-reflecting outlet boundary is applied 1.14 pitch lengths downstream from the trailing edge. The blade is extruded by 5% chord length along the span with periodic BCs applied in the spanwise, as well as the pitchwise direction. The entropy-conserving split-form variant by Chandrashekar [16] is used together with the corresponding Riemann solver and a uniform polynomial order of $N = 3$ (4th order accuracy in space) ensures that resolution jumps within the subcell-divided element are moderate.

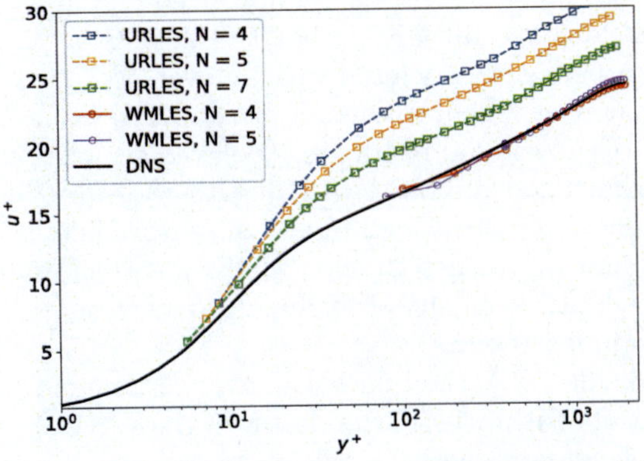

Fig. 1. Velocity profiles of URLES (dashed) and WMLES (solid) for the turbulent channel flow at $Re_\tau = 2003$. The velocity sampling location is indicated by solid markers. The reference DNS by Hoyas & Jiminez [13] is given as solid, black line.

Two meshes are employed in this study: the wall-refined grid G1 and the coarsened grid G2 whose first cell height corresponds to 0.2δ (δ = boundary layer thickness) measured at 10% of the absolute chord length c. Note that the streamwise element size of G2 is slightly lower than in G1 in order to follow the recommendation for elements with low aspect ratio for WMLES [6]. Three cases are compared: low-Reynolds wall treatment, wall-function treatment (with van-Driest compressibility correction) and a mixed BC, where the latter employs low-Reynolds treatment over the laminar section from the leading edge to the laminar separation bubble (LSB) and wall-function BC otherwise. An overview of the simulations in this work is given in Table 1:

Table 1. Overview of simulations.

Case	Wall BC	Grid	DOFs	Δt^*	y^+_{LE}	$\overline{y^+}$	t^*/CPUh
WRLES	LR	G1	145.7×10^6	7.0×10^{-6}	3.6	0.9	1.1×10^{-5}
URLES	LR	G2	132.8×10^6	1.8×10^{-5}	18.2	4.5	3.5×10^{-5}
Mixed	LR & WF	G2	132.8×10^6	1.8×10^{-5}	18.3	4.7	3.6×10^{-5}
WMLES	WF[a]	G2	132.8×10^6	1.8×10^{-5}	10.3	4.8	3.4×10^{-5}

LR = Low-Reynolds BC; WF = wall-function BC; DOF = degree of freedom; [a] = van-Driest corrected; $\overline{y^+} = \frac{1}{s}\int_0^s y^+(\hat{s})\mathrm{d}\hat{s}$; $t^* = t \cdot \overline{u}_0/c$; $\overline{u}_0$ = mean inflow velocity.

The surface isentropic Mach number, plotted in Fig. 2a, shows good agreement across all cases up to a near-constant offset to the experiments on the suction side. Only the case with wall-function treatment in the laminar region (WMLES) deviates from the reference WRLES and shows a less pronounced plateau upstream of the shock. The locally improved match with the experimental data is deemed coincidental, given that the deviation from the WRLES is increased. The discrepancy is associated with an over-prediction of the wall-shear stress in the laminar flow region upstream of the LSB, as seen in the skin friction coefficient C_f (Fig. 2b), and results in a delayed separation of the boundary layer and overall shortened bubble. Switching to wall-function BC within the LSB (mixed case) shows that the wall model under-predicts the transition peak in the shear stress and reattaches the flow slightly earlier than in the WRLES. Under-resolution (URLES), on the other hand, over-estimates the C_f peak but matches the reattachment point.

On the pressure side (PS), the boundary layer separates right at the LE and reattaches immediately within 5% of the chord. In the WRLES, spanwise vortices shed from the bubble and gradually transition to turbulence on the pressure side. This shedding behavior is not captured by any of the cases on the coarser grid G2, which immediately transition to a fully turbulent boundary layer and yield a higher friction coefficient within the first 20% of the blade's chord length.

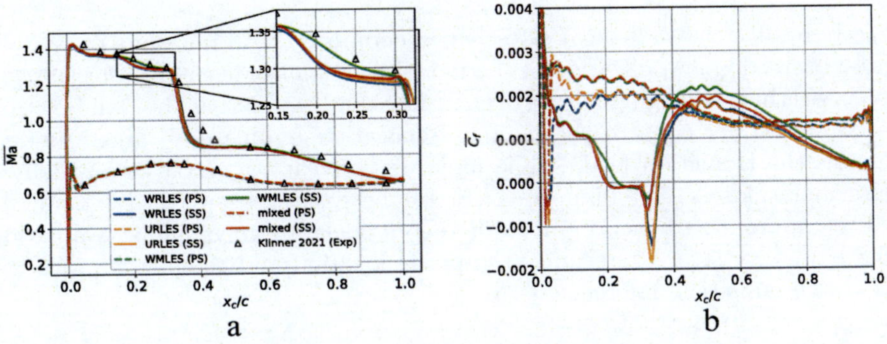

Fig. 2. (a) Time- and span-averaged isentropic Mach number. (b) Time- and span-averaged skin friction coefficient

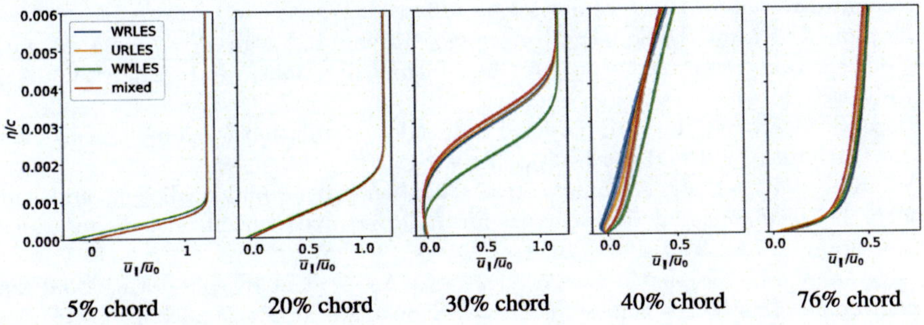

Fig. 3. Boundary layer profiles (normalized by the mean inflow velocity $\overline{u}_0$) on the suction side at various chord-wise locations.

The boundary layer shape is given for the different wall treatments at various locations in Fig. 3 and shows that all profiles, except for the case with full wall-function treatment (WMLES), collapse in the laminar flow region. The mismatch is caused by the over-estimation of the wall-shear stress at the LE because the velocity is probed too far away from the wall at the edge of the boundary layer, as the probing location is kept constant in this work. Within the separation bubble, the case with mixed BCs shows to have a slightly increased bubble height over the WRLES case, while delayed separation of the WMLES results in a significantly reduced bubble height. Downstream in the fully turbulent boundary layer, not surprisingly, all profiles converge again more closely. In all figures, the envelope of the curves indicates the 99% confidence interval of the averages.

4 Conclusion

We implemented and tested a simple wall function BC for DGSEM to be used for WMLES. The model is validated on the turbulent channel flow and shows that WMLES agrees well with reference DNS data, while under-resolved LES fails to correctly resolve the wall-shear stress when computed with the same number of degrees of freedom. Application of the model to a transonic compressor cascade emphasizes the challenging nature of such flows and shows that the wall function fails to accurately capture the laminar separation bubble and largely underestimates the transition peak, while under-resolved low-Reynolds walls deliver results more closely matching WRLES. In the future, we plan to extend the WMLES approach to incorporate different models and a dynamic evaluation of the boundary layer to optimize the probe location and thereby improve the match with numerical reference data.

References

1. Choi, H., Moin, P.: Grid-point requirements for large eddy simulation: Chapman's estimates revisited. Phys. Fluids **24**, 011702 (2012)
2. Larsson, J., Kawai, S., Bodart, J., Bermejo-Moreno, I.: Large eddy simulation with modeled wall-stress: recent progress and future directions. Mech. Eng. Rev. **3**(1), 15-00418 (2016)
3. Bose, S.T., Park, G.I.: Wall-modeled large-eddy simulation for complex turbulent flows. Annu. Rev. Fluid Mech. **50**, 535–561 (2018)
4. Winters, A.R., et al.: A comparative study on polynomial dealiasing and split form discontinuous Galerkin schemes for under-resolved turbulence computations. J. Comput. Phys. **372**, 1–21 (2018)
5. Bergmann, M., Morsbach, C., Ashcroft, G.: Assessment of split form nodal discontinuous Galerkin schemes for the LES of a low pressure turbine profile. In: García-Villalba, M., Kuerten, H., Salvetti, M.V. (eds.) DLES 2019. ES, vol. 27, pp. 365–371. Springer, Cham (2020). https://doi.org/10.1007/978-3-030-42822-8_48
6. Frère, A., de Wiart, C.C., Hillewaert, K., Chatelain, P., Winckelmans, G.: Application of wall-models to discontinuous Galerkin LES. Phys. Fluids **29**, 085111 (2017)
7. Gassner, G.J., Winters, A.R., Kopriva, D.A.: Split form nodal discontinuous Galerkin schemes with summation-by-parts property for the compressible Euler equations. J. Comput. Phys. **327**, 39–66 (2016)
8. Bassi, F., Rebay, S.: A high-order accurate discontinuous finite element method for the numerical solution of the compressible Navier–Stokes equations. J. Comput. Phys. **131**, 267–279 (1997)
9. Bergmann, M., Gölden, R., Morsbach, C.: Numerical investigation of split form nodal discontinuous Galerkin schemes for the implicit les of a turbulent channel flow. In: Proceedings of the 7th European Conference on Computational Fluid Dynamics (2018)
10. Hennemann, S., Rueda-Ramírez, A.M., Hindenlang, F.J., Gassner, G.J.: A provably entropy stable subcell shock capturing approach for high order split form DG for the compressible Euler equations. J. Comput. Phys. **426**, 109935 (2021)
11. Fernandez, P., Nguyen, N.-C., Peraire, J.: A physics-based shock capturing method for large-eddy simulation (2018)
12. Reichardt, H.: Vollständige darstellung der turbulenten geschwindigkeitsverteilung in glatten leitungen. ZAMM - Z. Angew. Math. Mechanik **31**(7), 208–219 (1951)
13. Hoyas, S., Jiménez, J.: Scaling of the velocity fluctuations in turbulent channels up to reτ=2003. Phys. Fluids **18**, 011702 (2006)

14. Klinner, J., Hergt, A., Grund, S., Willert, C.E.: High-speed PIV of shock boundary layer interactions in the transonic buffet flow of a compressor cascade. Exp. Fluids **62**, 1–19 (2021)
15. Schlüß, D., Frey, C., Ashcroft, G.: Consistent non-reflecting boundary conditions for both steady and unsteady flow simulations in turbomachinery applications. In: ECCOMAS Congress 2016 VII European Congress on Computational Methods in Applied Sciences and Engineering. Crete Island, Greece (2016)
16. Chandrashekar, P.: Kinetic energy preserving and entropy stable finite volume schemes for compressible Euler and Navier-Stokes equations. Commun. Comput. Phys. **14**, 1252–1286 (2013)

Numerical Investigation of the Transonic Non-ideal Gas Flow Around a Circular Cylinder at High Reynolds Number

C. Matar[1](✉), P. Cinnella[1], X. Gloerfelt[2], S. Sundermeier[3], L. Hake[3], and S. aus der Wiesche[3]

[1] Institut Jean le Rond dAlembert, Paris, France
{camille.matar,paola.cinnella}@dalembert.upmc.fr
[2] École Nationale Supérieure Arts et Métiers, Paris, France
xavier.gloerfelt@ensam.eu
[3] Muenster University of Applied Sciences, Münster, Germany
{stephan.sundermeir,leander.hake,Wiesche}@fh-muenster.de

1 Introduction

In the recent years, increased attention has been given to Organic Rankine cycle (ORC) systems which find their use in small scale applications such as waste-heat recovery [1]. ORC machines rely on the Rankine cycle and work with organic fluids which are molecularly complex compounds with high molar mass, and the operating conditions prevent the use of the calorically and thermally perfect gas model. These non-ideal thermodynamic states are typically located in the dense vapour region, defined by a fundamental derivative of gas dynamics $\Gamma < 1$ [2]. Moreover, due to the required reduced number of parts of such machines, leading to small number of stages in the turbine expander, the flow regimes are often transonic or supersonic [3]. As we know, compressibility strongly affects the flow around the blades, especially the developing boundary layer or the detached free mixing layers near rounded trailing edges [4]. It is now known that the trailing edge can typically account for 1/3 of the total blade mechanical losses in air [5]. The circular cylinder consists in a good prototype for bluff bodies in general [6], and is here considered immersed in a dense gas flow.

The incompressible high Reynolds number flow of a perfect gas around a smooth circular cylinder features complex phenomena such as boundary layer development in both favourable and adverse pressure gradients, flow separation, free shear layer transition to turbulence, and unsteady wake development, and it has been the subject of intensive research [7–9]. On the other hand, the compressible regime encompassing the drag divergence remains much less investigated and is home to strong coupling between flow compressibility and turbulence development [10]. Experimental measurements of pressure drag show relatively large disparities near the critical Mach number M_{cr} [11–13]. Steady and unsteady

C. Marchioli et al. (Eds.): DLES 2023, ERCO 31, pp. 164–169, 2024.
https://doi.org/10.1007/978-3-031-47028-8_25

solutions have been observed to exist intermittently at fixed Mach number, posing another challenge to numerical methods development [14]. Numerical works show systematic earlier onset of drag divergence linked to a delayed boundary layer separation compared to experimental measurements [15,16]. The non-ideal behaviour of dense gas can further affect the development of the complex shock systems around the cylinder, as well as the main loss mechanisms leading to drag divergence, and it has not been investigated until recently. Experiments are being conducted in the closed-loop wind tunnel (CLOWT) at University of Muenster, using a dense organic vapor (Novec^{TM}649) as the working fluid, and wall pressure measurements around a cylindrical rotatable Pitot tube up to high-subsonic conditions could be obtained for the first time [17], and cross validated with [19] steady and unsteady Reynolds-Averaged Navier Stokes (RANS) simulations. Although a reasonable agreement between the experimental and RANS/URANS results was obtained for low subsonic conditions, the discrepancy became larger as the Mach number was increased and dense gas effects became stronger. Furthermore, a Background Oriented Schlieren (BOS) technique has been developed to obtain visualisations of the compressible dense gas flow of Novec^{TM}649 around a circular cylinder [18]. This is a corner stone in the cross-validation of numerical methods and experiments, paving the way to investigations of other ORC-relevant flows.

To support experimental observations and further investigate the effect of the non-ideal gas behaviour on the global aerodynamic performance of the cylinder, we perform wall-resolved implicit Large Eddy Simulation (iLES) of the same inlet flow conditions, as well as explore the drag divergence regime. The results are compared with the preceding (U)RANS and experimental results, as well as recent Backward-Oriented Schlieren (BOS) visualizations obtained in the CLOWT facility.

2 Methods

Compressible LES are carried out using the in-house MUSICAA finite-difference solver. The inviscid fluxes are discretized by means of 10th-order centred differences whereas 4th-order is used for viscous fluxes. The scheme is supplemented with a 10th-order selective filter to eliminate grid-to-grid oscillations, along with a low-order shock capturing term activated locally by a combination of Jameson's shock sensor and Ducros' dilatation/vorticity sensor. A four-stage Runge-Kutta algorithm is used for time integration and high-order implicit residual smoothing [20] is applied to relax stability constraints on the time step. The filter also acts as a regularization term draining energy at subgrid scales, so that no explicit subgrid-scale model is used (implicit LES).

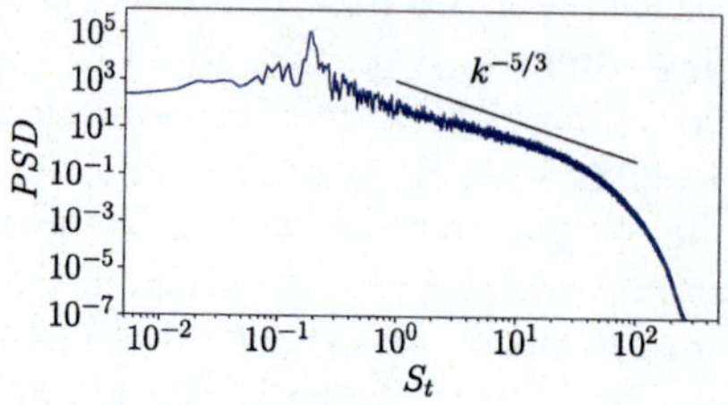

Fig. 1. Power spectrum in the near wake.

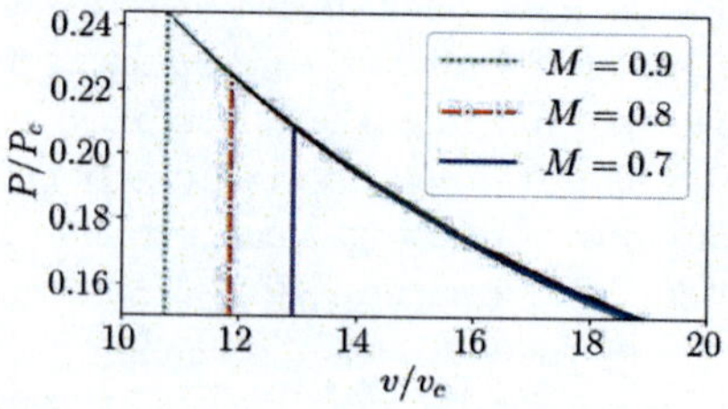

Fig. 2. P–v explored in the computations.

The numerical domain is a structured curvilinear grid combining H- and O-type topologies and extends over 30 D, 20D and 1D in the streamwise, transverse and spanwise directions, respectively, with D the cylinder diameter. The resolution has been set based on previous experience and literature [6,8] and ensures a wall non-dimensional unit $y^+_{avg} \leq 1.3$ to fully resolve the boundary layer. Fine details of the near wake flow are made possible by imposing a low cell growth rate from the wall, as well as a high number of points in the azimuthal direction $N_\theta = 1,500$ relative to other works in the literature [15,16]. Figure 1 gives the power spectral density of the transverse velocity component v against Strouhal number $St = fD/U_{in}$ computed at $(1D, 0)$ for $M_{in} = 0.7$. The power peak is associated with the vortex shedding frequency $St_{vs} \approx 0.19$. Resolution is beyond $St \geq 200$, indicating a wide spectrum of the inertial range when compared to similar wall-resolved LES [6,8]. Finally, adiabatic no-slip conditions are applied at the wall, and non-reflecting Tam & Dong's conditions are imposed at the boundaries. Periodicity is enforced in the spanwise direction.

Due to the typically high density of the gas, the thermodynamic conditions reached experimentally are limited to diluted conditions where the reduced volumes $v_r = v/v_c$ are high compared to the strongly non-ideal dense gas region. However, $\Gamma \leq 1$ in the freestream and in most of the flow, implying a reverse speed of sound behavior in isentropic perturbations compared to perfect gases. The inlet quantities considered in the following simulations are given in Table 1: we fix numerically the inlet value of Γ and Re_D and explore Mach numbers $M = 0.7, 0.8, 0.9$. The regions of the P–v space covered during the computations are shown in Fig. 2 and remain in the diluted gas region. The vertical lines indicate the lowest v_r for each case.

Table 1. Inlet Novec^{TM}649 conditions.

Fluid	p_{in}/p_c	ρ_{in}/ρ_c	Z_{in}	Γ_{in}	M_{in}	Re_D
Novec^{TM}649	0.166	0.060	0.880	0.938	0.7–0.9	2×10^5

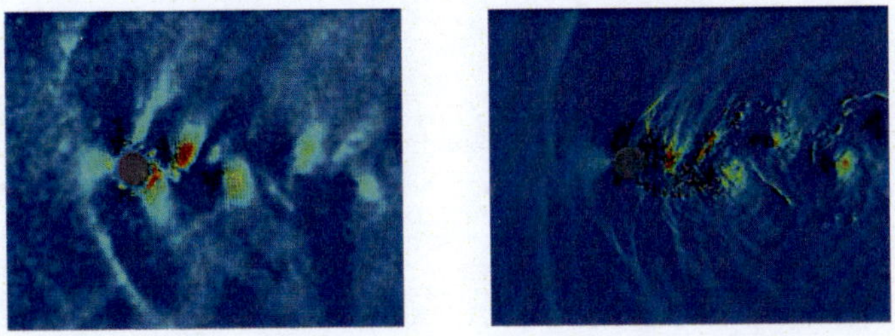

Fig. 3. BOS image (Left) and numerical Schlieren (Right) at $M = 0.7$.

3 Results and Discussion

The experimental and numerical Schlieren visualisations of the $M = 0.7$ flow are given in Fig. 3. Besides the vortex street, the present LES predicts upstream travelling acoustic waves as obtained in the experiments. Moreover, the Strouhal number measured in the experiments is also $St_{vs} \approx 0.19$. Strong compression waves are generated at the boundary layer separation point and at the end of the shear layers due to the locally supersonic zones. An indication of coupling between compressibility effects and the turbulent wake is the oscillation of the separation point which accompanies the vortex shedding, and covers a large angular range [10]. Figure 4 shows a conditional average of the wall pressure coefficient C_p only when lift is positive, corresponding to times where separation is likely to occur at the front and aft sections of the top and bottom of the cylinder, here shown for $0^0 \leq \theta \leq 180^0$ and $180^0 \leq \theta \leq 360^0$ respectively. As M increases, separation is delayed due to the stronger supersonic expansion at the wall. Moreover, the wake thickness $\Delta\theta$ decreases when M increases, in agreement with the onset of drag divergence.

We now inspect the C_p distribution on the cylinder at $M = 0.7$ and the mean pressure drag coefficient $C_d = -\int_0^\pi \overline{C_p}(\theta)\mathrm{d}\sin\theta$ against M in Fig. 5, where a previously performed URANS computation of the same dense gas flow is given. Computations and measurements for air from the literature, also reported in the figure, show higher aft section C_p compared to the present dense gas results. This is due to the lower equivalent isentropic exponent $\gamma_{Nov} \leq 1.04 \leq \gamma_{air}$ inducing higher Novec^{TM}649 compressibility and a stronger initial expansion. Furthermore, the mean back pressure coefficient $C_{p,b} = \overline{C_p}(180^0)$ in the present dense gas LES is much lower than the perfect gas value, which largely contributes to increase C_d. This is due to locally re-accelerating flow as the vortices roll-up and reach supersonic speeds, inducing low pressure at the cylinder back. Because a similar trend is observed in the URANS computation for Novec^{TM}649 and not in the data for air, this suggests that this is also a result of the lower γ_{Nov}. Finally, we notice the absence of a minimum of C_p near $\theta \approx 90^0$ in the present LES. This results from the large oscillations of the separation point which, upon time-averaging, essentially flattens the pressure curve. This element is also reported by the experimental measurements, while the URANS and literature numerical data

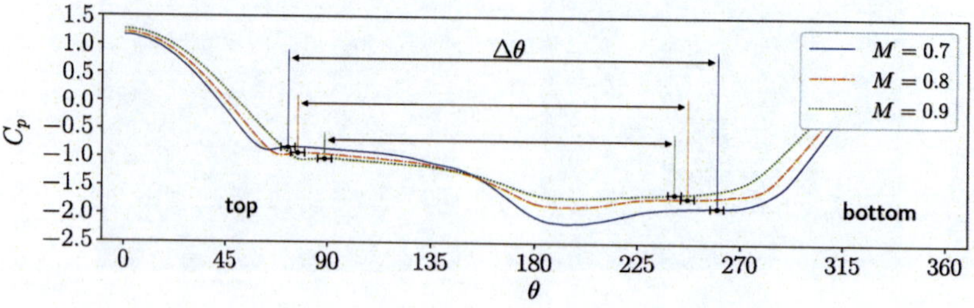

Fig. 4. Wall pressure conditional average.

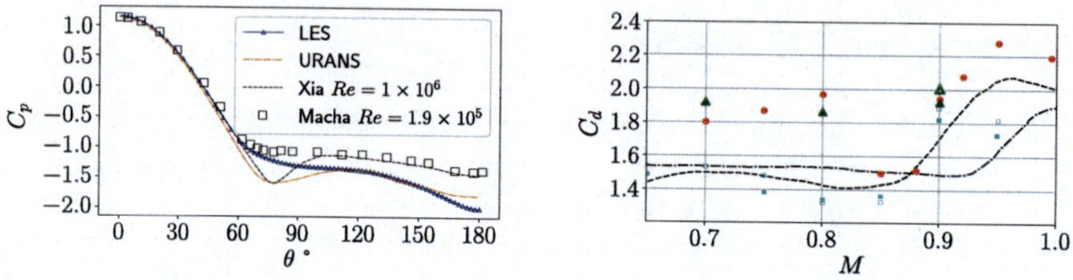

Fig. 5. (Left) Wall C_p distribution at $M = 0.7$ and (Right) C_d against M. ***Triangles*** present LES, air; ***filled triangles*** present LES, Novec^{TM}649; ***dots*** present URANS, Novec^{TM}649; ***dashed line*** data from flight tests of Welsh [23]; ***dash-dotted line*** mean line through data of Macha [12]; ***filled squares*** Xu *et al.* $Re_D = 4 \times 10^4$; ***squares*** Xu *et al.* $Re_D = 1 \times 10^6$.

show a pronounced local C_p minimum. We hypothesize that the high azimuthal resolution of the present LES allows to capture intricate details of the flow, such as boundary layer separation point oscillations.

The mean pressure drag coefficient is compared with the literature and the URANS results. The disparities within the literature are large; Hirsch [14] mentions the non uniqueness of these transonic flows and was supported by experimental measurements (Dyment [10]). Moreover, different wind tunnel blockage ratio or cylinder surface condition may strongly affect the development of shock waves and boundary layer respectively. The numerical works from Xu *et al.* indicate an earlier M_{cr} compared to the experiments, supporting the present LES for air. Most importantly, the present LES and URANS clearly predict a higher pressure drag coefficient for the dense gas within the $M \in [0.7; 0.9]$ range, as a direct result of the lower $C_{p,b}$.

4 Conclusions

The present study reported for the first time high-Reynolds number LES of transonic dense gas flows past cylinders. The results are in good agreement with companion BOS visualizations at Mach 0.7 in a closed-loop dense-gas wind tunnel. Dense gas effects enhance supersonic expansions around the cylinder, leading to lower values of the back pressure and to increased drag compared to air. Low-fidelity URANS simulations are found to capture well such qualitative behavior, although they miss the correct unsteady shock dynamics. In the future, similar works will be performed on the flow around a supersonic turbine blade profile with Novec^{TM}649 as working fluid in typical ORC operating conditions, supported by experimental measurements in the CLOWT facility.

Acknowledgements. This work was granted access to the HPC resources of IDRIS and TGCC under the allocation A0122A13457 made by GENCI. Financial support was provided by the ANR-DFG Grant REGAL-ORC, N. ANR-20-CE92-0019-01.

References

1. Colonna, P., et al. J. Eng. Gas Turbines Power **137** (2015)
2. Thompson, P.A.: Phys. Fluids **14**, 1843–1849 (1971)
3. Wheeler, A.P.S., Ong, J.: J. Eng. Gas Turbines Power **135** (2013)
4. Passmann, M.G.: PhD thesis (2021)
5. Denton, J.D.: J. Turbomach. (1993)
6. Lehmkuhl, O., Rodriguez, I., Borrel, R.: Phys. Fluids **26**, 1–22 (2014)
7. Norberg, C.: J. Fluids Struct. **17**, 57–96 (2003)
8. Cheng, W., Pullin, D.I., Samtaney, R., Zhang, W., Gao, W.: J. Fluid Mech. **820**, 121–158 (2017)
9. Demartino, C., Ricciardelli, F.: Eng. Struct. **137**, 76–114 (2017)
10. Dyment, A.: Vortex Motion, pp. 18–30 (1982)
11. Gowen, F.E., Perkins, E.W.: pp. 1–26 (1953)
12. Macha, J.M.: Eng. Notes **14**, 605–907 (1977)
13. Rodriguez, O.: AIAA **22**, 1713–1718 (1984)
14. Hirsch, C.: pp. 37–39 (2001)
15. Chen, L.W., Lu, X.Y.: Chin. Sci. Bull. (2009)
16. Xia, Z., Xiao, Z., Shi, Y., Chen, S.: AIAA **54** (2016)
17. Reinker, F., Wagner, F., Hake, L., aus der Wiesche, S.: Exp. Fluids**62** (2021)
18. Hake, L., et al.: NICFD2022, London (2022)
19. Matar, C., Cinnella, P., Gloerfelt, X., Reinker, F., aus der Wiesche, S.: Energy (2023)
20. Cinnella, P., Content, C.: J. Comput. Phys. **326**, 1–29 (2016)
21. Sciacovelli, L., Cinnella, P., Gloerfelt, X.: J. Fluid Mech. **821**, 153–199 (2017)
22. Gloerfelt, X., Robinet, J. -C., Sciacovelli, L., Cinnella, P., Grasso, F.: J. Fluid Mech. **893** (2020)
23. Welsh, C.J.: NASA report (1953)

Turbulent Boundary Layer in a 3-Element High-Lift Wing: Coherent Structures Identification

R. Montalà[1(✉)], B. Eiximeno[2], A. Miró[2], O. Lehmkuhl[2], and I. Rodríguez[1]

[1] Universitat Politècnica de Catalunya, Barcelona, Spain
{ricard.montala,ivette.rodriguez}@upc.edu
[2] Barcelona Supercomputing Center, Barcelona, Spain
{benet.eiximeno,arnau.mirojane,oriol.lehmkuhl}@bsc.es

Abstract. A large eddy simulation (LES) of the flow past a 30P30N high-lift three-element wing is performed at the angle of attack $AoA = 9°$ and a Reynolds number of $Re = 750,000$, based on the nested chord. Results are validated against experimental data available in the literature, proving their reliability. The present work is focused on the boundary layer development along the main suction side, including the analysis of the evolution of the thickness, the friction Reynolds number, the shape factor and the Clauser pressure-gradient parameter, as well as the Reynolds stresses at given chord locations. Finally, a proper orthogonal decomposition (POD) is performed, which allows visualizing the main turbulent coherent structures present in a turbulent boundary layer.

1 Introduction

The reduction of fuel consumption in airplanes has always stand as a very appealing topic for the aeronautical industry. Economical savings and a lower environmental impact are the main benefits that companies aim to reach with it. In this sense, the breakthroughs in computational sciences, as well as the major advances in flow analysis, have allowed to disentangle the complexity of turbulence and gain insight into drag physics. However, these techniques has been usually limited to the analysis of canonical fluid flows, such as plane turbulent boundary layers or channel flows.

In the present work, we propose to address the physics of a more complex configuration, closer to the geometry of a real wing. Specifically, computational predictions on the 30P30N three-element high-lift wing are conducted, which has been used as a reference case for the AIAA Workshop on Benchmark Problems for Airframe Noise Computations (BANC). Consequently, many numerical contributions can be found on the literature [1,3,14], as well as experimental studies [4,8,9]. However, most of these works are focused on the aeroacoustic noise and, more precisely, on the flow mechanisms occurring at the slat cove.

Therefore, this work aims to extend the knowledge on high-lift wings and assess, not only the noise generation, but also identify the main causes of aerodynamic drag. The present results are the partial conclusions of a much larger

C. Marchioli et al. (Eds.): DLES 2023, ERCO 31, pp. 170–176, 2024.
https://doi.org/10.1007/978-3-031-47028-8_26

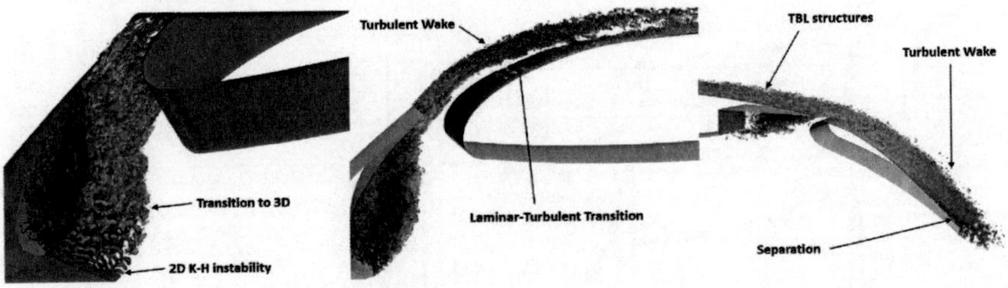

Fig. 1. Vortical structures represented by Q-isosurfaces

comprehensive analysis which targets to identify the most energy-containing coherent structures and frequencies, and relate them to the Reynolds stresses and drag. The considered Reynolds number and angle of attack are $Re_c = 750,000$ and $AoA = 9°$, respectively.

2 Numerical Method

In the present work, a wall-resolved large-eddy simulation (WRLES) is conducted on the 30P30N high-lift wing geometry employing the finite-element code Alya. In Alya, the convective operator of the equations is approximated by a low-dissipation scheme [5]. In this approach, the energy, momentum and angular momentum are preserved at the discrete level, providing enhanced results. The set of LES equations is time-advanced using an energy conserving third-order Runge-Kutta explicit method combined with an eigenvalue based time-step estimator. A non-incremental fractional-step method is used to stabilise the pressure. For the turbulence modelling, the Vreman [13] eddy-viscosity model is considered.

3 Results

The 2D geometry of the airfoil is positioned within a circular computational mesh that extends a radius of $R = 10C$ along the x-y plane and $L_z = 0.1C$ along the z direction, as recommended by Lockard and Choudhari (2009) [6], being C the stowed chord. This direction is assumed periodic and is discretized using 128 planes. As the inflow boundary condition, a uniform velocity profile is applied, whereas zero-gradients are imposed at the outflow regions. A total of 58 million grid points are employed and a structured-like inflation layer is considered around the airfoil, allowing to achieve a non-dimensional near wall distance of $\Delta y^+ \approx 1$. Here, a no-slip boundary condition is prescribed. Along the remaining directions, the maximum near wall distance is $\Delta x^+_{max} = 80$ and $\Delta z^+_{max} = 50$.

The predicted Q-criterion isosurfaces are depicted in Fig. 1. As can be observed, the fluid flow past a three-element airfoil exhibits a wide variety of

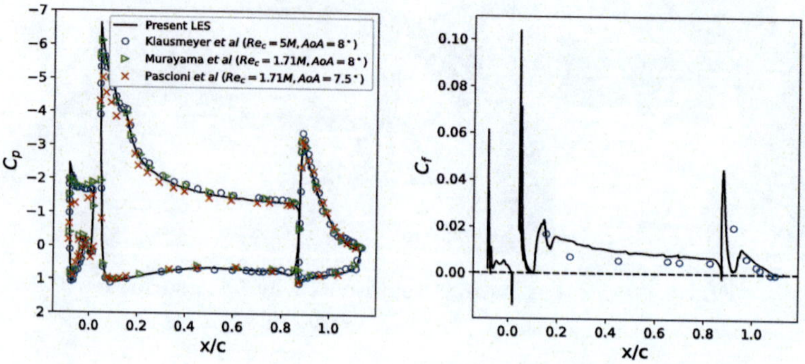

Fig. 2. Pressure C_p (left) and skin-friction C_f (right) coefficients. Skin friction represented only along the suction side. Present LES (black solid line) compared to the literature: Klausmeyer et al. (blue circles) [4], Murayama et al. (green triangles) [8] and Pascioni et al. (red crosses) [9].

phenomena: Shear layers, a laminar-turbulent transition, bounded and wake turbulence, among others flow physics. The pressure and skin friction coefficients are plotted in Fig. 2, along with selected results from the literature at comparable Reynolds numbers. Despite those experiments were performed at different flow conditions (Re_c and AoA), results show reasonably good agreement. The computed lift coefficient is equal to $C_L = 3.0879$, which lies in the range of values obtained by Murayama et al. [8] ($C_L = 3.2428$) and Pascioni et al. [9] ($C_L = 3.0559$); whereas the drag coefficient has a value of $C_D = 0.0885$ and no experimental results are available. The present work is focused on the bounded turbulence along the suction side of the main element.

3.1 Boundary Layer Development at Main Suction Side

The turbulent boundary layer (TBL) development can be observed in Fig. 3. In this figure, the boundary layer thickness, the friction Reynolds number, the Clauser pressure-gradient parameter and the shape factor are shown. It can be detected that, despite being an adverse pressure gradient (APG), the Clauser parameter is relatively small. This yields to a moderate growth of the boundary layer. In fact, the shape factor tends to $H = 1.40$, which is the typical value observed in zero pressure gradient (ZPG) TBLs. Also notice that, near the trailing edge, H suddenly decreases to a negative value due to the acceleration experimented by the fluid across the main-flap gap, leading to a local favorable pressure gradient (FPG).

In Fig. 4, the streamwise and wall-normal Reynolds stresses are plotted at three different locations and compared with the computations of other authors. Please refer to Table 1 for further information about these works. Notice that the historical effects at $x/c = 0.25$ are visible, most likely caused by the turbulent wake coming from the slat and the laminar-turbulent transition occurring nearby (see Fig. 1). On the other hand, a better agreement with the literature for both

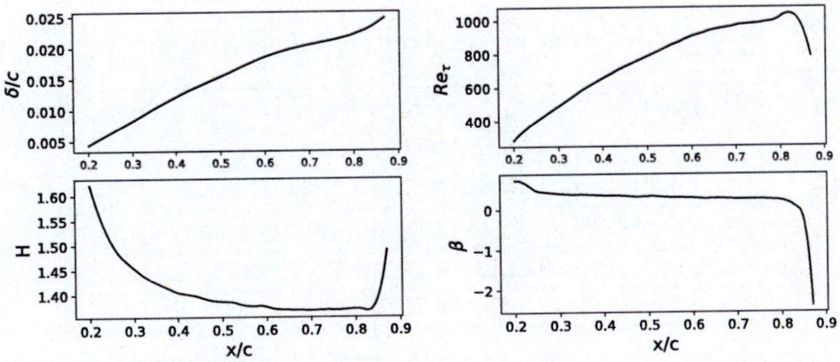

Fig. 3. Boundary layer thickness δ (top-left), friction Reynolds number Re_τ (top-left), shape factor H (bottom-left) and Clauser pressure-gradient parameter β (bottom-right) on the suction side of the wing.

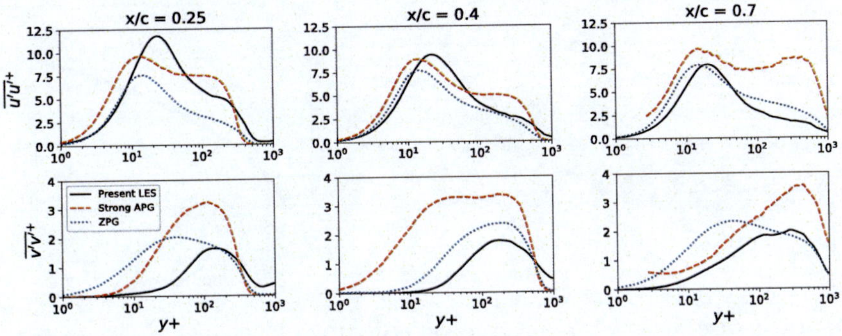

Fig. 4. Streamwise $\overline{u'u'}^+$ (top) and wall-normal $\overline{v'v'}^+$ (bottom) Reynolds Stresses at three different locations: $x/c = 0.25, 0.4$ and 0.7 (from left to right). Results compared with the literature at matched friction Reynolds number Re_τ: Present LES (black solid line); Strong APG (red dashed line) and ZPG (blue dotted line).

the magnitude and the shape of the stresses is observed at $x/c = 0.4$ and 0.7. In fact, as commented before, results resemble those of a ZPG TBL rather than a APG TBL. Thus, the inner $\overline{u'u'}$ peak can be easily identified, while the outer one is roughly present, which is a typical feature from APG TBL.

3.2 POD Analysis

In order to get more insight into the TBL structures present, a proper orthogonal decomposition (POD) [7] is conducted. For the study, a total number of 620 snapshots are taken every 250 simulation time steps. Only a small portion of the domain is considered for this analysis, comprising $x/C \in [0.55, 0.80]$, $y/C \in [0.04, 0.15]$ and the whole span-length.

Figure 5 shows the reconstruction of the fluid field employing the first one hundred modes, as well as the cumulative turbulent kinetic energy (TKE) distribution. The latter indicates a wide energy spread along high order modes,

Table 1. Summary of the literature employed in Fig. 4.

Location	Case	Re_τ	Re_θ	H	β	Ref
	Present LES	397	1,026	1.50	0.46	–
$x/c = 0.25$	Strong APG	373	1,722	1.74	4.10	[10]
	ZPG	492	1,421	1.43	0	[11]
	Present LES	655	1,626	1.40	0.34	–
$x/c = 0.4$	Strong APG	671	2,877	1.58	2.00	[12]
	ZPG	671	2,001	1.41	0	[11]
	Present LES	968	2,498	1.37	0.25	–
$x/c = 0.7$	Strong APG	1,070	–	–	2.4	[10]
	ZPG	1,139	3,600	–	0	[2]

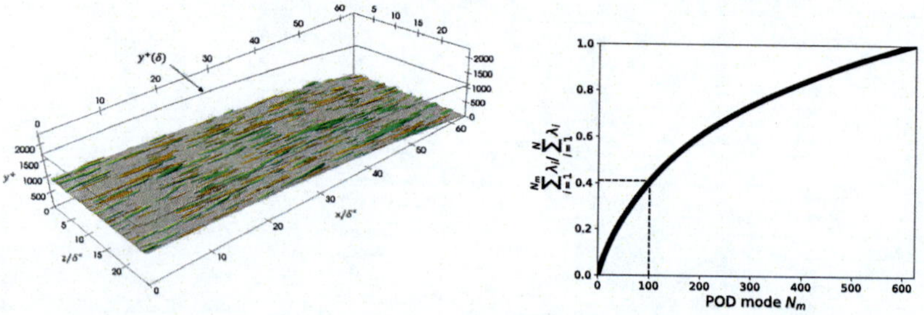

Fig. 5. Streamwise velocity u reconstruction using 100 POD modes (left) and cumulative sum of POD eigenvalues i.e. TKE (right).

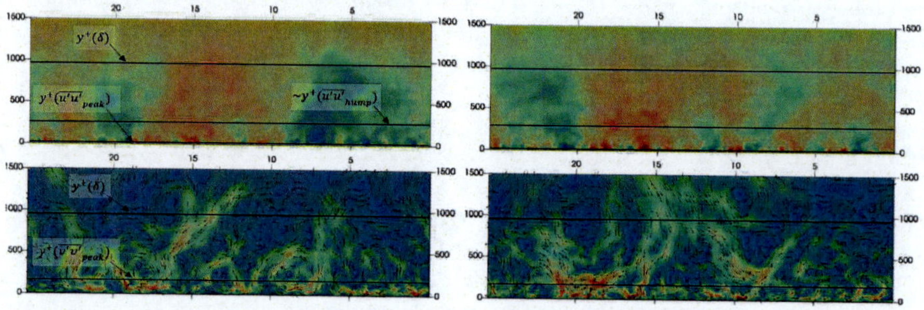

Fig. 6. ϕ_u (top) and $[0, \phi_v, \phi_w]_{mag}$ (bottom) isocontours at the transversal plane located at $x/c = 0.675$ for modes 1 and 3 (left and right, respectively).

requiring more than 360 modes to recover the 80% of the TKE. This may be explained due to the great amount of similar coherent structures present in a TBL, and that energy is homogeneously distributed, making the POD algorithm to fail in separating them efficiently. Despite this large energy spread, modes are capturing the most prominent structures, i.e. the boundary layer streaks.

In Fig. 6, the isocountours of the POD spatial modes ϕ along a transverse plane located at $x/c = 0.675$ are displayed. It is interesting to highlight how the most energetic ϕ_u structures are located at the height of $\overline{u'u'}_{peak}$, whereas the $[\phi_v, \phi_w]_{mag}$ structures are located nearby the region of $\overline{v'v'}_{peak}$. Also notice that for ϕ_u, there are also secondary structures around the primary ones that might be related to the outer $\overline{u'u'}$ hump.

4 Conclusions

A wall-resolved large-eddy simulation of the fluid around a 30P30N airfoil is performed at a Reynolds number of $Re_c = 750,000$ and an angle of attack of $AoA = 9°$. Results are validated with experiments carried out by other authors and further assessed in more detail along the suction side of the main element. Specifically, the boundary layer development is analysed, exhibiting a behaviour close to a ZPG TBL. Thus, the growth of the boundary layer is not pronounced and the outer peak of the streamwise Reynolds stresses is practically not present, which is a typical footprint of a APG TBL. A POD analysis is applied to a small portion of the TBL. In spite of the large energy spread to high order modes, TBL streaks are identified; and the location of the most energetic structures are associated to the peaks in the Reynolds stresses.

Acknowledgements. This work is partially financed by the Ministerio de Ciencia e Innovación of Spain (Ref. PID2020-116937RB-C21 / -C22). We also acknowledge the Red Española de Surpercomputación (RES) and the EuroHPC JU for granting us access to the MareNostrum IV machine based in Barcelona Supercomputing Center (Ref. IM-2022-3-0005) and to the HPC system Vega at the Institute of Information Science (Ref. EHPC-REG-2022R01-030), respectively. Finally, R. Montalá acknowledges the support of Departament de Recerca i Universitats de la Generalitat de Catalunya through a FI-SDUR graduate fellowship program.

References

1. Bodart, J., Larsson, J., Moin, P.: Large eddy simulation of high-lift devices. In: 21st AIAA Computational Fluid Dynamics Conference. San Diego (2013)
2. Eitel-Amor, G., Örlü, R., Schlatter, P.: Simulation and validation of a spatially evolving turbulent boundary layer up to $Re_\theta = 8300$. Int. J. Heat Fluid Flow **47**, 57–69 (2014)
3. Jin, Y., Liao, F., Cai, J.: Numerical simulation of 30p30n multi-element airfoil using delayed detached-eddy simulation. In: AIAA Aviation 2020 Forum (2020)
4. Klausmeyer, S., Lin, J.: An experimental investigation of skin friction on a multi-element airfoil. In: 12th AIAA Applied Aerodynamics Conference, Colorado (1994)
5. Lehmkuhl, O., Houzeaux, G., Owen, H., Chrysokentis, G., Rodriguez, I.: A low-dissipation finite element scheme for scale resolving simulations of turbulent flows. J. Comput. Phys. **390**, 51–65 (2019)
6. Lockard, D., Choudhari, M.: Noise radiation from a leading-edge slat. In: 15th AIAA/CEAS Aeroacoustics Conference, AIAA Paper 2009-3101 (2009)

7. Lumley, J.L.: The structure of inhomogeneous turbulent flows. Atmos. Turbul. Radio Wave Propag. 166–178 (1967)
8. Murayama, M., Nakakita, K., Yamamoto, K., Ura, H., Ito, Y.: Experimental study of slat noise from 30p30n three-element high-lift airfoil in JAXA hard-wall low-speed wind tunnel. In: 20th AIAA/CEAS Aeroacoustics Conference, Atlanta (2014)
9. Pascioni, K.A., Cattafesta, L.N.: Aeroacoustics measurements of leading-edge slat noise. In: 22nd AIAA/CEAS Aeroacoustics Conference, Lyon (2016)
10. Sanmiguel, V.C., Örlü, R., Vinuesa, R., et al.: Adverse-pressure-gradient effects on turbulent boundary layers: statistics and flow-field organization. Flow Turbul. Combust. **99**, 589–612 (2017)
11. Schlatter, P., Örlü, R.: Assessment of direct numerical simulation data of turbulent boundary layers. J. Fluid Mech. **659**, 116–126 (2010)
12. Vinuesa, R., Negi, P.S., Atzori, M., Hanifi, A., Henningson, D.S., Schlatter, P.: Turbulent boundary layers around wing sections up to $Re_c = 1,000,000$. Int. J. Heat Fluid Flow **72**, 586–99 (2018)
13. Vreman, A.W.: An eddy-viscosity subgrid-scale model for turbulent shear flow: algebraic theory and applications. Phys. Fluids **16**, 3670 (2004)
14. Zhang, Y., Chen, H., Wang, K., Wang, M.: Aeroacoustic prediction of a multi-element airfoil using wall-modelled large-eddy simulation. AIAA J. **55**(12), 4219–4233 (2017)

Aeroacoustic Source Terms from Turbulent Flow Through a 90° Pipe Bend Predicted by Large-Eddy Simulation

J. Tieber[1(✉)], H. Steiner[1], P. Maurerlehner[2], S. Schoder[2], M. Kaltenbacher[2], and G. Brenn[1]

[1] Institute of Fluid Mechanics and Heat Transfer, Graz University of Technology, Graz, Austria
{johannes.tieber,helfried.steiner}@tugraz.at
[2] Institute of Fundamentals and Theory in Electrical Engineering, Graz University of Technology, Graz, Austria
paul.maurerlehner@tugraz.at

1 Introduction

The large disparity of scales between the smallest relevant turbulent structures and the large acoustic wave lengths [1] makes a fully resolved Direct Numerical Simulation of both phenomena unfeasible for most technical applications. In Computational Aeroacoustics, (CAA) especially at low Mach numbers, so-called hybrid methods allow for a separate computation of the incompressible turbulent flow field, providing acoustic source terms, and the Acoustic Simulation (AS) of the sound propagation. The present study follows this approach using the Perturbed Convective Wave Equation (PCWE) [2] to predict the acoustic field inside a circular 90° pipe bend with relative curvature radius $R_m/D = 1.02$ at bulk Reynolds numbers $Re_\mathrm{B} = Du_\mathrm{B}/\nu = 5300$ and 12500. The respective bulk Mach numbers are $Ma_\mathrm{B} = u_\mathrm{B}/c_{ref} = 0.005$ and 0.011. The incompressible flow field is computed using Large-Eddy Simulation (LES) to predict acoustic source terms as input into the AS. The unresolved subgrid scale fluxes are computed from a Boussinesq-type eddy viscosity model introduced as the Coherent Structure Model by [3]. The PCWE is based on a perturbation Ansatz, where the flow field variables are decomposed into a mean, a fluctuating incompressible, and an acoustic component, generally written as $\phi = \langle\phi\rangle + \phi' = \langle\phi\rangle + \phi^\mathrm{ic} + \phi^\mathrm{a}$. This finally yields a convective wave equation for the acoustic potential ψ^a, called PCWE and is written as

$$\frac{1}{c_0^2}\frac{D^2\psi^\mathrm{a}}{Dt^2} - \frac{\partial^2\psi^\mathrm{a}}{\partial x_j \partial x_j} = \underbrace{-\frac{1}{\rho_0 c_0^2}\frac{Dp^\mathrm{ic}}{Dt}}_{\overline{Q}} = \underbrace{-\frac{1}{\rho_0 c_0^2}\frac{\partial \overline{p}^\mathrm{ic}}{\partial t}}_{\overline{Q}_t} \underbrace{-\frac{1}{\rho_0 c_0^2}\langle\overline{u}_i\rangle\frac{\partial \overline{p}^\mathrm{ic}}{\partial x_i}}_{\overline{Q}_c}, \quad (1)$$

providing the acoustic velocity $u_i^\mathrm{a} = -\partial\psi^\mathrm{a}/\partial x_i$ and pressure $p^\mathrm{a} = \rho_0 D\psi^\mathrm{a}/Dt$, with $D/Dt = \partial/\partial t + \langle\overline{u}_i\rangle\partial/\partial x_i$ beeing the material derivative. The source term on right-hand side of Eq. (1) is computed from the LES predicted instantaneous pressure fluctuations $\overline{p}^\mathrm{ic}$ and velocity field $\overline{u}_i$, where the latter is introduced into the convective component $\overline{Q}_c$ as temporal average denoted by the angular brackets. Unlike LES, AS basically requires

C. Marchioli et al. (Eds.): DLES 2023, ERCO 31, pp. 177–182, 2024.
https://doi.org/10.1007/978-3-031-47028-8_27

a coarser preferably uniform spatial resolution. The consequently applied conservative spatial interpolation of the source terms from the LES grid onto the AS grid may to some extent alter the high frequency content of the spectra [4]. The effect of the fine-to-coarse mapping on the finally predicted acoustics is still very small.

2 Simulation Setup

The computational domain is shown in Fig. 1a with the bend geometry in detail Fig. 1b. The operating fluid is air, which is assumed as a perfect gas.

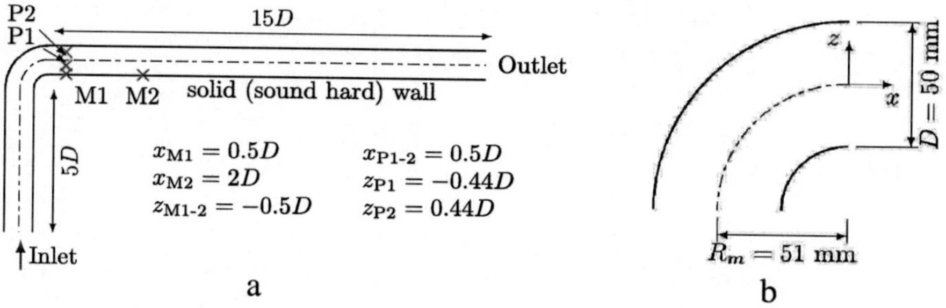

Fig. 1. a Computational domain with sensor positions P1 and P2. **b** Geometry of the pipe bend.

The LES equations were solved using a second order accurate finite volume method with an implicit first order accurate temporal discretization. The time step size was chosen sufficiently small ($\Delta t = 10^{-5}$ s) to ensure a insignificantly small discretization error also with the applied first order scheme. At the inlet, an instantaneous inflow boundary condition (BC) was prescribed, obtained from a precursor-LES of fully developed pipe flow. An averaged pressure BC was applied at the outlet, and the no-slip BC was imposed at the solid walls. After statistical convergence was reached, acoustic source terms were exported on the AS mesh for a period of $t = 0.2$ s. The PCWE was solved in the same domain using a finite element method, with a homogeneous von Neumann BC at the sound hard walls and adjacent Perfectly Matched Layer[5] (PML) regions at the inlet and outlet to avoid reflections.

3 Simulation Results

3.1 Flow Field

Figure 2a and Fig. 2b show LES results at $Re_B = 5300$ and 12500 compared against LES data and PIV measurements of Rütten et. al. [6] for $Re_B = 5000$ at $x = 1D$. Very good agreement is observed for the matching lower Reynolds number case. The higher Reynolds number case $Re_B = 12500$ shows similar profiles, however, with notably increased stresses in the highly turbulent region on the inner side.

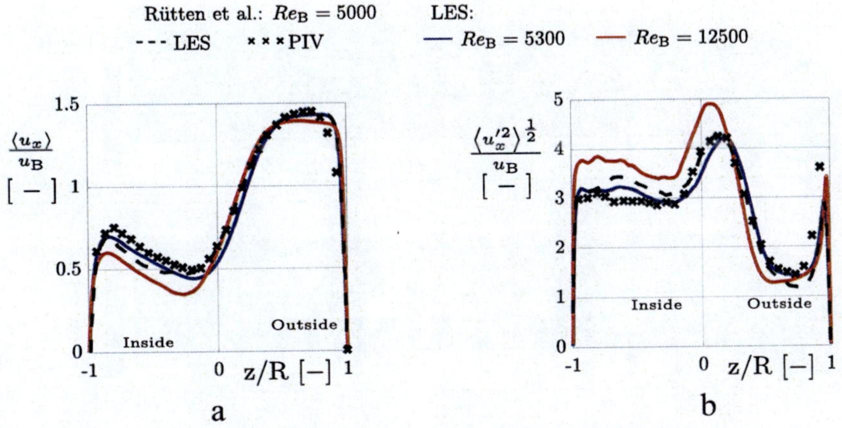

Fig. 2. **a** Profile of the mean x-velocity $x = 1D$ downstream of the bend exit. **b** Profile of the x-normal stresses $x = 1D$ downstream of the bend exit.

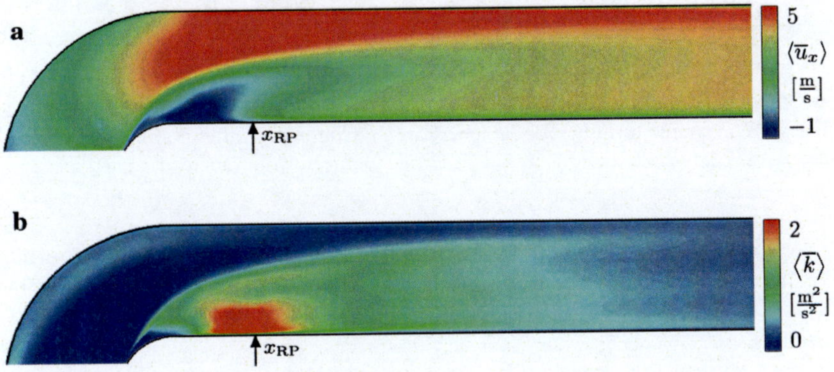

Fig. 3. **a** Contours of the mean x−velocity component $\langle \overline{u}_x \rangle$ for $Re_B = 12500$. **b** Contours of the turbulent kinetic energy $\langle \overline{k} \rangle$ for $Re_B = 12500$.

As seen from the contours in Fig. 3a, the flow detaches from the highly curved inner bend wall, producing a recirculation zone downstream of the bend exit, and a high-velocity region radially outside. The contours in Fig. 3b indicate a region of high turbulence kinetic energy extending near the reattachment point, which is located at $x_{RP} = 0.68D$ for $Re_B = 12500$. Further downstream the turbulence kinetic energy decays notably.

Figure 4a, Fig. 4b and Fig. 4c show contours of the instantaneous PCWE source term and its constitutive components as predicted from the incompressible LES. High values of both components are seen in the recirculation zone and in the highly sheared region. The source term components in Fig. 4a, Fig. 4b are negatively correlated, especially on

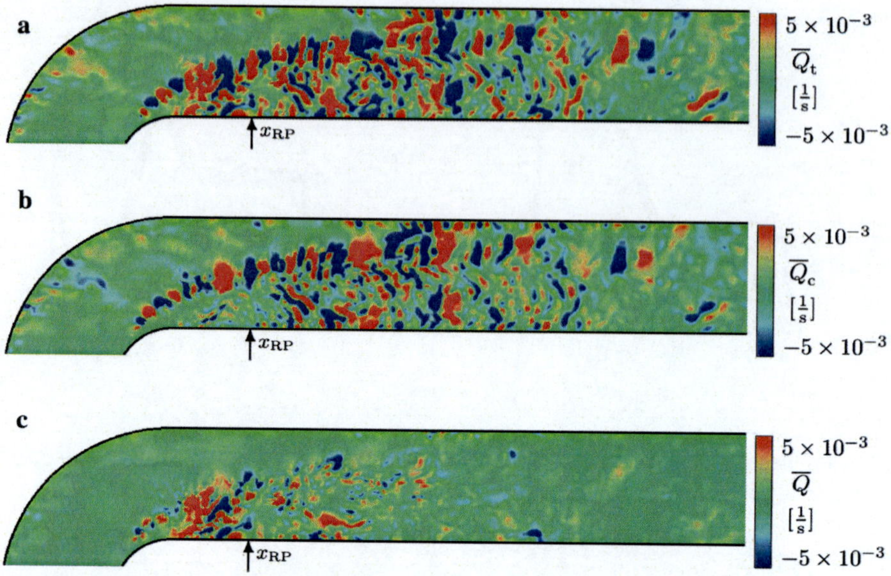

Fig. 4. **a** Contours of the transient PCWE source term $\overline{Q}_t$ for $Re_B = 12500$. **b** Contours of the convective PCWE source term $\overline{Q}_c$ for $Re_B = 12500$. **c** Contours of the total PCWE source term $\overline{Q}$ for $Re_B = 12500$.

the high-velocity side of the shear layer, which effectively reduces the total term $\overline{Q}$ in Fig. 4c. The highest total source terms are observed in the region near the reattachment point.

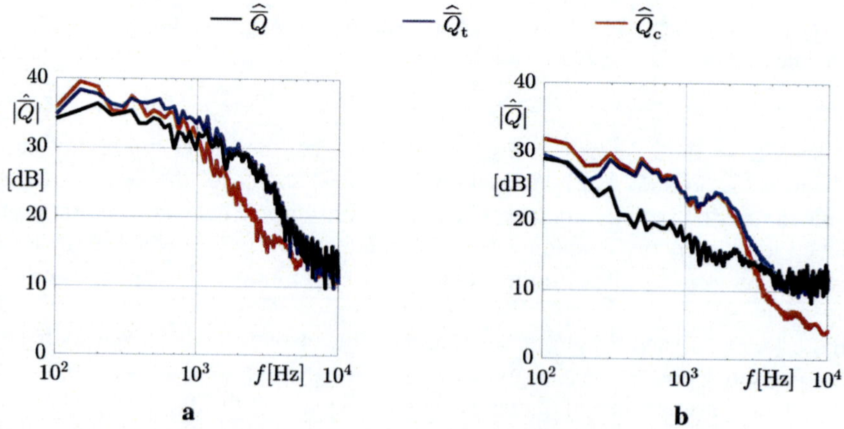

Fig. 5. **a** Spectra of the time derivative, convective and total PCWE source terms, at position P1 for $Re_B = 12500$. **b** Spectra of the time derivative, convective and total PCWE source terms, at position P2 for $Re_B = 12500$.

Figure 5a and Fig. 5b present spectra of the PCWE source terms at the two positions P1 and P2 (see Fig. 1a) downstream of the bend near the radially inner and outer side, respectively. $|\hat{\Phi}| = (\hat{\Phi}\hat{\Phi}^*)^{1/2}$ represents the amplitudes of the Discrete Fourier Transform of the considered time signal $\hat{\Phi} = DFT\{\Phi(t)\}$, which have been averaged over ten non-overlapping $\Delta t = 0.02s$ long subsegments taken from the full record, following Welch's method [7]. For the first position P1, which is located in the highly turbulent region inside the recirculation zone, the time derivative term is dominating, especially at higher frequencies. For the other position P2 shown in Fig. 5b, being located in the high axial velocity region, the amplitudes of the total PCWE source term is significantly smaller than its components. This indicates mutual cancellation of the two source term components over a wide range of frequencies.

3.2 Acoustic Simulation and Comparison Against Measurements

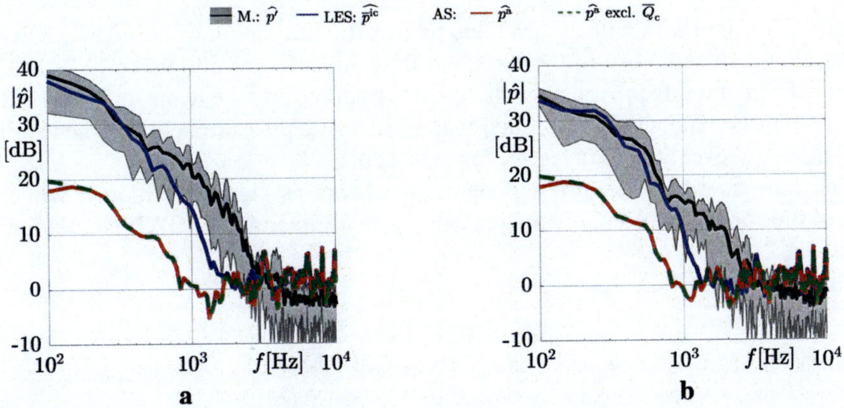

Fig. 6. a Pressure spectra from LES and AS against measurements at position M1 for $Re_B = 12500$. **b** Pressure spectra from LES and AS against measurements at position M2 for $Re_B = 12500$.

Figure 6a and Fig. 6b show spectral amplitudes of the AS predicted acoustic pressure together with LES predicted incompressible pressure amplitudes for $Re_B = 12500$ and results from dedicated microphone measurements [8]. The grey shaded area indicates the bandwidth of the experimental spectra obtained for the selected $\Delta t = 0.02\,\text{s}$ long subsegments spanning the considered $\Delta t = 0.2\,\text{s}$ of the investigated experimental record. The collapse of the predictions for $|\widehat{p^a}|$ with and without the contribution of the convective component of the PCWE source term indicates the negligible effect of this component on the acoustic field, despite its lowering effect on the total source term in the high axial velocity region, as seen in physical and spectral space in Fig. 4c and Fig. 5b, respectively. Overall, the acoustic pressure $|\widehat{p^a}|$ is significantly smaller than the incompressible pressure $\left|\widehat{p^{ic}}\right|$. It is also similar in amplitude at both positions, while the underlying incompressible pressure fluctuation exhibits higher amplitudes in the low frequency range at M1, which is located closer to the highly turbulent region downstream the bend exit. The predicted sound pressure level from the AS is $SPL = 20\log_{10}(p^{\mathrm{a}}_{\mathrm{rms}}/p_{\mathrm{ref}}) \sim 25\,\mathrm{dB}$ at M1, where $p_{\mathrm{ref}} = 2\cdot 10^{-5}\,\mathrm{Pa}$. The incompressible LES results generally show good agreement with the measurements for the lower frequencies, the high frequency content is underpredicted. The increasing discrepancy between the LES predictions and the measured pressure spectra for the higher frequencies may be attributed to the underresolution of the turbulent pressure fluctuations as well as the turbulent normal stresses inherent to LES. Including an appropriate subgrid-scale model contribution into the acoustic source terms should bring improvements here.

4 Conclusions

A hybrid CAA approach with a PCWE based acoustic simulation was used to predict the internal flow-induced noise of the turbulent flow downstream of a 90° pipe bend. The applied LES method is proven as well-suited for supporting the acoustic PCWE simulation with instantaneous sources, such that the acoustic pressure field and the resulting sound pressure level are predicted reasonably well.

The transient source term component was shown as clearly dominant, while the effect of the convective component appeared as negligible in the PCWE simulation.

References

1. Schoder, S., Kaltenbacher, M.: Hybrid aeroacoustic computations: state of art and new achievements. J. Theor. Comput. Acoust. **27**(04), 1950020 (2019)
2. Schoder, S., et al.: Aeroacoustic sound source characterization of the human voice production-perturbed convective wave equation. Appl. Sci. **11**, 2614 (2011)
3. Kobayashi, H.: The subgrid-scale models based on coherent structures for rotating homogeneous turbulence and turbulent channel flow. Phys. Fluids **17**, 045104 (2005)
4. Schoder, S., et al.: Application limits of conservative source interpolation methods using a low mach number hybrid aeroacoustic workflow. J. Theor. Comput. Acoust. **29**, 2050032 (2021)
5. Kaltenbacher, B., Kaltenbacher, M., Sim, I.: A modified and stable version of a perfectly matched layer technique for the 3-d second order wave equation in time domain with an application to aeroacoustics. J. Comput. Phys. **235**, 407–422 (2013)
6. Rütten, F., Schröder, W., Meinke, M.: Large-eddy simulation of low frequency oscillations of the Dean vortices in turbulent pipe bend flows. Phys. Fluids **17**(3), 035107 (2005)
7. Welch, P.D.: The use of fast Fourier transform for the estimation of power spectra: a method based on time averaging over short, modified periodograms. IEEE Trans. **15**(2), 70–73 (1967)
8. Maurerlehner, P., Floss, S., Tieber, J., Steiner, H., Brenn, G., Kaltenbacher, M.: Validation setup for the investigation of aeroacoustic and vibroacoustic sound emission of confined turbulent flows. In: INTER-NOISE and NOISE-CON Congress and Conference Proceedings. Institute of Noise Control Engineering, pp. 519–525 (2021)

The Effect of Wing-Tip Vortices on the Flow Around a NACA0012 Wing

S. Toosi[1,2,3](✉), A. Peplinski[1,2], P. Schlatter[3,1,2], and R. Vinuesa[1,2]

[1] FLOW, Engineering Mechanics, KTH Royal Institute of Technology, Stockholm, Sweden
{adam,rvinuesa}@mech.kth.se
[2] Swedish e-Science Research Centre (SeRC), Stockholm, Sweden
[3] Institute of Fluid Mechanics (LSTM), Friedrich-Alexander-Universität (FAU) Erlangen-Nürnberg, 91058 Erlangen, Germany
{siavash.toosi,philipp.schlatter}@fau.de

1 Introduction

The wing-tip vortices are formed near the tip of finite-span, lift-generating surfaces as a result of the pressure difference between the pressure and suction sides of the wing. The most significant impact of the tip vortices on the lift is due to their imposed downwash which reduces the effective angle of attack of the airfoil and causes a drop in lift compared to an infinite-span airfoil at the same geometric angle of attack. The other impact of the imposed downwash of the tip vortices is that the lift that the airfoil generates, which is normal to the free-stream direction perceived by the airfoil, has a non-zero component in the flight direction; i.e., a new component of drag. This is referred to as the induced drag (also lift-induced drag), or the vortex drag [2,7]. In a subsonic, commercial airliner in cruise conditions, it is estimated that around 40% of the total drag is due to vortex drag [6]. This ratio in even higher in the take-off, climb, and landing phases of the flight [2].

The goal of the present work is to perform a systematic study of the formation of wing-tip vortices and their interaction with and impact on the surrounding flow in more details. This is done by high-fidelity simulations of two wing geometries, i.e., infinite-span (periodic) and three-dimensional (wing-tip) wings, at a chord-based Reynolds number of $Re_c = cU_\infty/\nu = 200,000$ (where c, U_∞, and ν are the chord, free-stream velocity, and fluid kinematic viscosity, respectively) and at three different angles of attack: $\alpha = 0°, 5°, 10°$. Of particular interest is the interaction of these vortices with wall turbulence and the turbulent wake. However, the results presented here focus on the vortex formation and early development.

2 Numerical Method

To ensure of the accuracy of the results, we perform well-resolved, high-resolution large-eddy simulations (LES), at resolutions that are close to direct numerical simulation (DNS), where only the smallest scales (e.g., $\leq 6\eta$ in the wake, where η is the Kolmogorov length scale) are modeled by the subgrid scale (SGS) model. All simulations are performed by the high-order incompressible Navier–Stokes solver Nek5000 [4]

C. Marchioli et al. (Eds.): DLES 2023, ERCO 31, pp. 183–188, 2024.
https://doi.org/10.1007/978-3-031-47028-8_28

with adaptive mesh refinement (AMR) capabilities developed at KTH [10]. The AMR version adds the capability of refining each element individually, where solution continuity at non-conforming interfaces is ensured by interpolating from the "coarse" side onto the fine side.

The velocity field is expanded by a polynomial of order $p = 7$ on $p+1$ Gauss–Lobatto–Legendre (GLL) points while the pressure is represented on $p-1$ Gauss-Legendre (GL) points following the $P_N - P_{N-2}$ formulation [4]. The nonlinear convective term is overintegrated to avoid (or reduce) aliasing errors. Time stepping is performed by an implicit third-order backward-differentiation scheme for the viscous terms and an explicit third-order extrapolation for the nonlinear terms. Additional details on the high-resolution LES setup can be found in Refs. [11, 12].

Wings have a no-slip, no-penetration boundary condition. The computational domain has a rectangular cross-section in the normal-to-span direction that extends $20c$ upstream (c denoting the chord length), $30c$ downstream and $20c$ in the positive and negative vertical (*i.e.*, normal to chord) directions. The boundary layers are tripped on both the suction and pressure sides of the wings for all six cases.

The production grids are generated by iterative refinement of the initial grids, where at each iteration the elements with the highest contribution to solution error [3], based on solution on that grid, are selected for refinement. The spectral error indicator of Mavriplis (1990) is used for this purpose. The convergence process is accelerated by some manual input from the user. The adaptation process is terminated based on the resolution criteria available in the literature [11, 12]. Table 1 summarizes some information about the production grids used in this work, and Fig. 1 shows the spectral elements of grid RWT-5 from Table 1.

3 Results and Discussion

Figure 2 shows the mean streamwise vorticity of the RWT-0, RWT-5, and RWT-10 cases near the tip region at a few locations along the chord. Regarding the initial vortex development of the RWT-10 case (*e.g.*, at $x'/c = 0.6$), it is visually clear that the tip vortex of the RWT-10 case initially starts as a result of flow separation from the surface. Additionally, the combination of separation, imposed shear, and pressure gradients results in the formation of secondary vortices characterized by small regions of positive/negative streamwise vorticity between the primary vortex and the wall. This is qualitatively consistent with the observations of Ref. [5] (although for a different setup). The RWT-5 setup does not yet show a clear vortex core at this location. It should also be noted that the RWT-10 case has a larger, stronger primary vortex compared to RWT-5, and this vortex has a larger vertical separation from the wing surface (for instance at the trailing edge), but nearly the same spanwise location. Lastly, the zero-lift RWT-0 case does not exhibit any wing-tip vortices, but instead, a set of two counter-rotating vortices at the trailing edge, as previously observed in Ref. [5].

Table 1. Summary of the production grids used in this study and their near-wall resolution. Here, N_{GLL} denotes the number of GLL points, $\Delta_{\mathrm{mean}} = \delta/p$ is the average resolution of the element, where δ and p are the element size and polynomial order. y_1 shows the distance of the first off-wall GLL point to the wall. Wall resolutions are normalized by the viscous length $\delta_v = \nu/\sqrt{\tau_w/\rho}$. The P-* grids denote periodic cases, whereas RWT-* stands for rounded wing-tip cases. The number next to each grid shows the angle of attack, α.

Grid	N_{GLL}	$(\Delta x^+_{\mathrm{mean}}, y^+_1, \Delta z^+_{\mathrm{mean}})$
P-0	376×10^6	(10.3,0.72,8.7)
P-5	381×10^6	(10.5,0.73,8.5)
P-10	435×10^6	(12,0.8,9)
RWT-0	950×10^6	(10.3,0.7,5.5)
RWT-5	1.58×10^9	(10.5,0.75,5.7)
RWT-10	2.16×10^9	(12,0.8,6)

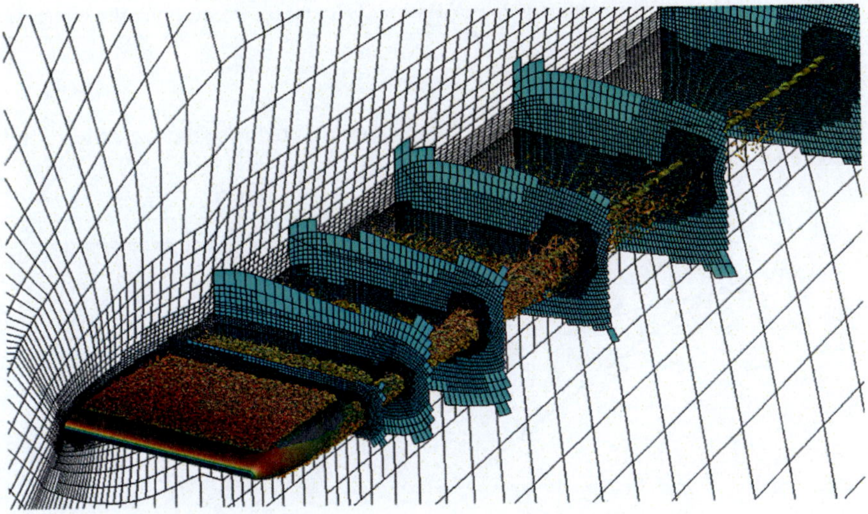

Fig. 1. Spectral-element grid for case RWT-5 with 3.1 million spectral elements. The grid is generated using the h-adaptation capabilities of the AMR version of Nek5000. We also show the instantaneous vortical structures represented by isosurfaces of $\lambda_2 c^2/U_\infty^2 = -100$ colored by streamwise velocity ranging from (blue) low to (red) high. (Color figure online)

Figure 3 shows the vertical and spanwise development of the core of the wing-tip vortex downstream of the wing. It is visually clear that the vertical development of the tip vortices is a function of the angle of attack: the vortex of the RWT-10 case encounters a larger downwash, as a result of the higher lift, and therefore, a larger downward angle. This was observed in previous experimental studies as well (cf. Ref. [5]). Despite the difference in their vertical development, the two tip vortices of RWT-5 and RWT-10 cases develop at nearly identical rates in the spanwise direction. This is slightly different

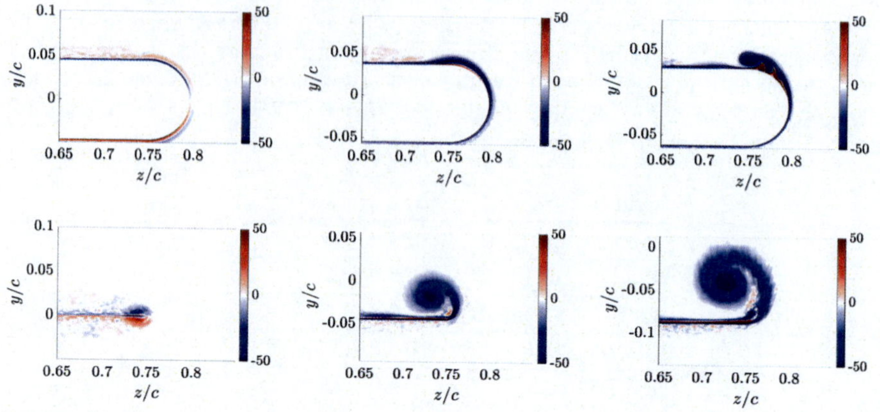

Fig. 2. Pseudocolor plots of streamwise vorticity $\omega_x c/U_\infty$ in the vicinity of the wing-tip for (left) RWT-0, (middle) RWT-5, and (right) RWT-10 at different locations along the chord: (top) $x'/c = 0.6$ and (bottom) $x'/c = 1.0$.

from the experimental results of Refs. [1,5], which could be due to the Reynolds number and turbulence effects as was suggested by Ref. [9].

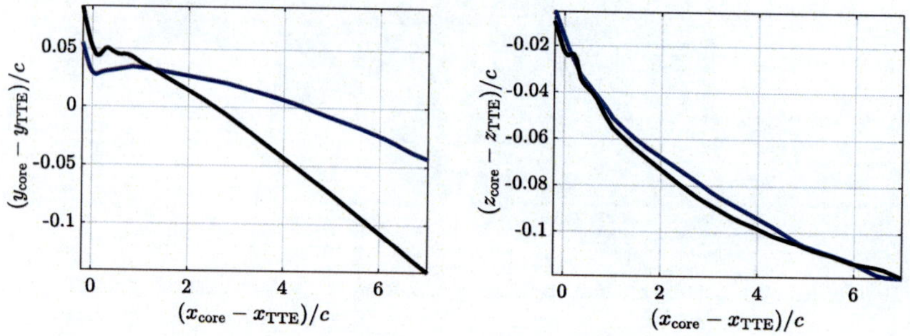

Fig. 3. Streamwise development of the vortex core, characterized by its vertical y_{core} (left) and spanwise z_{core} (right) location as function of the distance from the tip trailing edge, located at $(x_{TTE}, y_{TTE}, z_{TTE})$, for RWT-5 (blue) and RWT-10 (black) configurations.

It is also interesting to note that the two cases exhibit a completely different behavior in terms of their core velocity (see Fig. 4): the vortex core in the RWT-5 case has a persistent deficit compared to the freesteam velocity (in other words, a wake-like behavior) with a gradual trend towards the free-stream value (qualitatively consistent with the findings of Ref. [8]). However, the core of the RWT-10 case has a velocity surplus (i.e., a jet-like structure) near the trailing edge of the wing (again qualitatively consistent with the findings of Ref. [8]), which is decreased downstream and the core starts to exhibit a wake-like behavior for $x/c \geq 4$. This difference in wake development is primarily due to the different pressure level at the core (Fig. 4) and the different role that the pressure gradient term plays in the streamwise momentum equation. In the RWT-5 case, the pressure gradient term (essentially the slope of the line plot of pressure) is less significant compared to viscous diffusion, leading to a gradual increase in the streamwise velocity of the core towards U_∞. On the other hand, in the RWT-10 case there is a much larger pressure variation in the streamwise direction, which leads to a dominant pressure-gradient term that decelerates the core to values below U_∞. Based on the discussion made for the RWT-5 case (i.e., the dominance of viscous diffusion for lower pressure gradients), the core velocity of the RWT-10 case is also expected to approach the free-stream velocity as it develops further downstream and the pressure gradient term becomes less dominant; however, the downstream extent of the current simulations is not sufficient to observe this behavior. Also note that this behavior of RWT-10 was not observed previously in experimental results, but can be explained by the thicker turbulent boundary layers and wake in our setup.

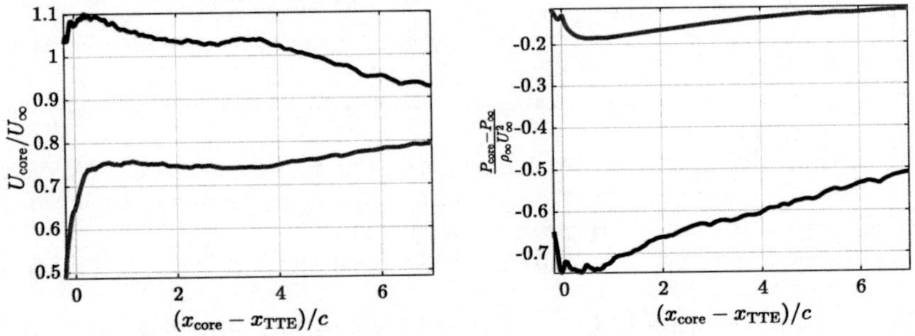

Fig. 4. Streamwise development of the vortex core, characterized by its vertical y_{core} (left) and lateral z_{core} (right) location as functions of distance from the tip trailing edge, located at $(x_{TTE}, y_{TTE}, z_{TTE})$, for RWT-5 (blue lines) and RWT-10 (black lines) configurations. ρ_∞, p_∞, and U_∞ are the free-stream density, pressure, and velocity, respectively. (Color figure online)

4 Summary and Conclusion

This work leverages the AMR version of Nek5000 to perform high-resolution large-eddy simulations of the flow around periodic and three-dimensional NACA0012 wings to investigates the impact of the wing-tip vortices on the flow. Our results show that during the vortex-formation process there are a number of secondary vortices that are formed in addition to the primary wing-tip vortex. The streamwise development of the wing-tip vortex appears to be strongly impacted by the angle of attack, which determines whether the vortex core has a velocity deficit or a velocity surplus compared to free-stream, and whether it can maintain its jet-like structure or loose momentum and become wake-like shortly downstream of the wing.

Additional work on this subject is needed, especially in terms of the interaction of the wing-tip vortex with the turbulent boundary layers. This will be investigated in the future using the available database from our simulations.

Acknowledgements. This work was supported by the European Research Council (ERC), the Swedish Research Council (VR), and the Knut and Alice Wallenberg Foundation. RV acknowledges the financial support from the ERC Grant No. "2021-CoG-101043998, DEEPCONTROL". Computing time for the production runs was provided by the Partnership for Advanced Computing in Europe (PRACE). The computations and postprocessing were enabled by resources provided by the Swedish National Infrastructure for Computing (SNIC), partially funded by the Swedish Research Council through grant agreement no. 2018-05973.

References

1. Devenport, W.J., Rife, M.C., Liapis, S.I., Follin, G.J.: The structure and development of a wing-tip vortex. J. Fluid Mech. **312**, 67–106 (1996)
2. Federal Aviation Administration: Pilot's handbook of aeronautical knowledge (2016)
3. Fidkowski, K.J., Darmofal, D.L.: Review of output-based error estimation and mesh adaptation in computational fluid dynamics. AIAA J. **49**(4), 673–694 (2011)
4. Fischer, P.F., Lottes, J.W., Kerkemeier, S.G.: NEK5000: open source spectral element CFD solver (2010). http://nek5000.mcs.anl.gov
5. Giuni, M., Green, R.B.: Vortex formation on squared and rounded tip. Aeros. Sci. Technol. **29**, 191–199 (2013)
6. Hills, D: The Airbus Challenge. EADS Engineering Europe. Budapest 9–10th May 2008
7. Houghton, E.L., Carpenter, P.W., Collicott, S.H., Valentine, D.T.: Aerodynamics for Engineering Students. Butterworth-Heinemann, Oxford (2013)
8. Lee, T., Pereira, J.: Nature of wakelike and jetlike axial tip vortex flows. J. Aircraft **47**(6), 1946–1954 (2010)
9. McAlister, K.W., Takahashi, R.K.: NACA 0015 Wing Pressure and Trailing Vortex Measurements. NASA Tech. Pap. 3151 (1991)
10. Offermans, N., Peplinski, A., Marin, O., Schlatter, P.: Adaptive mesh refinement for steady flows in Nek5000. Comps Fluids **197**, 104352 (2020)
11. Tanarro, A., Mallor, F., Offermans, N., Peplinski, A., Vinuesa, R., Schlatter, P.: Enabling adaptive mesh refinement for spectral-element simulations of turbulence around wing sections. Flow Turbul. Combust. **105**, 415–436 (2020)
12. Vinuesa, R., Negi, P.S., Atzori, M., Hanifi, A., Henningson, D.S., Schlatter, P.: Turbulent boundary layers around wing sections up to $Re_c = 1{,}000{,}000$. Int. J. Heat Fluid Flow **72**, 86–99 (2018)

LES Modeling and Hybrid LES-RANS

Subgrid-Scale Modeling for Remeshed Vortex Methods

M. de Crouy-Chanel[1(✉)], C. Mimeau[1], I. Mortazavi[1], and M. V. Salvetti[2]

[1] M2N, CNAM, Paris, France
marthe.de-crouy-chanel@lecnam.net
[2] Department of Civil and Industrial Engineering, Università di Pisa - UNIPI, Pisa, Italy

Abstract. In this study, we evaluate various Large Eddy Simulation (LES) models applied to turbulent flows in the framework of remeshed or semi-Lagrangian Vortex Methods. The models are tested on the Taylor Green Vortex and the decay of homogeneous isotropic turbulence. We find a good performance when applying the Smagorinsky model to the smallest resolved scales of the vorticity field.

1 Introduction

Vortex methods [1,2] are a type of particle methods in which the vorticity is discretized on numerical particles following the fluid dynamics. In this work, we use a semi-Lagrangian approach where particles are repositioned on a fixed mesh after being transported. This "hybridization" through remeshing allows for the use of Eulerian methods in an initially Lagrangian algorithm, which allows to solve the Poisson equation for the velocity field as well as the diffusion more efficiently than purely Lagrangian methods [3]. Semi-Lagrangian approaches have some positive features: by keeping the Lagrangian framework for the transport of vorticity, they do not require a CFL condition constraining the advection time step to the grid size, they are less dispersive and less diffusive compared with non-spectral Eulerian methods (see for example a comparison with the Lattice Boltzmann Method [4]) and show similar accuracy compared to pseudo-spectral methods [5].

This method has proved to be efficient for a number of laminar and transitional flows [2], illustrating the flexibility provided by the optimal coupling between Lagrangian and Eulerian schemes. As for turbulent flows, this method has been mainly used for direct numerical simulation until now. According to the features cited above (especially the low diffusivity property), the present approach represents a legitimate candidate to perform large eddy simulations.

Following the pioneer, but very few, works on turbulent models for vortex methods [6–8] and semi-Lagrangian vortex methods [9] in the context of large eddy simulations (LES), the present study investigates and compares different closure models. The models implemented are tested for two classical benchmarks: the Taylor-Green Vortex and the decay of homogeneous isotropic turbulence.

C. Marchioli et al. (Eds.): DLES 2023, ERCO 31, pp. 191–196, 2024.
https://doi.org/10.1007/978-3-031-47028-8_29

2 Remeshed Vortex Methods

Vortex methods are Lagrangian methods. They are based on the vorticity(ω)-velocity($\mathbf{u}$) formulation of the incompressible Navier-Stokes equations:

$$\partial_t \omega + (\mathbf{u}\cdot\nabla)\omega - (\omega\cdot\nabla)\mathbf{u} = \frac{1}{Re}\Delta\omega \qquad \Delta\mathbf{u} = -\nabla\times\omega \tag{1}$$

where $\omega = \nabla\times\mathbf{u}$ and where $(\mathbf{u}\cdot\nabla)\omega$ and $(\omega\cdot\nabla)\mathbf{u}$ denote respectively the advection and stretching terms. The Poisson equation $\Delta\mathbf{u} = -\nabla\times\omega$ allows to recover the velocity field $\mathbf{u}$ from the vorticity field ω. The vorticity field is discretized on a set of numerical particles with position $\mathbf{x}_p$ and volume $\mathbf{v}_p$. The resolution of the governing equations is based on a fractional step algorithm. One time step of such algorithm is decomposed as follows: first the particles, carrying the vorticity field, are convected in a Lagrangian way by solving the system of ODEs

$$\frac{d\mathbf{x}_p}{dt} = \mathbf{u}(\mathbf{x}_p(t),t) \qquad \frac{d\omega_p}{dt} = 0 \tag{2}$$

Then, in order to avoid the vorticity field distortion, the vorticity ω_p carried by each particle p is distributed on the neighboring points of an underlying Cartesian mesh in a "remeshing" step

$$\omega_i(\mathbf{x}) = \sum_p \omega_p(\mathbf{x})\Lambda\left(\frac{\mathbf{x}_p - \mathbf{x}_i}{\Delta\mathbf{x}}\right) \tag{3}$$

where Λ is a piecewise polynomial with compact support. At this stage, the whole vorticity field has been redistributed on the mesh and the stretching equation, the diffusion equation and the Poisson equation are finally discretized on the grid using Eulerian schemes: the stretching is discretized through 4th order finite differences and RK time integration scheme and the Poisson and diffusion equations are solved using fast Fourier transform, with implicit time integration for the diffusion.

3 Turbulence Modeling

In order to perform large eddy simulations in the context of the remeshed vortex method, we write the filtered Navier-Stokes equations in their velocity-vorticity formulation

$$\partial_t\overline{\omega} + \nabla\cdot(\overline{\omega}\otimes\overline{\mathbf{u}} - \overline{\mathbf{u}}\otimes\overline{\omega}) = \frac{1}{Re}\Delta\overline{\mathbf{u}} - \nabla\cdot\mathbf{R} \tag{4}$$

where $\overline{\omega}$ and $\overline{\mathbf{u}}$ denotes the large scales of the vorticity and velocity field and where the subgrid scale vorticity stress to be modeled is expressed as

$$\mathbf{R} = \overline{\omega\otimes\mathbf{u}} - \overline{\omega}\otimes\overline{\mathbf{u}} - \overline{\mathbf{u}\otimes\omega} + \overline{\mathbf{u}}\otimes\overline{\omega} \tag{5}$$

The tensor can be decomposed in two parts, since the vorticity-velocity formulation has two non-linear terms: the subgrid scales produced by convection $\mathbf{R}_C = \overline{\omega\otimes\mathbf{u}} - \overline{\omega}\otimes\overline{\mathbf{u}}$ and by stretching $\mathbf{R}_S = \overline{\mathbf{u}\otimes\omega} - \overline{\mathbf{u}}\otimes\overline{\omega}$.

Smagorinsky Model

A first approach to model **R** in the vorticity-velocity formulation is to take the curl of $\nabla \cdot \tau_{SGS}$ where $\tau_{SGS} = \nu_{SGS}(\nabla \overline{\mathbf{u}} + \nabla \overline{\mathbf{u}}^T)$ denotes the usual form of the Smagorinsky subgrid scale stress tensor in velocity-pressure formulation: $\nabla \times (\nabla \cdot (\nu_{SGS}(\nabla \overline{\mathbf{u}} + \nabla \overline{\mathbf{u}}^T))) \approx \nabla \times (\nu_{SGS} \nabla \cdot (\nabla \overline{\mathbf{u}} + (\nabla \overline{\mathbf{u}})^T)) = -\nabla \times (\nu_{SGS} \nabla \times \overline{\omega})$. One can show that it is equivalent to

$$\mathbf{R}_{SGS} = \nu_{SGS}(\nabla \overline{\omega} - \nabla \overline{\omega}^T) \tag{6}$$

We note that the tensor $\mathbf{R}_{SGS}$ is anti-symmetric, as the tensor **R** to be modeled. A symmetric formulation can be found in [9]. The coexistence of those seemingly different versions of Smagorinsky model in literature can be explained by numerical experiments in [8] showing that the transposed gradient term is negligible compared to the gradient term. The subgrid-scale viscosity is modelled by $\nu_{SGS} = (C\bar{\Delta})^2|\overline{\mathbf{S}}|$ where C is a coefficient to be chosen, $\bar{\Delta}$ taken to be equal to the grid size, $\overline{\mathbf{S}} = \frac{1}{2}(\nabla \overline{\mathbf{u}} + \nabla \overline{\mathbf{u}}^T)$ and $|\overline{\mathbf{S}}| = \sqrt{2\overline{\mathbf{S}}_{ij}\overline{\mathbf{S}}_{ij}}$ is the magnitude of $\overline{\mathbf{S}}$.

"Small" Model

Following [9, 10], we explore a second model from the variational multiscale (VMS) family of models. VMS models aim to circonvent the over dissipation of the largest scales in classical artificial viscosity models by applying the artificial viscosity only to the smallest of the resolved scales and thus require an additional explicit small-scale filtering. Let $\overline{f}$ be some resolved field, we define, in Fourier space, the largest resolved scales of $\bar{f}$ by $\widehat{\overline{f}}(k) = \widehat{G}(k)\overline{f}(k)$ where $\widehat{G}$ is some test filter, and the small resolved scales $\overline{f}_S$ by $\overline{f}_S = \overline{f} - \widehat{\overline{f}}$. One therefore defines the second model by $\mathbf{R}_{SGS} = \nu_{SGS}(\nabla \overline{\omega}_S - (\nabla \overline{\omega}_S)^T)$ where $\overline{\omega}_S$ denotes the small scales of the resolved vorticity field.

Dynamic Model

The most widely used LES model in practical applications is the dynamic version of the Smagorinsky model in which the coefficient is adapted in time and space (with usually an averaging procedure along homogeneous directions). Following [7, 8] we adapt this procedure to the vorticity-velocity formulation. The Leonard tensor writes $\ell_i = \widehat{\overline{\mathbf{u}}_j \partial_j \overline{\omega}_i} - \widehat{\overline{\omega}_j \partial_j \overline{\mathbf{u}}_i} - \widehat{\overline{\mathbf{u}}}_j \partial_j \widehat{\overline{\omega}}_i + \widehat{\overline{\omega}}_j \partial_j \widehat{\overline{\mathbf{u}}}_i$ and $m = \nabla \cdot (\widehat{\bar{\Delta}}^2 |\widehat{\bar{S}}|(\nabla \widehat{\overline{\omega}} - \nabla \widehat{\overline{\omega}}^T) - \widehat{G} * (\bar{\Delta}^2 |\bar{S}|(\nabla \overline{\omega} - \nabla \overline{\omega}^T)))$ where $\widehat{G}$ denotes the test filter, $\widehat{\bar{\Delta}}$ its size and $\widehat{\bar{\omega}}$ and $\widehat{\bar{\mathbf{u}}}$ the test-filtered vorticity and velocity fields. We define the dynamic coefficient as $C_d = \langle \ell_i m_i \rangle / \langle m_i m_i \rangle$ where the brackets represent an averaging over homogeneous directions.

Gradient Model

We also explored a structural model that aims at directly modeling the subgrid-scale tensor (as opposed to functional models which aim at reintroducing the dissipation at the smallest scales). We consider the gradient model [11], based on a Taylor expansion

$$\mathbf{R}^g_{ij} = \frac{\bar{\Delta}^2}{12}(\partial_k \overline{\omega}_i \partial_k \overline{u}_j - \partial_k \overline{\omega}_j \partial_k \overline{u}_i) \tag{7}$$

This model is known for its low cost and its good performance in a priori tests but has the drawback of being not dissipative enough. To circumvent this problem, we propose the following clipping procedure, based on the enstrophy transfer between resolved and modelled scales $\mathbf{R}^g_{ij}\Omega_{ij}$ where $\Omega_{ij} = \frac{1}{2}(\nabla\overline{\omega} + \nabla\overline{\omega}^T)$.

$$\mathbf{R}^{g,clip}_{ij} = \frac{\bar{\Delta}^2}{12}(\partial_k\overline{\omega}_i\partial_k\overline{u}_j - \partial_k\overline{\omega}_j\partial_k\overline{u}_i) \times \begin{cases} 1 & \text{if } \mathbf{R}^g_{ij}\Omega_{ij} < 0 \\ -1 & \text{otherwise.} \end{cases} \tag{8}$$

4 Numerical Experiments

We tested the models presented above for two classical benchmarks: the Taylor Green Vortex (TGV) test case at $Re = 5000$ and the decay of homogeneous turbulence (HIT). In both test cases, the simulations are performed in a periodic cubic box of side length $L = 2\pi$. The Smagorinsky coefficient was set at $C_S^2 = 0.027$ for the simple Smagorinsky model and $C_S^2 = 0.0563$ for the "small" model, which are close to the ones found in literature [9, 10]. The filtering of the small scales was performed using the 6th order filter described in [9].

The Taylor-Green vortex is a well suited benchmark to study transition to turbulence and the progressive formation of small scales. A Reynolds number of 5000 has been used recently in literature [12] and has been chosen as it is more turbulent than the transitional and classical $Re = 1600$ test case, such that a LES does not work without a model, but is still low enough for a DNS to be computationally accessible. The reference DNS was performed at a resolution of 512^3 and the LES were performed at a coarser resolution of 96^3.

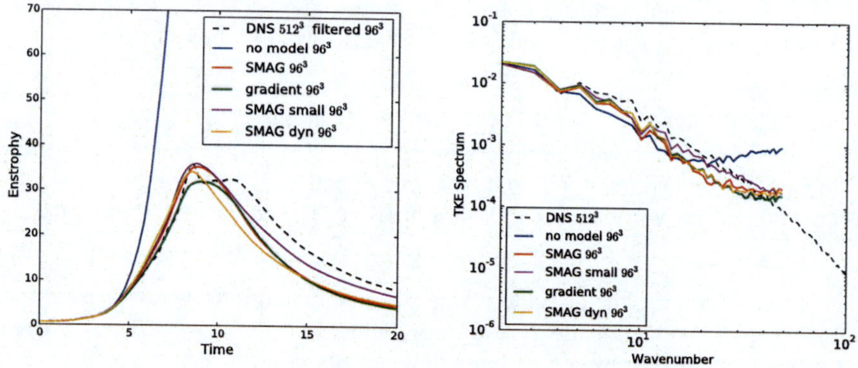

Fig. 1. Time history of enstrophy (left) and kinetic energy spectrum at $t = 8.5$ (right) for the TGV at $Re = 5000$

In Fig. 1, the time history of the enstrophy and the energy spectrum at $t = 8.5$ are used to evaluate the different models. The overestimation of the energy at the small scales by a LES with no additional model (in blue) shows the need for a subgrid-scale model in our case. The simple Smagorinsky model is too diffusive after the peak of enstrophy but its performance is improved when filtering the small scales, showing good agreement with the DNS data. The gradient model, a novel approach in the context of our method, also shows a good performance and the clipping has the expected effect of reducing backscatter. The dynamic Smagorinsky shows results similar to the gradient and Smagorinsky models but for a higher computational cost as it requires many additional filtering operations.

To initialize the decay of HIT, the forcing was done using the procedure described in [13] on the reference DNS at 512^3. Once the spectrum converged, the turbulent state was interpolated on a coarse 32^3 grid, used as the initial state for the LES, and a fine 256^3 grid. The results obtained are coherent with our conclusions for the TGV (Fig. 2).

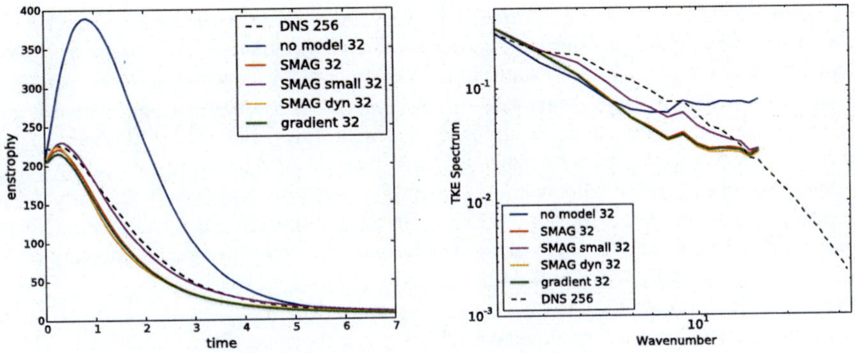

Fig. 2. Time history of enstrophy (left) and kinetic energy spectrum at $t = 1$ (right) for the decay of homogeneous isotropic turbulence

5 Conclusion and Perspectives

A benchmark analysis of various subgrid-scale models was performed in the context of a remeshed vortex method. This method presents the advantage, compared to most methods classically used in LES, of its semi-Lagrangian nature (showing low dissipation) and the use the vorticity, a field closely related to the growth of small-scale turbulence, as the main variable. Tests on two well known benchmarks showed that the model focusing on the smallest scales of the resolved vorticity ("small" model) gives the best results and the adaptation of the gradient model to our present context shows some promising results as well. Future works will focus on improving the existing models, as well as on developing Implicit Large Eddy Simulation (ILES) for our method and applying our models to test cases with walls.

References

1. Cottet, G.-H., Koumoutsakos, P.D.: Vortex Methods: Theory and Practice, vol. 8. Cambridge University Press, Cambridge (2000)
2. Mimeau, C., Mortazavi, I.: A review of vortex methods and their applications: from creation to recent advances. Fluids **6**, 68 (2021)

3. Cottet, G.-H., Poncet, P.: Advances in direct numerical simulations of 3D wall-bounded flows by vortex-in-cell methods. J. Comput. Phys. **193**(1), 136–158 (2004)
4. Mimeau, C., Mari, S., Mortazavi, I.: A comparison of semi-Lagrangian vortex method and lattice Boltzmann method for incompressible flows. Comput. Fluids **224**, 104946 (2021)
5. van Rees, W.M., Leonard, A., Pullin, D.I., Koumoutsakos, P.: A comparison of vortex and pseudo-spectral methods for the simulation of periodic vortical flows at high reynolds numbers. J. Comput. Phys. **230**, 2794–2805 (2011)
6. Cottet, G.-H.: Artificial viscosity models for vortex and particle methods. J. Comput. Phys. **127**(2), 299–308 (1996)
7. Winckelmans, G.S., Lund, T.S., Carati, D., Wray, A.A.: A priori testing of subgrid-scale models for the velocity-pressure and vorticity-velocity formulations. In: Summer Program, Center for Turbulence Research, Stanford University and NASA Ames, pp. 309–328 (1996)
8. Mansfield, J.R., Knio, O.M., Meneveau, C.: A dynamic LES scheme for the vorticity transport equation: formulation anda PrioriTests. J. Comput. Phys. **145**, 693–730 (1998)
9. Cocle, R., Dufresne, L., Winckelmans, G.: Investigation of multiscale subgrid models for LES of instabilities and turbulence in wake vortex systems. In: Kassinos, S.C., Langer, C.A., Iaccarino, G., Moin, P. (eds.) Complex Effects in Large Eddy Simulations. Lecture Notes in Computational Science and Engineering, vol 56, pp. 141–159. Springer, Heidelberg (2007). https://doi.org/10.1007/978-3-540-34234-2_11
10. Hughes, T.J.R., Mazzei, L., Oberai, A.A., Wray, A.A.: The multiscale formulation of large eddy simulation: decay of homogeneous isotropic turbulence. Phys. Fluids **13**, 505–512 (2001)
11. Vreman, A.: Direct and large-eddy simulation of the compressible turbulent mixing layer. Ph.D. thesis, University of Twente, Netherlands (1995)
12. Chapelier, J.B., Wasistho, B., Scalo, C.: A Coherent vorticity preserving eddy-viscosity correction for Large-Eddy Simulation. J. Comput. Phys. **359**, 164–182 (2018)
13. Rogallo, R.: Numerical Experiments in Homogeneous Turbulence, vol. 81315. National Aeronautics and Space Administration (1981)

Wavelet-based Adaptive LES for Compressible Flows

G. De Stefano[1](✉) and O. V. Vasilyev[2]

[1] Engineering Department, University of Campania, 81031 Aversa, Italy
giuliano.destefano@unicampania.it

[2] Keldysh Institute of Applied Mathematics, Russian Academy of Sciences, Moscow 125047, Russia

1 Introduction

Fully-developed turbulence is characterised by the existence of intermittent coherent energetic eddies at all scales, including the smaller ones. Therefore, traditional lowpass filter-based large-eddy simulation (LES) methods with *a priori* defined mesh resolution are inherently not able to investigate the multi-resolution features of coherent structures and intermittency of turbulent flows. Wavelet-based adaptive LES (WA-LES) is an extension of the LES methodology, where wavelet threshold filtering (WTF) is employed to separate resolved (more energetic) from residual (less energetic) turbulent flow structures, regardless of their size [1]. The effect of unresolved coherent structures on the resolved flow fields is approximated by means of subgrid-scale (SGS) models, with the turbulence resolution of the filtered solution being dictated by the WTF level that is prescribed.

The WA-LES governing equations are numerically solved by means of the anisotropic adaptive wavelet collocation (AWC) method, where the wavelet compression represents an integral part of the numerical algorithm, such that the solution is obtained with an optimal number of grid points for a given level of accuracy [2]. By separating the computational and the physical spaces, while introducing a suitable mapping between them, the structured rectilinear assembly of wavelet collocation points in the computational space can be retained. This way, the use of computationally efficient adaptive discrete wavelet transforms and derivative approximations is allowed, while anisotropic curvilinear spatial meshes can be employed for complex geometry flows.

In this work, the WA-LES approach to compressible turbulent flows is presented and discussed, with a particular focus on wall-bounded flows, while testing two different closure procedures that were recently introduced. The performance of the method is assessed by conducting adaptive numerical simulations of fully-developed compressible flow in a plane channel, which represents a well-established benchmark for this kind of flows [3].

C. Marchioli et al. (Eds.): DLES 2023, ERCO 31, pp. 197–202, 2024.
https://doi.org/10.1007/978-3-031-47028-8_30

2 Compressible WA-LES

The WA-LES governing equations for compressible flows are expressed in terms of wavelet-based density-weighted Favre-filtered variables [4]. Formally, these equations can be obtained by replacing the spatial low-pass filtering in the (traditional) compressible LES equations with multi-resolution nonlinear WTF. Denoting wavelet-filtered and Favre-filtered variables by using the over-bar and tilde symbols, respectively, the compressible WA-LES equations read:

$$\frac{\partial \overline{\rho}}{\partial t} = -\frac{\partial (\overline{\rho}\widetilde{u}_j)}{\partial x_j}, \tag{1}$$

$$\frac{\partial (\overline{\rho}\widetilde{u}_i)}{\partial t} = -\frac{\partial}{\partial x_j}\left(\overline{\rho}\widetilde{u}_i\widetilde{u}_j + \overline{p}\delta_{ij} - \hat{\sigma}_{ij}\right) - \frac{\partial \tau_{ij}}{\partial x_j} + \overline{\rho} f_i, \tag{2}$$

$$\frac{\partial (\overline{\rho}\widetilde{e})}{\partial t} = -\frac{\partial}{\partial x_j}\left[\left(\overline{\rho}\widetilde{e} + \overline{p}\right)\widetilde{u}_j - \hat{\sigma}_{ij}\widetilde{u}_i + \hat{q}_j\right] - \frac{\partial}{\partial x_j}\left(Q_j - D_j + J_j\right) + \overline{\rho} f_j \widetilde{u}_j. \tag{3}$$

While the detailed derivation of these equations can be found, for instance, in [5], let us simply recall some definitions: $\hat{\sigma}_{ij}$ and τ_{ij} are the resolved and SGS stresses; $\hat{q}_i$ and Q_i are the resolved and SGS heat fluxes; D_i and J_i are the SGS viscous and turbulent diffusion fluxes, and f_i represents a uniform body force sustaining the flow.

In this work, the unclosed terms appearing in the above equations are approximated by two different SGS models that were originally introduced for traditional non-adaptive LES, namely, the anisotropic minimum dissipation (AMD) model [6], and the dynamic k-equation model [7]. The AMD model, which has been further developed for WA-LES in [4], seeks to approximate the least eddy-viscosity and eddy-diffusivity magnitudes necessary to prevent the formation of turbulent structures smaller than the mesh size, while imposing the balance of production and dissipation of SGS energy at the local grid scale. Differently, the localized dynamic kinetic-energy (LDK) model, which has been recently extended for compressible WA-LES [5], involves the explicit solution of the transport equation for the SGS turbulent kinetic-energy. The various residual terms appearing in the wavelet-filtered governing equations are individually modeled, with the model coefficients being determined by means of either Germano-like or Bardina-like dynamic procedures. This way, differently from other SGS modeling approaches involving the use of a priori prescribed model coefficients, this approach leads to a self-closed modeling procedure. However, it is worth noting the higher computational cost of the LDK model compared to the AMD model, both in memory and complexity. The model equations, which are not reported here for brevity, can be found in Refs. [4,5].

Since the AWC method employs the same thresholding level (say ϵ) that is used for deriving the filtered governing equations, this user-defined parameter actually controls both the turbulence resolution and the numerical accuracy of the WA-LES solution. For statistically steady turbulent flows, this constraint is partially relaxed by adopting the split adaptation formulation, where two different thresholds for the resolved time-averaged and fluctuating components are employed [8]. Practically, in addition to the instantaneous balanced variables, the computational grid also adapts to the corresponding time-averages, but with a much lower threshold (say ϵ'). This allows to enhance the efficiency of the simulation, given the available computational resources.

3 Channel Flow Simulation

To demonstrate the WA-LES method for compressible flows, the simulations of isothermal-wall turbulent channel flow at various Mach and Reynolds numbers, matching the conditions of the pioneering direct numerical simulations (DNS) [3], are carried out. The results reported in this paper correspond to the flow case at $\mathrm{Ma_{bulk}} = 3$ and $\mathrm{Re_{bulk}} = 4880$, where WA-LES is performed with the WTF levels of $\epsilon' = 0.01$ and $\epsilon = 0.05$, for the mean and fluctuation components, respectively. Four adaptive collocation grids are employed in the spatial domain $4\pi\delta \times 2\delta \times 4\pi\delta/3$, with the finest resolution corresponding to a non-adaptive $384 \times 513 \times 384$ mesh.

Figures 1 and 2 show the adapted grids at two different time instants, say t_1 and $t_1 + \delta/U_{\mathrm{bulk}}$, with X and Y representing the streamwise and wall-normal directions, respectively. The instantaneous grids, colored by the thermal field (normalized by the wall temperature), are depicted for five equispaced cross-sections of the channel. Apparently, the evolution of dominant flow structures is predicted by means of dynamic grid adaptation in both space and time, as is confirmed by looking at the corresponding streamwise vorticity contours that are reported. It is worth noting that the finest spatial mesh, which is occasionally used in the wall region, practically corresponds to DNS-like resolution [3], and the present one can be referred to as wall-resolved WA-LES.

Table 1. Grid compression and mean flow results for channel flow at $\mathrm{Ma_{bulk}} = 3$.

Case	Compression	Re_τ	Ma_τ	$-B_q$	$\widetilde{T}_c/T_{\mathrm{wall}}$
WA-LES (AMD)	95%	458	0.107	0.13	2.69
WA-LES (LDK)	96%	442	0.113	0.13	2.37
LES [9]	0%	459	0.112	0.14	2.67
DNS [3]	–	451	0.116	0.137	2.49

The average grid compression that is achieved in this case is reported in Table 1, along with some mean flow results, which are the friction Reynolds (Re_τ) and Mach (Ma_τ) numbers, the heat flux coefficient ($-B_q$), and the normalized temperature at the midplane ($\widetilde{T}_c/T_{\mathrm{wall}}$). Due to the AWC method, less

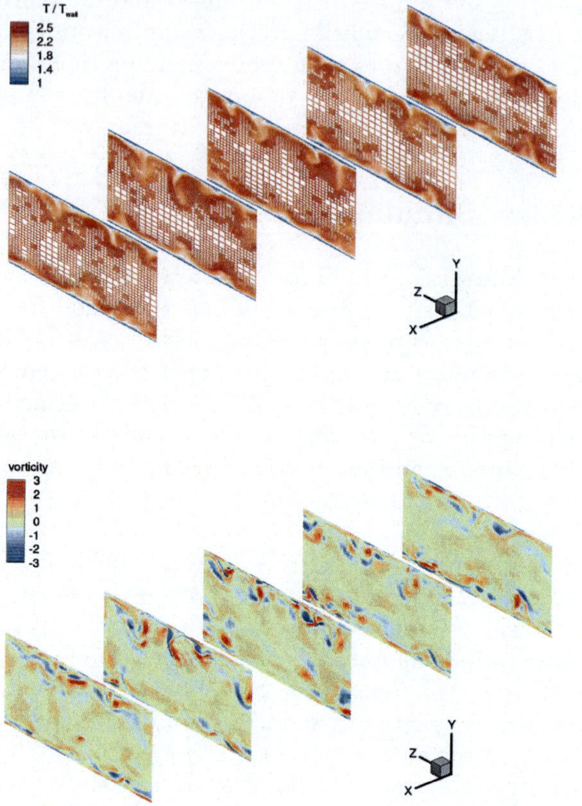

Fig. 1. Adapted mesh colored by temperature (top) and streamwise vorticity contours (bottom) at a given time instant, say t_1.

than 6% of the potentially available grid points are actually employed by the various calculations, confirming the good performance already obtained in the past for incompressible WA-LES [1]. The results of the statistical analysis, in terms of both mean flow variables and turbulent fluctuations, are found to be fully consistent with classical non-adaptive LES [9] and reference DNS [3]. For instance, in Fig. 3, the normalized mean temperature and the corresponding fluctuation root-mean-square are reported against the normalized wall distance $|Y|/\delta$. The good predictions of the present simulations, supplied with either LDK or AMD model, demonstrate that WA-LES represents a viable alternative to the traditional LES approach for wall-bounded attached compressible turbulent flows.

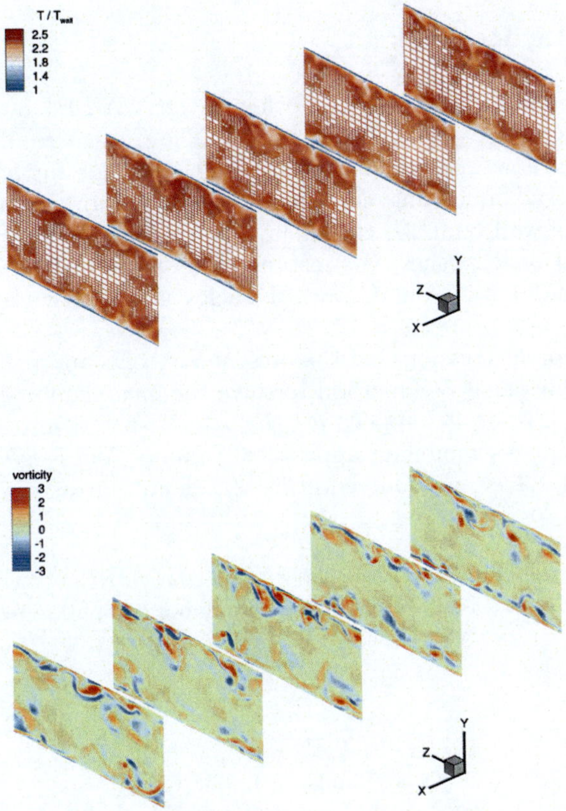

Fig. 2. Same as previous figure, at the time instant $t_1 + \delta/U_{bulk}$.

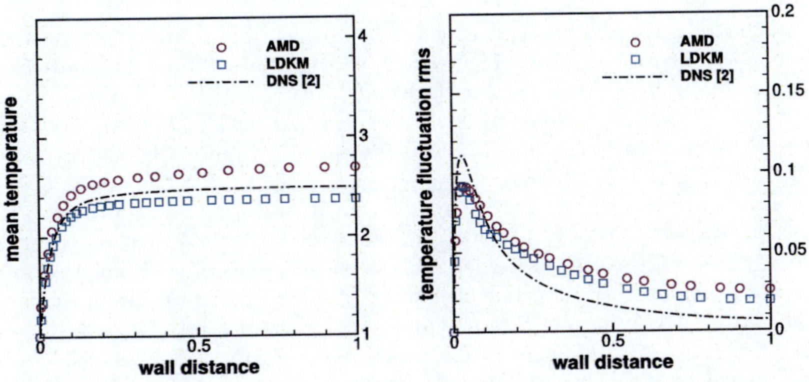

Fig. 3. Normalized mean temperature (left) and corresponding fluctuation rms (right).

4 Concluding Remarks

Originally developed for incompressible flows, the WA-LES method has been recently extended to the compressible regime. The present results obtained for turbulent channel flow are very encouraging and call for further development of the wavelet-based modelling and simulation framework, where fully adaptive simulations of wall-bounded turbulent flows can be effectively performed on optimal computational meshes. For instance, the numerical prediction of compressible turbulent flows around solid obstacles is an active area of ongoing research [10].

Finally, it is worth stressing that the present WA-LES approach is completely different from using the AWC method to solve the mean compressible flow governing equations [11], as it happens for the wavelet-based unsteady Reynolds-averaged Navier-Stokes modeling approach. Actually, WTF is employed there by the numerical solver, without entering the definition of the balanced flow variables [12].

Acknowledgements. The authors acknowledge the CINECA award under the ISCRA initiative for the availability of high performance computing resources (Project HP10CTESFT).

References

1. De Stefano, G., Vasilyev, O.V.: Fluids **6**(2), 83 (2021)
2. Brown-Dymkoski, E., Vasilyev, O.V.: J. Comput. Phys. **333**, 414–426 (2017)
3. Coleman, G.N., Kim, J., Moser, R.D.: J. Fluid Mech. **305**, 159–183 (1995)
4. De Stefano, G., Brown-Dymkoski, E., Vasilyev, O.V.: J. Fluid Mech. **901**, A13 (2020)
5. De Stefano, G., Dymkoski, E., Vasilyev, O.V.: Phys. Rev. Fluids **7**, 054604 (2022)
6. Rozema, W., Bae, H.J., Moin, P., Verstappen, R.: Phys. Fluids **27**, 085107 (2005)
7. Chai, X., Mahesh, K.: J. Fluid Mech. **699**, 385–413 (2012)
8. Ge, X., De Stefano, G., Hussaini, M.Y., Vasilyev, O.V.: Fluids **6**(9), 331 (2021)
9. Brun, C., Petrovan Boiarciuc, M., Haberkorn, M., Comte, P.: Theor. Comput. Fluid Dyn. **22**, 189–212 (2008)
10. Kasimov, N., Dymkoski, E., De Stefano, G., Vasilyev, O.V.: Fluids **6**(8), 293 (2021)
11. Ge, X., Vasilyev, O.V., De Stefano, G., Hussaini, M.Y.: Wavelet-based adaptive unsteady Reynolds-averaged Navier-Stokes computations of wall-bounded internal and external compressible turbulent flows. AIAA Paper 2018-0545. In: Proceedings of the 56th AIAA Aerospace Sciences Meeting, Kissimmee, FL, USA, 8–12 January 2018 (2018). https://doi.org/10.2514/6.2018-0545, ISBN: 978-162410524-1
12. Ge, X., Vasilyev, O.V., De Stefano, G., Hussaini, M.Y.: AIAA J. **58**(4), 1529–1549 (2020)

High-Reynolds Compressible Flows Simulation with Wall-Modeled LES and Immersed Boundary Method

F. De Vanna[1](✉), G. Baldan[2], F. Picano[1], and E. Benini[1]

[1] Department of Industrial Engineering, University of Padova, Via Venezia 1, 35131 Padua, Italy
francesco.devanna@unipd.it

[2] Department of Aerospace Science and Technology, Politecnico di Milano, Via La Masa 34, 20156 Milan, Italy
giacomo.baldan@polimi.it

1 Introduction

High-Reynolds compressible flows over complex geometries represent a real challenge for contemporary engineering. Fully resolved Direct Numerical Simulations (DNS) are still not feasible in realistic contexts; therefore, accurate turbulence models are required. In this scenario, Large Eddy Simulation (LES) represents a superior approach compared to more standard Reynolds Averaged Navier-Stokes (RANS) strategies since the dynamics of the energy-dominant and flow-dependent large eddies are directly resolved on the computational mesh rather than being modelled [1]. However, even if Wall-Resolved LES (WRLES) can be highly accurate in predicting such flows when applied to high-Reynolds conditions, they are still too computationally demanding due to the resolution requirements imposed by the presence of the solid boundary. Thus, the technique does not fit the industrial needs and, for this reason, Wall-Modelled LES (WMLES) have taken charge during the last decades [2]. The approach aims to resolve the most significant flow portions with standard LES techniques, while a wall-stress model considers the near-wall regions. Compared to other hybrids/zonal numerical discretization of the Navier-Stokes system of equations, the WMLES approach has proven to be more accurate in capturing near-wall dynamics, which is critical for enhanced predictions of unsteady problems at any applicative level in engineering.

In addition, treating complex geometries in a computationally efficient way is not trivial, and one of the most promising strategies in this path is the Immersed Boundary Method (IBM). The strategy allows the body surface to cut the computational cells so that a Cartesian mesh can be employed independently of the geometry complexity. Thus, a highly efficient and massive parallel Cartesian solver can be used for aerodynamics applications. The problem with IBM lies in resolving the near-wall regions since the mesh does not fit the body locally. For this reason, the combination of WMLES, which aims at placing the first off-the-wall point as far as possible, with an IB strategy looks promising, combining the possibility of dealing with high Reynolds number flows in a framework that is efficient from a computational point.

C. Marchioli et al. (Eds.): DLES 2023, ERCO 31, pp. 203–208, 2024.
https://doi.org/10.1007/978-3-031-47028-8_31

The present work proposes an innovative technique that unifies the IBM and the WMLES approaches according to the procedure proposed in [3–5]. The method is validated in high-Reynolds canonical flows consisting of turbulent channel and pipe flows whose walls are modeled using immersed bodies. Thus, the proposed methodology represents a real turning point for the aerodynamic characterization of complex geometries in application fields with high Mach and Reynolds numbers.

The present document is organized as follows: Sect. 2 provides the essential information concerning the numerical model. Section 3 presents the results of the analyses, while Sect. 4 states the conclusions and the final remarks.

2 Governing Equation and Numerical Methods

The present study is carried out with URANOS (Unsteady Robust All-around Navier-Stokes Solver), a fully compressible Navier-Stokes solver developed at the Industrial Engineering Department of the University of Padova [6–9]. The solver deals with the filtered Navier-Stokes system of equations in a conservative formulation which introducing both the Reynolds ($\phi = \bar{\phi} + \phi'$) and Favre ($\phi = \tilde{\phi} + \phi'', \tilde{\phi} = \overline{\rho\phi}/\bar{\rho}$) decompositions, reads as:

$$\frac{\partial \bar{\rho}}{\partial t} = -\frac{\partial \bar{\rho}\tilde{u}_j}{\partial x_j} \tag{1a}$$

$$\frac{\partial \bar{\rho}\tilde{u}_i}{\partial t} + \frac{\partial \bar{\rho}\tilde{u}_i\tilde{u}_j}{\partial x_j} = -\frac{\partial \bar{p}_i\delta_{ij}}{\partial x_j} + \frac{\partial \bar{\tau}_{ij}}{\partial x_j} - \frac{\partial T_{ij}^{SGS}}{\partial x_j} \tag{1b}$$

$$\frac{\partial \bar{\rho}\tilde{E}}{\partial t} + \frac{\partial \bar{\rho}\tilde{u}_j\tilde{E}}{\partial x_j} = -\frac{\partial \bar{p}\tilde{u}_j}{\partial x_j} + \frac{\partial \tilde{u}_j\bar{\tau}_{ij}}{\partial x_j} - \frac{\partial \bar{\mathcal{J}}_j}{\partial x_j} - \frac{\partial E_j^{SGS}}{\partial x_j} \tag{1c}$$

The SubGrid-Scale (SGS) terms are modeled via the canonical Boussinesq's hypothesis evaluating the turbulent viscosity, μ_{SGS} through the Wall-Adaptive Large-Eddy (WALE) viscosity model [10].

As far as the wall model, the one proposed by [2] is employed so that the wall shear-stress value, $\tau_{w,wm}$, and the wall heat flux, $q_{w,wm}$, are fed as boundary conditions for the external flow according to the procedure proposed by [4]. Concerning IBM, the one proposed by [3] is used.

3 Results

To validate the present methodology, a turbulent channel flow configuration and a fully developed pipe flow system are addressed as leading benchmarks. Thus, Figs. 1a and 1b show some snapshots of the instantaneous velocity field inside the channel and the pipe, respectively. From such qualitative representations, it can be observed how immersed blocks model wall surfaces. Simulation parameters are provided in Table 1, while, in the following, a detailed description of the models is provided.

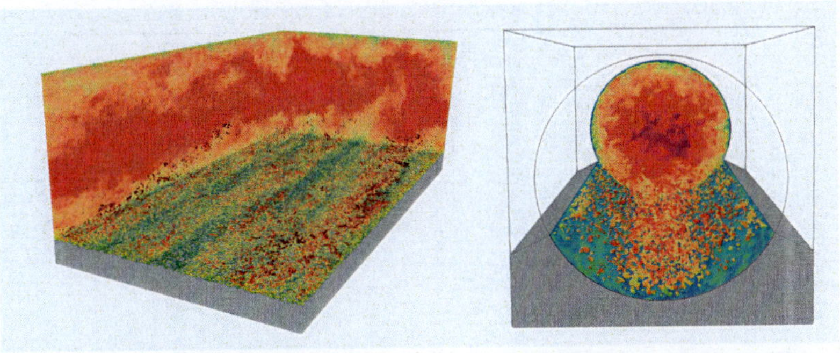

(a) Channel flow at $Re_\tau = 10k$ (b) Pipe flow at $Re_\tau = 6k$

Fig. 1. Representation of the domain of the turbulent channel flow, with the solid IBM entities clearly visible at the bottom. Top immersed block is not represent.

Table 1. Channel and pipe flow simulations parameters.

Case	$Re_{\tau,0}$	Re_τ	$N_x \times N_y \times N_z$	$\Delta x^+ \times \Delta y^+ \times \Delta z^+$
tch1	590	596	$48 \times 78 \times 24$	$77.2 \times 7.6 \times 77.2$
tch2	1000	1010	$80 \times 78 \times 40$	$78.5 \times 16.7 \times 78.5$
tch3	2000	1942	$160 \times 156 \times 80$	$78.5 \times 16.7 \times 78.5$
tch4	5200	5376	$160 \times 156 \times 80$	$204.2 \times 43.3 \times 204.2$
tch5	10000	9623	$320 \times 312 \times 160$	$196.3 \times 41.7 \times 196.3$
tch6	20000	19244	$640 \times 624 \times 320$	$196.3 \times 41.7 \times 196.3$
Case	$Re_{\tau,0}$	Re_τ	$N_x \times N_y \times N_z$	$\Delta r^+ \times \Delta \theta^+ \times \Delta z^+$
tpi1	500	551	$96 \times 96 \times 50$	$13.8 \times 13.8 \times 69.2$
tpi2	1140	1217	$128 \times 128 \times 96$	$22.7 \times 22.7 \times 79.2$
tpi3	3000	3183	$128 \times 128 \times 96$	$59.7 \times 59.7 \times 208.3$
tpi4	6000	6398	$150 \times 150 \times 96$	$102.4 \times 102.4 \times 418.8$

3.1 Turbulent Channel with Immersed Walls

Starting from the channel flow simulation, the latter is carried out by prescribing a bulk Mach number, $M_b = u_b/c_w$, of 0.1, while the friction Reynolds number, $Re_\tau = \rho_w u_\tau h/\mu_w$, is enforced up to $20k$. Here is not worthless to mention that such a Reynolds condition is still very far from any DNS or WRLES approach. The computation is carried out in a tridimensional box with size $L_x \times L_y \times L_z = 2\pi h \times 2h \times \pi h$ along the x, y and z coordinates, respectively, h being the channel half-height. A uniform mesh spacing is applied in all three Cartesian directions.

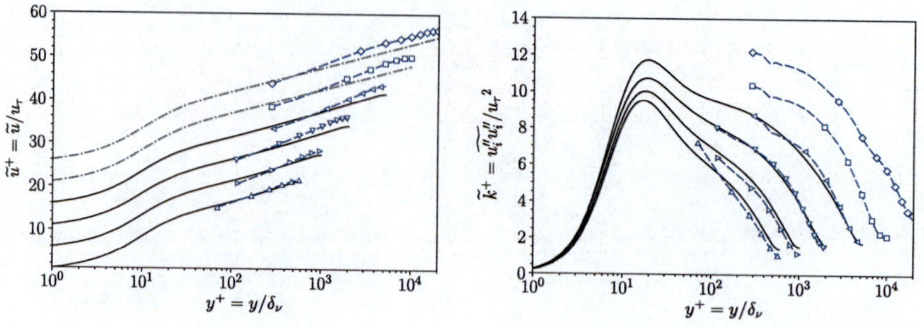

(a) Velocity profile

(b) Turbulent kinetic energy

Fig. 2. Mean streamwise velocity profiles, $u^+ = \tilde{u}/u_\tau$, and inner-scaled turbulent kinetic energy, $\tilde{k}^+ = \widetilde{u_i''u_i''}/u_\tau^2$, as a function of the inner scaled wall distance, $y^+ = y/\delta_\nu$, for channel flow cases at $M_b = 0.1$ and Re_τ up to 20k. Present results (blu dots) are compared with DNS data (solid black lines) by [11–13] and analytical reference according to the Reichardt's law [14].

Figure 2 reports the results of the computations. In particular, Fig. 2a show the mean scaled velocity profiles, $u^+ = \tilde{u}/u_\tau$, while Fig. 2b reports the inner-scaled turbulent kinetic energy, $\tilde{k}^+ = \widetilde{u_i''u_i''}/u_\tau^2$, as a function of the internal wall distance, $y^+ = y/\delta_\nu$. For greater clarity, velocity profiles are offset in the vertical direction by five wall unit steps. Overall, the results agree well with DNS references by [11–13]. Data are also provided for a nominal Re_τ of 10k and 20k. The latter are compared with Reichardt's law [14] since no DNS/experimental data are available for such Reynolds conditions. The trend is confirmed even for such extreme configurations.

3.2 Turbulent Pipe Inside an Immersed Block

The pipe flow configuration allows us to validate the current approach in a non-conformal geometrical arrangement. Thus, the computational domain is made up of a rectangular box of dimensions $2.4R \times 2.4R \times 2\pi R$ being R the pipe radius and is discretized with a uniform grid in all directions according to the description reported in Table 1. The system is forced flowing along the z-direction setting a bulk Mach number, $M_b = w_b/c_w$, is still enforced to 0.1. Concerning the boundary conditions, some distinctions must be made compared to the channel flow setup. In particular, the IB block is modeled as a no-slip and isothermal boundary condition enforcing the wall-shear stress and heat flux according to the WMLES procedure. Periodicity is enforced along with the z-direction, while no particular boundaries are associated with the edge of the computational domain since they do not affect the flow.

Figure 3 reports the results of the computations in terms of mean scaled velocity profiles, $w^+ = \tilde{w}/w_\tau$, (Fig. 3a) and turbulent kinetic energy, $\tilde{k}^+ = \widetilde{u_i''u_i''}/w_\tau^2$ (Fig. 3a). For greater clarity, velocity profiles are still offset in the vertical direction by five wall unit steps. Overall, the results agree well with DNS references by [15].

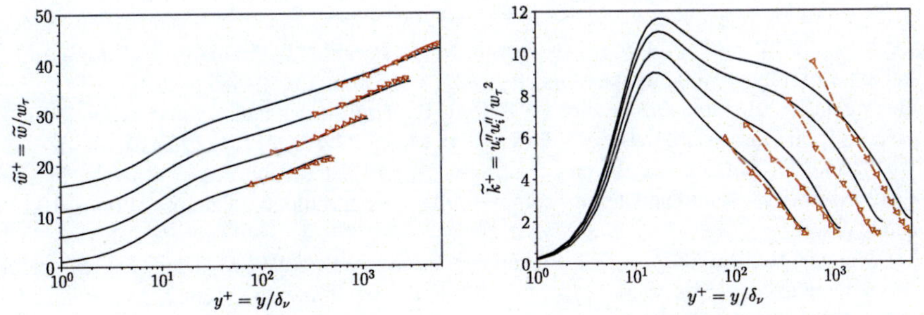

(a) Velocity profile (b) Turbulent kinetic energy

Fig. 3. Mean streamwise velocity profiles, $w^+ = \tilde{w}/w_\tau$, and inner-scaled turbulent kinetic energy, $\tilde{k}^+ = \widetilde{u_i''u_i''}/w_\tau^2$, as a function of the inner scaled wall distance, $y^+ = y/\delta_\nu$, for pipe flow cases at $M_b = 0.1$ and Re_τ up to 6k. Present results (red dots) are compared with DNS data (solid black lines) by [15].

4 Conclusions

The present work describes the numerical results obtained by coupling a sharp interface immersed boundary method with an innovative wall-modeled LES approach. The method is designed to address the simulation of complex geometries in high-Reynolds/high-Mach number conditions and is implemented within the URANOS solver framework, a massive-parallel CFD solver for compressible flow simulations. The approach is tested in canonical turbulent channel and pipe configurations.

The two tests see the solid walls modeled by immersed blocks and can enforce the correct shear stress thanks to the embedded WMLES reconstruction. Comparison with the highest-Reynolds DNS data available in the literature and analytical reference from wall-turbulence theory confirm the quality of the proposed methodology in a wide range. Future studies aim to investigate more complex flow configurations by characterizing complex geometries with higher compressibility levels.

References

1. Bose, S.T., Park, G.I.: Wall-modeled large-eddy simulation for complex turbulent flows. Annu. Rev. Fluid Mech. **50**(1), 535–561 (2018)
2. Kawai, S., Larsson, J.: Wall-modeling in large eddy simulation: length scales, grid resolution, and accuracy. Phys. Fluids **24**(1) (2012)
3. De Vanna, F., Picano, F., Benini, E.: A sharp-interface immersed boundary method for moving objects in compressible viscous flows. Comput. Fluids **201**, 104415 (2020)
4. De Vanna, F., Cogo, M., Bernardini, M., Picano, F., Benini, E.: Unified wall-resolved and wall-modeled method for large-eddy simulations of compressible wall-bounded flows. Phys. Rev. Fluids **6**(3), 034614 (2021)
5. De Vanna, F., Cogo, M., Bernardini, M., Picano, F., Benini, E.: A straightforward strategy to unify WR/WMLES approaches for compressible wall-bounded flows, p. 0181 (2022)

6. De Vanna, F., Picano, F., Benini, E., Quinn, M.K.: Large-eddy simulations of the unsteady behavior of a hypersonic intake at mach 5. AIAA J. **59**(10), 3859–3872 (2021)
7. De Vanna, F., Benato, A., Picano, F., Benini, E.: High-order conservative formulation of viscous terms for variable viscosity flows. Acta Mech. **232**(6), 2115–2133 (2021)
8. De Vanna, F., Baldan, G., Picano, F., Benini, E.: Effect of convective schemes in wall-resolved and wall-modeled les of compressible wall turbulence. Comput. Fluids 105710 (2022)
9. De Vanna, F., Bernardini, M., Picano, F., Benini, E.: Wall-modeled les of shock-wave/boundary layer interaction. Int. J. Heat Fluid Flow **98**, 109071 (2022)
10. Nicoud, F., Ducros, F.: Subgrid-scale stress modelling based on the square of the velocity gradient tensor. Flow Turbul. Combust. **62**(3), 183–200 (1999)
11. Vreman, A., Kuerten, J.G.: Statistics of spatial derivatives of velocity and pressure in turbulent channel flow. Phys. Fluids **26**(8), 085103 (2014)
12. Bernardini, M., Pirozzoli, S., Orlandi, P.: Velocity statistics in turbulent channel flow up to $Re_\tau = 4000$. J. Fluid Mech. **742**, 171–191 (2014)
13. Lee, M., Moser, R.D.: Direct numerical simulation of turbulent channel flow up to $Re_\tau = 5200$. J. Fluid Mech. (2015)
14. Reichardt, H.: Vollständige Darstellung der turbulenten Geschwindigkeitsverteilung in glatten Leitungen. J. Appl. Math. Mech. (1951)
15. Pirozzoli, S., Romero, J., Fatica, M., Verzicco, R., Orlandi, P.: One-point statistics for turbulent pipe flow up to. J. Fluid Mech. **926** (2021)

Stochastic Data-Driven POD-Based Modeling for High-Fidelity Coarsening of Two-Dimensional Rayleigh-Bénard Turbulence

S. R. Ephrati[1(✉)], P. Cifani[1,2], and B.J. Geurts[1,3]

[1] University of Twente, P.O. Box 217, 7500 AE Enschede, The Netherlands
{s.r.ephrati,p.cifani,b.j.geurts}@utwente.nl
[2] Gran Sasso Science Institute, Viale F. Crispi, 7 67100 L'Aquila, Italy
[3] Eindhoven University of Technology, P.O. Box 213, 5600 MB Eindhoven, The Netherlands

Abstract. A stochastic data-driven based model is proposed to augment coarse-grid numerical simulations of two-dimensional Rayleigh-Bénard convection. Proper orthogonal decomposition (POD) is used to decompose fine-grid reference data into spatial fields and temporal coefficients. The latter are modeled as stochastic processes, yielding a model with few tunable parameters. A relatively small number of POD modes is found to be sufficient to obtain accurate Nusselt number estimates at Rayleigh number 10^{10}.

1 Introduction

The aim of LES is that of performing accurate numerical simulations at a moderate computational cost. The challenge is to devise a significant coarsening of the model while retaining an adequate level of reliability. In recent years, due to ever-growing computational resources, it has become possible to perform high-resolution direct numerical simulations (DNS) from which data are used to build coarse-grid fluid models. The latter allows for high-fidelity simulation of the dynamics of large-scale flow structures over long time horizons, otherwise unreachable by DNS. In particular, data is used to find LES closure models [3] or to alter governing equations [2]. In addition, observational data can be used as an alternative or to complement a DNS dataset. Data assimilation, in fact, has been used for several decades in the context of geophysical fluid dynamics [1,7] and is of particular interest to this work.
Here we propose a data-driven approach to augment the coarse-grained numerical predictions of two-dimensional Rayleigh-Bénard convection. Proper orthogonal decomposition (POD) is used to decompose fine-grid reference data into spatial fields and temporal coefficients. These coefficients are modelled as stochastic processes whose variance equals the POD spectrum of the flow. We found that a comparably small number of POD modes is sufficient to attain an accurate estimation of the Nusselt number even at high Rayleigh number, thereby enabling studies in the range of 'ultimate turbulence' [8,9]. This strong coarsening makes the model efficient and accurate over long time.

C. Marchioli et al. (Eds.): DLES 2023, ERCO 31, pp. 209–214, 2024.
https://doi.org/10.1007/978-3-031-47028-8_32

2 Data Generation and Processing

Rayleigh-Bénard convection is described by the Navier-Stokes equations coupled to temperature transport under the Boussinesq approximation. The governing equations, in non-dimensional form, are the following:

$$\frac{\partial u}{\partial t} + u \cdot \nabla u = \sqrt{\frac{Pr}{Ra}} \nabla^2 u - \nabla p + T e_z, \tag{1}$$

$$\nabla \cdot u = 0, \tag{2}$$

$$\frac{\partial T}{\partial t} + u \cdot \nabla T = \frac{1}{\sqrt{PrRa}} \nabla^2 T. \tag{3}$$

Here, we restrict to two spatial dimensions. We adopt u, p and T to denote velocity, pressure and temperature, respectively. Buoyancy effects due to temperature gradients are included by means of the term Te_z, where e_z is the vertical unit vector. The dimensionless parameters that determine the flow are the Rayleigh number $Ra = g\beta\Delta L_y^3/(\nu\kappa)$ and the Prandtl number $Pr = \nu/\kappa$, set to 10^{10} and 1, respectively. Here g is the gravitational acceleration, β the thermal expansion coefficient, Δ the temperature difference between the boundaries, ν the kinematic viscosity and κ the thermal diffusivity. The computational domain is a rectangle of size $L_x = 2$, $L_y = 1$. Periodic boundary conditions are imposed in the horizontal direction while the top and bottom boundary are no-slip walls. The temperature is prescribed at 1 at the top wall and 0 at the bottom wall.

The time integration is carried out by employing an RK3 method. An energy-conserving finite difference method is employed for the spatial discretization [10] and parallelized as in [5]. A uniform grid spacing is adopted along the periodic direction while grid refinement close to the wall is realized by means of a hyperbolic tangent profile to properly resolve the boundary layer.

The DNS simulation is performed over a 4096×2048 grid. This resolution has been shown to be adequate for the selected Ra number, as indicated in [11]. Coarse-grid runs are carried out on a 64×32 grid. The rather strong coarsening allows to identify clear differences between the proposed model and the solution of system (1-3) solved on the fine grid without model, as shown in Fig. 1.

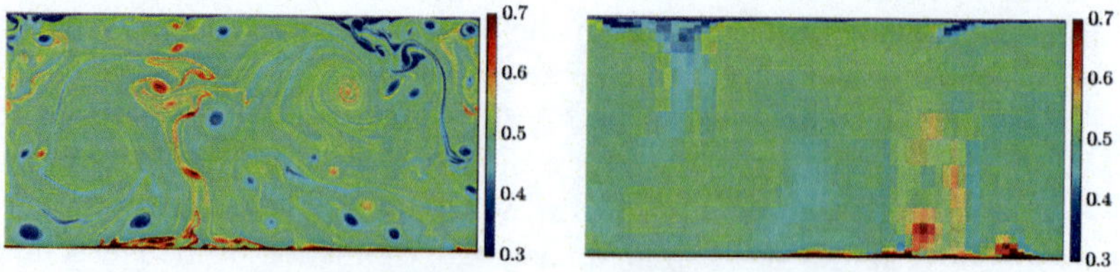

Fig. 1. Snapshots of the temperature field of the DNS (left) and the numerical simulation on the coarse grid using 100 POD modes (right).

Proper orthogonal decomposition (POD) is applied to a sequence of equispaced and uncorrelated DNS snapshots. The spatial and temporal coefficients, indicated by $\sigma_i(x)$ and $a_i(t)$, respectively, follow from the minimization problem

$$\min \int_\Omega \int_T \left(f(x,t) - \sum_{i=1}^{N} a_i(t)\sigma_i(x) \right)^2 \mathrm{d}t\mathrm{d}\Omega, \tag{4}$$

where $f(x,t)$ is either the velocity or the temperature field. The fraction of energy related to each of the modes is given by the variance of the corresponding temporal coefficient. The solution of (4) is given by the eigenvectors of the spatial-spatial correlation tensor:

$$\mu_i a_i(t) = \int_T \langle f(x,t) f(x,t') \rangle_\Omega a_i(t')\mathrm{d}t', \tag{5}$$

where $\langle\cdot\rangle_\Omega$ denotes integration over the spatial domain and μ_i is the standard deviation of the time series $a_i(t)$. The spatial profiles $\sigma_i(x)$ then follow from orthogonality:

$$\sigma_i(x) = \frac{\langle f(x,t) a_i(t) \rangle_T}{\langle a_i(t)^2 \rangle_T}, \tag{6}$$

where $\langle\cdot\rangle_T$ indicates integration over time. For an in-depth description of this procedure the reader is referred to [4].

3 Stochastic Modelling

3.1 Model Description

The momentum and temperature Eqs. (1, 3) can be written as a system of ODEs for the temporal coefficients by projection onto the POD basis. For each POD mode one has

$$\frac{\mathrm{d}a_{i,f}}{\mathrm{d}t} = L_f(\{a_{j,f}(t)\}), \quad i = 0, \ldots, N, \tag{7}$$

where L_f is formally the spectral representation of $\mathbf{u} \cdot \nabla$, ∇, ∇^2 and the source term appearing in (1–3). This system of ODEs involves all coefficients. In actual models we truncate this formulation to include only a certain number N of modes.
Representing the fine-scale dynamics in the coarsened model using the POD time series has been shown to be an efficient approach to improve coarse-grid simulations [6]. In the same spirit, here we regard the evolution of the coefficients as a stochastic process. First, Eq. (7) is integrated in physical space to find an intermediate value $\tilde{a}_{i,f}$. Subsequently, after projection, the temporal coefficients at the new time level $n+1$ are updated by

$$a_{i,f}^{n+1} = \tilde{a}_{i,f}^{n+1}\left(1 - \frac{\Delta t}{\tau_i}\right) + \mu_i \sqrt{1 - \left(1 - \frac{\Delta t}{\tau_i}\right)^2}\, r_i^n \tag{8}$$

where μ_i and τ_i are the standard deviation and time correlation of the temporal coefficients computed from the DNS dataset. Moreover, r^n is a random variable drawn from a standard normal distribution. The formulation (8) is a first-order autoregressive model with drift coefficient $1 - \Delta t / \tau_i$ and variance μ_i^2. Once $a_{i,f}^{n+1}$ is computed from (8), the full spatial field $f(x,t)$ is reconstructed in physical space and the time step is completed. The projection and reconstruction of the field from/to physical space to/from spectral space is a computationally intensive operation. However, as we will show in the next section, only a few POD modes are necessary to significantly improve the predictions of the flow statistics compared to adopting no model at all. Thus, the overall cost of the model is comparable with that of solving the coarse-grained PDE without the model.

3.2 Nusselt Number Predictions

First we verify that the simulated coarse-grid POD spectrum is well represented when using (8). Preliminary tests have shown that the temperature spectrum simulated without model matches the DNS spectrum well, in particular at large scales. However, the velocity spectrum showed considerable differences at all scales. Given these findings, we apply the model to the momentum equation only. In Fig. 2 and 3 the POD spectrum of both the velocity and the temperature is shown when using the primary 20 and 100 modes in the forcing. The velocity spectra match the reference spectrum up to the forced mode. Furthermore, an overall improvement is observed in the temperature spectra compared to the no-model case.

Next we predict the Nusselt number and compare it with its reference value obtained from DNS. We adopt the following definition

$$Nu = 1 + \sqrt{PrRa}\langle u_y T \rangle_\Omega \tag{9}$$

where u_y denotes the vertical velocity and $\langle\ ,\rangle_\Omega$ denotes the volume average. This definition is particularly suitable for the computation of the Nusselt number since it does not involve gradients, which would introduce an additional source of error when discretized on a coarse grid. Figure 4 shows the time averages of the Nusselt number as measured on the reference fine grid, on the coarse grid without model and on the coarse grid with the stochastic model. Increasing the number of POD modes clearly results

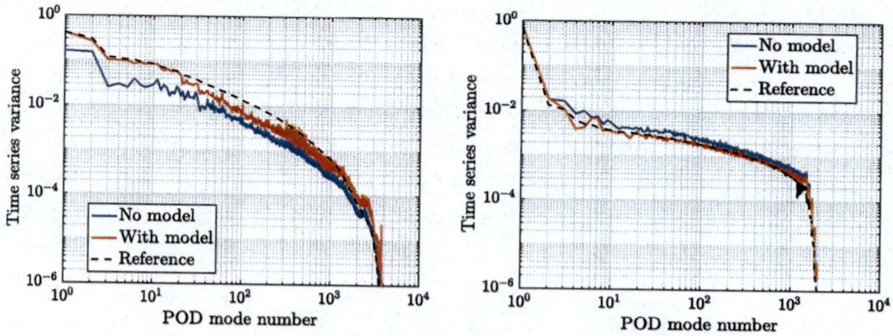

Fig. 2. POD spectra of the velocity (left) and temperature (right). The spectra of the reference solution (dashed line), the no-model (blue line) and the stochastic model using 20 POD modes (red line) are shown (Color figure online).

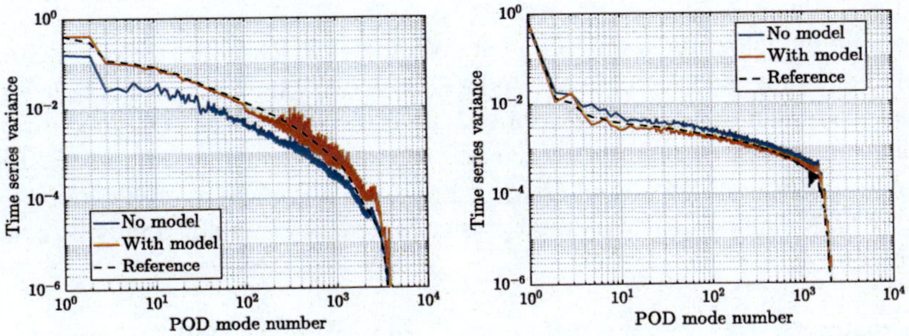

Fig. 3. POD spectra of the velocity (left) and temperature (right). The spectra of the reference solution (dashed line), the no-model (blue line) and the stochastic model using 100 POD modes (red line) are shown (Color figure online).

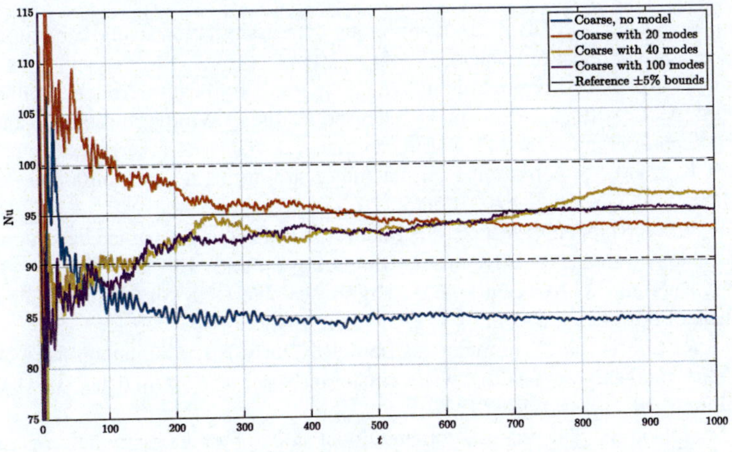

Fig. 4. Rolling time averages of the Nusselt number obtained from the no-model and the stochastic model using 20, 40 and 100 POD modes. The reference DNS result is represented by the solid line together with a 5% uncertainty interval.

in a more accurate estimation of Nu. Even at the lowest number of POD modes simulated, a significant improvement is clearly visible. This suggests that a correction on the large-scale spectrum of velocity is beneficial for the correct prediction of mean heat transfer.

4 Conclusions

In this work, a stochastic data-driven model for coarse-grained numerical simulations of two-dimensional Rayleigh-Bénard turbulence has been proposed. The model is based on the proper orthogonal decomposition (POD) of high-resolution numerical data. In particular, parameters are extracted from the temporal coefficients of the POD basis and are used to augment the coarse-grained numerical predictions via stochastic forcing.

The resulting model requires only two parameters per POD mode, making it straightforward to interpret and to implement. In fact, these two parameters, i.e., (τ_i, μ_i) for the time correlation and the standard deviation, can be computed per POD mode from the DNS data and are not free parameters. It was found that the Nusselt number estimates on the coarse grid with stochastic forcing improved significantly compared to those obtained without a model. The method presented here is general and may be extended to different flows.

Acknowledgments. This work was performed in the context of the SPRESTO project, funded by the Dutch Science Foundation (NWO) in their TOP1 program.

References

1. Azouani, A., Olson, E., Titi, E.S.: Continuous data assimilation using general interpolant observables. J. Nonlinear Sci. **24**(2), 277–304 (2014)
2. Bai, Z.: Data-driven methods in fluid dynamics: sparse classification from experimental data. In: Pollard, A., Castillo, L., Danaila, L., Glauser, M. (eds.) Whither turbulence and big data in the 21st century?, pp. 323–342. Springer, Cham (2017)
3. Beck, A., Kurz, M.: A perspective on machine learning methods in turbulence modeling. GAMM-Mitteilungen **44**(1), e202100002 (2021)
4. Cazemier, W., Verstappen, R.W.C.P., Veldman, A.E.P.: Proper orthogonal decomposition and low-dimensional models for driven cavity flows. Phys. Fluids **10**(7), 1685–1699 (1998)
5. Cifani, P., Kuerten, J.G.M., Geurts, B.J.: Highly scalable DNS solver for turbulent bubble-laden channel flow. Comput. Fluids **172**, 67–83 (2018)
6. Ephrati, S.R., Luesink, E., Wimmer, G., Cifani, P., Geurts, B.J.: Computational modeling for high-fidelity coarsening of shallow water equations based on subgrid data. SIAM Multiscale Model. Simul. **20**(4), 1468–1489 (2022)
7. Gesho, M., Olson, E., Titi, E.S.: A computational study of a data assimilation algorithm for the two-dimensional Navier-Stokes equations. Commun. Comput. Phys. **19**(4), 1094–1110 (2016)
8. Johnston, H., Doering, C.R.: Comparison of turbulent convection between conditions of constant temperature and constant flux. Phys. Rev. Lett. **102**(6), 064501 (2009)
9. Stevens, R.J., Verzicco, R., Lohse, D.: Radial boundary layer structure and Nusselt number in Rayleigh-Bénard convection. J. Fluid Mech. **643**, 495–507 (2010)
10. Vreman, A.W.: The projection method for the incompressible Navier-Stokes equations: the pressure near a no-slip wall. J. Comput. Phys. **263**, 353–374 (2014)
11. Zhu, X., Mathai, V., Stevens, R.J., Verzicco, R., Lohse, D.: Transition to the ultimate regime in two-dimensional Rayleigh-Bénard convection. Phys. Rev. Lett. **120**(14), 144502 (2018)

Boundary Conditions for Wall-Modelled Large-Eddy Simulation Using Spectral Element Discretization

T. Mukha(✉), G. Brethouwer, and P. Schlatter

Department of Engineering Mechanics, KTH Royal Institute of Technology, Stockholm, Sweden
{tmu,geert,pschlatt}@mech.kth.se

1 Introduction

Complementing large-eddy simulation (LES) with wall-modelling is, perhaps, the most straight-forward way to enable high-fidelity simulations at high Reynolds numbers. At the same time, high-order methods offer the benefits of high computational efficiency and potentially faster convergence with respect to mesh refinement even outside the asymptotic regime. Solvers based on the spectral-element method (SEM) in particular, are also well-suited for novel computing architectures such as graphics processing units, due to the high level of locality of the performed numerical operations [3]. This warrants the current interest in implementing wall-modelled LES (WMLES) in SEM solvers, such as Nek5000 [2,7].

This is also the topic of the current contribution. We consider the so-called 'wall-stress' class of wall models [4], in which the effect of inner-layer dynamics are modelled by providing the correct local value of the wall shear stress. Such wall-models can be split into three subcomponents.

1. Sampling the outer-layer solution to the wall model.
2. Determining the value of the wall shear stress, τ_w, based on the sampled quantities.
3. Applying the predicted stress as a boundary condition at the wall.

Here, the focus is on the third item, and we test two alternative boundary conditions for prescribing the wall stress, both implemented in Nek5000. The two approaches are discussed in detail in the next section. To make the comparison, simulations of turbulent channel flow at $Re_\tau = 8000$ are conducted. We use Spalding's law [8] to predict the values of the wall shear stress.

2 Wall Boundary Conditions

It is instructive to begin with a mathematical derivation of how the wall stress vector, τ_w, enters the governing equations. To that end, we consider only the viscous stress term in the Navier-Stokes equations. Let σ_{ij} be the viscous stress tensor, then

$$\sigma_{ij} = \frac{\partial}{\partial x_j}\left(2(\nu + \nu_{sgs})S_{ij}\right), \quad \frac{\partial \sigma_{ij}}{\partial x_j} = 0, \quad i = 1,2,3. \tag{1}$$

C. Marchioli et al. (Eds.): DLES 2023, ERCO 31, pp. 215–220, 2024.
https://doi.org/10.1007/978-3-031-47028-8_33

Here, S_{ij} is the rate-of-strain tensor, ν is the kinematic viscosity, and ν_{sgs} the subgrid scale viscosity. The corresponding weak form is

$$\int_{\Omega} v_i \frac{\partial \sigma_{ij}}{\partial x_j} d\xi = 0, \tag{2}$$

where v_i are test functions, and Ω the computational domain. Now, using the following identity,

$$\frac{\partial v_i \sigma_{ij}}{\partial x_j} = \frac{\partial v_i}{\partial x_j}\sigma_{ij} + v_i \frac{\partial \sigma_{ij}}{\partial x_j}, \tag{3}$$

we obtain

$$\int_{\Omega} \frac{\partial v_i}{\partial x_j}\sigma_{ij} d\xi = \int_{\Omega} \frac{\partial v_i \sigma_{ij}}{\partial x_j} d\xi. \tag{4}$$

Applying the Gauss-Ostrogradsky theorem to the right-hand side we get

$$\int_{\Omega} \frac{\partial v_i}{\partial x_j}\sigma_{ij} d\xi = \int_{\partial\Omega} v_i \sigma_{ij} n_j d\xi. \tag{5}$$

Now let $\tau_i = \sigma_{ij} n_j$ and thus $\tau_{w_i} = \tau_i - (\tau_j n_j) n_i$, leading to

$$\int_{\Omega} \frac{\partial v_i}{\partial x_j}\sigma_{ij} d\xi = \int_{\partial\Omega} \left[v_i \tau_{w_i} + (\tau_j n_j) v_i n_i\right] d\xi. \tag{6}$$

The second term under the integral on the right-hand side is equal to zero because of the strongly enforced non-penetration condition, $v_i n_i = 0$. The first term depends on the boundary condition. In case of the standard no-slip condition, it vanishes since the test functions are zero-valued at the wall. In the case of a Neumann condition, it is non-zero.

The above shows a natural way of setting the predicted τ_w, namely using a Neumann boundary condition for the wall-parallel velocity components. At the same time, we see that in the SEM framework the value is enforced weakly and only enters as a source term on the right-hand side. While the global integral balance (6) will hold (supplemented by the remaining terms in the momentum equation), there is no guarantee that the local value of the wall stress at each wall node will be equal to that in the source term. In fact, these only converge towards each other asymptotically. Since, by the very definition of WMLES, the near-wall region is under-resolved, we cannot expect this convergence to take place.

Since the statistics of τ_w are an important simulation outcome, the question arises on what values to report: the ones predicted by the wall model or those actually observed at the wall. To avoid this issue, an alternative boundary condition can be used to set the stress.

In particular, the value of ν_{sgs} at the wall can be used to that end. Assume that the no-slip condition is used for the velocity field. Also, let x and z point in the wall-parallel directions, and y in the wall-normal. Then, given some predicted value of τ_w, we could set ν_{sgs} to

$$\nu_{sgs} = |\tau_w| / \left(2\sqrt{S_{xy}^2 + S_{zy}^2}\right) - \nu \tag{7}$$

in order to enforce this τ_w at the wall. Since all the boundary conditions are now Dirichlet, the stress is enforced strongly. Note, however, that since ν_{sgs} is a scalar, the direction of τ_w cannot be controlled, and is determined by the local values of S_{xy} and S_{zy}.

A comparison between these two boundary condition has been made in the context of finite-difference solvers in [1], with the viscosity condition leading to somewhat better results. The viscosity-based condition has also been successfully used with finite-volume-based codes [5]. However, both boundary conditions have not been compared yet in SEM.

2.1 Subgrid Scale Modelling

3 Case Setup

To test the boundary conditions, we consider channel flow, at $Re_b = 200400$. This corresponds to $Re_\tau \approx 8000$ according to the DNS data [9], which serves as a reference solution.

The size of the domain is $8\delta \times 2\delta \times 6\delta$, where δ is the half-height of the channel. The simulations are performed on a set of 3 meshes using polynomials of order 7. The first mesh M1 corresponds to node resolution $\approx \delta/30$ in all directions, although some modulation of this value is introduced by the Gauss-Legendre-Lobatto distribution of the nodes within each element. The mesh M2 is produced from M1 by doubling the number of elements in the wall-normal direction, and for M3 the same is also done for both wall-parallel directions. All three meshes can be considered fine by WMLES standards. More information about the meshes can be found in Table 1.

Table 1. Computational meshes for channel flow simulations. N_x, N_y, and N_z are the number of elements along the three coordinate axes. N_{gll} is the total number of nodes in the mesh.

Mesh	N_x	N_y	N_z	N_{gll}
M1	32	8	24	$3.14 \cdot 10^6$
M2	32	16	24	$6.29 \cdot 10^6$
M3	64	16	48	$25.17 \cdot 10^6$

As already mentioned in the introduction, Spalding's law used to predict the wall stress on both walls. Sampling is performed at a distance of 0.2δ, independent of the computational mesh. The Sigma model was used for subgrid scale modelling [6]. The simulation time step was selected to keep the Courant number below 0.5. The flow statistics were gathered over $40\delta/u_\tau$.

4 Results

We start by examining the accuracy of the wall-modelling. Table 2 shows the errors in the value of the mean friction velocity u_τ, both as predicted by the wall-model and as computed at the wall. For both boundary conditions, the wall-model predictions are extremely accurate. However, for the Neumann condition, the values at the boundary exhibit a large error. While not showing figures to keep this communication brief, we additionally note that the latter are also found to be highly sensitive to subgrid scale modelling. When using subgrid scale models that are more dissipative than Sigma in the near-wall region, the disparity between the wall-model predictions and the values at the boundary decreases. By contrast, for the viscosity condition, good consistency is observed independent of the model (see last two columns in Table 1). The fact that the values are not exactly the same can be attributed to the lack of control over the direction of τ_w, mentioned in Sect. 2.

Table 2. Relative errors in the predictions of the friction velocity u_τ in the simulations.

Mesh	Wall-model prediction, Neumann b.c	Value at the boundary, Neumann b.c.	Wall-model prediction, viscosity b.c	Value at the boundary, viscosity b.c
M1	0.77%	−15.48%	0.47%	−0.53%
M2	0.20%	−9.08%	0.57%	−1.95%
M3	−0.27%	−13.82%	0.11%	−1.49%

Figure 1 shows the mean velocity profiles in outer scaling, with U_b denoting the bulk velocity. The plot is split in two at the channel symmetry line, with the left part showing the profiles obtained with the Neumann boundary condition, and the right one those with the viscosity-based condition. In both cases, the effect of mesh resolution is marginal, yet the accuracy is better in the case of the Neumann condition.

Using the no-slip condition for velocity, combined with the coupling of the solution within the whole near-wall element, leads to an under-prediction of the mass-flux in the near-wall region. Since the total mass flux through the channel is enforced, this necessarily leads to an over-prediction in the core of the channel.

To assess the accuracy of second-order statistics the variance of the streamwise velocity is presented in Fig. 2. The other components of the Reynolds stress tensor exhibit a similar level of accuracy. A trend typical of WMLES is observed in the results. The fluctuations are too strong near the wall, but outside the inner layer an under-prediction with respect to the DNS is observed instead. Overall, the accuracy in the outer layer is good for a WMLES.

The influence of the grid is marginal when the Neumann condition is used, although on the finest grid a small improvement in accuracy can be discerned. With the viscosity condition, the dependency is stronger. Also, the near-wall wiggles extend further into the domain, particularly on the M1 grid.

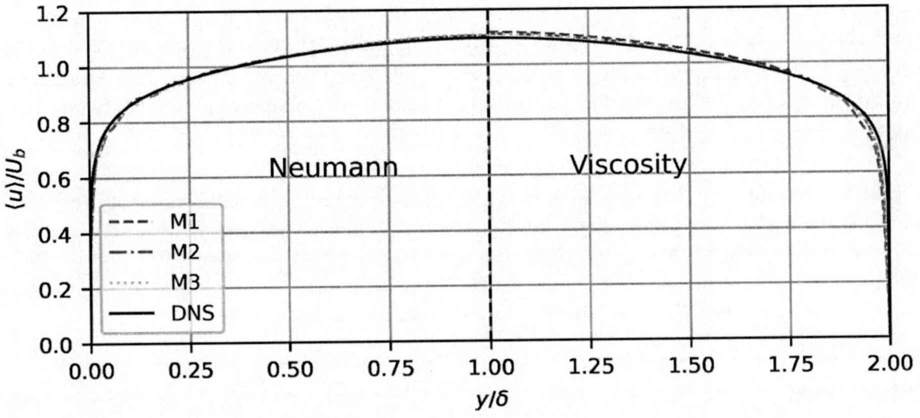

Fig. 1. Mean streamwise velocity profiles. Results from using the Neumann condition (left), and the viscosity-based condition (right).

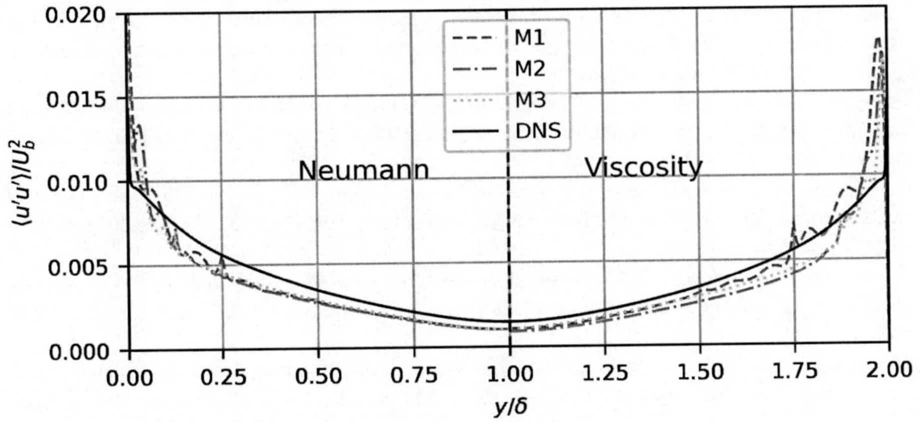

Fig. 2. Profiles of the variance of the streamwise velocity. Results from using the Neumann condition (left), and the viscosity-based condition (right).

5 Conclusions

Two alternatives for wall-stress boundary conditions have been tested in the spectral-element solver Nek5000. Since the Neumann condition is imposed weakly, poor agreement between the predictions of the wall model and the actual value found at the wall. Using the viscosity condition, this issue is mostly resolved, however, the profiles of the velocity statistics exhibit oscillations and deviate stronger from the reference data.

These results highlight the sensitivity of spectral element solvers to the simulation setup. The question remains, which boundary condition is to be recommended. Generally, accuracy of the mean velocity profile can be considered the most important outcome for a WMLES, thus favouring the Neumann condition. One could then report the τ_w values predicted by the wall model, arguing that the solution fields in the inner region can be disregarded in a WMLES. Nevertheless, we believe that additional simulations—of more involved flow configurations—may be necessary to arrive to final conclusions regarding this question.

Acknowledgements. The authors wish to thank Simon Sticko at Uppsala University for assistance in mathematical derivations and fruitful exchange of ideas. The computations were enabled by resources provided by the Swedish National Infrastructure for Computing (SNIC) at the PDC Center for High Performance Computing, KTH Royal Institute of Technology, partially funded by the Swedish Research Council through grant agreement no. 2018-05973.

References

1. Bae, H.J., Lozano-Durán, A.: Effect of wall boundary conditions on wall-modeled large-eddy simulation in a finite-difference framework. Fluids **6**, 112 (2021). https://doi.org/10.3390/fluids6030112
2. Gillyns, E., Buckingham, S., Winckelmans, G.: Implementation and validation of an algebraic wall model for LES in Nek5000. Flow Turbul. Combust. **109**, 1111–1131 (2022). https://doi.org/10.1007/s10494-022-00378-y
3. Jansson, N., Karp, M., Podobas, A., Markidis, S., Schlatter, P.: Neko: a modern, portable, and scalable framework for high-fidelity computational fluid dynamics. Preprint (2021). https://arxiv.org/abs/2107.01243
4. Larsson, J., Kawai, S., Bodart, J., Bermejo-Moreno, I.: Large eddy simulation with modeled wall-stress: recent progress and future directions. Mech. Eng. Rev. **3**, 15 (2016). https://doi.org/10.1299/mer.15-00418
5. Mukha, T., Rezaeiravesh, S., Liefvendahl, M.: A library for wall-modelled large-eddy simulation based on OpenFOAM technology. Comput. Phys. Commun. **239**, 204–224 (2019). https://doi.org/10.1016/j.cpc.2019.01.016
6. Nicoud, F., Toda, H.B., Cabrit, O., Bose, S., Lee, J.: Using singular values to build a subgrid-scale model for large eddy simulations. Phys. Fluids **23** (2011). https://doi.org/10.1063/1.3623274
7. Pal, P., Ameen, M.M., Xu, C: Near-wall modeling of turbulent flow and heat transfer using a spectral element CFD solver. In: AIAA Scitech 2022 Forum (2022). https://doi.org/10.2514/6.2022-1369
8. Spalding, D.B.: A single formula for the 'law of the wall'. J. Appl. Mech. **28**, 455–458 (1961)
9. Yamamoto, Y., Tsuji, Y.: Numerical evidence of logarithmic regions in channel flow at $Re_\tau = 8000$. Phys. Rev. Fluids **3**, 012602 (2018). https://doi.org/10.1103/PhysRevFluids.3.012602

Structural Models for Particles in LES: Assessment Through the Subfilter Enstrophy Field

M. Rajek and J. Pozorski(✉)

Institute of Fluid Flow Machinery, Polish Academy of Sciences,
Fiszera 14, 80-231 Gdańsk, Poland
{mrajek,jp}@imp.gda.pl

1 Introduction

The large-eddy simulations (LES) are increasingly often applied to turbulent flows with particles or droplets. One of the general questions refers to the impact of LES smoothing on the dispersed phase. A considerable body of evidence has been gathered over the years concerning the droplet preferential concentration, the average settling velocity in the gravity field, as well as the collision statistics [5,8,17]. The latter involve particle collective effects; another example is droplet evaporation or fuel particle devolatilisation at larger volume fractions of the dispersed phase. For such statistical quantities, the *one-point* models of the functional type, such as stochastic Lagrangian closures for the fluid velocity along the particle trajectories (the fluid "seen" by particles, see, e.g. [6,11]) are inadequate and some approximate reconstruction of the subgrid scale (SGS) fluid velocity *field* is sought for as an input to the particle equation of motion in the point-particle approximation. In the same vein, one-point closures prove sufficient for the long term turbulent dispersion, but two-point models are necessary to correctly predict relative dispersion.

The structural-type models are able to provide two- and multi-point statistics; the word "structural" simply means that the whole field (or a dynamically relevant part of it) is somehow generated or mimicked in a simplified manner. Several proposals for such closures have been put forward, including the kinematic simulations [13,15], approximate deconvolution [7], fractal interpolation [1,9], spectral enrichment [2], and hybrid variants of those [10,18]. They revealed to be partly successful, depending on the droplet inertia quantified by the Stokes number, see [12] for an overview. In the following, modeling proposals to account of the SGS flow effects on particles are referred to as "the SGS particle models" for brevity.

A prospective candidate for an SGS particle model to be put forward should preferably be autonomous, i.e. based on the resolved LES quantities only (like the dynamic Germano model in the case of a single-phase flow). For reasons of consistency, also the SGS stress tensor for the fluid motion closure should ideally be determined from the postulated small-scale velocity field. Another non-trivial criterion for the assessment of the SGS particle models is their computational complexity: clearly, any viable proposal has to be considerably less expensive than the full DNS.

C. Marchioli et al. (Eds.): DLES 2023, ERCO 31, pp. 221–226, 2024.
https://doi.org/10.1007/978-3-031-47028-8_34

In the present contribution, we attempt to tackle the issue of structural SGS particle closures, in particular through considering an interesting and physically-motivated model based on the multifractality (MF) of the enstrophy cascade in a turbulent flow [3]. The MF concepts used there served to derive the closure for the SGS tensor in single-phase flow; however, despite some plans [4], they have not been applied to conceive an SGS particle model (it would be cumbersome and computationally costly due to the Biot-Savart formulae present there).

2 A Priori LES and Multifractal Model

The pseudo-spectral code we use has already been applied for the purpose of DNS, LES and *a priori* LES studies, as documented in [14, 17]. As the departure point for analyses of general turbulent flows, we have simulated homogeneous isotropic turbulence (HIT) with a large-scale forcing (be it stochastic or deterministic), both in the presence of gravity and without it. In a companion paper [14] results are reported from a particle-laden HIT where small droplets, otherwise treated as point-particles, were allowed to collide. There, *a priori* LES was studied with the spectral cutoff of the highest wavenumbers. We supplemented the large-scale velocity field with kinematic simulations (KS, either frozen or time-dependent). As an important statistic, we have computed the increase of the mean droplet settling velocity w.r.t. the droplet terminal velocity in DNS. The results clearly indicate that frozen SGS fields are inadequate. We also determined the two-point statistics relevant for droplet collisions: the radial distribution function (RDF, a measure of particle concentration non-uniformity) and the radial relative velocity (RRV). Both are computed at contact (nearly touching droplets); see, e.g., [16] for details.

The present work is meant to provide hints for conceiving new SGS particle models, beyond the KS reported in [14]. Detailed flow and dispersed phase data (DNS together with the Lagrangian particle tracking) are essential for the development of an SGS particle model, the parameter space including the Reynolds number R_λ as well as the particle Stokes and Froude numbers (for gravity-affected cases). Along with the DNS, *a priori* LES have been performed as well as *a posteriori* (true) LES with the spectral eddy diffusivity.

The experimental and DNS evidence has amply shown that the small-scale turbulence is dominated by vortical structures and the velocity gradients are strongly non-Gaussian. A relevant measure of this flow aspect is the enstrophy $\Omega = |\omega|^2$ where $\omega = \nabla \times \mathbf{u}$ is the vorticity. Given that the solid particles of low inertia are strongly sensitive to the small-scale eddy structures, we argue that the SGS enstrophy field is a useful quantity to assess structural models for LES.

From among existing proposals we have chosen the multifractal model and the kinematic simulations. The former, exploiting the multifractal scale similarity of turbulence gradient fields, is of particular interest as it explictly reconstructs the vorticity magnitude and orientation cascades out of locally available large-scale data.

In the present *a priori* LES analysis, the starting point is the DNS of HIT. The necessary input to the multifractal model is the velocity field $\mathbf{u}_\Delta$ computed as the difference of two low-pass filtered fields at the scale Δ and 2Δ, respectively (Δ is the LES mesh size). The SGS enstrophy $\Omega_{\rm sgs}$, to be distributed over the fine-scale (DNS-size) cells, is determined as proportional to $\Omega_\Delta = |\nabla \times \mathbf{u}_\Delta|^2$, see [3] for the actual formulae.

The MF model control parameter is an assumed PDF $P(\mathscr{M})$ of a fractal multiplier, $\mathscr{M}_n(\mathbf{x},t)$, needed to construct the vorticity magnitude values at each subsequent refinement level $n = 1,\ldots,\mathscr{N}$, according to the formula [3]

$$|\omega_{\text{sgs}}|^2(\mathbf{x},t) = Q_{\text{sgs}}(2^{\mathscr{N}})^3 \prod_{n=1}^{\mathscr{N}} \mathscr{M}_n(\mathbf{x},t).$$

Another cascade has been proposed in [3] for the SGS vorticity orientation $\hat{\mathbf{e}}^n(\mathbf{x},t)$; subsequent levels are constructed in a local spherical coordinate system using azimuthal and polar decorrelation angles, ϕ and θ, respectively: $\hat{\mathbf{e}}^n = \hat{\mathbf{e}}^{n-1} + \mathbf{f}(\phi^n, \theta^n)$. Finally, the quantity $\omega^{\mathscr{N}}(\mathbf{x},t)$, assumed to play the role of the SGS vorticity field, is determined in all finest-level cells of size $\Delta/2^{\mathscr{N}}$ according to $\omega^{\mathscr{N}}(\mathbf{x},t) = |\omega_{\text{sgs}}|(\mathbf{x},t)\,\hat{\mathbf{e}}^{\mathscr{N}}(\mathbf{x},t)$; Fig. 1 provides some illustration of the process.

Out of $\omega^{\mathscr{N}}$, generated from a stochastic process recalled above, the corresponding SGS velocity field can could in principle be found from the Biot-Savart law. The structural model based on the multifractality of enstrophy would be made this way. In [3], after further simplifying assumptions adopted (some of them seem to be quite strong), the MF model served to derive a closure for the SGS stress tensor.

Here, our guiding idea has been to apply the synthetic SGS velocity field generated this way, and interpolated at the positions of inertial particles, as the fluid velocity "seen". However, before attempting to compute the resulting particle statistics, as done before for the kinematic simulations [14], we decided to look at the enstrophy field. There are two major difficulties. First, the computational cost of inverting vorticity into velocity through the Biot-Savart formula is rather high, the more so that the formula is global (not limited to a single LES cell). Second, as it transpires from the MF model, the stochastic $\omega^{\mathscr{N}}$ field is not guaranteed to be solenoidal. However, this is a necessary condition for this field to represent vorticity (which is divergence-free by definition). So, rigorously, before even attempting to find the SGS velocity out if it, the field of $\omega^{\mathscr{N}}$ would need to be projected on a div-free space, meaning another computational procedure (non-local in the flow domain, hence costly).

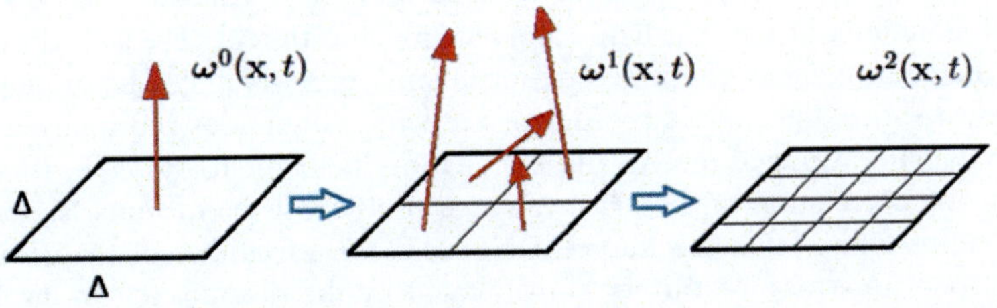

Fig. 1. A 2D scheme of the vorticity multifractal cascade. Left picture: the departure vorticity vector $\omega^0 = \nabla \times \mathbf{u}_\Delta$. Mid picture: first-level subfilter vorticity ω^1 reconstructed in all sub-cells of size $\Delta/2$. Right picture: second-level reconstruction in cells of $\Delta/4$ (vorticity vectors not shown). The LES cell (a cube of edge length Δ) is represented here in 2D only (as a square) for simplicity; the same for subsequent plots where 2^3 and 4^3 cells should appear (instead of squares). Note that the generated vorticity vectors (ω^1, ω^2, etc.) differ from ω^0 both in magnitude and orientation.

3 Results: SGS Enstrophy

We performed the DNS of HIT at $R_\lambda = 107$ on 256^3 grid. Using the low-pass filtering at two scales, we computed the velocity field $\mathbf{u}_\Delta$. Then, starting from the LES mesh of 64^3, we proceeded down to the DNS mesh of 256^3 (here, $\mathcal{N} = 2$). This means that the vorticity data at each LES cell serve to construct the vorticity within $2^{3\mathcal{N}}$ fine-scale DNS cells.

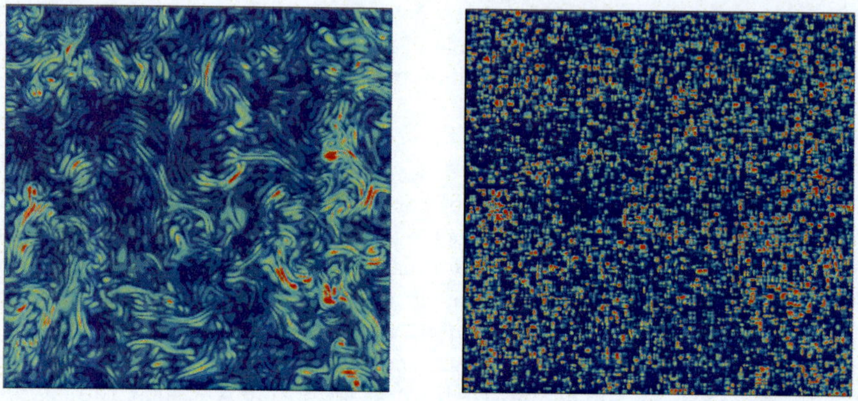

Fig. 2. Subfilter vorticity magnitude in forced isotropic turbulence: computed in DNS (left panel) and generated from multifractal enstrophy cascade (right panel). Both quantities are normalised by the respective domain-averaged values of $|\omega_{\mathrm{sgs}}|$.

Figure 2 shows the snapshots of: (i) the SGS enstrophy (precisely: the vorticity magnitude ω_{sgs}) computed using the pseudospectral DNS code and high-pass velocity filtering and (ii) the SGS vorticity magnitude generated from the MF cascade. The positive observation is that the MF model acts locally (through $\mathbf{u}_\Delta$) and this is confirmed by the maxima and minima approximately coinciding in both panels of Fig. 2. However, the spatial structures of the two fields are very much different. The naked-eye observation of the color maps is confirmed by results in Fig. 3 which can be more precisely interpreted; it shows the sample profiles of vorticity magnitudes (normalised by their respective domain-averaged mean value), including those for KS as well. It is noticed that the spatial correlation of the MF-generated profile will be of much shorter range than the one resulting from the filtered DNS; also the variability of the enstrophy is way too strong; this may be further corroborated by the discrepancy in the PDFs of enstrophy in the computational domain (not shown).

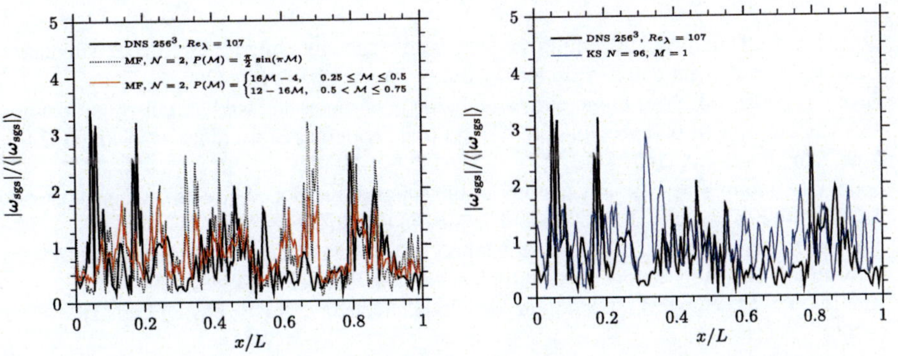

Fig. 3. Subfilter vorticity magnitude in forced isotropic turbulence computed in DNS on a 256^3 mesh, generated from filtered DNS data on 64^3 mesh using a two-level ($\mathcal{N} = 2$) multifractal enstrophy cascade (MS, left panel) and using the kinematic simulation (KS [14], right panel). Two PDFs $P(\mathcal{M})$ of fractal multipliers were considered: the original one [3], and the sine function [4].

4 Conclusion

In this paper, we have briefly recalled the issue of SGS particle models. Then, we focused on *a priori* LES to gain insights in the SGS enstrophy, or the magnitude of vorticity, as this quantity plays a vital role in the subfilter dynamics of low inertia particles (the most problematic ones for all structural models proposed to date). We have extracted the SGS enstrophy from the DNS of forced isotropic turbulence using high-pass spectral filtering. To assess the viability of the multifractal model [3] for dispersed flows, we generated a synthetic SGS enstrophy field using the multifractal cascade of vorticity magnitude. It reveals that the two SGS fields (high-pass DNS and the MF model) considerably differ, which seems to be rather discouraging as far as the use of the MF procedure to feed the SGS particle model is concerned. The proposed methodology can be applied to assess other structural models in the same way; in particular, for the kinematic simulations the resulting enstrophy profiles have been found to be in a qualitative agreement with the DNS reference data.

Acknowledgements. The work was supported by the National Science Centre (NCN, Poland), through the research project Sheng 2018/30/Q/ST8/00341. The authors are grateful to Prof. B. Rosa (IMGW-PIB Warsaw, Poland) for insightful discussions on the pseudo-spectral DNS code applied in this work. The HPC resources were provided by the Interdisciplinary Centre for Mathematical and Computational Modelling (ICM, University of Warsaw), grant G87-1145.

References

1. Akinlabi, E.O., Wacławczyk, M., Malinowski, S.P., Mellado, J.P.: Fractal reconstruction of sub-grid scales for large eddy simulation. Flow Turbul. Combust. **103**, 293–322 (2019)
2. Bassenne, M., Esmaily, M., Livescu, D., Moin, P., Urzay, J.: A dynamic spectrally enriched subgrid-scale model for preferential concentration in particle-laden turbulence. Int. J. Multiphase Flow **116**, 270–280 (2019)

3. Burton, G.C., Dahm, W.J.A.: Multifractal subgrid-scale modeling for large-eddy simulation. I. Model development and a priori testing. Phys. Fluids **17**, 075111 (2005)
4. Fayed, H.E., Ragab, S.A.: Large-eddy simulation of particles and bubbles collisions in homogeneous isotropic turbulence. In: 43rd AIAA Fluid Dynamics Conference and Exhibit, 24–27 June, San Diego, CA (2013)
5. Fede, P., Simonin, O.: Numerical study of the subgrid turbulence effects on the statistics of heavy colliding particles. Phys. Fluids **17**, 045103 (2006)
6. Knorps, M., Pozorski, J.: Stochastic modeling for subgrid-scale particle dispersion in large-eddy simulation of inhomogeneous turbulence. Phys. Fluids **33**, 043323 (2021)
7. Kuerten, J.G.M.: Subgrid modeling in particle-laden channel flows. Phys. Fluids **18**, 025108 (2006)
8. Marchioli, C.: Large-eddy simulation of turbulent dispersed flows: a review of modelling approaches. Acta Mech. **228**, 741–771 (2017). https://doi.org/10.1007/s00707-017-1803-x
9. Marchioli, C., Salvetti, M.V., Soldati, A.: Appraisal of energy recovering sub-grid scale models for large-eddy simulation of turbulent dispersed flows. Acta Mech. **201**, 277–296 (2008)
10. Michałek, W.R., Kuerten, J.G.M., Liew, R., Zeegers, C.H., Pozorski, J., Geurts, B.J.: A hybrid deconvolution stochastic model for LES of particle-laden flow. Phys. Fluids **25**, 123202 (2013)
11. Minier, J.-P.: Statistical descriptions of polydisperse turbulent two-phase flows. Phys. Rep. **665**, 1–122 (2016)
12. Pozorski, J.: Models of turbulent flows and particle dynamics. In: Minier, J.-P., Pozorski, J. (eds.) Particles in Wall-Bounded Turbulent Flows: Deposition, Re-Suspension and Agglomeration. CISM, vol. 571, pp. 97–150. Springer (2017). https://doi.org/10.1007/978-3-319-41567-3_3
13. Pozorski, J., Rosa, B.: The motion of settling particles in isotropic turbulence: filtering impact and kinematic simulations as subfilter model. In: Salvetti, M.V., Armenio, V., Fröhlich, J., Geurts, B.J., Kuerten, H. (eds.) Direct and Large-Eddy Simulation XI. ERCOFTAC Series, vol. 25, pp. 215–220. Springer (2019). https://doi.org/10.1007/978-3-030-04915-7_29
14. Rajek, M., Pozorski, J.: Kinematic simulations as a subgrid-scale model for particle motion: a priori LES of homogeneous isotropic turbulence. J. Phys. Conf. Ser. **2367**, 012029 (2022)
15. Ray, B., Collins, L.R.: A subgrid model for clustering of high-inertia particles in large-eddy simulations of turbulence. J. Turb. **15**, 366–385 (2014)
16. Rosa, B., Kopeć, S., Ababaei, A., Pozorski, J.: Collision statistics and settling velocity of inertial particles in homogeneous turbulence from high-resolution DNS under two-way momentum coupling. Int. J. Multiphase Flow **148**, 103906 (2022)
17. Rosa, B., Pozorski, J.: Impact of subgrid fluid turbulence on inertial particles subject to gravity. J. Turb. **18**, 634–652 (2017)
18. Zhou, Z., Wang, S., Yang, X., Jin, G.: A structural subgrid-scale model for the collision-related statistics of inertial particles in large-eddy simulations of isotropic turbulent flows. Phys. Fluids **32**, 055106 (2020)

DNS-Based Turbulent Closures for Sediment Transport Using Symbolic Regression

Y. Stöcker[1], C. Golla[1(✉)], R. Jain[1], J. Fröhlich[1], and P. Cinnella[2]

[1] Institute of Fluid Mechanics, Technische Universität Dresden, Dresden, Germany
christian.golla@tu-dresden.de

[2] Institut Jean le Rond d'Alembert, Sorbonne Université, Paris, France

1 Introduction

Particle-laden flows occur in many ways in natural and technological situations. Jain et al. [4] presented four DNS studies of sediment transport with varying particle shapes. However, in industrial applications, the RANS equations are commonly used for simulations of turbulent flows.

Chauchat et al. [2] developed a closure model in their two-phase CFD solver for sediment transport applications, SedFoam-2.0. The mean flow quantities are represented well. However, SedFoam-2.0 relies on modelling assumptions for the fluid-particle interactions which are often not satisfied, in particular for the cases simulated with DNS considered. Among others, sediment does not always act as purely dissipative, but can also enhance turbulence by its roughness. The aim of the present work, therefore, is to improve the turbulence modelling in RANS simulations utilising the DNS data.

Based on the high-fidelity data from DNS conducted with spherical particles, a machine learning (ML) algorithm based on symbolic regression was used to improve the model. To this end, the error of the modelled turbulence transport equation served as the target data for the deterministic symbolic regression technique SpaRTA introduced by Schmelzer et al. [6]. The learnt models were implemented in the CFD solver for cross-validation to assess their predictive performance.

2 RANS Model

The phase-weighted average of a generic flow value ψ as defined in Burns et al. [1] is denoted by a tilde, and the fluctuations can be obtained as $\psi^{\mathrm{p}\prime} = \psi - \widetilde{\psi}^{\mathrm{p}}$, where p indicates the phase, f or s. The volume fraction of the fluid, ϕ, and the sediment, α, are expressed as the spatially averaged respective phase indicator function χ^{p} [1].

The RANS equations feature the fluid Reynolds stress tensor $\tilde{R}_{ij}$, which in the present work is modelled based on the Boussinesq assumption, using the eddy viscosity ν_{t} and the deviatoric part of the mean fluid strain rate tensor $\tilde{S}_{ij}$ [2]. The eddy viscosity is determined using a k-ε-turbulence model with $\nu_{\mathrm{t}} = C_\mu k^2/\varepsilon$, with $C_\mu = 0.09$, since this is the approach in the cited reference. The transport equation for the turbulent kinetic energy of the fluid k reads

$$D_t k = P - \varepsilon + T + I. \tag{1}$$

C. Marchioli et al. (Eds.): DLES 2023, ERCO 31, pp. 227–231, 2024.
https://doi.org/10.1007/978-3-031-47028-8_35

The turbulent production P and transportation T are modelled similar to the clear fluid closure and the dissipation ε is computed from its own transport equation, which is analogous to Eq. (1). The interfacial term I accounts for sediment-fluid interaction and is split into two parts. The first is modelled as a drag-induced turbulence damping effect, the second results from a density stratification which, for upwards decreasing sediment concentrations, also provides turbulence attenuation [2].

3 Setup and DNS Data

The data basis is formed by high-fidelity simulations of particle laden flows, described in [4]. It is an incompressible turbulent open-channel flow with a rough sediment bed using various particle shapes. The domain is of size $108d \times 23d \times 36d$ in the stream-wise (x), wall-normal (y) and span-wise (z) direction, respectively, where d is the particle diameter. The present work is based on the case with spherical particles (Fig. 1).

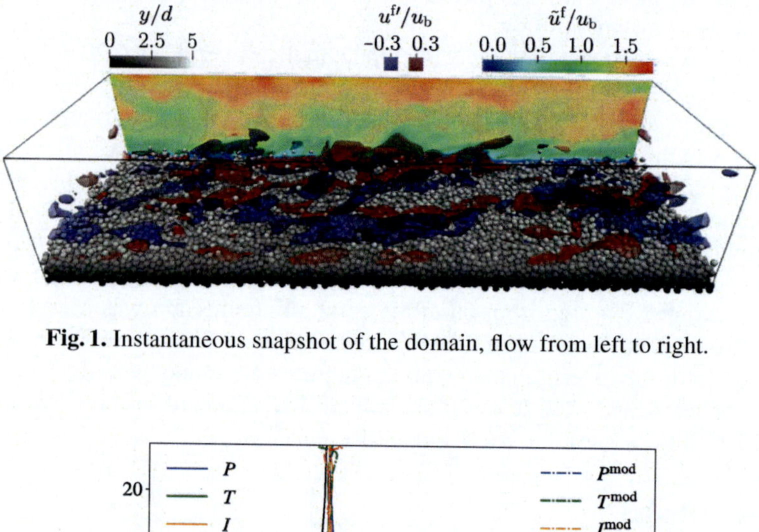

Fig. 1. Instantaneous snapshot of the domain, flow from left to right.

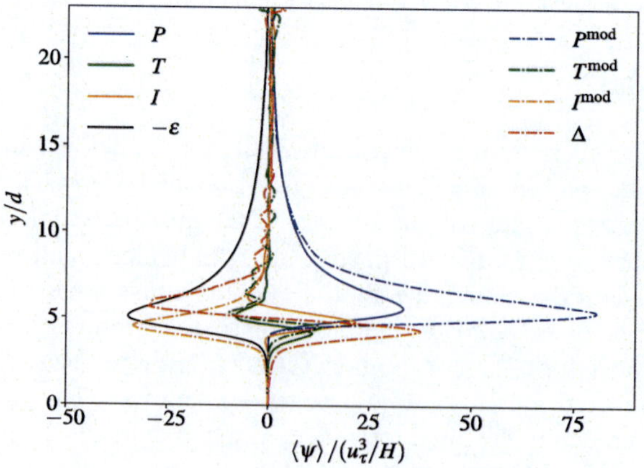

Fig. 2. Wall-normal profiles for the exact RHS of the k-transport equation (1) using DNS data by Jain et al. [4] juxtaposed to the corresponding model term of the RANS model, evaluated from the DNS data. All quantities are defined in the text.

The terms of Eq. (1), both exact and modelled, can be evaluated employing the high-fidelity data for u_i^{f}, u_i^{s}, k, $\tilde{R}_{ij}$, ε and χ^{f}. The term on the left-hand side is assumed to be negligible when averaging in time and periodic directions. Figure 2 shows the resulting wall-normal profiles for all terms of the right-hand side (RHS) of Eq. (1). Compared to the exact terms, the modelled production is largely overestimating and the interfacial term is of the wrong sign. Hence, the model terms do not satisfy Eq. (1), a residual Δ is obtained, which is also displayed in Fig. 2.

4 SpaRTA Approach

The basic idea of the SpaRTA approach [6] is to add corrective terms R_{ij}^{Δ} and $\mathscr{R}$ to the Reynolds tensor and to the RHS of the modelled k-transport equation, respectively, which are represented through ML ansatzes trained from data. A counterpart of the corrective term for k is also included in the ε-equation as in [6].

In this work, only the transport equations are corrected. The aim of the study was to find a data-driven model $\mathscr{R}$ for the k-transport equation that best fits the residual Δ. Therefore, first a functional space is defined, constituted by all linear combinations of a large number of candidate terms, i.e., a function "dictionary". Afterwards, sparse regression is used to select the terms that best fit the data and to infer the corresponding parameters.

Here, two different modelling approaches were considered: the turbulent production analogy, expressing $\mathscr{R}$ as [6]

$$\mathscr{R} = 2k\left(b_{ij}^{\mathscr{R}} - 1/3\,\delta_{ij}\right)\partial_j \tilde{u}_i^{\mathrm{f}}, \tag{2}$$

and a newly introduced dissipation analogy, with

$$\mathscr{R} = c\,\varepsilon. \tag{3}$$

For the first one, a tensorial model for $b_{ij}^{\mathscr{R}}$ has to be found, and for the second one, a functional expression for the scalar c is needed. It is assumed that both depend on the mean strain-rate tensor $\widetilde{S}_{ij}$, the mean rotation-rate tensor $\widetilde{\Omega}_{ij}$, and a time scale $\tau = C_\mu k/\varepsilon$.

For both approaches, polynomial model expressions for the target values were assembled from a predefined set of non-linear candidate functions. These candidates depend on the invariants that form the minimal integrity basis [3] of the normalized tensors $\tau\widetilde{S}_{ij}$ and $\tau\widetilde{\Omega}_{ij}$. Additionally, for considering sediment-phase related quantities in the models, the sediment volume fraction α was included and an additive constant was also considered. The scalar c was assembled via

$$c\left(\tau\tilde{S}_{ij}, \tau\tilde{\Omega}_{ij}, \alpha\right) = f\left(I_1, \ldots, I_m, \alpha\right), \tag{4}$$

with the m scalar invariants I_m of the minimal integrity basis and f a linear combination of monomials of the arguments up to degree 6. For building $b_{ij}^{\mathscr{R}}$, the monomial functions of the dictionary were additionally multiplied with each base tensor $\mathscr{T}_{ij}^{(n)}$

$$b_{ij}^{\mathscr{R}}\left(\tau\tilde{S}_{ij}, \tau\tilde{\Omega}_{ij}, \alpha\right) = \sum_{n=1}^{10} f_n\left(I_1, \ldots, I_m, \alpha\right) \cdot \mathscr{T}_{ij}^{(n)}. \tag{5}$$

For two-dimensional flows only three base tensors are needed to represent $b_{ij}^{\mathscr{R}}$ and only two invariants are independent [5].

The model selection and inference procedures following [6] were employed. Elastic net regularisation [7] was used to identify relevant candidates. It adopts a sparsity constraint, i.e., it allows only few non-zero model coefficients. To enforce small coefficients, ridge regression [7] was employed. The overall goal is to find models balancing error and complexity and to avoid overfitting. This increases the likelihood of yielding models with a good predictive performance that are stable when implemented in a CFD solver. As a result of the process, numerous symbolic model expressions M for $b_{ij}^{\mathscr{R}}$ or c were obtained that regress the target data.

5 Cross-Validation

The models obtained by the above procedure were implemented in SedFoam-2.0 to evaluate their predictive quality. The simulation setup used to conduct the RANS simulations employs the same geometry, bulk Reynolds number, and total particle concentration as the DNS. The main goal was to improve the modelled k-transport equation. However, other quantities should not be deteriorated, so that the cross-validation was performed not only on k, but also on the stream-wise velocity of the fluid $\tilde{u}^{\mathrm{f}}$ and of the sediment $\tilde{u}^{\mathrm{s}}$, the total shear stress τ_{tot} (as defined in [4]) and the fluid volume fraction ϕ. The performance $\mathscr{P}$ of a model on a single flow quantity ψ was evaluated using the mean-squared error E with respect to the high-fidelity data via $\mathscr{P}_{\psi} = 1 - \mathrm{E}(\psi^{M})/\mathrm{E}(\psi^{0})$. The index M denotes that the value is obtained from the CFD simulations employing the machine-learnt model, and 0 indicates the result of the unmodified solver. The total performance $\mathscr{P}_{\mathrm{tot}}$ of a model was assessed by the arithmetic mean of the single performances on all five quantities.

The total performance $\mathscr{P}_{\mathrm{tot}}$ of all stable models is shown in Fig. 3. The results are split into models that are obtained from the production analogy Eq. (2) and dissipation analogy Eq. (3), and whether the volume fraction of the sediment, α, is used in the models. Additionally, the single performances $\mathscr{P}_{\psi}$ for the respective model with the best total performance is displayed.

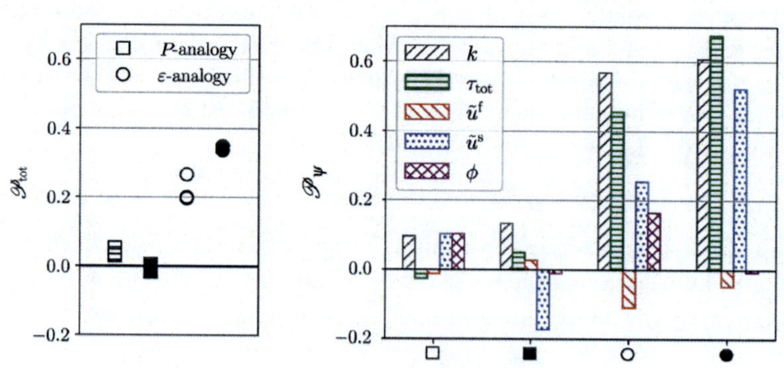

Fig. 3. Performance of the models, split into models from the production analogy Eq. (2) and dissipation analogy Eq. (3). Filled symbols: same with α incorporated. Left: total performance, right: single performances of the best performing models.

In general, the models of the dissipation analogy performed better than the ones from the production analogy, and the inclusion of α yielded an additional improvement, resulting in a total performance up to 35%. No model was capable of improving all quantities at the same time, but particularly the turbulent fluctuations were better captured, thereby reducing the error on k and τ_{tot} up to 65%.

6 Conclusions

In this work, the applicability of the SpaRTA algorithm of Schmelzer et al. [6] to multiphase flows is demonstrated. The newly introduced dissipation analogy showed significantly better results than the production analogy, yielding an approximately twenty percent higher total performance. However, the improvements are not as good as in the separating single-phase flows evaluated in [6], where the objective was to improve predictions of separation and reattachment of the flow along a given geometry. Multiphase flow, as considered here, is a substantially different target.

The dissipation analogy enables the possibility to easily include new values such as the gradient of the phase volume fraction in the models. One can also consider to incorporate the sediment velocity. In a further study, the available DNS data obtained with different particle shapes will be used, since they exhibit substantial differences from the ones for spherical particles, e.g. a different shape of the sediment bed. Most likely, this will require different correction models. Other basic RANS models could also be employed with the same methodology.

References

1. Burns, A.D., Frank, T., Hamill, I., Shi, J.-M.: The Favre averaged drag model for turbulent dispersion in Eulerian multi-phase flows. In: ICMF 2004, p. 392 (2004)
2. Chauchat, J., Cheng, Z., Nagel, T., Bonamy, C., Hsu, T.-J.: SedFoam-2.0: a 3-D two-phase flow numerical model for sediment transport. Geosci. Model Dev. **10**, 4367–4392 (2017). https://doi.org/10.5194/gmd-10-4367-2017
3. Gatski, T.B., Jongen, T.: Nonlinear eddy viscosity and algebraic stress models for solving complex turbulent flows. Prog. Aerosp. Sci. **36**(8), 655–682 (2000). https://doi.org/10.1016/S0376-0421(00)00012-9
4. Jain, R., Tschisgale, S., Fröhlich, J.: Impact of shape: DNS of sediment transport with non-spherical particles. J. Fluid Mech. **916**, A38 (2021). https://doi.org/10.1017/jfm.2021.214
5. Pope, S.: A more general effective-viscosity hypothesis. J. Fluid Mech. **72**(2), 331–340 (1975). https://doi.org/10.1017/S0022112075003382
6. Schmelzer, M., Dwight, R.P., Cinnella, P.: Discovery of algebraic Reynolds-stress models using sparse symbolic regression. Flow, Turbul. Combust. **104**, 579–603 (2019). https://doi.org/10.1007/s10494-019-00089-x
7. Zou, H., Hastie, T.: Regularization and variable selection via the elastic net. J. R. Stat. Soc. Series B Stat. Methodol. **67**(2), 301–320 (2005). https://doi.org/10.1111/j.1467-9868.2005.00503.x

Uncertainty Quantification of LES for Buoyancy-Driven Mixing Processes Using PCE-HDMR

P. J. Wenig[1(✉)], S. Kelm[2], and M. Klein[1]

[1] Institute of Applied Mathematics and Scientific Computing, University of the Bundeswehr Munich, Werner-Heisenberg-Weg 39, 85577 Neubiberg, Germany
{philipp.wenig,markus.klein}@unibw.de

[2] Institute of Energy and Climate Research, Forschungszentrum Jülich GmbH, 52425 Jülich, Germany
s.kelm@fz-juelich.de

1 Introduction

The investigation of uncertainties in high-dimensional and computationally expensive engineering applications requires Uncertainty Quantification (UQ) methods that efficiently provide reliable High-Dimensional Model Representations (HDMR) of stochastic results with a justifiable number of calculation runs. Therefore, the promising PCE-HDMR technique, which is based on Cut-HDMR in conjunction with Polynomial Chaos Expansions (PCE), is applied for the qualification of suitable UQ methods for large-scale CFD-applications. The method enables to achieve a high level of accuracy and provides great flexibility in stochastic model construction. The methodology is applied for the UQ analysis of a buoyancy-driven mixing process between two miscible fluids within the Differentially Heated Cavity (DHC) of aspect ratio 4. Result quantities or responses, which are presented, allow for the comprehensive description of the mixing process. Uncertainties in these responses, which originate from input uncertainties, need to be quantified. Models for stochastic processes at hand are derived from Large Eddy Simulations (LES) through PCE-HDMR and the Sparse Grid Method (SGM), which serves as a reference for the results from PCE-HDMR. The presented results include the statistics of modeled time-dependent stochastic processes and indicate that PCE-HDMR yields very good agreement with the SGM reference and is proven a promising UQ method for large-scale computational applications distinguished through flexible adaptation, extension and improvement of existing stochastic models.

2 Methods

Numerical techniques for the prediction of underlying flow phenomena are covered in Sect. 2.1. The applied UQ methodology is detailed in Sect. 2.2.

C. Marchioli et al. (Eds.): DLES 2023, ERCO 31, pp. 232–237, 2024.
https://doi.org/10.1007/978-3-031-47028-8_36

2.1 CFD Methods

Low-Mach number flow of two Newtonian viscous fluids within the DHC of aspect ratio 4 is investigated by means of LES. The DHC is filled with air and 40 vol.% of helium in the upper third, as shown in a schematic sketch for the case setup in Fig. 1. Calculations are performed by using the finite volume code OpenFOAM v2006. The Prandtl numbers are $Pr_{Air} = 0.71$ and $Pr_{He} = 0.66$. The Rayleigh number Ra, which is derived for the undisturbed free circulation of air, is equal to 2×10^9. Pressure-velocity coupling is addressed by the use of the PIMPLE algorithm. The convective momentum flux was evaluated by a second-order linear upwind scheme. The remaining convective fluxes and diffusive fluxes were evaluated by the limited linear scheme. Temporal advancement was achieved by blending of 10% Euler and 90% Crank–Nicolson scheme. It has been ensured that the CFL number is always below the value of 0.5. For modeling of the sub-grid scale effects in the LES, the WALE-model and the gradient flux approach with a turbulent Prandtl- and Schmidt-number $Pr_t = Sc_t = 0.85$ are applied. The spatial grid resolution is chosen according to [1]. The uncertain input parameters Q comprise the temperature difference between the left and right wall ΔT, the top and bottom wall temperatures $T_{b/t}$, the vertical temperature gradient at the left and right wall $T_y\big|_{l,r}$, the initial helium stratification via ΔX and the molecular diffusion coefficient D. A detailed description of the definition of the uncertain input parameters can be found in [1,2].

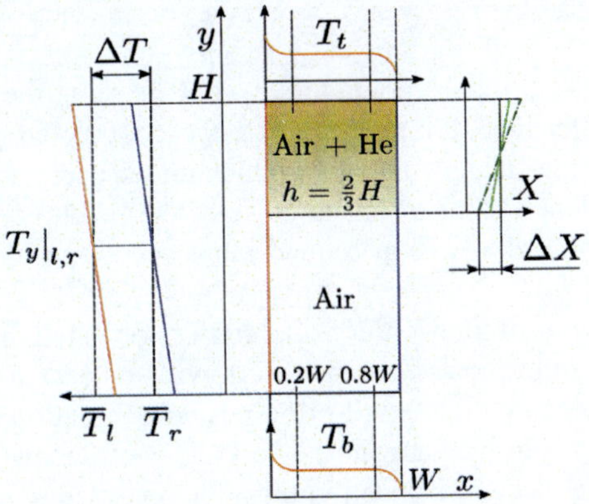

Fig. 1. Schematic sketch of the DHC with uncertain parameters.

2.2 UQ Methods

Non-intrusive PCEs together with Karhunen-Loeve Expansions (KLEs) [3,4] were utilized for the random field approximation $R^{P,K}(\mathbf{x},\mathbf{Q})$ through the discrete KLE method [5] with K terms and numerically generated orthogonal polynomials $\Psi_\beta(\mathbf{Q})$ with corresponding field-dependent expansion coefficients $\alpha_\beta^K(\mathbf{x})$:

$$R^{P,K}(\mathbf{x},\mathbf{Q}) = \mu_R(\mathbf{x}) + \sum_{\beta \in \mathscr{A}^{p,n}} \left[\alpha_\beta^K(\mathbf{x})\right] \Psi_\beta(\mathbf{Q}) \, , \quad \mathscr{A}^{p,n} = \{\beta \in \mathscr{N}^n : |\beta| \leq p\} \, , \quad (1)$$

where $\beta = (\beta_1, \cdots, \beta_n)$ with $\beta_i \geq 0$ is an n-dimensional index and $\mu_R(\mathbf{x})$ is the expectation of the random field. For computational purposes, the PCE is truncated by retaining a total polynomial degree of $|\beta| = \sum_{i=1}^{n} \beta_i \leq p$, which leads to a total number of P terms of the PCE. The PCE coefficients for the SGM reference results are computed using discrete projection through SGM with sparse grid level of 2. The KLE eigenvalue decay rate is 0.02. In the present work, the field variable $\mathbf{x}$ refers to time t and consequently time-dependent stochastic processes are considered.

Response functions $R(\mathbf{Q})$ can be represented by a unique HDMR or Sobol decomposition, which in practice is usually truncated after the second-order term [6]:

$$R(\mathbf{Q}) \approx R_0 + \sum_{i=1}^{n} R_i(Q_i) + \sum_{1 \leq i < j \leq n} R_{ij}(Q_i, Q_j) \,. \tag{2}$$

Cut-HDMR or anchored HDMR proposes the determination of the series component functions of the HDMR through the evaluation of the response function along cut lines, planes, and hyperplanes through a reference point $\mathbf{q}^*$. When using PCE-HDMR, metamodels of the component functions within the Cut-HDMR are simultaneously constructed through PCEs [7]:

$$R(\mathbf{Q}) \approx R(\mathbf{q}^*) + \sum_{i=1}^{n} \sum_{\beta \in \mathscr{A}^{p,1}} \alpha_\beta \Psi_\beta \left(Q_i, \mathbf{q}^*_{\sim i}\right) + \sum_{1 \leq i < j \leq n} \sum_{\beta \in \mathscr{A}^{p,2}} \alpha_\beta \Psi_\beta \left(Q_i, Q_j, \mathbf{q}^*_{\sim ij}\right) , \tag{3}$$

where $\mathbf{q}^*_{\sim i}$ indicates that all input parameters except q_i are at the reference point, and $\mathbf{q}^*_{\sim ij}$ analogously indicates that all input parameters except q_i and q_j are at the reference point. Q_i or Q_j denote the random input variables under current consideration. In the process, a suitable reference point $\mathbf{q}^*$ has to be selected. Its choice is crucial and it was defined at the center point of sparse grid quadrature according to recommendations in [8].

Prior analysis by means of total-order Sobol indices revealed that the uncertain input parameters show negligible interaction behavior with respect to considered random responses [2]. For this reason, the second-order terms in Equation (3) were omitted and univariate effects were taken into account. The PCE coefficients for the PCE-HDMR were computed by using discrete projection through Gaussian quadrature with quadrature order of 5.

3 Results

The UQ methods are applied for the stochastic process representation of Quantities of Interest (QoI), that characterize the underlying buoyancy-driven mixing process. This includes the mixture uniformity σ_X, which is the volume-weighted standard deviation of the mole fraction X from the homogeneous equilibrium state mole fraction X_{eq} over the whole fluid volume V. It represents the progress of the mixing process and is determined through the expression

$$\sigma_X = \sqrt{V^{-1} \int_V (X - X_{eq})^2 \, \mathrm{d}V} \,. \tag{4}$$

In addition, the spatially averaged Nusselt number Nu over the left wall is considered, which provides information about the convective heat transfer inside the cavity in dimensionless form. It is determined by the expression

$$Nu_l = \frac{1}{A_l} \int_{A_l} \frac{\left.\frac{\partial T}{\partial \mathbf{n}}\right|_{wall} H}{\Delta T} \mathrm{d}A \,, \tag{5}$$

where $\mathbf{n}$ denotes the wall-normal unit vector and A_l denotes the face area of the left wall. The response variables described above are time-dependent stochastic processes and are shown in the following over dimensionless time represented by the Fourier number $Fo = D_0\, t\, /H^2$, where D_0 is the nominal molecular diffusion coefficient.

Stochastic Models, from which statistics of the random response variables were derived, were computed by means of PCE-HDMR and SGM. Results from both methods are compared in Fig. 2. As a reference for the PCE-HDMR method, results from SGM are employed. The determined statistics, which are shown in Fig. 2, include PDFs $f_R(r)$ of the random responses at different times as well as the visualization of the cumulative distribution function $F_R(r)$ or quantiles using a color map. At the top of Fig. 2 the statistics regarding the mixture uniformity σ_X, are depicted. The mixing process starts with high values for σ_X, which represent the initial inhomogeneous state, and ends with $\sigma_X = 0$, which represents the homogeneous state. At the beginning of the mixing process a distribution of σ_X is recognizable, which is due to the definition of the uncertain initial distribution of the helium. This initial distribution turns into a normal distribution, which gradually widens during the mixing process. After reaching the homogeneous state of the mixture, the probability gradually accumulates completely around $\sigma_X = 0$, which is indicated by high values for the PDFs $f_R(r)$. The PDFs were truncated at $\sigma_X = 0$ in order to ensure the visibility of the initial distributions. Since σ_X reaches zero at different points in time, it is evident, that the duration of the mixing process is subject to uncertainty. This is due to the definition of uncertain input parameters that slow down or accelerate the mixing process. The comparison of obtained statistics with SGM and PCE-HDMR indicates that PCE-HDMR achieves a very good reproduction of the reference results obtained by SGM, while the cost is reduced by 75%. Statistics for the left wall Nusselt number Nu_l are depicted at the bottom of Fig. 2. Nu_l shows a stagnant phase at the beginning of the mixing process, which is also characterized by a normal distribution. This is followed by a transient transition to the new quasi steady state, which takes place at different points in time and thus entails a larger uncertainty band for the values of Nu_l. In addition, this transition is evident from the gradual decrease in probability at low Nu_l and the gradual increase at higher Nu_l over time. Subsequently, the new state is reached and yields a normal distribution again. Slight differences between PCE-HDMR and SGM are only visible in the region, where transient transition to the quasi steady state occurs. In the remaining temporal domain both methods provide good agreement in the derived statistics.

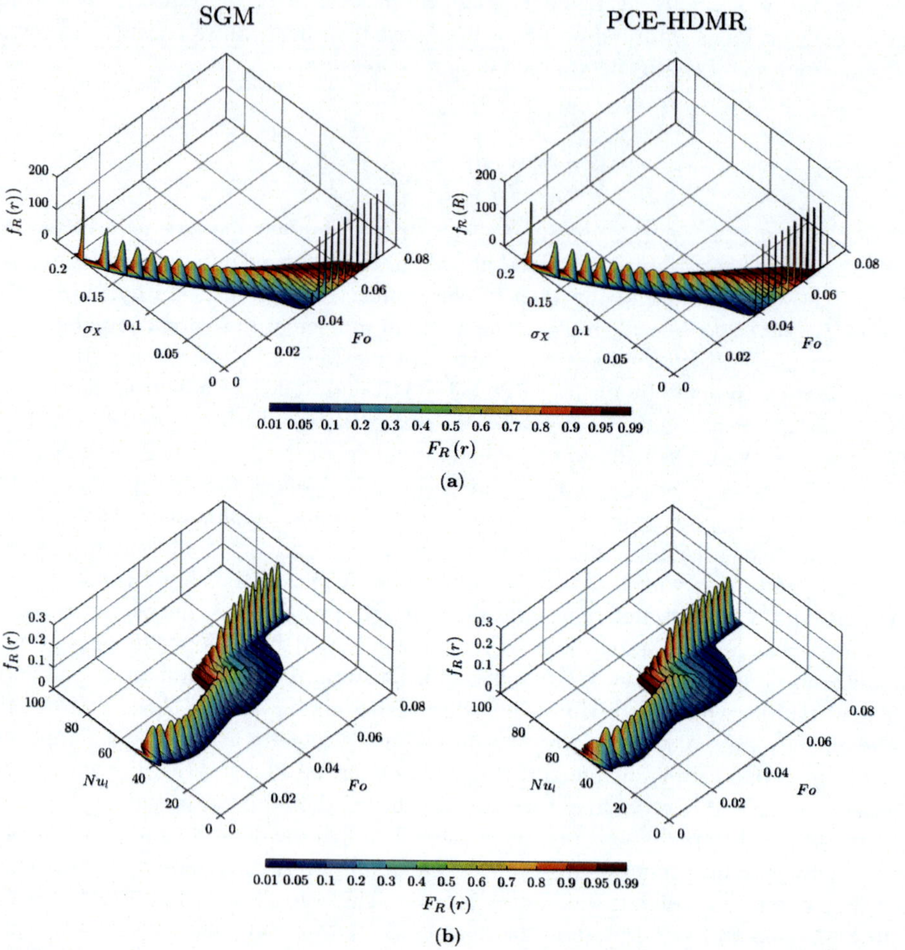

Fig. 2. Statistics of stochastic models from SGM on the left and PCE-HDMR on the right for the time-dependent stochastic processes over time via dimensionless Fourier number Fo: (**a**) Mixture uniformity σ_X; (**b**) left-wall Nusselt number Nu_l.

4 Conclusions

Uncertainty of CFD results was approximated via PCEs and KLEs of LES for buoyancy-driven mixing processes employing PCE-HDMR and SGM. The stochastic model statistics from both approaches revealed very good agreement. Therefore, PCE-HDMR is proven a promising UQ-method for large-scale CFD applications, since it allows for the investigation of a large number of random inputs and further model extensions at will. In addition, the stochastic model construction by means of PCE-HDMR is distinguished through its efficiency by taking advantage of the high convergence rate of numerical integration methods like Gaussian quadrature for individual low-dimensional random input spaces. This reduced the number of 105 simulation runs for SGM for the reference UQ results to 26 simulation runs for PCE-HDMR, which originate from five simulations per random input variable and the calculation at the reference point. PCE-HDMR facilitates tremendous savings of computing resources while preserving high accuracy and is therefore suitable for further investigations of industrial scale CFD applications.

Acknowledgements. Parts of the work presented in this paper are funded by the German Federal Ministry for the Environment, Nature Conservation, Nuclear Safety and Consumer Protection (BMUV, grant No. 1501595) on the basis of a decision by the German Bundestag. The computational resources for this project have been provided by the Gauss Centre for Supercomputing/Leibniz Supercomputing Centre under the project name pn29ce.

References

1. Wenig, P.J., Ji, R., Kelm, S., Klein, M.: Towards uncertainty quantification of LES and URANS for the buoyancy-driven mixing process between two miscible fluids - differentially heated cavity of aspect ratio 4. Fluids **6**, 161 (2021). https://doi.org/10.3390/fluids6040161
2. Wenig, P.J., Kelm, S., Klein, M.: CFD uncertainty quantification using stochastic spectral methods - exemplary application to a buoyancy-driven mixing process. Nucl. Eng. Des. **409**, 112317 (2023)
3. Jivani, A., et al.: Uncertainty quantification for a turbulent round jet using Multifidelity Karhunen-Loeve Expansions. In: AIAA Scitech 2021 Forum (2021). https://doi.org/10.2514/6.2021-1367
4. Huan, X., et al.: Uncertainty propagation using conditional random fields in large-eddy simulations of scramjet computations. In: AIAA Scitech 2019 Forum (2019). https://doi.org/10.2514/6.2019-0724
5. Schenk, C.A., Schuëller, G.I.: Uncertainty Assessment of Large Finite Element Systems, Springer, Heidelberg (2002). https://doi.org/10.1007/11673941
6. Sobol, I.M.: Sensitivity estimates for nonlinear mathematical models. Math. Model. Comput. Exp. **1**, 407 (1993)
7. Yue, X., Zhang, J., Gong, W., Luo, M., Duan, L.: An adaptive PCE-HDMR metamodeling approach for high-dimensional problems. Struct. Multidiscip. Optim. (2021). https://doi.org/10.1007/s00158-021-02866-7
8. Gao, Z., Hesthaven, J.S.: On ANOVA expansions and strategies for choosing the anchor point. Appl. Math. Comput. **217**, 3274–3285 (2010)

Multiphase and Reactive Flows

DNS of Drag Reduction in Dispersed Bubbly Taylor-Couette Turbulence

A. D. Franken[1(✉)], S. R. Ephrati[1], P. Cifani[1,2], and B. J. Geurts[1,3]

[1] Mathematics of Multiscale Modeling and Simulation, University of Twente, Enschede, The Netherlands
a.d.franken@utwente.nl

[2] Gran Sasso Science Institute, Viale F. Crispi 7, 67100 L'Aquila, Italy

[3] Multiscale Physics, Center for Computational Energy Research, Department of Applied Physics, Eindhoven University of Technology, PO Box 513, 5600 Eindhoven, MB, The Netherlands

1 Introduction

Draining of energy due to friction has been a major issue in fluid transport over many decades. Such drain of energy is ubiquitous in naval transportation, process technology, transportation of liquified natural gas and other industries. Historically, many ideas have been proposed to reduce this energy drain, both passively and actively. The particular idea of introducing gas bubbles to fluid flow in order to reduce drag has been around for at least a century. This mode of flow modification with the aim of drag reduction is investigated in this paper. We present a DNS methodology that enables the simulation of large numbers of deformable bubbles in water, thereby resolving all flow features inside and around each of the bubbles. It is shown that the multiphase flow reorganises itself in such a way that the total drag is indeed reduced. This phenomenon is illustrated for bubbles at a Weber number of 8, i.e., bubbles with considerable deformability, displaying reduced friction of several percent.

Latorre [1] reports that patents on using air lubrication to reduce the drag on ship hulls were filed as early as 1880. Early implementations of this idea focused mostly on the addition of polymers to the flow, which showed promising results in turbulent pipe flow [2]. During the last four decades, however, many theoretical, numerical and experimental studies have been performed on the lubricating effect of air bubbles in liquid flows, which show promising results in reducing the frictional energy loss in various flow configurations (e.g., [3] and references therein). The magnitude of drag reduction or of reduction of the driving force in two-phase flows compare to single-phase flows can be quite substantial. It was shown that adding dispersed bubbles at a gas volume concentration of just 4% can lead to a drag reduction of up to 40% in Taylor-Couette flows [4]. Also in many other application areas, the reduction of skin friction can improve the performance of key processes, e.g., in the transport of chemicals in industrial equipment.

Much research on the mechanism behind drag reduction is focussed on the study of Taylor-Couette (TC) flow taking place between two concentric cylinders. This is a closed flow with clearly defined exact global balances that characterize the dynamics.

C. Marchioli et al. (Eds.): DLES 2023, ERCO 31, pp. 241–246, 2024.
https://doi.org/10.1007/978-3-031-47028-8_37

The flow is usually driven by the externally controlled rotation of the cylinders. Statistically stationary states in the Taylor-Couette system enable a detailed evaluation of the energy balances [5]. Despite the research efforts in the past decades, there is still no solid understanding of the physical mechanisms behind the observed drag reduction in turbulent flows. Various theories have been proposed to explain the origin of this phenomenon. Among these mechanisms, bubble compressibility and deformability, as well as bubble size play an important role in the realisation of drag reduction. This indicates the important role of bubble deformability in hindering the exchange of angular momentum between the boundary layer and bulk in TC flow [6].

Direct numerical simulation (DNS) of the Navier-Stokes equations describing a gas-liquid flow of dispersed deformable bubbles in water enables quantitative characterization of the modulation of turbulence arising from an immersed phase. This paper adopts the volume-of-fluid method as implemented in the TBFSolver (https://github.com/cifanip/TBFsolver) to simulate dispersed bubbly turbulence, at considerably higher global gas volume fractions compared to literature, using high-performance computing [7]. Attention is given to (i) the fundamental resolution of bubble-bubble and bubble-wall interactions and (ii) to the clustering of bubbles in Taylor-Couette turbulence.

The organization of this paper is as follows. The computational model for solving two-phase TC flow is presented in Sect. 2. The results on drag reduction in a simulation of turbulent bubbly Taylor-Couette flow are presented in Sect. 3.

2 Computational Model

The numerical technique used in the TBFsolver is based on the volume-of-fluid (VOF) method in which the one-fluid formulation is adopted, i.e., a single set of equations is solved on the entire domain. The VOF method adopts the Piecewise Linear Interface Calculation (PLIC) as a way to geometrically reconstruct the fluid-bubble interface. To describe the dispersed embedded flow discontinuous material properties and interfacial terms associated with the bubbles are accounted for using a marker function f_i for each bubble i. This marker function f_i equals 1 in cells where bubble i fully occupies the cell, 0 in cells that are not occupied by the bubble, and a value between 0 and 1 indicates that the cell contains the bubble interface. The value of the marker function is also referred to as the volume fraction. Given N bubbles, the marker functions are advected via

$$\frac{\partial f_i}{\partial t} + \mathbf{u} \cdot \nabla f_i = 0 \qquad \text{for } i = 1, \ldots, N. \tag{1}$$

The nondimensionalized incompressible Navier-Stokes equations and continuity equation are used to describe the flow:

$$\rho \frac{D\mathbf{u}}{Dt} = -\nabla p + \frac{1}{\mathrm{Fr}^2}\rho \hat{\mathbf{g}} + \frac{1}{\mathrm{Re}}\nabla \cdot (2\mu \mathbf{S}) + \frac{1}{\mathrm{We}} k\mathbf{n}\delta_s \tag{2}$$

$$\nabla \cdot \mathbf{u} = 0 \tag{3}$$

Here $D\mathbf{u}/Dt$ is the material derivative of the velocity $\mathbf{u}$ with t the time. Moreover, p is the pressure, ρ the mass density, μ the viscosity, k is the curvature of the bubble-fluid interface, $\hat{\mathbf{g}}$ is the normalized gravity vector, $\mathbf{S}$ the deformation tensor, $\mathbf{n}$ is the normal vector to the interface, and δ_s is a surface Dirac-δ function that is nonzero only at the interface. The dimensionless numbers are the Froude number $\mathrm{Fr} = U/\sqrt{gL}$, the Reynolds number $\mathrm{Re} = UL\rho_1/\mu_1$, and the Weber number $\mathrm{We} = LU^2\rho_1/\sigma$. Here L and U denote a characteristic length and a characteristic velocity, respectively. The subscript 1 denotes the continuous phase mass density.

The mass density and viscosity at a certain point follow from the marker functions and the properties of the continuous and dispersed phases. For instance, a cell with volume fraction c would have a density and a viscosity of

$$\rho = \rho_1\,(1-c) + \rho_2\,c, \qquad \mu = \mu_1\,(1-c) + \mu_2\,c. \tag{4}$$

were $c = \max_i f_i$. The continuous surface force (CSF) method is used to model the surface tension term. This method replaces the delta function $\delta(n)\mathbf{n}$ by a smooth term, which is computationally easier but may induce spurious currents. Reducing such spurious currents can effectively be done by accurately computing the curvature of the interface and hence the surface tension term, for which a generalized height function method is adopted.

A three-dimensional Cartesian grid is used to discretize the domain and a staggered arrangement of the variables was selected. The spatial discretization of the convective term is based on the finite volume approach. The fully conservative discretization scheme of the Navier-Stokes equation by [8] is adopted. Additionally, the QUICK interpolation scheme is implemented to avoid unphysical oscillations that may occur as the spatial resolution is too low. A second-order finite difference scheme is employed for the diffusive term. Finally, a third-order Runge-Kutta scheme is used to discretize the convection and diffusion terms of the Navier-Stokes equations, while a Crank-Nicolson scheme is employed for the surface tension term. Also for incompressible multiphase flow the Poisson equation is a primary computational challenge - here we adopt a fast Fourier transform method to solve for the variable density problem. The implementation and high-performance TBFSolver code that harnesses these methods is directly available online via open-source access.

3 Bubbly Taylor-Couette Turbulence

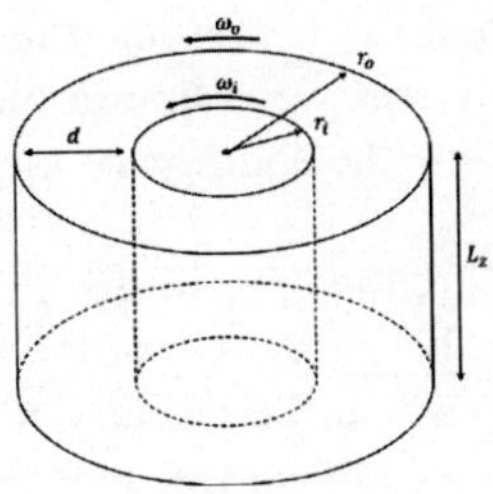

Fig. 1. Domain for the Taylor-Couette flow.

We concentrate on turbulence in Taylor-Couette (TC) flow as this presents a finite domain in which all conditions can be controlled with high fidelity. To simulate the turbulent flow in this configuration the TBFsolver was extended to cylindrical coordinates. We focus on a high-Reynolds number TC flow with $Re = 5 \cdot 10^3$ on a grid with $N_\theta \times N_r \times N_z = 768 \times 192 \times 384$ intervals in the circumferential, radial and vertical directions respectively. We investigate a case with 120 bubbles with deformability characterized by a Weber number $We = 8$ and a bubble diameter $d_b = 0.1(r_o - r_i)$ in terms of the outer and inner radii of the TC domain. The bubble volume fraction is taken as $\alpha = 0.8\%$ and the mass density ratio $\rho_f/\rho_b = 20$ in our simulations.

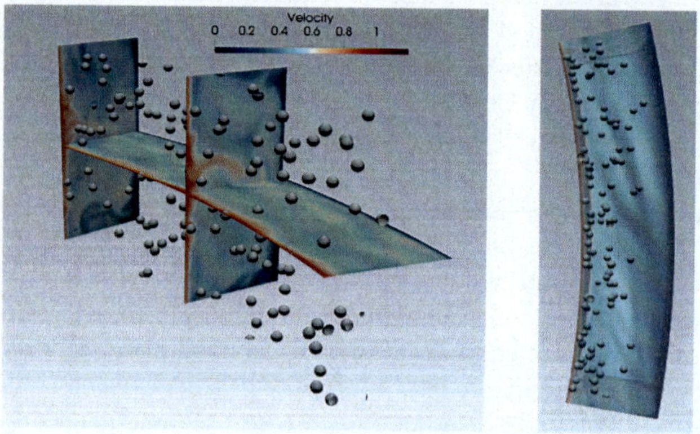

Fig. 2. Snapshot of the velocity magnitude in two-phase TC flow - red/blue indicates high/low velocity. Bubbles are seen to cluster near the inner wall as seen clearly in the top-view (right figure).

Figure 1 shows the domain for this experiment. We initialize the flow by first allowing the system to evolve in the absence of bubbles. The flow is driven by the fixed constant angular velocity of the inner cylinder, while the outer cylinder is kept stationary. After the single-phase flow reaches a statistically stationary state, the bubbles are injected into the flow and we further allow the flow to develop until a new statistically stationary state is reached.

A snapshot of the flow with bubbles is presented in Fig. 2. We observe strong clustering in the bubble concentration, corresponding to the large-scale circulations in the single-phase flow being disrupted by the addition of the bubbles. This clustering is further quantified in Fig. 3.

In a statistically stationary state, the drag in the TC system is given by the torque on the cylinder walls, which can be measured as the total shear stress on the walls. Following [9], the wall shear stress can be directly related to the radial transport of angular velocity. In a direct analogy to heat transport in Rayleigh-Bénard convection, [9] introduces a Nusselt number Nu^ω based on the transport of angular velocity

$$\mathrm{Nu}^\omega = \frac{J^\omega}{J^\omega_{lam}}, \quad \text{where } \langle T \rangle = 2\pi L_z \rho J^\omega, \tag{5}$$

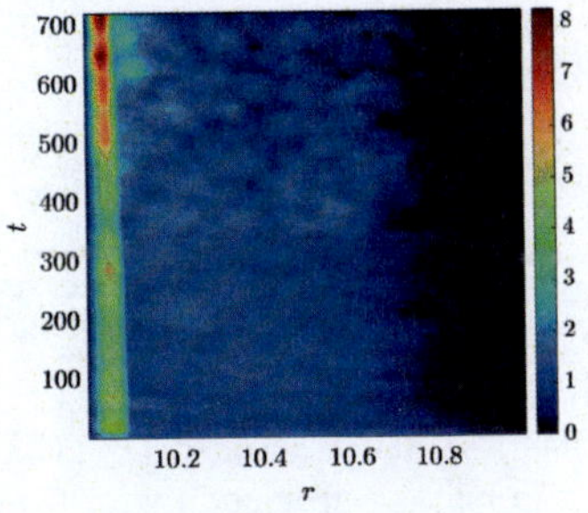

Fig. 3. Probability distribution function of the radial coordinates of the gas phase. PDFs are given as temporal averages over 2 flow-through times t_F, where $t_F = 4$.

where $\langle T \rangle$ is the time-averaged torque and J_{lam} is the angular velocity transport in the case of laminar flow, which allows us to measure drag reduction in terms of the Nusselt number.

In Fig. 4 the instantaneous torque on the inner and outer cylinder walls are plotted at $We = 8$. We observed that the flow reaches a statistically stationary state after a phase in which the initially regularly placed bubbles got mixed. The Nusselt number for this problem Nu_ω was found to be $8.48 \pm 0.05\%$ in case bubbles are entirely absent. From Fig. 4 we hence infer that there is a clear but small reduction in the drag due to the addition of bubbles.

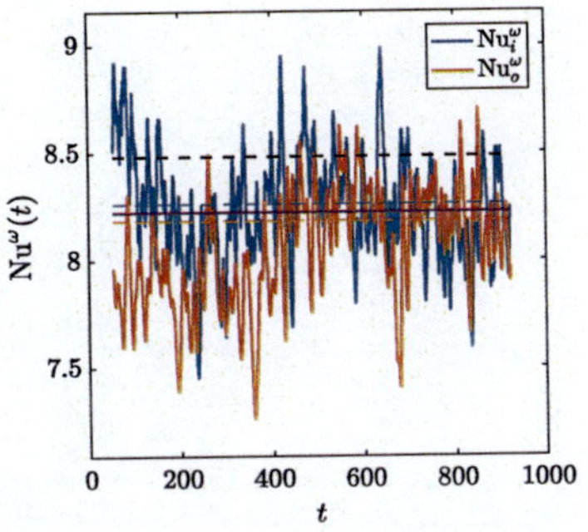

Fig. 4. Evolution of the torque on the inner and outer wall in turbulent TC flow as measured on the inner wall (blue) and outer wall (red). Indicated with the dashed black line is the Nusselt number in the absence of bubbles. The solid line indicates the average Nusselt number after reaching the statistically stationary state.

This new flow solver enables high-resolution DNS of bubbly Taylor-Couette turbulence. The simulations allow us to study the effects of buoyancy and deformability of large numbers of highly deformable bubbles in turbulent TC flow, which appear to be crucial ingredients for drag reduction.

References

1. Latorre, R.: Ship hull drag reduction using bottom air injection. Ocean Eng. **24**, 161–175 (1997)
2. Virk, P.S.: Drag reduction fundamentals. AIChE J. **21**, 625–656 (1975)
3. Verschoof, R.A., van der Veen, R.C.A., Sun, C., Lohse, D.: Bubble drag reduction requires large bubbles. Phys. Rev. Lett. **117**, 104502 (2016)
4. Van Gils, D.P.M., Narezo Guzman, D., Sun, C., Lohse, D.: The importance of bubble deformability for strong drag reduction in bubbly turbulent Taylor-Couette flow. J. Fluid Mech. **722**, 317–347 (2013)
5. van den Berg, T.H., Luther, S., Lathrop, D.P., Lohse, D.: Drag reduction in bubbly Taylor-Couette turbulence. Phys. Rev. Lett. **94**, 44501 (2005)
6. Spandan, V., Verzicco, R., Lohse, D.: Physical mechanisms governing drag reduction in turbulent Taylor-Couette flow with finite-size deformable bubbles. J. Fluid Mech. **849**, R31–R313 (2018)
7. Cifani, P., Kuerten, J.G., Geurts, B.J.: Highly scalable DNS solver for turbulent bubble-laden channel flow. Comput. Fluids **172**, 67–83 (2018)
8. Morinishi, Y., Vasilyev, O.V., Ogi, T.: Fully conservative finite difference scheme in cylindrical coordinates for incompressible flow simulations. J. Comput. Phys. **197**, 686–710 (2004)
9. Eckhardt, B., Grossmann, S., Peterson, W.H.: Torque scaling in turbulent Taylor-Couette flow between independently rotating cylinders. J. Fluid Mech. **581**, 221–250 (2007)

Analysis of Turbulent Kinetic Energy Budgets in Co-current Taylor Bubble Flow

E. M. A. Frederix(✉), S. Tajfirooz, and E. M. J. Komen

Nuclear Research and Consultancy Group, Westerduinweg 3, 1755 Petten, LE, The Netherlands
frederix@nrg.eu

1 Introduction

Boiling in a nuclear power plant reactor vessel may occur at normal operation or only in accident scenarios, depending on the reactor type. After a planned or emergency shutdown, the removal of residual heat, initially at hundreds of megawatts, is crucial. The rate of heat removal by the primary coolant in a reactor reduces dramatically when too much steam is generated, in turn increasing fuel temperatures which can lead to irreparable damage to the reactor. This drives the need for accurate two-phase models describing the complex turbulent flow dynamics inside reactor core sub-channels, in order to determine operational safety limits.

Whereas in the past, such models have mainly relied on spatial averaging (system codes) or full temporal averaging (two-fluid model), current research is dedicated to a new hybrid approach for the description of two-phase structures. The idea is to rely on unsteady Reynolds-Averaged Navier-Stokes (RANS) models for the description of the Reynolds stress tensor, but to apply a separation of two-phase scales treating small bubbles in an averaged or filtered sense while resolving large structures on the grid. Such an approach is called *multiple flow regime* modeling, and carries an obvious similarity with how eddies are treated in LES.

However, predictions of Turbulent Kinetic Energy (TKE) by traditional single-phase RANS models applied to two-phase scenarios are known to be much too high as a result of the presence of a resolved two-phase interfaces [2]. To make multiple flow regime models work, improved two-phase RANS models are therefore needed, tailored to two-phase flows incorporating surface tension effects and applying asymmetric TKE damping at the interface, such as the one of [1]. In support of this model development, high-fidelity simulation of relevant turbulent two-phase flow can play a pivotal role as a means of model training using machine learning or for validation.

In this work, we analyze two-phase TKE budgets using high fidelity simulation of turbulent co-current Taylor bubble flow. We use the simulation setup developed in [3], which leverages a Moving Frame of Reference (MFR) attached to the Taylor bubble, allowing to compute two-phase TKE budgets as given in [4] relative to the position of the volumetric center of the bubble. As compared to the single-phase wall-bounded flow, the two-phase Taylor bubble flow is much richer in physics, including laminarization of the liquid flow along the Taylor bubble towards its tail, a toroidal wake vortex, and turbulence interaction at the interface. Therefore, the appropriate mesh resolution for a direct numerical simulation of the situation is not known a priori, and a mesh convergence study of the TKE budgets is performed to investigate mesh sensitivity.

C. Marchioli et al. (Eds.): DLES 2023, ERCO 31, pp. 247–252, 2024.
https://doi.org/10.1007/978-3-031-47028-8_38

2 Model and Simulation Setup

The two-phase flow is modeled by the one-fluid Volume Of Fluid (VOF) method, in which the liquid volume fraction f adheres to

$$\partial_t f + u_j \partial_j f = 0 \tag{1}$$

with mixture velocity u_j. The momentum equation is given by

$$\partial_t u_i + \partial_j (u_i u_j) = \frac{1}{\rho}\left[-\partial_i p + \frac{\partial_j (2\mu S_{ij})}{\text{Re}} + \frac{\kappa \partial_i f}{\text{We}}\right] + \frac{g_i}{\text{Fr}^2} \tag{2}$$

with mixture density ρ, mixture viscosity μ, pressure p, interface curvature κ, strain rate tensor S_{ij}, gravity g_i, Reynolds number Re, Weber number We and Froude number Fr. All quantities in Eq. (1–2), including density and viscosity, are non-dimensional. The scaling of density and viscosity introduces two additional non-dimensional parameters, i.e., the liquid–gas viscosity ratio Γ_μ and liquid–gas density ratio Γ_ρ, according to

$$\rho = f + (1-f)/\Gamma_\rho \qquad \text{and} \qquad \mu = \frac{1}{f + (1-f)\Gamma_\mu}, \tag{3}$$

where a harmonic mixing of phase-specific viscosities is used in the mixture viscosity, which was identified to give better results for simple two-phase flow validation cases. The momentum equation is supplemented by the continuity equation which, as a result of assumed incompressibility, is given by $\partial_i u_i = 0$.

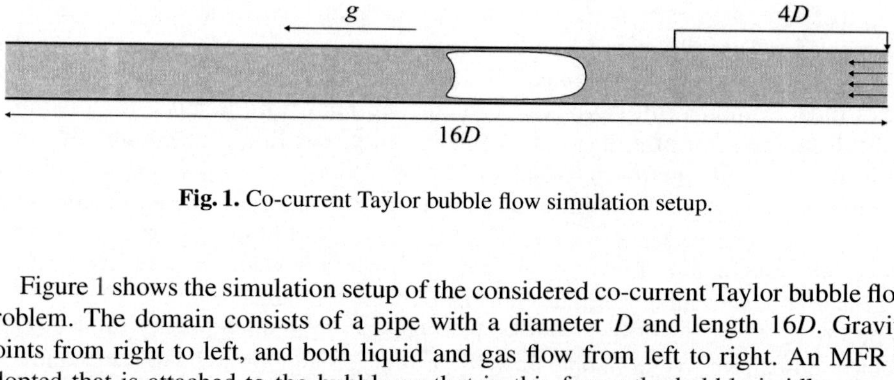

Fig. 1. Co-current Taylor bubble flow simulation setup.

Figure 1 shows the simulation setup of the considered co-current Taylor bubble flow problem. The domain consists of a pipe with a diameter D and length $16D$. Gravity points from right to left, and both liquid and gas flow from left to right. An MFR is adopted that is attached to the bubble so that in this frame the bubble is idle, conveniently allowing for statistical averaging of data relative to the bubble's center of mass. Because the bubble rises faster than the liquid due to buoyancy, in the MFR, liquid moves on average from right to left, meaning that the inlet (as seen by the bubble) is above the bubble. In order to have developed single-phase turbulent flow ahead of the bubble, the velocity field is recycled over a distance of $4D$ as indicated in Fig. 1. Simulations are initialized from a fully developed single-phase flow into which a cylindrical bubble with a diameter $0.85D$ and length $2D$ with an attached hemispherical head is imposed, in such a way that its nose is at roughly $6D$ from the inlet. After some initial development phase, data is averaged in time and space (circumferentially), and across four different realizations of the same flow that are started with four statistically independent initial flow fields. This ensemble averaging is needed, because the bubble slowly loses void by break-up, thus limiting the temporal averaging window beyond which the flow topology has changed measurably.

Preliminary simulations have identified suitable values for the five parameters of the problem. First, the Reynolds number (based on the liquid bulk velocity) is set to 5300, allowing for comparison with single-phase DNS databases of the flow ahead of the bubble. The Weber number is set to 40 and the Froude number to 1.5, allowing for significant relative velocity (about 32% of the liquid bulk velocity) without too much bubble break-up, which would break the statistical steady state. The density ratio does not affect the Taylor bubble flow much, and is moderately set to 10, which improves numerical convergence. Likewise, the viscosity ratio is also set to 10 so that the liquid and gas-based Reynolds numbers are comparable. The Taylor bubble is driven by its buoyancy, so that its motion is determined by three non-dimensional numbers: the ratio of inertia over buoyancy, surface tension over buoyancy and viscosity over buoyancy. In terms of the non-dimensional numbers defined above, these quantities are Fr^2, $\mathrm{Fr}^2/\mathrm{We}$ and $\mathrm{Fr}^2/\mathrm{Re}$, respectively. Using the proposed values, we observe that $\mathrm{Fr}^2 > \mathrm{Fr}^2/\mathrm{We} > \mathrm{Fr}^2/\mathrm{Re}$, meaning that the bubble is in the inertial regime [5]. In turn, the bubble velocity relative to the liquid is proportional to $\sqrt{gD}$, showing that the choice for viscosity and density does not influence the motion of the bubble for the selected values of Fr, We and Re.

The problem is simulated using the second-order collocated finite-volume solver *Basilisk*, in which a uniform Cartesian mesh is employed with cubic cells mapping a rectangular cuboid with dimensions $16D \times D \times D$. The wall of the pipe, which intersects cells, is modeled using Basilisk's 'embedded boundaries' approach. Simulations are performed on 'Mesh 11', having $2^{11} = 2048$ cells along the pipe length, and 'Mesh 12', having $2^{12} = 4096$ cells along the pipe length. Mesh 11 and 12 have 31M and 247M cells, respectively. On these meshes, single-phase turbulent pipe flow simulations are first performed and compared with reference data, showing excellent agreement for Mesh 12. More details on the method and mesh can be found in [3].

Two-phase TKE budgets are calculated using the TKE equation derived by Vreman & Kuerten [4] which reads, for each phase k,

$$\partial_t(\alpha_k K_k) = -C_k + \mathscr{P}_k - \varepsilon_k + D_k + S_k + \mathrm{PTI}_k = 0, \tag{4}$$

with phase k mean volume fraction α_k, TKE convection C_k, production $\mathscr{P}_k$, dissipation ε_k, viscous diffusion D_k, surface tension S_k, and combined Pressure diffusion – Transport – Interfacial budgets PTI_k. This last budget is computed indirectly, by leveraging the fact that the solution is in a statistical steady state such that $\partial_t(\alpha_k K_k) = 0$, allowing PTI_k to be computed as the negative sum of the other budgets. It was found that a direct calculation of PTI_k is complicated by its dependence on $\partial_j f$, a term that approaches infinity as the grid size decreases, meaning that it does not converge. The indirect calculation does not suffer from this difficulty. Moreover, in RANS models, the contributions to the PTI budget are almost always lumped into a single term too, and approximated using a gradient-diffusion model. Thus, their indirect lumped representation may offer a validation base for RANS models.

3 Results

Fig. 2. Coherent turbulent structure visualization using iso surfaces (green) of the λ_2 criterion of the instantaneous velocity field. The bubble is represented by an iso surface (blue) at $f = 0.5$.

Figure 2 shows a visualization of coherent turbulent structures in the flow around the Taylor bubble, using iso-surfaces of the λ_2 criterion. It can be seen that vortices in the liquid ahead of the Taylor bubble are stretched as they are entrained by the film enclosing the bubble. At the tail of the bubble, standing ring vortices are observed. In the wake of the bubble there is much turbulence production by the liquid ejected at relatively high velocity from the film. The turbulence is mostly contained by the toroidal vortex that flows downward (with respect to the bubble) at large radii close to the wall and upward at small radii, where the liquid wake turbulence induces a turbulent flow inside the lower part of the bubble (not visible in Fig. 2).

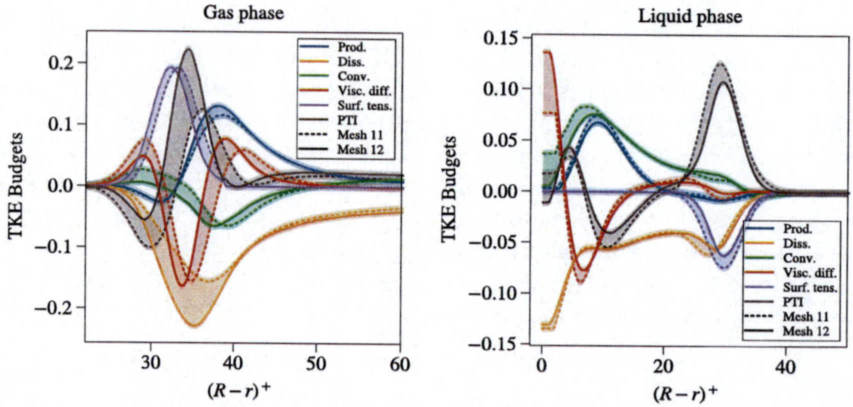

Fig. 3. Gas (left) and liquid (right) TKE budgets calculated across the Taylor bubble on Mesh 11 and 12. Data is in wall units. The variable r is the radial coordinate. The area between the same budget on Mesh 11 and 12 is shaded, to better highlight the grid sensitivity.

Figure 3 shows two-phase TKE budgets (per unit mass) in the gas phase (left) and liquid phase (right), across a radial line through the bubble at a distance of roughly D below its nose. The position where the average void fraction $\overline{f}$ crossed 0.5 is at roughly $(R-r)^+ = 30$. Shown are the production, dissipation, convection, viscous diffusion, surface tension and PTI budgets on both Mesh 11 and 12, and the area between the same budget on either mesh is shaded, to better highlight the grid sensitivity.

Three important conclusions can be drawn from Fig. 3. First, it is observed that the TKE budgets are still quite sensitive to the mesh, most notably for the dissipation in the gas phase, suggesting that these results cannot be regarded as true DNS. However, the mesh is such that Δy^+ values for the considered liquid Re are spatially uniform and close to unity, indeed providing excellent agreement with reference data for the single-phase TKE budgets ahead of the bubble (not shown here). Thus, the two-phase nature of the problem induces additional requirements on the mesh that are not yet met, or may never be met due to $\partial_i f \rightarrow \infty$ as the mesh is refined. After inspection of the behavior of turbulent structures in the wake of the Taylor bubble using the λ_2 criterion (as shown in Fig. 2), it can be concluded that the liquid turbulent motion entrapped by the Taylor bubble wake toroidal vortex induces a turbulent flow in the bubble itself, transferred across the bubble's tail interface. The tail turbulence inside the bubble possesses smaller length scales than what the current meshes (designed based on single-phase wall-bounded flow arguments) can accommodate, resulting in significant mesh dependence.

Second, it is observed that surface tension plays a measurable role in the production and destruction of TKE. In the liquid phase, surface tension is seen to damp turbulence but appears to be mostly balanced by the PTI budget. In the gas phase, the surface tension budget it is seen to produce turbulence. This is an effect that has not been considered in any conventional RANS model and gives rise to further research in this direction, e.g., by study of the influence of We on the surface tension TKE budget.

Third, it can be observed that production and dissipation in the gas phase close to the interface behave much like how they would in single-phase wall-bounded flow, while in the liquid phase, there is negligible TKE production close to the interface. This suggests that two-phase RANS model improvement using wall functions in the gas phase may have merit. Moreover, as for example discussed in [1], suitable two-phase RANS models should possess an asymmetric damping term, acting mostly in the lighter phase.

4 Conclusions

In summary, this work explores the behavior of two-phase TKE budgets near large scale resolved interfaces using high-fidelity simulation, allowing for a better understanding of the behavior of separate terms affecting TKE. In turn, RANS models may be improved from the insight generated by this data, to be potentially harvested by modern machine learning tools. The results establish the relevance of surface tension in the modeling of mean TKE, and suggest an asymmetry in the modeling of production and dissipation on either side of the interface. However, the used meshes that are designed based on single-phase wall-bounded turbulent flow arguments was found to be insufficient to resolve the smallest length scales present in the flow, such that further refinement is needed for the development of a true DNS of the simulated setting.

References

1. Dong, Z., Bürgler, M., Hohermuth, B., Vetsch, D.F.: Density-based turbulence damping at large-scale interface for Reynolds-averaged two-fluid models. Chem. Eng. Sci. **247**, 116975 (2022)
2. Frederix, E.M.A., Mathur, A., Dovizio, D., Geurts, B.J., Komen, E.M.J.: Reynolds-averaged modeling of turbulence damping near a large-scale interface in two-phase flow. Nucl. Eng. Des. **333**, 22–130 (2018)
3. Frederix, E.M.A., Tajfirooz, S., Hopman, J.A., Fang, J., Merzari, E., Komen, E.M.J.: Two-phase turbulent kinetic energy budget computation in co-current Taylor bubble flow. Nucl. Sci. Eng. Accepted for publication (2022)
4. Vreman, A.W., Kuerten, J.G.M.: Turbulent channel flow past a moving array of spheres. J. Fluid Mech. **856**, 580–632 (2018)
5. Wallis, G.B.: One-Dimensional Two-phase Flow. McGraw-Hill, New York (1969)

Taylor Bubble Breakup in Counter-Current Turbulent Flow Using Algebraic and Geometric Volume of Fluid Method

J. Kren[1,2](✉), E. M. A. Frederix[3], and B. Mikuž[1]

[1] Reactor Engineering Division, Jožef Stefan Institute, Jamova cesta 39, Ljubljana, Slovenia
{jan.kren,blaz.mikuz}@ijs.si

[2] Faculty of Mathematics and Physics, University of Ljubljana, Jadranska ulica 19, 1000 Ljubljana, Slovenia

[3] Nuclear Research and Consultancy Group, Westerduinweg 3, 1755 Petten, LE, The Netherlands
frederix@nrg.eu

1 Introduction

Turbulent flow of two-phase fluids occurs in many industrial systems. One of the manifestations of the two-phase flow in a pipe is slug flow, which consist of large gas bubbles (i.e. Taylor bubbles) separated from each other by liquid slugs. We are particularly interested in the phenomena of coalescence and breakup in the slug flow regime and the development of numerical models for the Large Eddy Simulation (LES) method as they are still in early stages of development for two-phase flows [1].

The counter-current Taylor bubble flow regime is subjected to bubble breakup due to several effects, such as breakup due to turbulent fluctuations, shearing-off process or due to viscous shear forces [2]. Comparison between LES simulations and experimental measurements [3] show that the simulations overpredict the bubble breakup. The rate of artificial numerical breakup is attributed mainly to the used numerical method for interface capturing. In this work we present results obtained with improved geometric Piecewise Linear Interface Capturing (PLIC) method in the OpenFOAM computer code.

2 Simulation of Taylor Bubble in Turbulent Counter-Current Flow

A two-phase mixture of gas and liquid is modeled using the one-fluid formulation of the Navier-Stokes equations with the Volume Of Fluid (VOF) approach for interface capturing. In this method a void fraction α is introduced and the advection equation for this quantity is the following:

$$\partial_t \alpha + \mathbf{u} \cdot \nabla \alpha = 0, \tag{1}$$

with partial time derivative ∂_t, velocity vector $\mathbf{u}$ and gradient operator ∇. The fluid mixture is described by the incompressible Navier-Stokes equations, i.e.,

$$\nabla \cdot \mathbf{u} = 0 \tag{2}$$

C. Marchioli et al. (Eds.): DLES 2023, ERCO 31, pp. 253–258, 2024.
https://doi.org/10.1007/978-3-031-47028-8_39

and

$$\partial_t(\rho\mathbf{u}) + \nabla\cdot(\rho\mathbf{u}\mathbf{u}) = -\nabla p + \rho\mathbf{g} + \nabla\cdot(2\mu_{eff}\mathbf{D}) + \sigma\kappa\delta(n)\mathbf{n} \quad (3)$$

with mass density ρ, pressure p, gravitational acceleration $\mathbf{g}$, deformation tensor $\mathbf{D}$, effective mixture viscosity μ_{eff}, with the harmonic expression of viscosity, interface curvature κ, interface normal unit vector $\mathbf{n}$ and interface Dirac delta function $\delta(n)$.

The described equations were solved in the open-source CFD code OpenFOAM v9 [4]. We have used a modified interFoam solver, which enables the usage of the Diagonally Implicit Runge-Kutta (DIRK) time integration schemes integrated with PLIC geometric reconstruction. This solver was based on a solver developed before in OpenFOAM v4 by Frederix et al. [5]. Turbulence at the sub-grid scales was modeled by Vreman model [6]. In the current case, the fluid properties were selected to mimic water-air mixture. Taylor bubble motion has been simulated for two different cases, i.e. in a pipe with a diameter of 12.4 mm (the laminar case) and in a pipe with a diameter of 26 mm (the turbulent case). In each case the total length of the domain is 20 Dh. Results have been compared with the measurements in an experiment [7] performed at THELMA laboratory of Reactor Engineering Division of Jožef Stefan Institute.

In order to achieve a fully developed turbulent flow a recycling boundary condition was used at the inlet, which in this case is upstream of the Taylor bubble. The flowrate was adjusted at every timestep so that the bubble‘s bouyancy is balanced by hydrodynamic drag and the bubble stays at approximately constant position throughout the simulation. This procedure is demonstrated in Fig. 1.

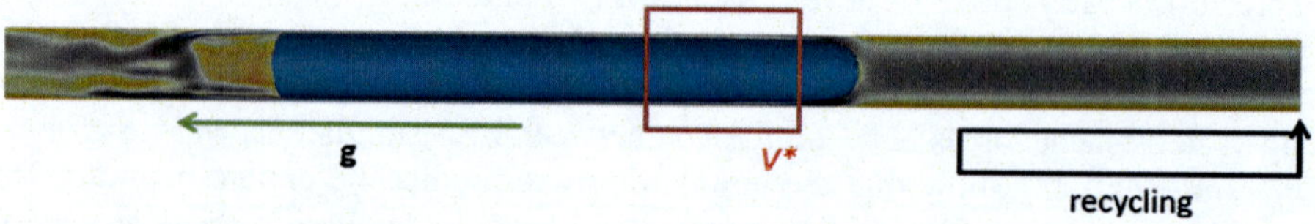

Fig. 1. Visualisation of the recycling procedure. The bubble motion was calculated for every timestep in the volume V^* and the flow rate has been adjusted accordingly.

The same numerical mesh with approximately 1 million cells and mesh refinement towards the wall has been used for all cases as can be seen in Fig. 2.

The key to an algebraic VOF method is to correctly compute the numerical fluxes to update the function of void fraction [9]. Originally, a donor-acceptor formulation was used with flux limiters to ensure the boundendness of the α function. Algebraic capturing is in OpenFOAM v9 performed through Multidimensional Universal Limiter for Explicit Solution (MULES) framework, that uses Flux Corrected Transport (FCT) with an interface compression scheme. As opposed to the algebraic VOF method, a geometric VOF method can be divided into reconstruction and the advection of the interface [10]. Firstly, the approximation of the interface is built from the information on the volume fraction. This can be done in various manners such as PLIC interface reconstruction methods. Second, the reconstructed interface is advected by the given velocity field. For PLIC methods the reconstruction is a two-step procedure. In any

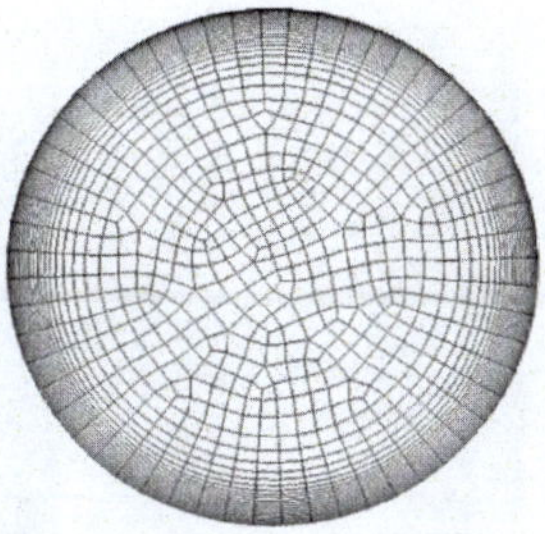

Fig. 2. Cross-section of the mesh.

given cell the normal is first determined from the known α function in this cell and in the neighboring ones. Geometrically, the line is then moved in the normal direction so that the interface position in each cell corresponds to the local void fraction value. In OpenFOAM, the PLIC method is used to compute the fluxes, then the α equation is solved algebraically.

3 Results

Figure 3 shows isosurfaces at $\alpha = 0.5$ of the instantenous gas void fraction in the laminar case for the two different interface capturing methods at $t = 5$ s, $t = 10$ s and $t = 15$ s. We observe void shedding for the algebraic VOF case, whereas it is not present for the geometric VOF. It should be noted that in the experiment almost no void shedding has been observed for Taylor bubbles of similar lengths in laminar flow (in few hours the bubble sheds). A significant difference is visible also in the velocity fields in the wake region of the bubble in Fig. 3. The reason for the observed difference in the velocity field is attributed mainly to the amount of small bubbles downstream the Taylor bubble. This shows that better capture of the interface is crucial for correct determination of all fluid flow physical properties and especially the bubble breakup and coalescence.

Figure 4 shows isosurfaces $\alpha = 0.5$ of the instantenous gas void fraction in the turbulent case for the two different interface capturing methods at time $t = 2.2$ s after the initial state. Clearly, at the skirt of the bubble the velocities are the largest which essentialy slowly breaks the bubble. For the algebraic capturing the breakup is much more significant and faster as signified by the many smaller bubbles in the wake of the Taylor bubble.

Figure 5 shows the surface averaged α at the outlet of the domain. It is shown that breakup as predicted by the geometric method is much more moderate than that by the algebraic method, giving better agreement with the experimental value of about 0.00034 [8]. For the PLIC method the bubble was still present in the simulation after 15 s, whilst with the algebraic capturing the bubble completely broke down in less than 6 s. This shows that accurate determination of the bubble shape is extremely important and has high impact on the breakup and coalescence rate. Additional models for the breakup and coalescence should be developed after the usage of more accurate interface capturing schemes.

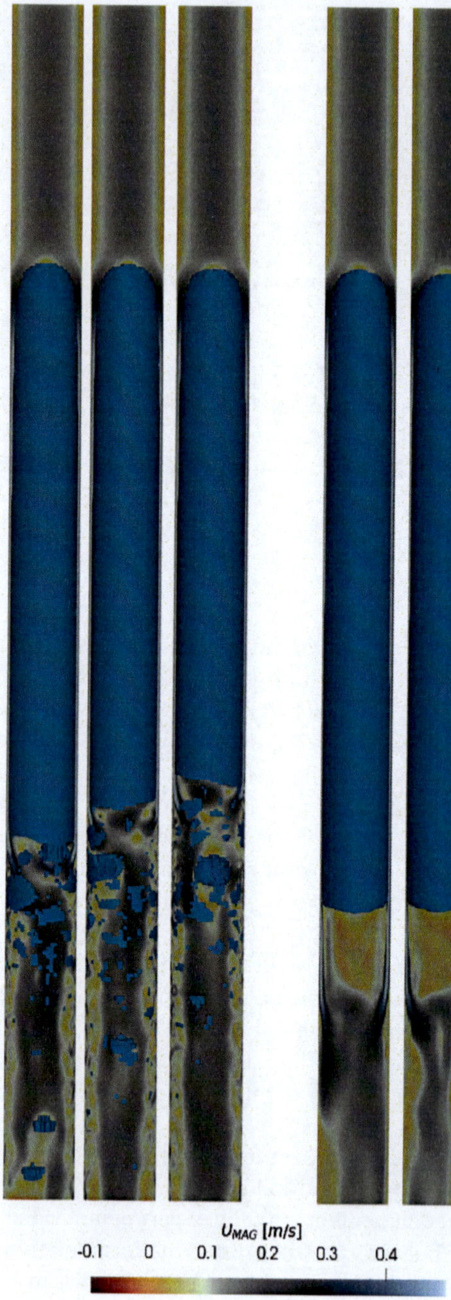

Fig. 3. Isosurfaces of the instantaneous gas void fraction (blue color) at three different timesteps ($t = 5$ s, $t = 10$ s and $t = 15$ s) obtained with standard algebraic interface capturing (left) and PLIC reconstruction (right) for laminar case ($Re = 1400$). The color scheme in the liquid phase represents the velocity magnitude.

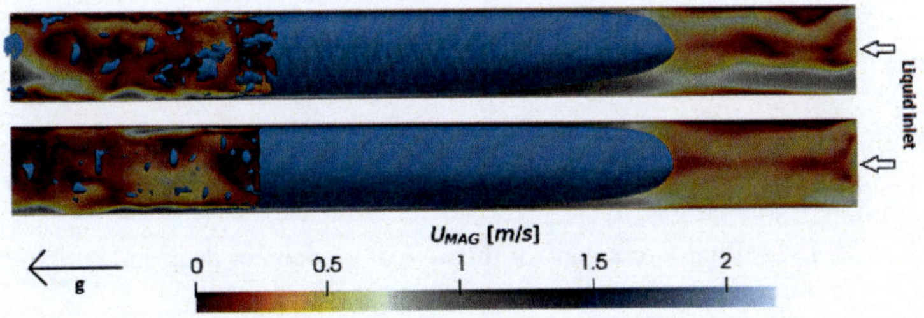

Fig. 4. Isosurfaces of the instantaneous gas void fraction 2.2 s after the initial state obtained with standard algebraic interface capturing (top) and PLIC reconstruction (bottom) for turbulent case ($Re = 5520$). The color scheme in the liquid phase represents the velocity magnitude. (Color figure online)

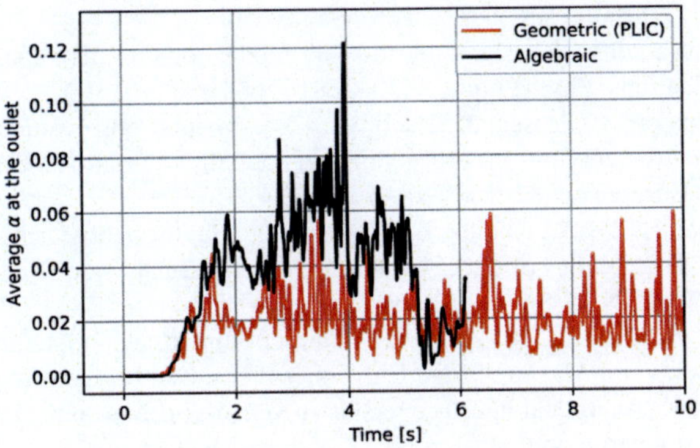

Fig. 5. Comparison of the surface averaged α at the pipe outlet between PLIC and algebraic interface capturing method. The average experimental value for α at the outlet is significantly smaller −0.000034.

4 Summary

In this work we presented a comparison between the performance of algebraic and geometric interface capturing methods for the VOF solver with Runge-Kutta time-integration. The methods were developed, tested and validated in the OpenFOAM v9. Current results of the Taylor bubble in the counter-current turbulent flow case show that the usage of the geometric interface reconstruction scheme PLIC significantly reduces artificial numerical breakup of the bubble and makes it closer to the experimental results. This confirms that the PLIC method is superior to algebraic capturing. It also suggests that bubble breakup depends significantly on the shape of the bubble interface. This is confirmed in particular with the simulation results of the laminar case, where PLIC method shows significantly improved results with no void shedding.

References

1. Klein, M., Ketterl, S., Hasslberger, J.: Large eddy simulation of multiphase flows using the volume of fluid method: part 1-governing equations and a priori analysis. Exp. Comput. Multiph. Flow **1**, 130–144 (2019)
2. Liao, Y., Lucas, D.: A literature review of theoretical models for drop and bubble breakup in turbulent dispersions. Chem. Eng. Sci. **15**, 3389–3406 (2009)
3. Mikuž, B., Frederix, E.M.A., Komen, E.M.J., Tiselj, I.: Taylor bubble behaviour in turbulent flow regime. In: Proceedings of the Conference Computational Fluid Dynamics for Nuclear Reactor Safety (CFD4NRS-8) (2020)
4. The OpenFOAM Foundation : OpenFOAM | Free CFD Software (2021). https://openfoam.org/
5. Frederix, E.M.A., Komen, E.M.J., Tiselj, I., Mikuž, B.: LES of turbulent co-current Taylor bubble flow. Flow Turbul. Combust. **2**, 471–495 (2020)
6. Vreman, A.W.: An eddy-viscosity subgrid-scale model for turbulent shear flow: algebraic theory and applications. Phys. Fluids, 3670–3681 (2004)
7. Mikuž, B., Kamnikar, J., Prošek, J., Tiselj, I. : Experimental observation of Taylor bubble disintegration in turbulent flow. In: Proceedings of the 28th International Conference Nuclear Energy for New Europe, pp. 87–106 (2019)
8. Mikuž, B., Frederix, E.M.A., Komen, E.M.J., Tiselj, I.: Taylor bubble behaviour in turbulent flow regime. In: Proceedings of the Conference Computational Fluid Dynamics for Nuclear Reactor Safety (CFD4NRS-8), vol. 12 (2020)
9. Tryggvason, G., Scardovelli, R., Zaleski, S.: Direct Numerical Simulations of Gas-Liquid Multiphase Flows, vol. 324. Cambridge University Press, Cambridge (2011)
10. Dai, D., Tong, A.Y.: Analytical interface reconstruction algorithms in the PLIC-VOF method for 3D polyhedral unstructured meshes. Int. J. Numer. Meth. Fluids **5**, 213–227 (2019)

Turbulent Kinetic Energy Budget in Turbulent Flow of a Dilute Polymer Solution

F. Serafini(✉), F. Battista, P. Gualtieri, and C. M. Casciola

Department of Mechanical and Aerospace Engineering, Sapienza University of Rome, Via Eudossiana 18, 00184 Rome, Italy
francesco.serafini@uniroma1.it

1 Introduction

The addition of a small number of polymers to a Newtonian solvent can drastically reduce the drag in wall-bounded turbulent flows [1] due to the mechanical interaction between the polymer chains and the velocity fluctuations [2]. Drag reduction can reach huge amounts, even up to 80% [3]. The phenomenon is still not clearly understood and still debated, see [4,5] for an elastic or a viscous interpretation of drag reduction. Direct Numerical Simulation (DNS) has been extensively used to investigate the drag reduction phenomenon through viscoelastic constitutive models to account for the polymer phase. The most famous, and most used, model is the FENE-P (Finitely Extensible Nonlinear Elastic with Peterlin's approximation) [6], whose results show qualitative accordance with the experimental investigations. However, the FENE-P model has well-known deficiencies in describing polymer dynamics [7] and is based on an unrealistic assumption of the polymer conformation [8]. The previously mentioned issues did not allow for a match of the model parameters with actual polymer data, e.g. polymer concentration, in numerical simulations of viscoelastic flows with the FENE-P model. DNS of realistic polymer solutions can be performed if a Lagrangian description of the polymers is adopted. Such description allows to overcome at once all the limitations and assumptions of the Eulerian FENE-P model, see [9,10] where numerical simulations of realistic DNA polymer solutions are performed.

In this paper, direct numerical simulation (DNS) is used to investigate the turbulent kinetic energy budget of a drag-reducing polymer solution at moderate Reynolds $Re_\tau = 180$ and Weissenberg $Wi = 10$ numbers. Two different models are considered: the Eulerian-Eulerian FENE-P model and the Eulerian-Lagrangian FENE dumbbell model. Both models capture an apparent modification of production and dissipation of turbulent kinetic energy. In particular, the production is reduced by the polymer action, while the dissipation is reduced at the wall. With respect to the Lagrangian FENE dumbbell model, the FENE-P model overestimates the effect of the polymers on the turbulent kinetic energy budget; the decrease of production is overestimated and a decrease of the dissipation above $y^+ > 25$ is predicted, something that does not find confirmation in the more accurate FENE dumbbell simulation.

C. Marchioli et al. (Eds.): DLES 2023, ERCO 31, pp. 259–263, 2024.
https://doi.org/10.1007/978-3-031-47028-8_40

2 Methodology

The incompressible Navier-Stokes system, momentum-coupled with the polymer suspension, namely

$$\begin{aligned} \nabla \cdot \mathbf{u} &= 0 \\ \frac{\partial \mathbf{u}}{\partial t} + \nabla \cdot (\mathbf{u} \otimes \mathbf{u}) &= -\nabla p + \frac{1}{\mathrm{Re}} \nabla^2 \mathbf{u} + \mathbf{F}, \end{aligned} \tag{1}$$

has been numerically solved in the geometry of a pipe. In Eq. (1) $\mathbf{u}$ is the fluid velocity, p the hydrodynamic pressure, $\mathbf{F}$ is the polymer feedback on the Newtonian solvent, and $\mathrm{Re} = U_b^* \ell_0^* / \nu^*$ the bulk Reynolds number. Equation (1) is reported in dimensionless form using as reference quantities (denoted by an asterisk): the pipe radius ℓ_0^* as reference length scale, the bulk velocity $U_b^* = Q_b^* / (\pi \ell_0^{*2})$ of the Newtonian case as reference velocity (Q_b^* is the volumetric flow rate), and the kinematic viscosity ν^*.

Simulations are performed with the no-slip condition at the wall, periodic boundary conditions in the axial direction, and at imposed pressure gradient – corresponding to $Re_\tau = u_\tau^* \ell_0^* / \nu^* = 180$ (where $u_\tau^* = \sqrt{\tau_w^* / \rho^*}$ is the friction velocity). In this condition, the drag reduction appears as a flow rate increase. Finally, the Weissenberg number (the polymer relaxation time ratio to the fluid time scale ℓ_0^* / U_b^*) is fixed to $\mathrm{Wi} = 10$.

For the Lagrangian FENE model, the $(j)^{th}$ polymer dumbbell is tracked according to Newton's equations,

$$\begin{aligned} \frac{d\mathbf{x}_c^{(j)}}{dt} &= \frac{\check{\mathbf{u}}_1^{(j)} + \check{\mathbf{u}}_2^{(j)}}{2} + \sqrt{\frac{r_{eq}^2}{3\mathrm{Wi}}} \frac{\mathbf{q}_1^{(j)} + \mathbf{q}_2^{(j)}}{2} \\ \frac{d\mathbf{r}^{(j)}}{dt} &= \check{\mathbf{u}}_2^{(j)} - \check{\mathbf{u}}_1^{(j)} - \frac{1}{\mathrm{Wi}} \frac{\mathbf{r}^{(j)}}{1 - \|\mathbf{r}\|^{(j)2} / L^2} + \sqrt{\frac{r_{eq}^2}{3\mathrm{Wi}}} \left(\mathbf{q}_2^{(j)} - \mathbf{q}_1^{(j)} \right), \end{aligned} \tag{2}$$

where $\mathbf{x}_c^{(j)} = (\mathbf{x}_1^{(j)} + \mathbf{x}_2^{(j)})/2$ is the polymer centre, $\mathbf{r}^{(j)} = \mathbf{x}_2^{(j)} - \mathbf{x}_1^{(j)}$ is the link vector, r_{eq} is the equilibrium size of the chain (set in a quiet solvent by the Brownian forces $\mathbf{q}_{1/2}^{(j)}$), L the polymer contour length, and $\check{\mathbf{u}}_{1/2}^{(j)}$ is the velocity of the fluid at the position of the beads denoted by the subscripts $_{1,2}$. The expression of polymer forcing on the fluid encompasses the sum on all the N_p dumbbells $\mathbf{F} = -\sum_{j=1}^{N_p} \mathbf{D}_1^{(j)} \delta(\mathbf{x} - \mathbf{x}_1^{(j)}) + \mathbf{D}_2^{(j)} \delta(\mathbf{x} - \mathbf{x}_2^{(j)})$ where $\mathbf{D}_{1/2}^{(j)} = \gamma(\check{\mathbf{u}}_{1/2}^{(j)} - d\mathbf{x}_{1/2}^{(j)}/dt)$ are the friction forces exchanged with the Newtonian solvent, and γ is the friction coefficient of the beads. The polymer force is singular and is physically regularised using the Exact Regularised Point Particle method (ERPP) [11–13].

For the Eulerian-Eulerian FENE-P model, the polymers are modelled according to the conformation tensor $\mathrm{C} = \langle \mathbf{r} \otimes \mathbf{r} \rangle$ that represents the covariance of the local polymer population. The corresponding equation for C is

$$\frac{d\mathrm{C}}{dt} = \mathrm{C} \cdot \nabla \mathbf{u} + (\mathrm{C} \cdot \nabla \mathbf{u})^T - \frac{2}{\mathrm{Wi}} \left[\frac{L^2 - 3 r_{eq}^2}{L^2 - \mathrm{Tr(C)}} \mathrm{C} - \frac{r_{eq}^2}{3} \mathbf{I} \right], \tag{3}$$

where it has been assumed a uniform polymer concentration c_o, the velocity of the beads $\check{\mathbf{u}}^{(j)}_{1/2}$ has been linearised around the polymer centre $\mathbf{x}_c$, the diffusion of the centre of mass has been neglected, and the Peterlin's approximation has been used to constrain the average polymer extension Tr(C) to the polymer contour length L. The polymer forcing is evaluated as the divergence of the extra-stress tensor, $\mathbf{F} = \nabla \cdot T_p$, whose expression is

$$\mathrm{T}_p = \frac{c_o \gamma}{2\mathrm{Wi}} \left[\frac{L^2 - 3\,r_{eq}^2}{L^2 - \mathrm{Tr(C)}} \mathrm{C} - \frac{r_{eq}^2}{3} \mathbf{I} \right] . \tag{4}$$

An illustrative example of the flow configuration is reported in Fig. 1 where the instantaneous field of the turbulent kinetic energy is shown as a contour plot for the FENE and FENE-P simulations.

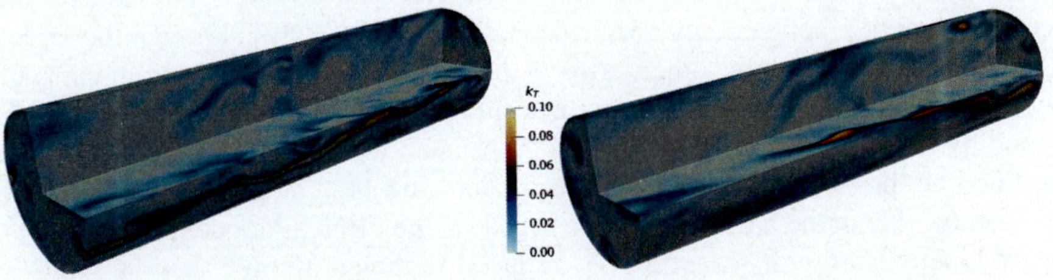

Fig. 1. Example of the flow configuration and snapshot of turbulent kinetic energy field for the Eulerian-Lagrangian FENE simulation (left) and for FENE-P simulation (right). The dimensionless domain is $(2\pi \times 1 \times 6\pi)$, the number of grid points are $[N_\theta \times N_r \times N_z] = [812 \times 129 \times 2436]$ for the FENE simulation, and $[N_\theta \times N_r \times N_z] = [128 \times 129 \times 384]$ for the FENE-P simulation.

3 Results

As widely documented in the literature, the back-reaction of the polymers reduces the Reynolds stress. The reduced "momentum mixing" is believed to be at the origin of the corresponding reduction of the friction at the wall. In this paper we investigate how the polymers modify the budget of turbulent kinetic energy $k_T = \langle \mathbf{u}'^2 \rangle / 2$, described by the following equation:

$$\nabla \cdot \Phi_T = \mathscr{P} \quad - \varepsilon_T + \langle \mathbf{F}' \cdot \mathbf{u}' \rangle \tag{5}$$

$$\Phi_T = \langle \mathbf{u} \rangle k_T + \langle \mathbf{u}' p' \rangle \quad - \frac{1}{\mathrm{Re}} \nabla k_T + \frac{1}{2} \langle \|\mathbf{u}'\|^2 \mathbf{u}' \rangle . \tag{6}$$

On the left-hand side of Eq. (5) we find the divergence of the spatial fluxes Φ_T, whose expressions are given in Eq. (6). On the right-hand side of Eq. (5) we find the production term $\mathscr{P} = -\langle \mathbf{u}' \otimes \mathbf{u}' \rangle : \nabla \langle \mathbf{u} \rangle$, the turbulent dissipation term $\varepsilon_T = \langle \nabla \mathbf{u}' : \nabla \mathbf{u}' \rangle / \mathrm{Re}$, and the energy sink/source term due to the fluid-polymer momentum coupling $\langle \mathbf{F}' \cdot \mathbf{u}' \rangle$. Note that the divergence of the flux $\nabla \cdot (\langle \mathbf{u} \rangle k_T)$ is statistically zero due to the symmetry of the pipe flow.

Figure 2 reports the turbulent kinetic energy budget, for the FENE dumbbell model, panel (a), and for the FENE-P model, panel (b). Both models capture a qualitatively similar scenario. Turbulent fluctuations take energy from the mean flow via the production term $\mathscr{P}$ whose peak is significantly lower than the corresponding one for the Newtonian flow (dashed-blue line). The production peaks in the buffer layer, and the peak is shifted away from the wall, especially in the FENE-P simulation. Part of this energy injected in the buffer layer is locally dissipated, while the remaining part is transferred towards the wall by the fluxes, namely $\langle p'\mathbf{u}'\rangle$, $\nabla k_T/\mathrm{Re}$, $\langle \|\mathbf{u}'\|^2\mathbf{u}'\rangle/2)$, where is dissipated by ε_T. It is worth noting that the turbulent dissipation at the wall in the polymer case is lower than in the Newtonian case (dashed red line), and again the effect is amplified in the FENE-P simulation. The remaining part of the available energy is transferred towards the pipe centre by the turbulent flux $\langle \|\mathbf{u}'\|^2\mathbf{u}\rangle/2)$. The correlation between the polymer forcing and the velocity fluctuation $\langle \mathbf{u}'\cdot\mathbf{F}'\rangle$, behaves as a local source of turbulent kinetic energy very close to the wall $y^+ < 10$, although the effect is small when compared with the other terms in the budget. In contrast, far from the wall at $y^+ > 10$, the correlation $\langle \mathbf{u}'\cdot\mathbf{F}'\rangle$ behaves as a sink of turbulent kinetic energy; the effect predicted by the FENE model is small and decays above $y^+ = 70$. Instead, this effect is amplified by the FENE-P model, which predicts an apparent larger amplitude of the correlation $\langle \mathbf{u}'\cdot\mathbf{F}'\rangle$ in the buffer layer. Furthermore, the FENE-P model predicts that the turbulent kinetic energy dissipation ε_T is reduced in the entire pipe section, whilst the more accurate Lagrangian FENE dumbbell model does not predict significant changes of dissipation in the region $y^+ > 25$, with respect to the Newtonian case.

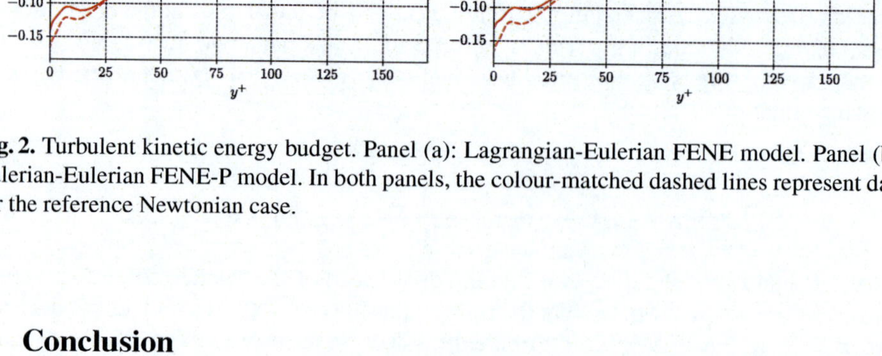

Fig. 2. Turbulent kinetic energy budget. Panel (a): Lagrangian-Eulerian FENE model. Panel (b): Eulerian-Eulerian FENE-P model. In both panels, the colour-matched dashed lines represent data for the reference Newtonian case.

4 Conclusion

The turbulent kinetic energy budget is addressed for a drag-reducing polymer solution at moderate friction Reynolds number $\mathrm{Re}_\tau = 180$ and Weissenberg number $\mathrm{Wi} = 10$.

Two different models are considered for the analysis, i.e. the Eulerian-Eulerian FENE-P model and the Eulerian-Lagrangian FENE dumbbell model. Both models capture a significant modification of the turbulent kinetic energy budget; the turbulent production is importantly reduced by the polymers, thus less energy feed the turbulent fluctuations with respect to the Newtonian case. The modification of the production term is slightly overestimated by the FENE-P model. The dissipation of turbulent kinetic energy is significantly reduced at the wall, and again this effect is amplified in the FENE-P simulation, especially in the bulk where a lower dissipation with respect to the Newtonian case is predicted. Finally, the direct polymer contribution to the budget behaves as a weak source of turbulent fluctuation in the near wall region ($y^+ < 10$), and as a sink elsewhere.

Acknowledgments. We acknowledge the CINECA award under the ISCRA initiative, Iscra B number HP10B0F5V3, for the availability of high-performance computing resources. This work was partially supported by Sapienza 2021 Funding Scheme, Project #RG12117A66DC803E.

References

1. Toms, B.A.: Proceedings of the 1st International Congress on Rheology, vol. II, North Holland, Amsterdam, p. 135 (1949)
2. Xi, L.: Turbulent drag reduction by polymer additives: fundamentals and recent advances. Phys. Fluids **31**(12), 121302 (2019)
3. Procaccia, I., L'vov, V.S., Benzi, R.: Colloquium: theory of drag reduction by polymers in wall-bounded turbulence. Rev. Mod. Phys. **80**(1), 225 2008
4. Tabor, M., De Gennes, P.G.: A cascade theory of drag reduction. EPL (Europhys. Lett.) **2**(7), 519 (1986)
5. Lumley, J.L.: Drag reduction in turbulent flow by polymer additives. J. Polym. Sci.: Macromol. Rev. **7**(1), 263–290 (1973)
6. Sureshkumar, R., Beris, A.N., Handler, R.A.: Direct numerical simulation of the turbulent channel flow of a polymer solution. Phys. Fluids **9**(3), 743–755 (1997)
7. Graham, M.D.: Drag reduction in turbulent flow of polymer solutions. Rheol. Rev. **2**(2), 143–170 (2004)
8. Keunings, R.: On the Peterlin approximation for finitely extensible dumbbells. J. Nonnewton. Fluid Mech. **68**(1), 85–100 (1997)
9. Serafini, F., Battista, F., Gualtieri, P., Casciola, C.M.: Drag reduction in turbulent wall-bounded flows of realistic polymer solutions. Phys. Rev. Lett. **129**(10), 104502 (2022)
10. Serafini, F., Battista, F., Gualtieri, P., Casciola, C.M.: Drag reduction in polymer-laden turbulent pipe flow. Fluids **7**(11), 355 (2022)
11. Gualtieri, P., Picano, F., Sardina, G., Casciola, C.M.: Exact regularized point particle method for multiphase flows in the two-way coupling regime. J. Fluid Mech. **773**, 520–561 (2015)
12. Battista, F., Mollicone, J.P., Gualtieri, P., Messina, R., Casciola, C.M.: Exact regularised point particle (ERPP) method for particle-laden wall-bounded flows in the two-way coupling regime. J. Fluid Mech. **878**, 420–444 (2019)
13. Motta, F., Battista, F., Gualtieri, P.: Application of the exact regularized point particle method (ERPP) to bubble laden turbulent shear flows in the two-way coupling regime. Phys. Fluids **32**(10), 105109 (2020)

Elongated Non-spherical Particles in Turbulent Channel Flow Using Euler/Lagrange Approach

Manuel A. Taborda(✉) and Martin Sommerfeld

Multiphase Flow Systems (MPS), Institute for Process Engineering,
Otto-Von-Guericke-University Magdeburg, Hoher Weg 7B, 06120 Halle (Saale), Germany
manuel.taborda@ovgu.de

Abstract. In numerical simulations of dispersed multiphase flows by computational fluid dynamics (CFD) methods the most usual assumption is that the particles are spherical. This is however a very crude assumption when considering flows laden with non-spherical particles, such as fibres or plates. This applies to the relevant fluid dynamic forces and to the interactions between particles as well as particle-wall collisions. Consequently, such numerical results are not reflecting the full truth yielding to an unrealistic modelling of the problem. Numerical simulations of fibre-laden flows using a point-particle Euler/Lagrange approach are based on tracking these elongated particles with respect to translation and rotation, where the centre of gravity position and the orientation of the fibres for each Lagrangian calculation are known. Here, an LES-Euler/Lagrange approach for elongated, inertial fibres was developed based on first principles and implemented in the open-source platform OpenFOAM®. The validation of the numerical approach was done considering a particle-laden turbulent channel flow which was thoroughly studied using DNS. Special attention was directed to the fibre-wall interaction model. For that purpose, a wall-impact model considering the particle orientation and the point of wall contact was implemented. The obtained numerical results were compared to those using a crude centre of mass specular reflection for the fibres. Both simulations yielded completely different results regarding the fibre orientation and concentration near the wall. The latter was considerably lower for the realistic wall collision model.

Keywords: Non-spherical particles · Large Eddy Simulation · Euler/Lagrange approach · Wall Collision Model

1 Introduction

Many industries such as paper making and pulp production involve particle transport in gas- or liquid-flow devices, such as mixing vessels, conveying in pipes or channels and particle separators. The particles found in these processes are however not spherical but have a fibre-like shape. Moreover, many of the mentioned multiphase flows are wall bounded and are typically highly turbulent. Although for many years, research is related to the behavior of regular non-spherical particles in turbulent flows, a conclusive

C. Marchioli et al. (Eds.): DLES 2023, ERCO 31, pp. 264–272, 2024.
https://doi.org/10.1007/978-3-031-47028-8_41

description is, not yet available. In available design rules as well as in numerical computations of dispersed multiphase flows of technical and industrial relevance the most common assumption is that the particles are spherical. This applies to the relevant fluid dynamic forces in a confined flow and to the interactions between particles as well as the particle-wall interactions (i.e., collisions and lubrication), which of course yields a non-realistic simplification of the problem. In this contribution, the LES-Euler/Lagrange framework using the point-particle assumption was applied to a turbulent channel flow [7] laden with elongated non-spherical particles using especially a recently developed wall collision model for non-spherical particles [9].

2 Numerical Modelling

A CFD model using an LES-Euler/Lagrange approach was developed and implemented in the open-source platform OpenFOAM®, elaborating a custom solver. This solver extends the classical point-particle Lagrangian tracking approach for spherical particles, considering additional properties for non-spherical particles, such as orientation angle and particle rotational velocity. The gas phase (continuous) aerodynamics is calculated using the volume-average Navier-Stokes equations comprising continuity and momentum conservation laws. Large Eddy Simulation (LES) approach was used to model the flow field and turbulence of the gas phase requiring a fully transient solution of the transport equations, as follows:

$$\frac{\partial(\rho_c)}{\partial t} + \nabla \cdot (\rho_c \boldsymbol{u_c}) = 0 \tag{1}$$

$$\frac{\partial(\rho_c \boldsymbol{u_c})}{\partial t} + \nabla \cdot (\rho_c \boldsymbol{u_c} \boldsymbol{u_c}) = -\nabla p - \nabla \cdot (\rho_c \boldsymbol{\tau_c}) \tag{2}$$

$$\boldsymbol{\tau_c} = -\mu_{eff} \left\{ \left(\nabla \boldsymbol{u_c} + (\nabla \boldsymbol{u_c})^{\mathrm{T}} \right) - \frac{2}{3} I (\nabla \cdot \boldsymbol{u_c}) \right\} \tag{3}$$

where, $\mathbf{u}_c$ corresponds to the velocity vector of the carrier phase, p represents the pressure, and ρ_c is the density of the carrier phase. In the Eq. (3), the effective viscosity term $\mu_{eff} = \mu_c + \mu_{t,c}$ which is included in the momentum equation, comprises two contributions, the molecular viscosity μ_c and the turbulent viscosity $\mu_{t,c}$. In this work, the turbulent viscosity, which accounts for the contribution of the sub-grid scale (SGS) turbulence, is described using the wall adapted local eddy-viscosity (WALE) approach [1].

In addition, the numerical calculation of particle-laden flows by point-particle Euler/Lagrange approach is based on tracking each non-spherical particle's translational and rotational motion, so that the centre position and the orientation of the particles is known. For this purpose, three coordinate systems can be defined (Fig. 1): **x** $_{[x, y, z]}$ the inertial or laboratory frame of reference, commonly referred to as world space, **x'** $_{[x', y', z']}$ the particle coordinate system centred at the particle centre of mass with the axis being parallel to the major axis of the particle, commonly referred to as body space, and **x''** $_{[x'', y'', z'']}$ the co-moving particle coordinate system also centred at the particle mass centre but with its axis being parallel to the laboratory coordinate system. For non-spherical particles the translational motion is calculated in world space, **x**, frame and the rotational motion in body space, **x'**, in the following way:

$$m_p \frac{d\boldsymbol{u}_p}{dt} = \boldsymbol{F}_d + \boldsymbol{F}_{LS} \tag{4}$$

$$I_{x'} \frac{d\omega_{x'}}{dt} - \omega_{y'}\omega_{z'}\left(I_{y'} - I_{z'}\right) = T_{H,x'} \tag{5}$$

$$I_{y'} \frac{d\omega_{y'}}{dt} - \omega_{z'}\omega_{x'}\left(I_{z'} - I_{x'}\right) = T_{H,y'} \tag{6}$$

$$I_{z'} \frac{d\omega_{z'}}{dt} - \omega_{x'}\omega_{y'}\left(I_{x'} - I_{y'}\right) = T_{H,z'} \tag{7}$$

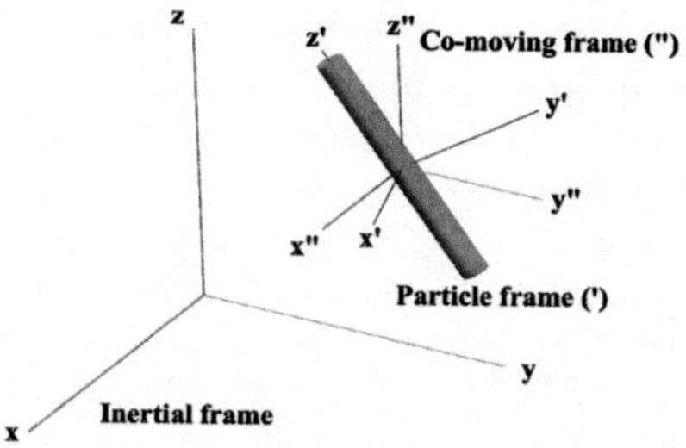

Fig. 1. Definition of coordinate systems; inertial coordinate system **x** = (x, y, z) and particle coordinate systems **x'** = (x', y', z') and **x"** = (x", y", z").

Here, m_p, $\boldsymbol{F}_d$, $\boldsymbol{F}_{LS}$ are the mass of the particle, the aerodynamic drag, and the profile or shape lift, which includes the effect of the angle of incidence on the respective coefficient [2, 3]. $I_{x'}$, $I_{y'}$, and $I_{z'}$ correspond to the moments of inertia about the particle principal axis [i.e., **x'**] and $T_{H,x'}$, $T_{H,y'}$ and $T_{H,z'}$ are the components of the aerodynamic torques [2]. As the calculation is conducted in different coordinate systems, a transformation between **x'** and **x"** is required through a transformation matrix [4]. This transformation matrix may be expressed in terms of the Euler angles or the Euler's four parameters or very effectively by quaternions. Unfortunately, a singularity exists when evaluating the temporal change of the Euler angles [4] wherefore, mostly the Euler parameters [5] or Quaternions [6] are applied instead. These can also be used to determine the particle orientation angles.

3 Simulation Setup and Results

Using the models mentioned above, a validation of the Euler/Lagrange point-particle framework for elongated, inertial fibres was accomplished comparing the obtained results with the numerical validation data obtained by Marchioli's research group [7], in which a periodic turbulent channel flow (considering DNS), was studied (Fig. 1a and 1b) for a specific friction Reynolds number of $Re_\tau = hu_\tau/\nu = 150$, based on the channel half-height h. The density and viscosity of the fluid were considered as $\rho = 1.3$ kg/m^3 and $\nu = 1.57 \times 10^{-5}$ m^2/s, respectively. The shear velocity is calculated as $u_\tau = \sqrt{\tau_w/\rho}$, where τ_w correspond to the wall shear stress. The normalized wall distance is determined as $z^+ = zu_\tau/\nu$. The computational domain is sized as $2\pi \times 2\pi h \times 4\pi h$, discretizes with $128 \times 128 \times 129$ numerical control volumes. For the LES calculations, it is known that for obtaining a turbulent flow, many flow-through times are needed before turbulence has developed through the wall boundary layer. However here, an artificial perturbation to initiate the turbulent process was applied based on the work of de Villiers [8]. Figure 2c shows the initial perturbated velocity field measured at $z^+ = 10$.

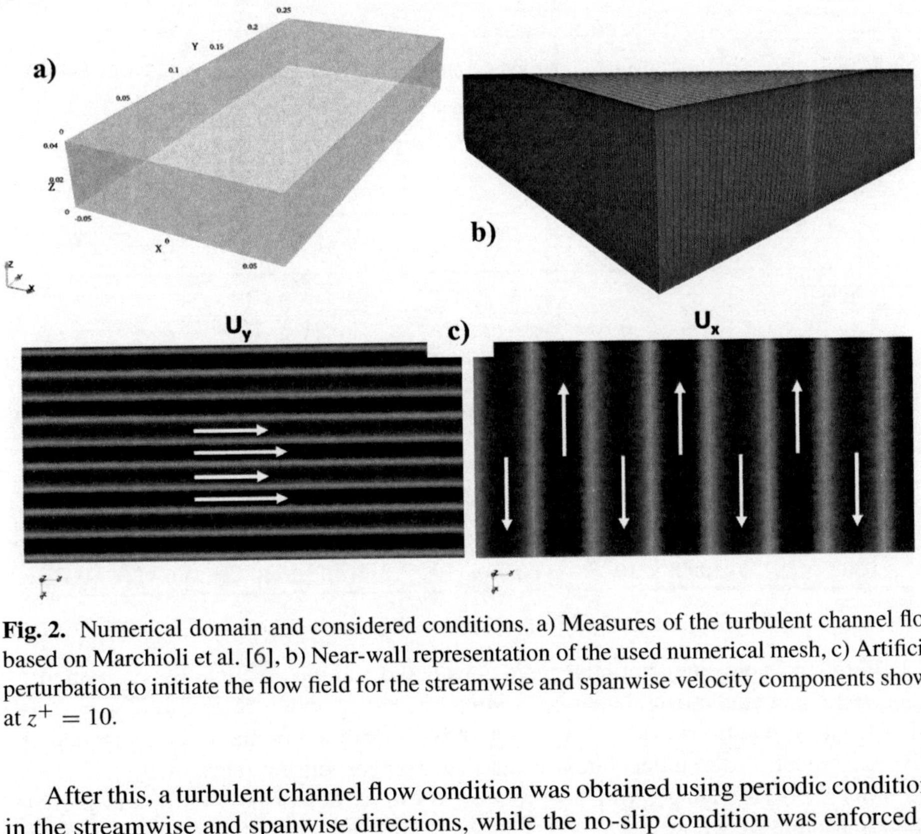

Fig. 2. Numerical domain and considered conditions. a) Measures of the turbulent channel flow based on Marchioli et al. [6], b) Near-wall representation of the used numerical mesh, c) Artificial perturbation to initiate the flow field for the streamwise and spanwise velocity components shown at $z^+ = 10$.

After this, a turbulent channel flow condition was obtained using periodic conditions in the streamwise and spanwise directions, while the no-slip condition was enforced at the walls. In order to get the flow statistics more than $t^+ = 13000$ wall units were simulated. Figure 3 shows the obtained LES statistical moments in the streamwise and the root mean squared (RMS) values of the velocity fluctuations in the wall-normal and spanwise directions of the turbulent channel solution in comparison with the DNS data [7]. A very good agreement was obtained for the considered sub-grid eddy viscosity approach. After arriving at a converged solution, this was further used for the analysis of the particle's behaviour specially in the near wall region.

The fibre behaviour is characterized using the non-dimensional particle relaxation time based on the viscous time scale ($\tau_f = \nu/u_\tau^2$) and the equivalent sphere volume diameter, defined as $\tau_{eq}^+ = \tau_{eq}/\tau_f = \phi_p d_{eq}^{+2}/18\nu$, where $d_{eq} = \left(6d_f^2 l_f/4\right)^{1/3}$, and $\phi_p = \rho_p/\rho \approx 833$. The fibre diameter and fibre length were selected such that the mass of particles with equal response times, τ_{eq}^+, were equivalent. Table 1 presents the parameters used for the fibres considered in the numerical simulations. Additionally, the time step of the particle tracking calculation is automatically adjusted along the trajectory by considering the time required for a particle to cross a control volume, and the particle response time. This means that a new value is calculated for every Lagrangian time-step choosing the minimum of these time-scales and making it one order of magnitude lower.

Table 1. Parameters of the considered particles, for spherical and elongated particles (fibres); d_f: fiber diameter, l_f: fiber length, d_{eq}: volume equivalent diameter, l_f^+: non-dimensional fiber length, d_{eq}^+: non-dimensional equivalent diameter.

τ_{eq}^+	$AR[l_c/d_c]$	d_f[m]	l_f[m]	d_{eq}[m]	l_f^+	d_{eq}^+
5	Sphere	-	-	4.533×10^{-5}	-	0.34
5	3	2.745×10^{-5}	8.237×10^{-5}	4.533×10^{-5}	0.62	0.34
5	10	1.838×10^{-5}	1.838×10^{-4}	4.533×10^{-5}	1.38	0.34
5	30	1.274×10^{-5}	3.823×10^{-4}	4.533×10^{-5}	2.87	0.34
25	Sphere	-	-	1.013×10^{-4}	-	0.76
25	3	6.137×10^{-5}	1.841×10^{-4}	1.013×10^{-4}	1.38	0.76
25	10	4.108×10^{-5}	4.108×10^{-4}	1.013×10^{-4}	3.08	0.76
25	30	2.848×10^{-5}	8.546×10^{-4}	1.013×10^{-4}	6.41	0.76

Different particle characteristics were considered and injected into the domain, differing in volume equivalent size ($\tau_{eq}^+ = 5$ and $\tau_{eq}^+ = 25$) and aspect ratio ($AR = 3$, 10 and 30). The simulations evolved with 100,000 fibres in the domain, ensuring statistically convergent results independent of particle number, during more than $t^+ = 4000$ wall units. Here, the shape, which is related to the particle aspect ratio, and particle orientation angle were considered, especially in the treatment of the interaction between particle and wall, where two models were studied, as follows. First a model that uses a centre of mass specular reflection for the wall collision of both spheres and fibres is considered [7], which however is completely unrealistic as the finite dimensions of the particles are not accounted for. Moreover, a more realistic wall-impact model considering the particle orientation, angular velocity, and point at wall contact is used. Hence, the recently proposed model by Quintero et al. [9] that accounts for the change of linear and rotational momentum of the non-spherical particles during a compression and recovery period was considered. The model takes into account the restitution coefficient e, and friction coefficient μ parameters, which were specified in terms of the impact angle as derived from experiments.

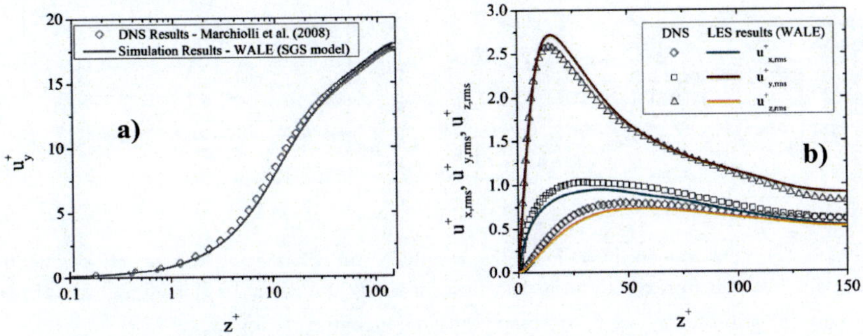

Fig. 3. Statistical moments of the turbulent channel flow in comparison with DNS solution [6], a) Mean streamwise velocity, ($u_y'^+$), b) Root mean square of velocity fluctuations of wall-normal ($u'^+_{z,rms}$), and spanwise ($u'^+_{x,rms}$) directions.

Figure 4 shows the results of the particle concentration in the region close to the wall for two different values of τ^+_{eq}, i.e., 5 and 25, varying the aspect ratio. In these simulations, only specular reflection was considered. In addition, the results for spherical particles are included. Figure 4a shows the results for $\tau^+_{eq} = 5$, where it is obvious that the concentration distribution strongly varies with the aspect ratio, especially for the region below $z^+ < 5$. For AR = 3 the maximum particle concentration is found just on the wall and then continuously decreases with a kind of parabolic shape. For the other two aspect ratios (i.e., AR = 10 and 30), the concentration profiles are very similar, having a lower value and at the wall and remaining constant until $z^+ \approx 1$. For $AR = 30$ the wall-concentration is only slightly lower than for $AR = 10$. Between $1.5 < z^+ < 8$ the concentration for the case with $AR = 3$ is remarkably lower in comparison with the two larger aspect ratios considered. The concentration distribution for spherical particles in the region between $1 < z^+ < 10$, has a completely different shape with a strong peak at $z^+ \approx 1$, followed by a rapid decay towards $z^+ \approx 4$. This implies that the effect of the orientation of the particles on the lift coefficient causes the particles to be able to move to other regions of the channel cross-section. Naturally, the developed particle concentration distribution for the different aspect ratios is not only related to the drag and lift forces but also to the torque experienced by the particle in this region, where a high velocity gradient is found. Although the particles have the same equivalent diameter, they differ in their angular velocity since the inertia of the particle changes depending on their aspect ratio.

On the other hand, when larger particles are considered ($\tau^+_{eq} = 25$) the behavior is completely different, which is reflected in the concentration of the particles near the wall (Fig. 4b). A direct comparison with smaller particles ($\tau^+_{eq} = 5$, see Fig. 4a), shows that the concentration of the fibre-like particles near the wall is much higher due to larger particle inertia, therefore this will affect the interaction with the wall, obtaining, for example, much more wall collision events. The lowest near-wall value is again found

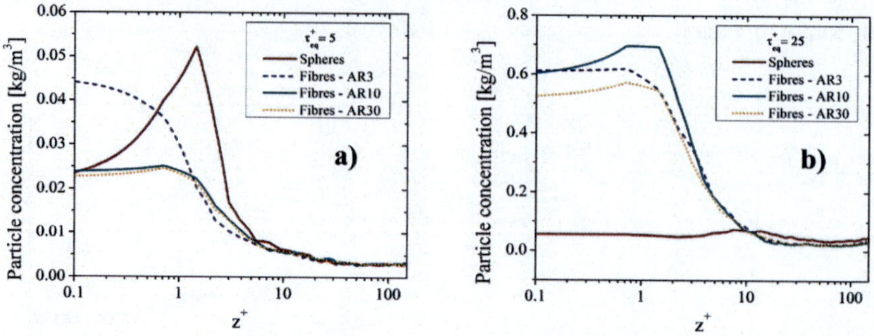

Fig. 4. Mean particle concentration in the near-wall region considering only specular reflection, including spheres and fibres with an aspect ratio of 3, 10 and 30 and $\tau_{eq}^+ = 5$ and 25. a) Mean particle concentration for $\tau_{eq}^+ = 5$, b) Mean particle concentration for $\tau_{eq}^+ = 25$.

for the $AR = 30$ fibers. For the small fibres, the concentration remains constant up to about $z^+ \approx 1$ and then decays. The larger fibres, however, show a small maximum of concentration in this region. For the spherical particles, the obtained concentration is much lower and is almost constant for $z^+ < 4$. Interestingly further up in the boundary layer, also a small concentration maximum develops, i.e., at $z^+ \approx 10$.

Figure 5 shows the particle concentration and fibre orientation in the near-wall region obtained with the two previously described wall-interaction models, considering a $\tau_{eq}^+ = 5$ and an aspect ratio of 10. Additionally, the result for the spherical particles with specular wall reflection is shown, where the particle concentration is highest and yields a peak for $z^+ \approx 1$. Interestingly, the fiber-like particles, considering the extended wall collision model [9], show the same shape of the concentration profile, however, with considerably lower values (Fig. 5a). On the other hand, the specular wall reflection of the fibres gave a constant concentration until $z^+ \approx 1$ followed by a continuous decay. For the extended wall collision model, this concentration profile is of course provoked by the possibility allowing an exchange between linear and rotational momentum depending on the locative location of contact point and fibre centroid. Thereby, a tilting of the fibre or an enhancement of rotation may be the result. Consequently, mainly lift and torque are modified. In terms of the wall coordinate, the volume equivalent diameter is $z^+ \approx 0.34$ and the length of the fibre corresponds to $z^+ \approx 1.38$ (see also Table 1).

In Fig. 5b the fibre orientation with respect to the wall is presented, revealing that both wall collision models yield the same orientation outside the wall layer (i.e., $z^+ > 1$), namely around 0° which implies an average alignment parallel to the wall in flow direction. For the consistent wall collision model [9] the average orientation angle in the near-wall region is only slightly higher, having about 10°. This may be caused by the convective flow and a preferred tilting of the fibres upon wall impact. When specular reflection is considered, the orientation of the elongated particles in the $z^+ < 1$ region shows a continuous increase of the orientation angle from 10° up to 70°, meaning that the fibres tend to move almost perpendicular to the wall. It is also clear that the extended wall collision model yields a change of the rotational velocity of the fibres, modifying the orientation in this region. Thereby also the fluid dynamic forces are changing in this

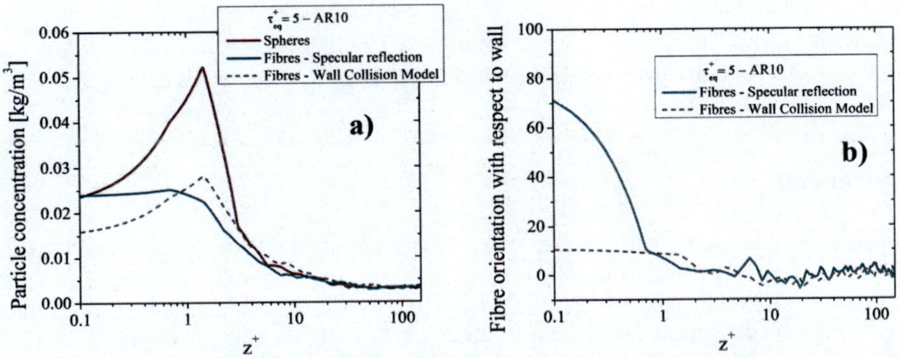

Fig. 5. Mean numerical results of the particle phase in the near-wall region for the two wall interaction models considered (specular reflection and wall collision model based on [9]), including spheres and fibres with a $\tau_{eq}^{+} = 5$ and aspect ratio of 10. a) Mean particle concentration, b) Fibre orientation angle with respect to the wall.

region. All these phenomena cannot be captured by the crude specular reflection model. Nevertheless, it should be noted that the coefficients of restitution and friction for the collision between the non-spherical particles and the wall presented by Quintero et al. [9], were developed for a wall collision of irregular-shaped non-spherical particles. In any case, however, the presented results reveal, that a crude centre of gravity specular reflection wall collision model does not capture the dynamics of a completely three-dimensional fibre-wall collision. The extended wall collision model predicts completely different distributions of particle concentration and orientation angle close to the wall.

4 Conclusions

An LES-Euler/Lagrange numerical computation of elongated non-spherical particles including the modelling of particle dynamic motion in the point-mass approximation, e.g., fibre orientation, was conducted successfully for a turbulent channel flow configuration. A very good agreement of the computed flow field with the DNS results presented by Marchioli et al. [7] was found. In the considered wall-bounded channel flow, of course the proper modelling of the fibre-wall impact is essential. Naturally, a specular wall refection model is not reproducing correctly the dynamic behaviour of fibres in the near-wall region. Therefore, a more realistic non-spherical particle wall-collision model was considered, which identifies the real conditions of the fibres upon wall impact, namely orientation and impact point. Hence, this model captures through the consideration of sliding and non-sliding collisions the exchange between translational and rotational momentum, whereby angular velocity is considerably altered through a wall collision. Consequently, the fibre concentration distribution and the mean fibre orientation in the near-wall region is predicted correctly. This realistic wall collision model yields an averaged fibre orientation in the near-wall region being almost parallel to the wall and in contrary the specular refection gives an orientation angle of 70° which is almost perpendicular.

Acknowledgements. Financial supports from the German Research Council (Deutsche Forschungsgemeinschaft) under contract number SO 204/53-1 is gratefully acknowledged.

References

1. Nicoud, F., Ducros, F.: Subgrid-scale stress modelling based on the square of the velocity gradient tensor. Flow Turbul. Combust. **62**(3), 183–200 (1999)
2. Zastawny, M., Mallouppas, G., Zhao, F., van Wachem, B.: Derivation of drag and lift force and torque coefficients for non-spherical particles in flows. Int. J. Multiph. Flows **39**, 227–239 (2012)
3. Hölzer, A., Sommerfeld, M.: New simple correlation formula for the drag coefficient of non-spherical particles. Powder Technol. **184**, 361–365 (2008)
4. Goldstein, H.: Classical Mechanics. 2nd Ed., Addison-Wesley, Reading 31 (1980)
5. Lain, S., Sommerfeld, M.: Kinematic simulations of non-spherical particle response behaviour. In: 8th International Conference on Multiphase Flow, ICMF 2013, Jeju, Korea, May 26–31 (2013)
6. Zhao, F., van Wachem, B.: A novel quaternion integration approach for describing the behaviour of non-spherical particles. Acta Mechanica 224(12), 3091–3109 (2013)
7. Marchioli, C., Fantoni, M., Soldati, A.: Orientation, distribution, and deposition of elongated, inertial fibers in turbulent channel flow. Phys. Fluids **22**, 033301 (2010)
8. de Villiers, Eugene, The Potential of Large Eddy Simulation for the Modelling of Wall Bounded Flows, PhD Thesis (2006)
9. Quintero, B., Lain, S., Sommerfeld, M.: Derivation and validation of a hard-body particle-wall collision model for non-spherical particles of arbitrary shape. Powder Technol. **380**, 526–538 (2021)

DNS of Magnetic Density Separation in the Wake of a Honeycomb

L. C. Thijs, J. G. M. Kuerten(✉), J. C. H. Zeegers, and S. Tajfirooz

Eindhoven University of Technology, Eindhoven, The Netherlands
{l.c.thijs,j.g.m.kuerten,j.c.h.zeegers}@tue.nl, tajfirooz@nrg.nl

1 Introduction

Magnetic density separation (MDS) is an innovative technique that separates different types of plastic particles according to their different mass densities. To this end an effective vertical mass density gradient in a liquid is created by applying a strong magnetic field to a ferrofluid. Particles transported by this liquid settle at the height where their mass density equals the effective fluid mass density. To obtain a high separation efficiency, it is important that the level of turbulence in the liquid is kept as small as possible. The liquid is first guided through a honeycomb in which the flow is relaminarized. However, in the wake of the honeycomb cell walls disturbances are created [5], which together with the effect of the particle motion may lead to turbulent flow further downstream. In this paper, we investigate this and its effects on particle separation using temporal point-particle DNS of the system, taking into account the full interaction between particles and fluid and particle collisions.

2 Governing Equations and Numerical Methods

In Fig. 1 the schematic of the MDS apparatus is given. The ferrofluid enters the system from the left and is guided through a honeycomb, whereas the particles to be separated can only enter the separation chamber via the top and bottom channels. In the separation chamber, a strong magnetic field is applied by magnets on the bottom and top (not indicated in the figure). The top and bottom walls of the separation chamber move with the mean fluid velocity. At the downstream end of the separation chamber, the separated particles are sorted by means of the separator blades. In all simulations shown here the magnetic field corresponds to the one-dimensional magnetic field between two identical magnetic Halbach arrays with their stronger sides facing each other.

We use an Euler-Lagrange approach, in which the flow around the spherical particles is not resolved, but a point-particle approach is applied instead. The effect of the magnetic field on the flow can be incorporated in a modified pressure [4]. Since the mass density of the particles is close to that of the fluid, the complete equation of motion is solved for each particle, including added mass force, force due to the undisturbed velocity field and history force, along with the equation for particle rotation [1]. It has been shown that the history force has a significant effect on the particle motion [4], which

C. Marchioli et al. (Eds.): DLES 2023, ERCO 31, pp. 273–278, 2024.
https://doi.org/10.1007/978-3-031-47028-8_42

could be expected since the mass density of the particles is close to that of the fluid. This force has been implemented in the efficient way described by Van Hinsberg et al. [2]. The reaction forces of the particles on the fluid are distributed over a volume equal to the particle volume. Particle collisions are taken into account by an efficient collision search method and taking into account conservation of linear and angular momentum and the coefficient of restitution for collisions of wet plastic particles. Experimental results of the collision of two spherical plastic particles in a fluid indicate that all collisions are in the sliding collision regime and not in the sticking collision regime and that the friction coefficient is very small.

The simulation domain is the region between the exit of the honeycomb and the separator blades, for which a temporal approach is applied with periodic conditions in streamwise and spanwise directions. The effect of the honeycomb is incorporated in the initial condition, which plays the role of the inflow condition in a temporal approach. A distinction is made between laminar and turbulent flow in the cells of the honeycomb, which are of rectangular shape. For laminar flow in the honeycomb cells the analytical solution is used as initial condition, on which perturbations are superposed with an amplitude that corresponds to experimental results at the same Reynolds number obtained by particle image velocimetry. The initial condition corresponding to turbulent flow in the honeycomb cells is obtained from a high-resolution DNS of the flow in a rectangular duct with the same aspect ratio and at the same Reynolds number. For laminar flow symmetry between the honeycomb cells is broken by using different perturbations in each cell. For turbulent flow, behind each cell the DNS solution at a different time is taken to break the symmetry. Directly behind the cell walls the initial velocity is set to zero.

A pseudo-spectral method is applied for the solution of the DNS with Fourier-Galerkin in the two periodic directions and Chebyshev-tau in the vertical direction [3]. The equations are solved in a frame which moves with the mean fluid velocity where the top and bottom wall are stationary and the honeycomb moves in the negative streamwise direction. A combination of a three-stage Runge-Kutta method and the implicit Crank-Nicolson method is used for time integration of the fluid equations. Forward Euler is

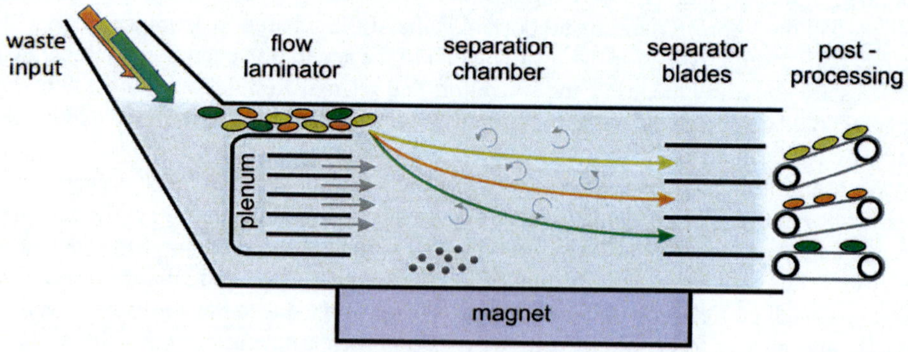

Fig. 1. Schematic of magnetic density separation apparatus.

applied for integration of the particle equations of motion to enable deterministic treatment of particle collisions [4].

In the results shown here we consider only one honeycomb consisting of 6×8 parallel rectangular ducts with height 15.2 mm, width 16.6 mm and wall thickness 3 mm. Above and below the honeycomb there are two channels in which the initial velocity has a Couette profile and where the particles are initially located. We consider three different Reynolds numbers in the honeycomb cells, 900, 1800 and 5600, all based on the hydraulic diameter of the cells and the mean velocity. Figure 2 shows the initial streamwise velocity in a plane perpendicular to the streamwise direction for the highest Reynolds number, where the flow in the honeycomb cells is turbulent. To fully resolve the honeycomb cells 576 Fourier and Chebyshev modes are applied in spanwise and vertical directions. The particle volume fraction is 2% and spherical particles with a diameter of 4 mm and ten different mass densities, both higher and lower than the fluid mass density, are used.

3 Results and Conclusions

In order to obtain good separation of particles of different mass density, particles should move towards the vertical position where the effective fluid mass density equals their own mass density as quickly as possible. To that end, disturbances in the fluid velocity should be kept as small as possible. Moreover, particle collisions generally result in a delay in the time at which they reach their equilibrium position. The purpose of the honeycomb is to eliminate velocity disturbances. However, the wakes of the walls between the honeycomb cells also generate disturbances, which may result in turbulent flow further downstream. Furthermore, the relative particle motion in vertical direction generates velocity disturbances as well. Comparison of results with and without two-way coupling shows the effect of the latter. The vertical motion of the particles with respect to the fluid moves the peak of the fluid velocity fluctuations that occurs downstream of the honeycomb in upstream direction.

In the results of the simulations in the cases where the flow in the honeycomb cells is laminar it is clearly visible that the individual channel profiles downstream of the honeycomb gradually develop into one uniform velocity profile. At the edges of the individual honeycomb channels instabilities arise and grow until the individual channel profiles break up. This break up corresponds with an increase in the velocity fluctuations which reaches its maximum and then starts to decay [5]. Particles with an equilibrium position at the center of the channel show a higher level of particle dispersion than particles at the top and bottom of the channel. Particles with their equilibrium positions in the central region of the channel have more interaction with other particles, which results in an increase of the settling time. Furthermore, in the center of the channel there is a lower magnetic field gradient, which results in slower vertical motion of these particles [4]. Also, the velocity fluctuations result in an additional increase in the level of particle dispersion.

The effect of a honeycomb-induced background flow on the separation error is investigated by comparing the results with a case with a uniform background flow and shown in Fig. 3. This is done for the case of laminar flow inside the honeycomb

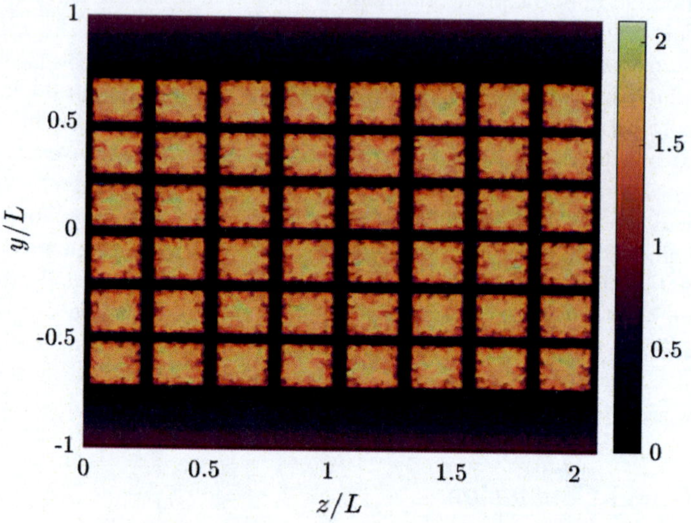

Fig. 2. Initial condition for turbulent flow in the honeycomb cells. The colors indicate the streamwise velocity component.

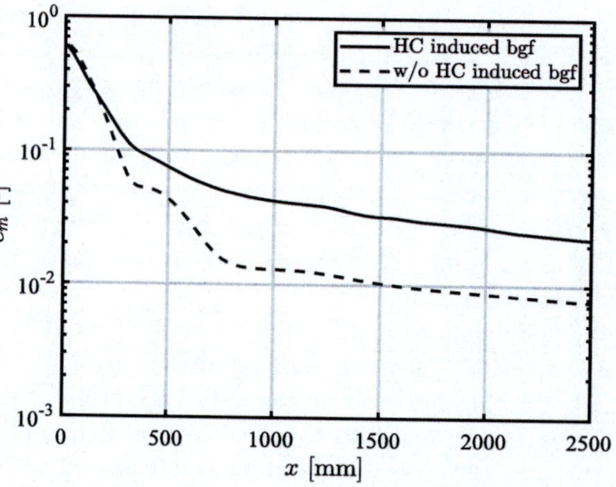

Fig. 3. Separation error as a function of streamwise coordinate x for a uniform background flow and a honeycomb-induced background flow at a honeycomb cell Reynolds number of 1800.

cells at a cell Reynolds number of 1800. The separation error is quantified by the root-mean square of the deviation of the actual vertical position of the particles and their equilibrium position. Until around $x = 200$ mm the difference between the two cases is relatively small. After $x = 200$ mm the difference is visible, background flow disturbances increase the levitation time of the particles and therefore result in a separation error which is almost three times larger.

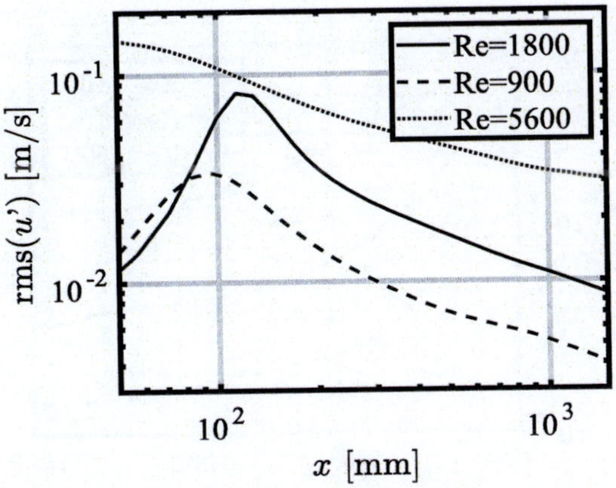

Fig. 4. Root-mean square of the streamwise velocity fluctuations averaged over the vertical and spanwise direction for Re = 900 (dashed), 1800 (solid) and 5600 (dotted).

Figure 4 shows the root-mean square of the streamwise fluid velocity component averaged over the vertical and spanwise direction as a function of the distance from the honeycomb exit, where the mean fluid velocity has been used to transform time to streamwise position. For the two laminar cases the peak position and maximum velocity fluctuation increase with increasing Reynolds number. The decay rate downstream of the peak is for both Reynolds numbers approximately the same. For the turbulent case, the peak in velocity fluctuations is located at the exit of the honeycomb and the fluctuations decay more slowly.

Figure 5 shows the resulting separation error, which is plotted as a function of the streamwise position and makes it possible to estimate the required length of the separation chamber to reach a certain separation error. The results correspond very well to the velocity fluctuations shown in Fig. 3. The higher the velocity fluctuations are, the larger the separation error is. These velocity fluctuations reach their maxima at a certain position, which is dependent on the Reynolds number, and then start to decay. The position where the RMS of velocity fluctuations reaches a maximum is related to the break-up of the individual channel profiles. For the turbulent case, the highest intensity occurs directly behind the walls which can be explained by the vortex street in the wake of the cell walls. Turbulent flow is not favorable for particle separation since it is accompanied by far larger velocity fluctuations than laminar flow. This might lead one to conclude that a lower Reynolds number always results in higher separation efficiency. However, a lower Reynolds number also implies a lower particle mass flow rate and the separation efficiency is a combination of separation quality and throughput. The optimal choice is therefore a trade-off between separation error and particle mass flow rate.

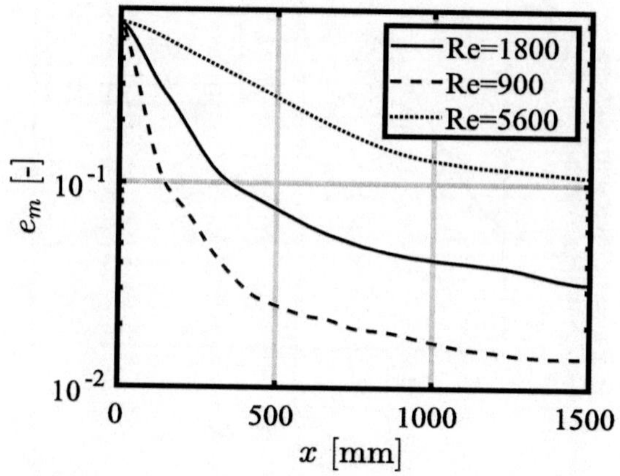

Fig. 5. Separation error as a function of the streamwise coordinate for Re=900 (dashed), 1800 (solid) and 5600 (dotted).

Acknowledgements. This work was sponsored by NWO Exacte en Natuurwetenschappen for the use of supercomputer facilities, with financial support from the Netherlands Organization for Scientific Research.

References

1. Dennis, S.C.R., Singh, S.N., Ingham, D.B.: The steady flow due to a rotating sphere at low and moderate Reynolds numbers. J. Fluid Mech. **101**, 257–279 (1980)
2. Van Hinsberg, M.A.T., ten Thije Boonkkamp, J.H.M. and Clercx, H.J.: An efficient, second order method for the approximation of the Basset history force, J. Comp. Phys., **230**, 1465–1478 (2011)
3. Kuerten, J.G.M., Van der Geld, C.W.M., Geurts, B.J.: Turbulence modification and heat transfer enhancement by inertial particles in turbulent channel flow. Phys. Fluids **23**, 123301 (2011)
4. Tajfirooz, S., Meijer, J.G., Dellaert, R.A., Zeegers, J.C.H., Kuerten, J.G.M.: Direct numerical simulation of magneto-Archimedes separation of spherical particles. J. Fluid Mech. **910**, A52 (2021)
5. Thijs, L.C., Dellaert, R.A., Tajfirooz, S., Zeegers, J.C.H., Kuerten, J.G.M.: Honeycomb-generated Reynolds-number dependent wake turbulence. J. Turbul. **22**, 535–561 (2021)

Volume Conservation Methods for VOF-Based Long-Term Simulations of Turbulent Bubble-Laden Flows on Coarse Grids

E. Trautner[1(✉)], J. Hasslberger[1], P. Cifani[2], and M. Klein[1]

[1] Institute of Applied Mathematics and Scientific Computing, Department of Aerospace Engineering, University of the Bundeswehr, Munich, Germany
{elias.trautner,josef.hasslberger,markus.klein}@unibw.de
[2] Faculty EEMCS, University of Twente, Enschede, The Netherlands
p.cifani@utwente.nl

1 Introduction

Turbulent bubble-laden flows are an integral part of numerous technical applications. Due to the computational cost of Direct Numerical Simulation, Large Eddy Simulation (LES) comes into focus for the design of such devices. One of the central challenges for two-phase LES is the under-resolution of the interface dynamics, which leads to a variety of issues with numerical algorithms for two-phase flows. This work discusses the violation of volume conservation in the context of Volume-of-Fluid (VOF)-based simulations of turbulent bubble-laden channel flows. Despite the good volume conservation properties of the VOF method, minimal volume errors on a time-step level can add up to severe errors when bubbly flows are simulated in periodic domains and on coarse grids. The problem is reinforced for long simulation intervals, which are indispensable to compute converged statistics.

This work firstly demonstrates the problem for a number of test cases. Subsequently, two volume conservation methods are proposed. Finally, the flow and bubble statistics and the individual bubble dynamics for both methods are compared.

2 Numerical Method and Flow Configuration

The study is performed using the finite-volume code "TBFsolver" [1], which is based on the one-fluid formulation of the incompressible Navier-Stokes equations. The mass and momentum conservation are given by

$$\frac{\partial u_i}{\partial x_i} = 0\,, \tag{1}$$

$$\rho\left(\frac{\partial u_i}{\partial t} + \frac{\partial u_i u_j}{\partial x_j}\right) = -\frac{\partial p}{\partial x_i} + \frac{\partial}{\partial x_j}\left[\mu\left(\frac{\partial u_i}{\partial x_j} + \frac{\partial u_j}{\partial x_i}\right)\right] + \sigma n_i \kappa \delta_S + (\rho - \rho_0) g_i\,. \tag{2}$$

Here, u_i, ρ, ρ_0, μ, p and g denote the i-th velocity component, density, average density, dynamic viscosity, pressure and gravitational acceleration, respectively. Using the

C. Marchioli et al. (Eds.): DLES 2023, ERCO 31, pp. 279–284, 2024.
https://doi.org/10.1007/978-3-031-47028-8_43

Continuous-Surface-Force method [2], the surface tension is computed from the surface tension coefficient σ, the interface normal n_i and the interface curvature κ. The latter is determined with a height function approach. The interface indicator function δ_S is approximated as $\delta_S = |\nabla f|$, where f denotes the marker function for the volume fraction of the gas phase in the framework of the VOF method. The local density and viscosity values are linearly interpolated based on the value of f. The code uses a geometrical reconstruction algorithm and a multiple-marker formulation [3], i.e., an individual marker function f_b is advected for every single bubble:

$$\frac{\partial f_b}{\partial t} + u_i \frac{\partial f_b}{\partial x_i} = 0\,. \tag{3}$$

The projection method is used to solve the introduced equations together with the Poisson equation for pressure. The QUICK scheme is used to discretize the convective term in Eq. 2, and the diffusive terms are treated with central differences. A staggered grid arrangement is used for the velocity components. Finally, the time step is adjusted corresponding to CFL $= 0.2 =$ const., and a fully explicit second-order Adams-Bashforth method advances the simulation in time.

The investigated configuration (see Fig. 1) is a vertical downflow channel of size $L_x = 8H$, $L_y = 2H$ and $L_z = 4H$, where H is the half width and x, y and z denote the stream-wise, the wall-normal and the span-wise direction. The x and z directions are periodic, whereas no-slip walls are prescribed in y-direction. The flow is controlled by a constant pressure gradient corresponding to a friction Reynolds number of $Re_\tau = (\sqrt{\tau_w/\rho_l H})/\nu_l = 590$, where τ_w is the average wall shear stress and ρ_l and ν_l are the density and the kinematic viscosity of the liquid phase, respectively. A total of 780 freely deformable bubbles with an initial diameter of $d = 0.25H$ are considered, leading to a gas volume fraction of 10%. The domain is discretized using an equidistant cubic mesh. The grid resolution Δ is varied such that it either corresponds to $d/\Delta = 12.5$ or $d/\Delta = 25$. Both the density and dynamic viscosity ratio are set to 20. The Eötvös

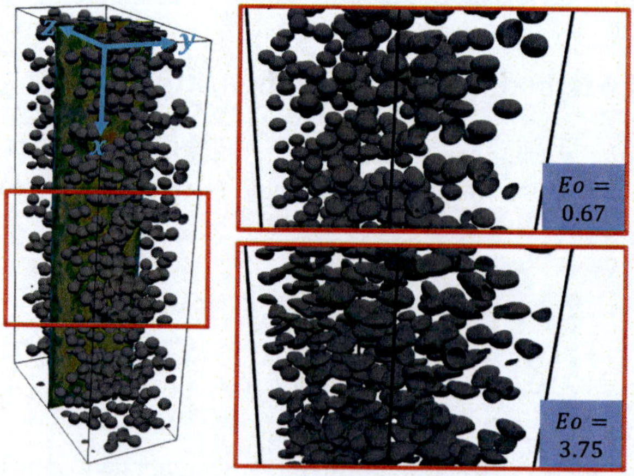

Fig. 1. Investigated channel configuration and bubble shapes for $Eo = 0.67$ and $Eo = 3.75$.

number $Eo = (gd^2\rho_l)/\sigma$ is varied between $Eo = 0.67$ and $Eo = 3.75$, which leads to nearly spherical bubbles for the lower, and to ellipsoidal and wobbling bubbles for the higher value of Eo.

3 Volume Conservation: Background and Methods

The implemented VOF method combines a multiple-marker formulation with a split-direction advection algorithm by Puckett et al. [4]. In this context, there are three root causes for the violation of volume conservation for the tracked gas phase:

1. Over- ($f_b > 1$) and undershoots ($f_b < 0$) of VOF values in single cells that are clipped to one or zero, respectively. This issue could be avoided by, e.g., using the VOF advection method proposed by Weymouth & Yue [5].
2. Resetting of cell VOF values to zero in cells with $0 < f_b < 1$, when no full gas cell with $f_b = 1$ is detected within a region of $5 \times 5 \times 5$ cells around the respective cell. This is a standard procedure in multiphase codes to avoid smearing.
3. Removal of small gas structures that can separate from the respective main bubble when high shear forces overcome the restoring surface tension forces.

The first scenario can lead to both an increase and a decrease of gas volume, while the latter two can only lead to volume losses.

For low Reynolds and Eötvös numbers and DNS-like resolutions, the observed volume errors are negligible [1]. For the configurations studied in this work, however, the small deviations on a time step level accumulate to significant volume errors. Figure 2 illustrates the global relative volume loss of the bubble collective for different setups over the dimensionless time. As the figure shows, the error is more severe for the higher Eötvös number and the lower resolution. Moreover, it is clearly demonstrated that solely using the VOF advection scheme by Weymouth & Yue ("WY") [5], which per construction avoids over- and undershoots, does not solve the issue, since it only avoids the first of the three origins of volume errors listed above. Therefore, the VOF advection scheme by Puckett et al. ("P") [4], which is the default in TBFsolver, is used in the following. It is important to note that the observed volume errors are not specific for the chosen solver or the multiple marker formulation, and similar issues have been observed using other two-phase codes.

In the following, two volume conservation methods are proposed. The underlying idea is to only use a single correction procedure to correct the volume error originating from all the three issues listed above in each time step. The methods rely on the given circumstance that the volume error is minimal for a single time step and only has an impact over a long simulation period. Moreover, the correction is performed for each bubble individually.

The first method (Method A) is also referred to as "simple method". To correct the volume error ΔV_b^t for bubble b in time step t, all N interface cells with $\varepsilon \leq f_b \leq (1-\varepsilon)$ are identified. Then, the volume error is corrected by adding or subtracting $\Delta V_b^t/(N \cdot V_{cell})$ to the VOF values of these cells. For the investigated setup, a tolerance of $\varepsilon = 1 \times 10^{-4}$ has been used. This avoids additional over- and undershoots in the correction step. The method is simple and free of kinematic assumptions, but, due to ε,

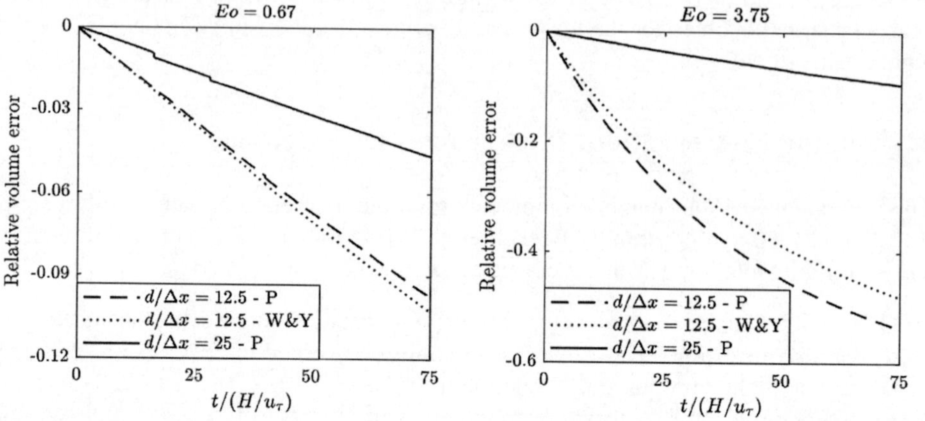

Fig. 2. Global relative volume loss $(V(t) - V_0)/V_0$ for $Eo = 0.67$ (left) and $Eo = 3.75$ (right) over the dimensionless time. The illustrated time interval corresponds to around 135 flow-through times.

it is not parameter-free. However, additional investigations have shown that the precise value of ε is of minor importance.

The second method (Method B) is termed "dilatation/contraction method". The evolution of a single bubble's volume can be expressed as

$$\frac{dV_b}{dt} = \int_{V_b} (\nabla \cdot \boldsymbol{u})\, dV. \tag{4}$$

Consequently, imposing a fictitious velocity field with $\nabla \cdot \boldsymbol{u} \neq 0$ allows to correct the volume error for the bubble. Assuming $\alpha = \nabla \cdot \boldsymbol{u}$, where α is a constant valid for one time step and a single bubble, Eq. 4 can be discretized as

$$\frac{V_b^{\text{corr.}} - V_b^t}{\Delta t} = \alpha V_b^t, \tag{5}$$

where $V_b^{\text{corr.}}$ is the correct bubble volume to be conserved. Since the volume error in Eq. 5 can be measured, α can be determined. Now, a velocity field with $\nabla \cdot \boldsymbol{u}^* = \alpha$ can be determined as $\boldsymbol{u}^* = (\alpha/3)(x - x_0, y - y_0, z - z_0)^T$, where (x_0, y_0, z_0) denotes the center of the bubble. This velocity field is now used to perform an additional reconstruction and advection step for the bubble's VOF field. This can be interpreted as a slight dilatation (contraction) of the bubble for $V_b^t < V_b^{\text{corr.}}$ ($V_b^t > V_b^{\text{corr.}}$). It can be shown that this procedure preserves the bubble volumes to machine precision.

4 Results

It is essential that the volume conservation methods do not exhibit a notable influence on the flow physics. The two proposed methods are verified against each other, i.e., if they lead to identical results they are considered reliable. Figure 3 shows the average gas fraction ($\langle f \rangle$), stream-wise velocity ($\langle u \rangle$) and turbulent kinetic energy (k) profiles for the setups at the coarser resolution ($d/\Delta = 12.5$). As the results show, the two methods lead to identical global flow statistics. For comparison, the respective simulations have been conducted without a volume conservation method. As the results demonstrate, there is a clear tendency towards single-phase statistics for this scenario, including a more pronounced peak of the wall-induced turbulence at $y^+ \approx 20$. This demonstrates the importance of the volume conservation methods.

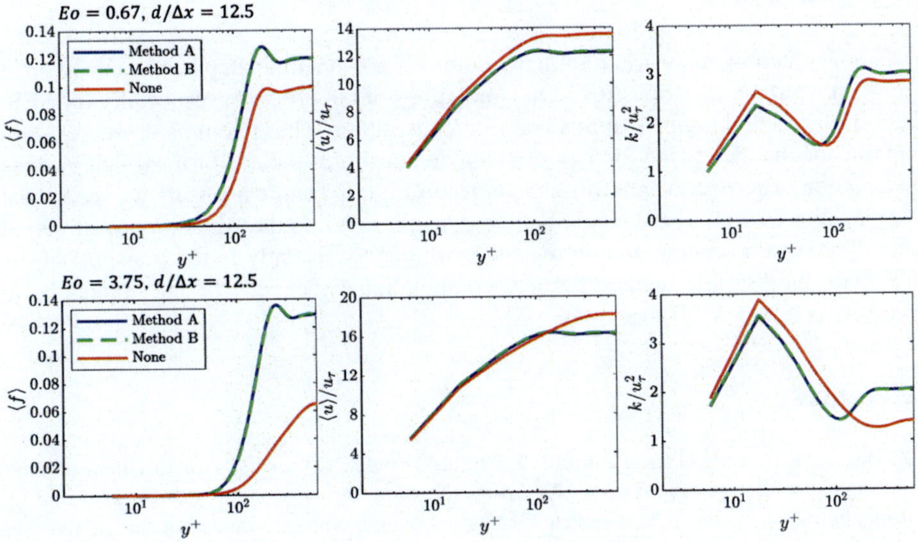

Fig. 3. Average gas fraction ($\langle f \rangle$), stream-wise velocity ($\langle u \rangle$) and turbulent kinetic energy (k) profiles over the dimensionless wall distance (y^+) for the setups at the coarser resolution ($d/\Delta = 12.5$). The first (second) row shows the results for $Eo = 0.67$ ($Eo = 3.75$). The blue (green) profiles represent the simulations performed with the "simple method" ("dilatation and contraction method"). The red profiles show the results for simulations without volume conservation after $t = 400(H/u_\tau)$.

In addition to the global flow statistics, the methods are also validated on the level of single bubble dynamics. For this purpose, the temporal behavior of the bubbles' interface areas is investigated. The surface area of a bubble b is numerically approximated as $\int_{V_b} |\nabla f_b| dV$ and normalized by the surface area of a volume-equivalent sphere to obtain the normalized surface area $A_b(t)$. From the latter, the time- and bubble-averaged mean deformation $\overline{A}$ and the Root Mean Square of the fluctuations A' around the mean value are computed. Table 1 provides the results for the setups at the lower resolution $d/\Delta = 12.5$. As the table demonstrates, the mean and fluctuation values of the bubbles' surface areas, and therefore the bubble dynamics, are quasi identical for the two methods. It is worth noting that the values of $\overline{A}$ for $Eo = 0.67$ are below the theoretical limit of one. This is simply due to the fact that the numerical approximation of the surface area is less accurate on coarse grids.

Table 1. Mean value $\overline{A}$ and fluctuation A' of the bubbles' surface areas for $d/\Delta = 12.5$.

Case	Method A ("simple")	Method B ("dilatation/contraction")
$Eo = 0.67$ $\overline{A}$ (A')	0.9846 (1.5600×10^{-2})	0.9846 (1.5603×10^{-2})
$Eo = 3.75$ $\overline{A}$ (A')	1.0559 (4.1447×10^{-2})	1.0561 (4.1676×10^{-2})

5 Conclusions

As demonstrated in this work, minimal volume errors on a time-step level can add up to significant volume deviations for VOF simulations of bubbly flows, especially for LES-like resolutions and high Reynolds and Eötvös numbers. These volume errors result in a significant change of flow statistics, which demonstrates the need for explicit volume conservation. The two proposed methods lead to quasi-identical results for the global flow statistics as well as the single bubble dynamics. While the "dilatation/contraction method" is parameter-free and corrects the volume consistently to the transport of the VOF field, the "simple method" is less computationally expensive and applicable to arbitrarily complex VOF structures.

References

1. Cifani, P., Kuerten, J.G.M., Geurts, B.J.: Highly scalable DNS solver for turbulent bubble-laden channel flow. Comp. Fluids **172**, 67–83 (2018)
2. Brackbill, J.U., Kothe, D.B., Zemach, C.: A continuum method for modeling surface tension. J. Comput. Phys. **100**, 335–354 (1992)
3. Coyajee, E., Boersma, B.J.: Numerical simulation of drop impact on a liquid-liquid interface with a multiple marker front-capturing method. J. Comput. Phys. **228**, 4444–4467 (2009)
4. Puckett, E.G., Almgren, A.S., Bell, J.B., Marcus, D.L., Rider, W.J.: A high-order projection method for tracking fluid interfaces in variable density incompressible flows. J. Comput. Phys. **130**, 269–282 (1997)
5. Weymouth, G.D., Yue, D.K.P.: Conservative volume-of-fluid method for free-surface simulations on cartesian grids. J. Comput. Phys. **229**, 2853–2865 (2010)

Hydrogen Jet Flame Global Instabilities and Their Control Using Suction

A. Wawrzak[1(✉)], K. Wawrzak[1], A. Boguslawski[1], A. Tyliszczak[1], and B. J. Geurts[2,3]

[1] Faculty of Mechanical Engineering and Computer Science, Czestochowa University of Technology, Czestochowa, Poland
{agnieszka.wawrzak,karol.wawrzak,andrzej.boguslawski, artur.tyliszczak}@pcz.pl

[2] Mathematics of Multiscale Modeling and Simulation, Faculty of Electrical Engineering, Mathematics and Computer Science, University of Twente, Enschede, The Netherlands
b.j.geurts@utwente.nl

[3] Multiscale Energy Physics, Eindhoven University of Technology, Eindhoven, The Netherlands

Abstract. This study employs large-eddy simulations (LES) to explore how counter-current co-axial flow influences global instability in hydrogen jet flames. It reveals that the critical velocity ratio for triggering global instability falls between 0.1 and 0.2. Additionally, applying strong suction ($I = 0.2$) around the fuel jet can effectively change the position and size of the flame.

1 Introduction

Flame control has been at the centre of interest for decades. It constitutes a subject of intensive research as it may improve the efficiency and safety of various technical devices. Numerous industrial applications, such as burners or combustion chambers, involve jet-type flames with fuel issuing from a nozzle into an oxidizer stream. An interesting phenomenon emerging in round jets, which may also be considered a flow control technique, is self-excited global instability triggered by absolutely unstable local flow regions [1]. Theoretical predictions showed that absolute instability can be induced in variable density and counter-current jet configurations [2,3]. Two absolutely unstable modes, Mode I and Mode II, have been analytically identified [3] and their occurrence in real life was confirmed by experimental and numerical works [4–7].

However, there are no investigations devoted to global instability in combustion problems such as jet flames, despite their significance for practical solutions. In the present paper, the effect of counter-current co-axial flow on the emergence of global instability in hydrogen jet flames is studied with the help of large-eddy simulations (LES).

2 Computations

The test case configuration is presented schematically in Fig. 1. It corresponds to the experimental set-up [8] used for hydrogen autoignition in a turbulent co-flow of heated

C. Marchioli et al. (Eds.): DLES 2023, ERCO 31, pp. 285–290, 2024.
https://doi.org/10.1007/978-3-031-47028-8_44

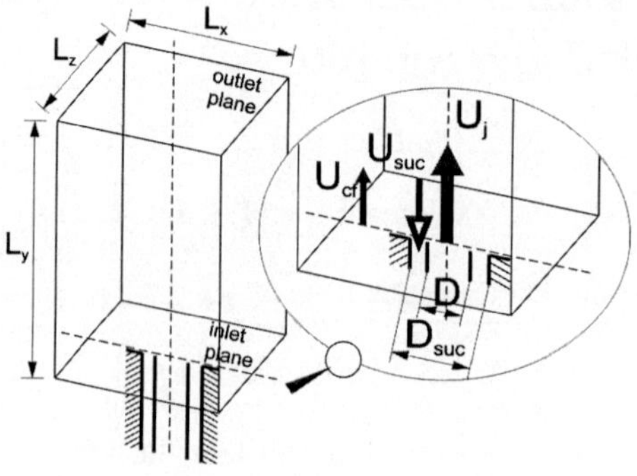

Fig. 1. Computational configuration

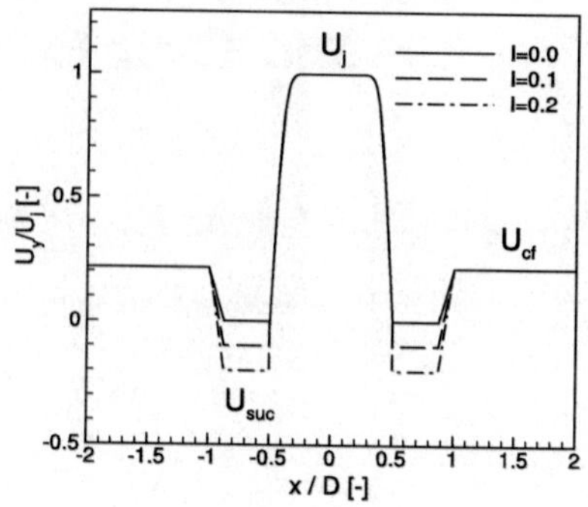

Fig. 2. The inlet velocity profiles for the three cases considered

Table 1. Parameters describing the simulated hydrogen jet flames

Fuel (jet)	Oxidiser (co-flow)	T_j [K]	T_{cf} [K]	U_j [m/s]	U_{suc} [m/s]	I [-]	U_{cf} [m/s]
0.13 H_2/0.87 N_2	0.23 O_2/0.77 N_2	691	1010	120	0, −12, −24	0, 0.1, 0.2	26

air. A mixture of hydrogen and nitrogen with mass fractions $Y_{H_2} = 0.13$ and $Y_{N_2} = 0.87$ is injected into heated air through a 2.25 mm (D) internal diameter pipe. Table 1 shows details of the temperature and velocity of the fuel jet as well as the co-flow stream. The density ratio $S = \rho_j/\rho_{cf}$, where ρ_j and ρ_{cf} denote density of the jet and co-flow, is slightly above the critical density ratio for which global oscillations emerge in a jet without counterflow for a given shear layer thickness characterized by the parameter $D/\theta = 40$ (θ - momentum thickness) [5]. The suction is applied through the annular nozzle ($D_{suc} = 2D = 4.5$ mm) which is placed around the main nozzle. It produces a counter-current region in the direct vicinity of the main jet. The strength of the counterflow is controlled by the velocity ratio $I = -U_{suc}/U_j$ (U_j - velocity of the jet, U_{suc} - velocity of the counter-current). To analyze the impact of the suction on the flame dynamics in two different flow regimes we consider the cases with $I = 0.1$ and $I = 0.2$ which correspond to values well below and well above the critical one observed previously [3,4,6,7]. The effect of the counterflows is assessed in relation to the case without suction ($I = 0$).

In the present work we do not consider the inner geometry of the nozzles. Instead, the computational domain is a rectangular box with dimensions $L_y = 30D, L_x = L_z = 15D$, where ‘$y$’ is the axial direction. The applied mesh counts $N_y = 288$, $N_x = N_z = 192$ nodes in the axial and radial directions, respectively. The inlet boundary conditions are specified in terms of the instantaneous velocity profile. The velocity fluctuations are computed according to the method proposed by Klein et al. [9] whereas the inlet mean velocity is described by the Blasius profile. The profiles of the mean axial velocity at the inlet plane of the computational domain for $I = 0, 0.1$ and 0.2 are displayed in Fig. 2.

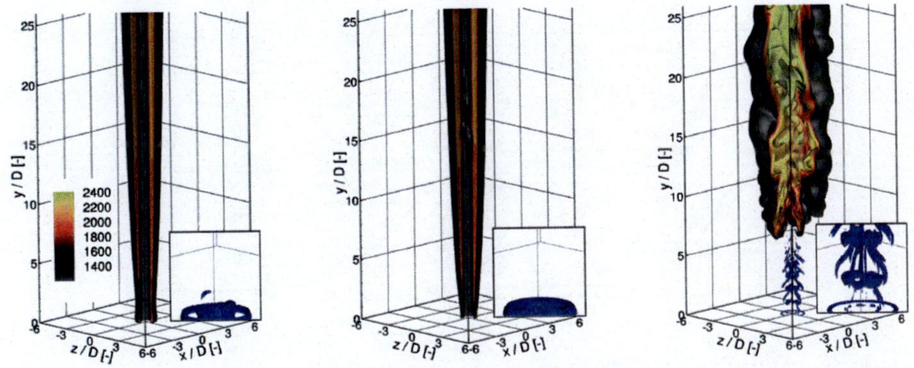

Fig. 3. Instantaneous iso-surfaces of the Q-parameter ($Q = 0.05\ \text{s}^{-2}$, blue) and temperature inside the flames at $I = 0$ (left), $I = 0.1$ (middle), $I = 0.2$ (right)

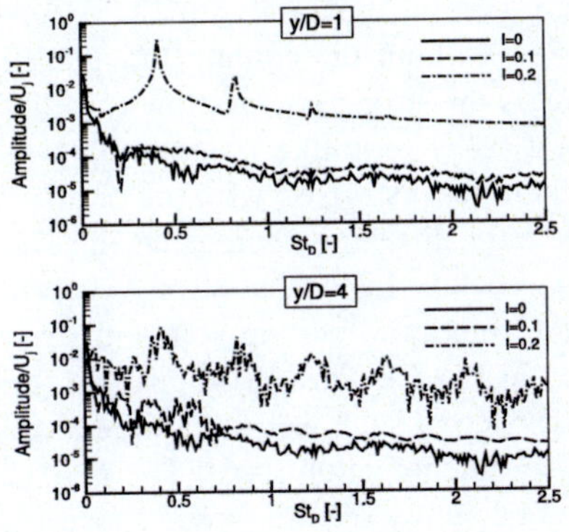

Fig. 4. Amplitude spectra of the axial velocity at two locations from the inlet plane

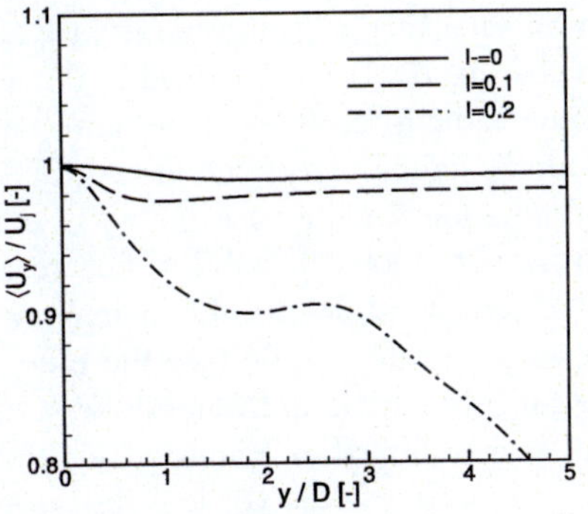

Fig. 5. Profiles of the time-averaged mean axial velocity along the jet axis

The LES code used in this study is an in-house high-order solver based on the low Mach number approximation. The Navier-Stokes and continuity equations are discretized using the sixth-order compact difference method on half-staggered meshes [10]. A sub-grid model of Vreman [11] is used to compute the sub-grid viscosity for the SGS-stress tensor. The chemical reactions are computed using the CHEMKIN interpreter with the help of a detailed mechanism of hydrogen oxidation [12] involving 9 species and 21 reactions. The calculations are conducted without any closure for turbulence/combustion interactions at the sub-grid level. Correspondingly, the reaction rates of species were obtained directly using the filtered variables.

3 Results

Figure 3 displays fully developed flames visualised by the instantaneous iso-surfaces of the temperature inside the flame, and the Q-parameter close to the inlet plane. It

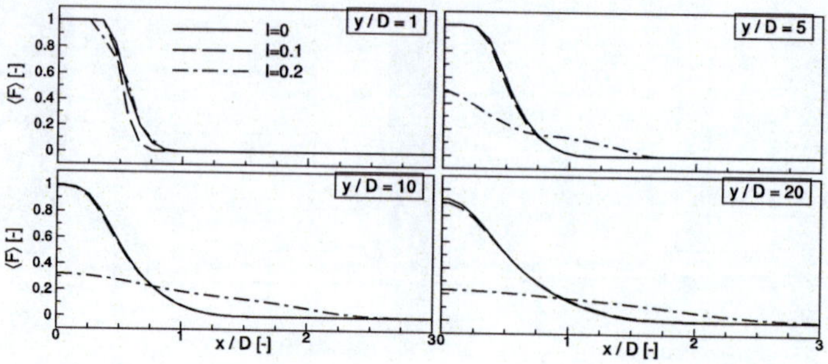

Fig. 6. Radial distributions of the time-averaged mixture fraction at different axial locations

appears that the flow pattern is similar for $I = 0$ and $I = 0.1$ while significant differences emerge when the strongest suction is applied, i.e., for the case with $I = 0.2$. As can be seen in Fig. 3, the flame in the configuration without the counterflow ($I = 0$) and also for $I = 0.1$ is attached to the nozzle, whereas the strong counterflow ($I = 0.2$) stabilizes the lifted flame. The flames at $I = 0$ and $I = 0.1$ look like a smooth laminar flame expanding across the thin shear layer where the fuel is mixed with the oxidizer. The situation for the case with $I = 0.2$ is significantly different. Increasing the suction triggers global instability before auto-ignition occurs. In fact, at strong suction, the velocity field drastically changes revealing the formation of strong coherent structures close to the nozzle (see the blue iso-surfaces in Fig. 3). These vortices pair and subsequently break up further downstream. The presence of such well-organised structures is recognized as a sign of the global instability phenomenon [4,6]. The generation and pairing of the vortices repeat in time. Figure 4 shows the amplitude spectra of the axial velocity expressed in terms of a non-dimensional frequency defined through the Strouhal number $St_D = fD/U_j$. The spectra were calculated on the basis of time signals registered at the jet centerline at downstream positions $y/D = 1$ and 4. Well-defined distinct peaks at $St_D = fD/U_j = 0.41$ are found immediately above the inlet plane for $I = 0.2$. These indicate self-excited global oscillations [4,6]. The characteristic frequency remains unchanged at further axial distances. Downstream of $y/D = 4$, the vortices break up. In contrast, the flow fields in the cases with $I = 0$ and $I = 0.1$ are ruled by completely different phenomena. The amplitudes of the oscillations are approximately two orders of magnitude smaller. Moreover, their spectra do not reveal any distinct peaks.

As can be seen in Fig. 5 showing velocity profiles along the jet axis the velocity decays close to the nozzle exit for the case with $I = 0.2$. In this situation, the mixing is drastically intensified directly behind the inlet plane. Figure 6 presents the radial profiles of time-averaged mixture fraction. Its values are reduced already at $y/D = 1$. Further downstream, at $y/D = 5$ a large amount of fluid is taken from the surroundings and transported toward the periphery of the jet by large vortices. This makes the mixture significantly leaner across the whole jet. The enhanced mixing of fuel and oxidizer does not necessarily mean more favourable regimes for flame development. The mixed gases have a lower temperature while local values of the scalar dissipation rate are higher. All this prevents upstream flame propagation and the flame lifts off.

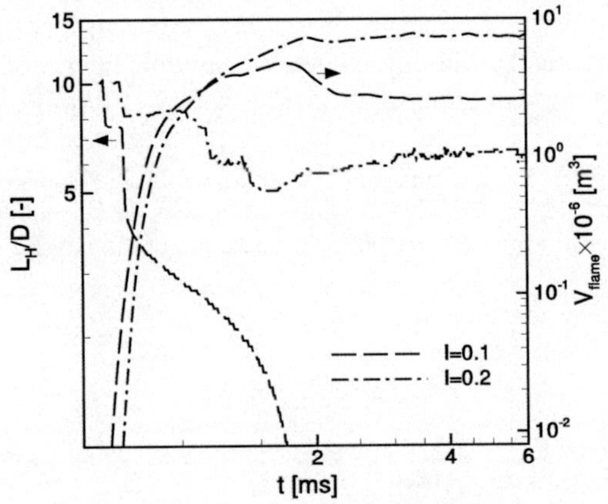

Fig. 7. Temporal evolution of the flame volume

The above results revealed that by applying sufficiently strong suction one can create a counter-current flow conducive to the generation of the global instability phenomenon. Significant differences in the instantaneous flame dynamics due to global instability are demonstrated in Fig. 7 presenting the temporal evolution of lift-off heights and the flames volumes for $I = 0.1$ and $I = 0.2$. The former was obtained from analysis of the temperature distribution in the main 'x-y' cross-section, i.e., $L_H = \min(y^{T \geq 1.01 T_{cf}})$. The latter was calculated as the sum of the volumes of the computational cells in which the temperature is larger than 1200 K ($V_{\text{flame}} = \sum V_{\text{cell}}^{T \geq 1200\text{K}}$). The auto-ignition is weakly dependent on the suction strength. As can be seen the V_{flame} starts to increase later for $I = 0.2$. Shortly after, the flames propagate quickly and attain their final position at $t \approx 2$ ms. The flame with $I = 0.1$ is attached to the nozzle whereas the flame at $I = 0.2$ anchors at the distance $y/D \approx 6$. The lift-off height changes slightly depending on the flow variations upstream of the flame front. Before both flames stabilize, the values of V_{flame} for $I = 0.1$ and $I = 0.2$ are very similar. Then, the profiles of V_{flame} start to deviate significantly and eventually become approximately constant. For $I = 0.1$, where the flame has a shape similar to a slightly widening tube (see Fig. 3), its volume is approximately three times smaller than the volume of the flame at $I = 0.2$. In this case, the flame lift-off is accompanied by growth in size of the flame. Hence, it can be said that global instability makes the mixing process very dynamic which ensures a much faster combustion process.

4 Conclusions

The present study shows that the critical velocity ratio for which global instability is triggered in the considered configuration is between 0.1 and 0.2. The global mode observed at sufficiently strong counter-current flow ($I = 0.2$) causes qualitative changes in the flame characteristics. It was demonstrated that the position and size of the flame can be effectively modified by the application of suction around the fuel jet. This may be desirable from a practical point of view since a suitable alteration of the flame would lead to a considerable improvement in efficiency, safety or pollution reduction.

Acknowledgments. This work was supported by the National Science Center in Poland (Grant 2020/39/B /ST8/02802) and the National Agency for Academic Exchange (NAWA) (Research project ANIMATE No. PPI/APM/2019/1/00062). The computations were carried out using PL-Grid Infrastructure.

References

1. Huerre, P., Monkewitz, P.: Local and global instabilities in spatially developing flows. Annu. Rev. Fluid Mech. **22**, 473–537 (1990)
2. Monkewitz, P., Sohn, K.: Absolute instability in hot jets. AIAA J. **26**, 911–916 (1988)
3. Jendoubi, S., Strykowski, P.: Absolute and convective instability of axisymmetric jets with external flow. Phys. Fluids **6**, 3000–3009 (1994)
4. Strykowski, P., Niccum, D.: The stability of countercurrent mixing layers in circular jets. J. Fluid Mech. **227**, 309–343 (1991)
5. Boguslawski, A., Tyliszczak, A., Wawrzak, K.: Large eddy simulation predictions of absolutely unstable round hot jet. Phys. Fluids **28**, 025108 (2016)
6. Wawrzak, K., Boguslawski, A., Tyliszczak, A.: A numerical study of the global instability in counter-current homogeneous density incompressible round jets. Flow Turbul. Combust. **107**, 901–935 (2021)
7. Wawrzak, K., Boguslawski, A.: Self-excited oscillations in variable density counter-current round jets. In: Twelfth International Symposium on Turbulence and Shear Flow Phenomena. Begel House Inc. (2022)
8. Markides, C., Mastorakos, E.: An experimental study of hydrogen autoignition in a turbulent co-flow of heated air. Proc. Combust. Inst. **30**, 883–891 (2005)
9. Klein, M., Sadiki, A., Janicka, J.: A digital filter based generation of inflow data for spatially developing direct numerical or large eddy simulations. J. Comput. Phys. **186**, 652–665 (2003)
10. Tyliszczak, A.: High-order compact difference algorithm on half-staggered meshes for low Mach number flows. Comput. Fluids **127**, 131–145 (2016)
11. Vreman, A.: An eddy-viscosity subgrid-scale model for turbulent shear flow: algebraic theory and applications. Phys. Fluids **16**, 3670–3681 (2004)
12. Mueller, M., Kim, T., Yetter, R., Dryer, F.: Flow reactor studies and kinetic modeling of the H_2/O_2 reaction. Int. J. Chem. Kinet. **31**, 113–125 (1999)

Numerical Methods

Improved Evaluation of Subfilter Stresses Using Approximate Deconvolution of Implicit Filters Induced by Numerical Schemes

A. Boguslawski[1(✉)], K. Wawrzak[1], and B. J. Geurts[2,3]

[1] Czestochowa University of Technology, Czestochowa, Poland
{andrzej.boguslawski,karol.wawrzak}@pcz.pl
[2] Mathematics of Multiscale Modeling and Simulation, Faculty EEMCS, University of Twente, PO Box 217, 7500 AE Enschede, The Netherlands
b.j.geurts@utwente.nl
[3] Multiscale Physics of Energy Systems, Center for Computational Energy Research, Faculty of Applied Physics, Eindhoven University of Technology, PO Box 513, 5600 MB Eindhoven, The Netherlands

1 Introduction

Large Eddy Simulation (LES) introduces a low-pass filtration of the Navier-Stokes equations distinguishing large resolved scales from small subfilter scales. Large scales are approximated by the solution of the filtered equations in which the influence of the small scales is introduced as a model for the subfilter stresses. However, one should distinguish between two types of filtering in LES [1]. Firstly, the numerical discretisation and mesh truncate the unknown exact solution, acting like an implicit filter with a cutoff lengthscale determined by the mesh resolution and the numerical scheme that is applied. Secondly, a filter operator with a cut-off length scale that can be explicitly applied to the discretised equations. Both types of filtration introduce additional terms to the filtered equations in the form of subfilter stresses. Subfilter stresses can be modeled in various ways, e.g., by the eddy viscosity and the resolved strain rate magnitude or using somehow estimated unfiltered velocity field. The present paper deals with the implicit filter induced by the numerical discretisation only. Knowing the explicit form of this numerical filter enables to partially recover by approximate deconvolution the corresponding small-scale dynamics in which the deconvolved velocity field can be used to evaluate subfilter stresses. As shown by Geurts and van der Bos [3] any finite difference scheme can be understood as an exact differentiation of the filtered function with a filter characterized by a filter kernel composed of top-hat filters. This reasoning can readily be extended to implicit compact differencing. Hence, for any explicit or implicit finite differencing scheme of arbitrary order we may apply a deconvolution procedure to recover part of the subfilter scales eliminated by the numerical scheme. In this work, we show the results of such a deconvolution for the sixth-order compact differencing method. As a test case, two-dimensional decaying turbulence is studied. Special attention will be given to the importance of the filter kernel discretization in relation to the quality of the deconvolution procedure.

C. Marchioli et al. (Eds.): DLES 2023, ERCO 31, pp. 293–298, 2024.
https://doi.org/10.1007/978-3-031-47028-8_45

2 Approximate Deconvolution of Induced Filters

Consider a general finite difference method for numerically approximating the first-order partial derivative of a function u in one spatial dimension given by

$$\delta_x u(x_i) = \frac{1}{h} \sum_{j=-n}^{m} a_j u_{i+j} \tag{1}$$

where we denoted $u_{i+j} = u(x_{i+j})$, i.e., the solution in the point x_{i+j} of the grid $\{x_k\}$. For convenience we restrict to discretization on a uniform grid with grid spacing h, and adopt a general stencil of $m+n+1$ nodes, defining the stencil $[-n,m]$. The implied filter L can be related to the discretization weights $\{a_j\}$ and expressed in terms of a series of explicit top-hat filters [3]. It can be expressed as a weighted average of skewed top-hat filters of width h

$$L(u(x_i)) = \sum_{j=-n+1}^{m} b_j \left(\frac{1}{h} \int_{x_i+(j-1)h}^{x_i+jh} u(\eta) d\eta \right) \tag{2}$$

where

$$b_j = \sum_{i=j}^{m} a_i, \quad j = -n+1, \ldots, m \tag{3}$$

To specify the implied filter (2) on a numerical mesh, the integral in (2) should be approximated. Several possible quadrature rules can be selected to approximate L - here we consider the trapezoidal rule, the Cavalieri-Simpson rule [4] and the Fourier series approximation [5], for periodic problems.

We start the analysis by selecting the trapezoidal rule for the numerical integration in (2), which implies

$$L(u(x_i)) = \sum_{j=-n+1}^{m} b_j \frac{u_{i+j-1} + u_{i+j}}{2} \tag{4}$$

Rearranging the sum one can arrive at the discrete form of the implied filter

$$\begin{aligned} &\textstyle\sum_{j=-n+1}^{m} b_j \frac{u_{i+j-1}+u_{i+j}}{2} \\ &= \tfrac{b_{-n+1}}{2} u_{i-n} + \textstyle\sum_{j=-n+1}^{m-1} \frac{b_j+b_{j+1}}{2} u_{i+j} + \frac{b_m}{2} u_{i+m} \equiv \sum_{j=-n}^{m} G_j^{(-n,m),tr} u_{i+j} \end{aligned} \tag{5}$$

with the discrete filter kernel defined as

$$\begin{aligned} G_{-n}^{(-n,m),tr} &= \tfrac{b_{-n+1}}{2} \\ G_j^{(-n,m),tr} &= \tfrac{b_j+b_{j+1}}{2} \quad for \quad j = -n+1, \ldots, m-1 \\ G_m^{(-n,m),tr} &= \tfrac{b_m}{2} \end{aligned} \tag{6}$$

Here, the superscript 'tr' is used to emphasize that the discrete filter adopts the trapezoidal rule for integration. Knowing the discrete form of the implied filter kernel its approximate inverse can be found. First, Wiener [7] type inverse filtering will be considered, focussing attention on periodic problems in the domain $0 < x < 1$. Taking Fourier series of the discrete filter kernel over the computational domain

$$\hat{G}_k^{(-n,m),tr} = \frac{1}{N}\sum_{i=0}^{N-1} G_i^{(-n,m),tr} e^{-2\pi \underline{i} k x_i}; \ k = -K, \ldots, K \tag{7}$$

where $\underline{i} = \sqrt{-1}$ and $N = 2K+1$ is the odd number of mesh points, the inverse filter kernel in Fourier space is defined as

$$\hat{Q}_k^{(n,m),tr} = \frac{1}{\hat{G}_k^{(-n,m),tr} N^2} \tag{8}$$

and the inverse filter kernel in physical space is found by the inverse Fourier transform

$$Q_i^{(-n,m),tr} = \sum_{k-K}^{K} \hat{Q}_k^{(-n,m),tr} e^{2\pi \underline{i} k x_i}; \ i = 0, \ldots, N-1 \tag{9}$$

An analogous procedure to discretize the filter kernel can be applied using more precise methods of integration in Eq. (2). Examples include Simpson-Cavalieri [4] with parabolic interpolation, higher order Lagrange polynomials or, in the periodic test case under consideration, Fourier series.

In the present paper trapezoidal, Simpson-Cavalieri and Fourier series methods are compared for the case of compact differencing of sixth order. The discrete form of the filter kernel induced by the numerical scheme and its inverse enables the subfilter stress tensor [2] to be expressed as

$$\tau_{ij} = G[Q(u_i)Q(u_j)] - u_i u_j \tag{10}$$

where u_i, u_j denote the resolved velocity components and G and Q general filtration in two dimensions with filter corresponding to the numerical scheme and its inverse, respectively. We will investigate LES that employs Eq. (10) as sub-filter stress tensor model.

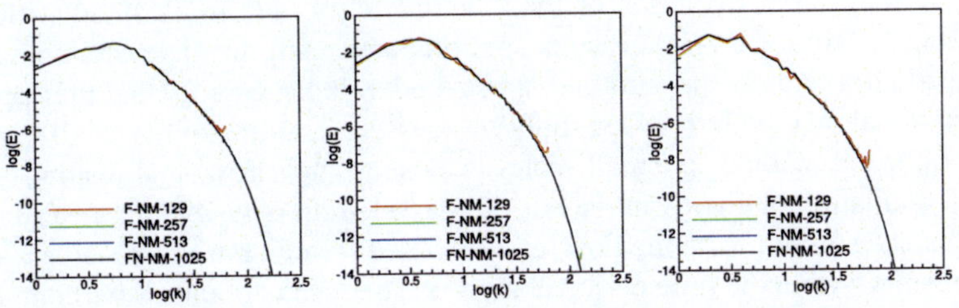

Fig. 1. Evolution of energy spectra, DNS, Re = 500, mesh density testing. Figures from left to right correspond to times t = 2, 4, 6

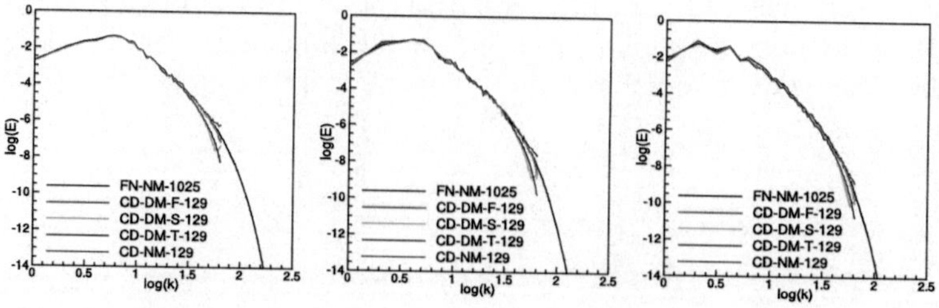

Fig. 2. Evolution of energy spectra. LES compared to DNS and the results with no model. Figures from left to right correspond to times t = 2, 4, 6. Abbreviations stand for: F-Fourier spectral method, CD-compact differencing of 6th order, NM-no model, DM - deconvolution model, T, S, F-trapezoidal, Simpson, Fourier integration, respectively.

3 Homogeneous Turbulence: Spectral and Vorticity Dynamics

The initial field has been obtained following the method proposed by San and Staples [6]. Two-dimensional freely decaying incompressible flow on a square domain of length 2π and with periodic boundary conditions has been taken into consideration. The initial energy spectrum is assumed to be

$$E(k) = \frac{a_s}{2}\frac{1}{k_p}\left(\frac{k}{k_p}\right)^{2s+1}\exp\left[-\left(s+\frac{1}{2}\right)\left(\frac{k}{k_p}\right)^2\right] \tag{11}$$

where s is a shape parameter assumed to be equal 3 and $a_s = \frac{(2s+1)^{s+1}}{2^s s!}$. The maximum value of the initial energy spectrum is obtained for $k_p = 12$. The calculations were performed for relatively low Reynolds number, Re = 500, applying sixth-order compact differencing and ADM models with trapezoidal, Cavalieri-Simpson, and Fourier methods for integration in Eq. (2). Figure 1 shows the evolution of kinetic energy spectra obtained with various mesh sizes using the Fourier spectral method. It is seen that for the mesh size 257×257 and finer all the scales are captured precisely, hence the results are considered as Direct Numerical Simulation (DNS). Only the coarsest mesh 129×129 appears to require LES and modelling. All further LES were performed on this coarsest mesh. The evolution of the kinetic energy spectra obtained with LES is shown in Fig. 2. The LES-ADM results are compared with the DNS and compact differencing solution without any subfilter stress model. It can be seen that the application of the trapezoidal method introduces slightly excessive damping of the small scales. The Cavalieri-Simpson method for small scales is close to the trapezoidal results, however, a small accumulation of energy at the cut-off wave-length is visible. In the range of the small scales, the solution with the Fourier series used for integration coincides best with the DNS results within the time period analyzed. The results obtained without any subfilter model are seen to overpredict energy at the small scales close to the cut-off wave number. The influence of the ADM method applied to the numerically induced filter on large-scale dynamics is illustrated by the vorticity fields. Figure 3 shows the vorticity field at time $t = 5$. The rectangle indicates two merging structures to be compared with LES-ADM results shown in Fig 4. Visual comparison of these two merging structures

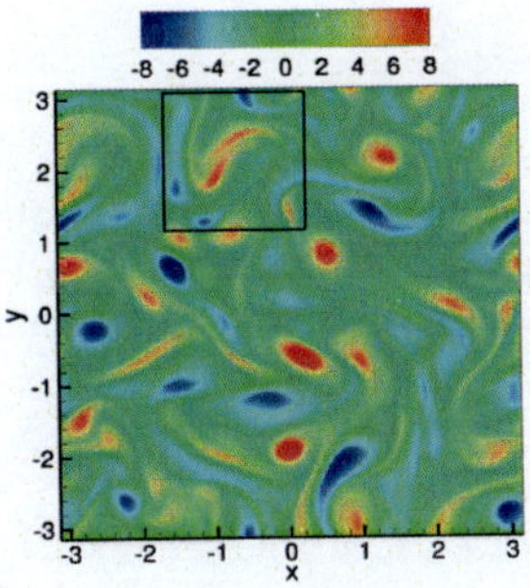

Fig. 3. DNS-vorticity, iso-contours, time instant t = 5

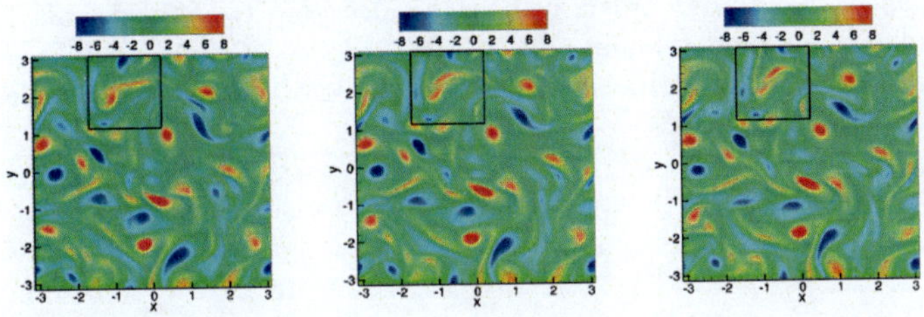

Fig. 4. LES-vorticity iso-contours, time instant t = 5. Figures from left to right correspond to the trapezoidal, Simpson, and Fourier methods applied for integration

predicted with various ADM versions shows that all methods perform accurately. On closer inspection, the trapezoidal results show these two structures are distorted somewhat when compared to the DNS results while Cavalieri-Simpson and Fourier ADM models predict the merging process in better agreement with the DNS.

4 Conclusions

A new deconvolution method to recover small scales filtered implicitly by the filter induced by the spatial discretisation scheme is proposed. It is shown that for each explicit and implicit finite difference method a dedicated filter kernel can be formed and its inverse can be applied to find an approximately unfiltered velocity field with which the subfilter stress tensor can be evaluated. The importance of the type of quadrature used to define the discrete filter kernel corresponding to the discretisation scheme was analyzed showing best results in combination with the Fourier spectral method. An extension to non-periodic problems is the subject of current research, as is an extension to 3D.

Acknowledgments. The research was supported by the Polish National Science Centre, Project No. 2018/29/B/ST8/00262, and by the National Agency for Academic Exchange (NAWA) within the International Academic Partnerships Programme No. PPI/APM/2019/1/00062.

References

1. Bull, J.R., Jameson, A.: Explicit filtering and exact reconstruction of the sub-filter stresses in large eddy simulation. J. Comput. Phys. **306**, 117–136 (2016)
2. Geurts, B.J.: Elements of Direct and Large-Eddy Simulation. R. T. Edwards Inc., Philadephia (2003)
3. Geurts, B.J., van der Boss, J.: Numerically induced high-pass dynamics in large-eddy symulation. Phys. Fluids **17**, 125103–125111 (2005)
4. Quarteroni, A., Sacco, R., Saleri, F.: Numerical Mathematics. Springer, Heidelberg (2007). https://doi.org/10.1007/b98885
5. Canuto, C., Hussaini, M.Y., Quarteroni, A., Zang, T.A.: Spectral Methods in Fluid Dynamics. Springer, Heidelberg (1988). https://doi.org/10.1007/978-3-642-84108-8
6. San, O., Staples, A.E.: High-order methods for decaying two-dimensional homogeneous isotropic turbulence. Comput. Fluids **63**, 105–127 (2012)
7. Wiener, N.: Extrapolation, Interpolation, and Smoothing of Stationary Time Series. The MIT Press, Cambridge (1949)

Seamless Interface Methods for Grey-Area Mitigation in Scale-Resolving Hybrid RANS-LES

M. Carlsson[1(✉)], S. Wallin[2], L. Davidson[1], S.-H. Peng[3], and S. Arvidson[4]

[1] Division of Fluid Dynamics, M2, Chalmers University of Technology, Gothenburg, Sweden
{magnus.carlsson,lada}@chalmerse.se

[2] KTH Royal Institute of Technology, Stockholm, Sweden
stefanw@mech.kth.se

[3] Swedish Defence Research Agency, Stockholm, Sweden
peng@foi.se

[4] Saab Aeronautics, Linköping, Sweden
sebastian.arvidson@saabgroup.com

1 Introduction

A new Grey-Area Mitigation (GAM) method for hybrid RANS-LES is presented. In regions of variable resolution, the transfer of turbulence energy between RANS-modeled and LES-resolved turbulence is quantified by a commutation residue term, as proposed for Partially-Averaged Navier Stokes (PANS) modeling [1]. Girimaji and Wallin [1] showed that the commutation residue term needs to be accounted for in regions with a variable PANS parameter, f_k, which is the ratio of modeled to total turbulence energy, in order to have a proper transition from modeled to resolved turbulence, and vice versa. A model was proposed for this term in PANS, of which an extension was also indicated for other hybrid RANS-LES formulation based on length scales [1].

Commutation residue stemming from length scale variation was quantified by Hamba [2] for fully developed turbulent channel flow, by a filtering operation with the filter width varying between RANS and LES length scales in different regions. The magnitude of the commutation term was found to be of the same order as the turbulence production term. This was further investigated by Arvidson et al. in their hybrid RANS-LES method based on the $K-\omega$ model [3]. It was shown that the commutation residue term can significantly accelerate the transition from modeled to resolved turbulence, and thus effectively mitigating the grey area.

In this work, an effort has been made to exploit the formulation proposed by Girimaji and Wallin and expand it further in DES-type modeling aiming for a more rapid transition between modeled and resolved turbulent scales in the vicinity of a RANS-LES interface.

2 Mathematical Formulation

The turbulence energy cascade in the spectral space is illustrated in Fig. 1 with $\kappa_c \propto 1/L_{ref}$ as the spectral cut-off wave number dividing the turbulence energy into the

C. Marchioli et al. (Eds.): DLES 2023, ERCO 31, pp. 299–305, 2024.
https://doi.org/10.1007/978-3-031-47028-8_46

resolved and modeled turbulence energy, denoted by K_r and K_u, respectively. It is noted further that L_{ref} is a reference length scale. For constant resolution, where κ_c is constant in space, the energy cascade σ is responsible for the energy exchange between the resolved and unresolved scales.

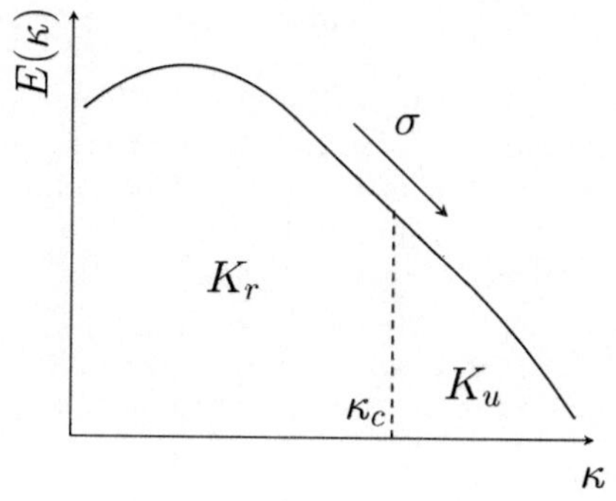

Fig. 1. Illustration of energy spectrum with cut-off wave number.

With variable L_{ref}, κ_c can vary in time or space and, correspondingly, the position of the interface between K_r and K_u will change. This leads to an additional mechanism of energy exchange between K_r and K_u. Following Girimaji and Wallin [1], this energy exchange can be described by integrating the unresolved turbulent kinetic energy in spectral space

$$\frac{\mathrm{d}K_u}{\mathrm{d}t} = \frac{\mathrm{d}}{\mathrm{d}t}\int_{\kappa_c}^{\infty} E(\kappa,t)\mathrm{d}\kappa = \int_{\kappa_c}^{\infty} \frac{\mathrm{d}E}{\mathrm{d}t}\mathrm{d}\kappa - E(\kappa_c,t)\frac{\mathrm{d}\kappa_c}{\mathrm{d}t} \tag{1}$$

The first term on the RHS represents the change of K_u considering non-varying κ_c. The second term on the RHS corresponds to energy exchange between K_r and K_u in the vicinity of κ_c, denoted by P_{Tr}. By assuming a standard spectrum in the Kolmogorov inertial subrange, $E(\kappa) = C\varepsilon^{2/3}\kappa^{-5/3}$, one can show that

$$P_{Tr} = -E(\kappa_c,t)\frac{\mathrm{d}\kappa_c}{\mathrm{d}t} = \ldots = -\frac{2}{3}\frac{K_u}{\kappa_c}\frac{\mathrm{d}\kappa_c}{\mathrm{d}t} \tag{2}$$

where $\frac{d}{dt}$ is the material derivative, i.e. advection by the resolved flow field. Using the relation $\kappa_c \propto 1/L_{ref}$, one can express the commutation residue in terms of the variation of the hybrid length scale L_{ref}:

$$P_{Tr} = \frac{2}{3}\frac{K_u}{L_{ref}}\frac{\mathrm{d}L_{ref}}{\mathrm{d}t} \tag{3}$$

The commutation residue term P_{Tr} in Eq. (3) acts as a source/sink term in the equation for K_u. If a $K-\omega$ based hybrid model is used, as in this work, it can be shown that a corresponding term has to be added to the ω-equation, namely $-(K_u/\omega_u)P_{Tr}$ [1].

In the case of decreasing L_{ref}, energy is transferred from the RANS-modeled to the LES-resolved scales and $P_{Tr} < 0$. Energy conservation dictates that the energy transfer rate, P_{Tr}, removed from the unresolved scales must be added to the resolved scales. The energy transfer to/from the resolved scales is modelled as a diffusion term in the momentum equation [1]

$$\nu_{Tr} = \frac{P_{Tr}}{S^2} \tag{4}$$

where $S = \sqrt{2S_{ij}S_{ij}}$. It should be emphasized that ν_{Tr} and ν_t represent different physics and should not be mixed up. In the momentum equations the total turbulent viscosity reads:

$$\nu^* = \nu_t + \nu_{Tr} \tag{5}$$

With increasingly refined resolution in space/time, energy is transferred from unresolved to resolved turbulence. This will bring up a negative ν_{Tr}. Decreasing resolution will, on the other hand, transfer energy from resolved to unresolved scales through positive ν_{Tr} and P_{Tr}. We limit $\nu_{Tr} \geq -\nu_t$ to ensure positive total turbulent diffusion for numerical stability. Note that Eqs. (3) and (4) are deployed in the whole computational domain.

3 Numerical Setup

The computations have been conducted using the M-Edge code, which is an edge- and node-based Navier-Stokes flow solver applicable for both structured and unstructured grids [4]. The compressible Navier-Stokes equations are discretized with a finite-volume approximation and integrated in time using a 2nd-order backward difference scheme. The viscous fluxes are estimated with a 2nd-order central scheme. The inviscid fluxes are based on the low-dispersive and low-dissipative (LD2) scheme, which combines a low-dissipative convection operator with a low-dispersive reconstruction of the face values [5].

All computations have been carried out using the $K-\omega$ SST DDES model [6,7] in combination with the proposed GAM formulation. The hybrid length scale in Eq. (3) is computed as

$$L_{ref} = L_{RANS} - f_d \max(0, L_{RANS} - L_{LES}) \tag{6}$$

where $L_{RANS} = \beta^* K_u \omega_u$ and $L_{LES} = C_{DES}\Delta$ are the RANS and LES length scales, respectively, and f_d is the DDES shielding function. The LES filter width is computed as $\Delta = \tilde{\Delta}_\omega$ [8], which adapts the subgrid-scale based on the local vorticity direction of the flow. Furthermore, another modification of the sub-grid model aiming at acceleration of the RANS-LES transition in separated shear layers for the DDES method is evaluated. The σ-DDES model [9,10] replaces the standard strain rate invariant S in the production term by a blended variant

$$S_{\sigma-DDES} = S - f_d \,\mathrm{pos}(L_{RANS} - L_{LES})(S - B_\sigma S_\sigma) \tag{7}$$

where $\mathrm{pos}(a) = 1$ if $a > 0$, otherwise 0. The subgrid model of SST σ-DDES then performs as the algebraic σ-model [11] instead of the Smagorinsky model in the LES domain, and the empirical constant $B_\sigma = 60$ is calibrated in decaying grid turbulence. In quasi two-dimensional flow regions, S_σ, which is the strain rate operator in the σ-model, is close to zero and ensures the decrease of the eddy viscosity in shear layers by rapidly reducing the production term in the LES mode of DDES.

4 Results and Discussion

The commutation term is evaluated and verified for the mixing shear layer. Simulations using the commutation term (i.e. Eqs. (3) and (4)) are indicated by CT. The baseline case is the SST DDES version with length scale $\tilde{\Delta}_\omega$, while the variations with GAM methods are made by incorporating $\sigma - DDES$ and/or the commutation term. Figure 2 presents the Q-criterion highlighting the resolved flow features.

4.1 Mixing Shear Layer

The free shear layer flow was investigated experimentally by Delville [13]. The computational domain includes a very thin flat plate, with turbulent boundary layers on each side, a mixing layer is formed in the region downstream of the flat plate trailing edge. The experimental boundary layer properties at the trailing edge are presented in Table 1. The focus region, i.e. the region from the flat plate trailing edge at $x = 0$ to $x = 1$ m, is meshed with $(n_x, n_y, n_z) = (640, 196, 96)$ cells. The grid is equidistant in the streamwise x-direction and spanwise z-direction, $\Delta x = \Delta z = 1.5625$ mm. The total number of hexahedral grid cells are 13.7 million. A timestep of $\Delta t = 2.5 \times 10^{-5}$ second is used. The upstream boundary layers are treated in RANS mode and the DDES automatically switches to LES after the trailing edge.

The growth of the mixing layer, the vorticity thickness, is presented in Fig. 3a. The baseline model clearly underpredicts the growth of the shear layer. With either of the two GAM methods the results are improved. By using a combination of both $\sigma - DDES$ and the commutation term the agreement with the experimental result is significantly improved for $x > 0.2$ m. However, it is observed that all tested models underpredict the initial growth of the mixing layer just behind the trailing edge, between $x \in [0, 0.2]$ m. The very initial developing shear layer is thin with small-scale turbulence. The grid in

Fig. 2. Q-criterion coloured by vorticity magnitude contours for mixing shear layer. Results are acquired using SST σ-DDES with commutation terms.

Table 1. Flow parameters of mixing-layer case. Data from experiment [13].

Measure	Notation	High vel. BL	Low vel. BL
Velocity	U_∞	41.54 m/s	22.40 m/s
Thickness	δ	9.6 mm	6.3 mm
Displ. thick	δ_1	1.4 mm	1.0 mm
Mom. thick	θ	1.0 mm	0.73 mm
Shape factor	H	1.35	1.37
Re based on θ	Re_θ	2900	1200
Turbulence level	u'/U_∞	~0.3%	~0.3%

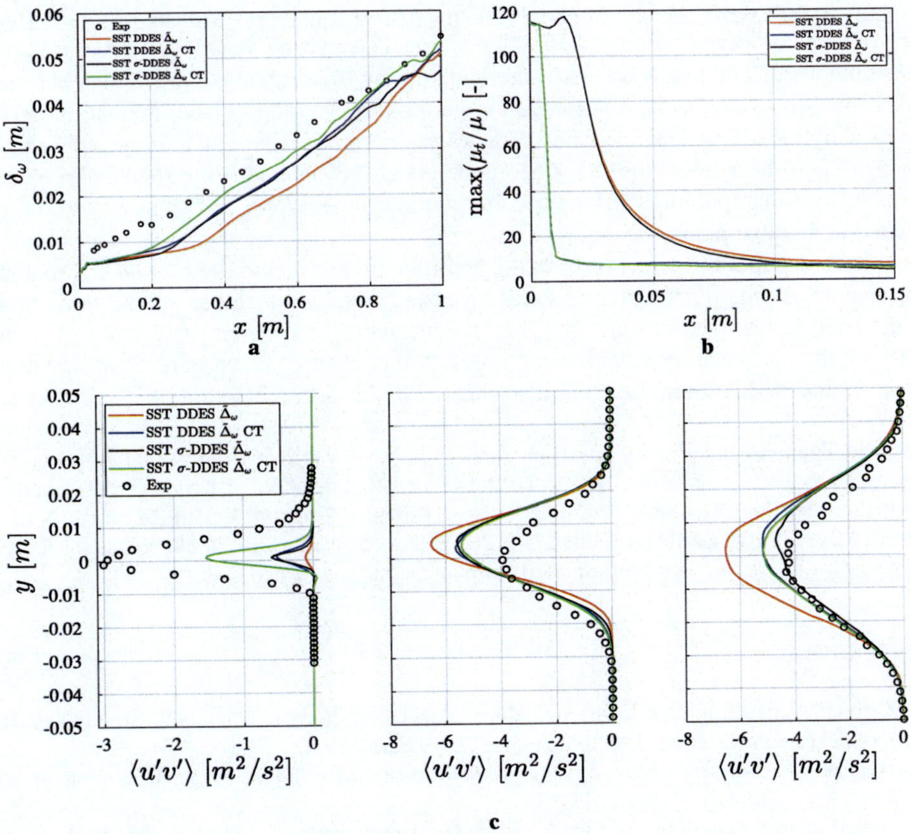

Fig. 3. (a) Vorticity thickness. (b) Maximum eddy viscosity. (c) Resolved shear stress at locations $x = 0.2$ m (left), $x = 0.65$ m (middle) and $x = 0.95$ m (right).

this region is most likely not fine enough to contain these scales, in particular in the spanwise direction. Moreover, some amount of synthetic turbulence might be needed initially for boosting the initial transition from the steady upstream RANS boundary layer. The maximum eddy viscosity at the initial part of the shear layer is shown in Fig. 3b. The effect of the commutation term is large, as indicated by a rapidly reduced eddy viscosity. Without the commutation term, such a reduction is relatively slow. A reduced eddy viscosity level contributes to a rapid growth of resolved turbulence, which is visualised in Fig. 3c. As shown, the resolved shear stress is better predicted at station $x = 0.2$ m, where the prediction incorporating both the commutation terms combined with σ-DDES gives the best result.

5 Conclusions

A new Grey-Area Mitigation (GAM) method is proposed in this paper. The work presents hybrid RANS-LES computations undertaken in the verification of GAM formulations incorporated into the SST DDES model. A mixing shear layer test case has been computed. The proposed GAM method, aiming to mitigate the grey area over the RANS-LES interface, is based on a commutation term stemming from the variation of the local hybrid length scale. Furthermore, a sub-grid model enabling to reduce the eddy viscosity in a shear layer, the σ-DDES model, is also evaluated. Both methods require no additional manipulation of, or explicitly defining, the RANS-LES interface and are applied in a global manner.

It is shown that the effect of both the commutation term and the $\sigma - DDES$ model improve the results for the mixing shear layer and are able to trigger a rapid transition of modeled to resolved turbulence. This is indicated by improved predictions of both vorticity thickness and resolved shear stress. Although not shown here, a verification made for the NASA hump flow has also shown similar GAM function.

Acknowledgement. This work has been funded by the Swedish Governmental Agency for Innovation Systems (VINNOVA), the Swedish Defence Materiel Administration (FMV) and the Swedish Armed Forces within the National Aviation Research Programme (NFFP, Contract Number 2017-04887) and Saab Aeronautics. The 2nd author (S. Wallin) would like to acknowledge all the early discussions with Prof. Sharath Girimaji on this and related topics.

References

1. Girimaji, S., Wallin, S.: Closure modeling in bridging regions of variable-resolution (VR) turbulence computations. J. Turbul. **14**(1), 72–98 (2014)
2. Hamba, F.: Analysis of filtered Navier-Stokes equation for hybrid RANS/LES simulation. Phys. Fluids **23**(1), 015108 (2011)
3. Arvidson, S., Davidson, L., Peng, S.: Interface methods for grey-area mitigation in turbulence-resolving hybrid RANS-LES. Int. J. Heat Fluid Flow **73**, 236–257 (2018)
4. Eliasson, P.: Edge, a Navier-Stokes solver for unstructured grids. Finite Vol. Complex Appl. **3**, 527–534 (2002)
5. Löwe, J., Probst, A., Knopp, T., Kessler, R.: Low-dissipation low-dispersion second-order scheme for unstructured finite-volume flow solvers. AIAA J. **54**, 2961–2971 (2016)

6. Spalart, P.R., Deck, S., Shur, M.L., et al.: A new version of detached-eddy simulation, resistant to ambiguous grid densities. Theoret. Comput. Fluid Dyn. **20**, 181 (2006)
7. Gritskevich, M., Garbaruk, A., Schütze, J., Menter, F.: Development of DDES and IDDES formulations for the k-ω shear stress transport model. Flow Turbul. Combust. **88**, 431–449 (2012)
8. Shur, M.L., Spalart, P.R., Strelets, M.K., Travin, A.K.: An enhanced version of DES with rapid transition from RANS to LES in separated flows. Flow Turbul. Combust. **95**(4), 709–737 (2015). https://doi.org/10.1007/s10494-015-9618-0
9. Mockett, C., et al.: Two non-zonal approaches to accelerate RANS to LES transition of free shear layers in DES Notes. Numer. Fluid Mech. Multidiscip. Des. **130**, 187–201 (2015)
10. Guseva, E.K., et al.: Assessment of two approaches to accelerate RANS to LES transition in shear layers in the framework of ANSYS-FLUENT. J. Phys.: Conf. Ser. **1038**, 012134, (2018)
11. Nicoud, F., Toda, H.B., Cabrit, O., Bose, S., Lee, J.: Using singular values to build a subgrid-scale model for large eddy simulations. Phys. Fluids **23**, 1–12 (2011)
12. Greenblatt, D., Paschal, K., Yao, C.S., Harris, J. :A separation control CFD validation test case part 2. Zero-efflux oscillatory blowing. In: AIAA 2005-485. 43rd AIAA Aerospace Sciences Meeting and Exhibit (2005)
13. Delville, J.: La décomposition orthogonale aux valeurs propres et l'analyse de l'organisation tridimensionnelle des écoulements turbulents ci saillés libre. Ph.D. thesis, Université de Poitier (1995)

On a Conservative Solution to Checkerboarding: Examining the Discrete Laplacian Kernel Using Mesh Connectivity

J. A. Hopman(✉), F. X. Trias, and J. Rigola

Heat and Mass Transfer Technological Center, Technical University of Catalonia, C/Colom 11, 08222 Terrassa, Spain
{jannes.hopman,francesc.xavier.trias,joaquim.rigola}@upc.edu

1 Introduction

CFD codes that are used for industrial applications commonly use a collocated grid arrangement to calculate the physical flow variables. When using a central differencing gradient (CDG) to discretise the continuous operators of the Navier-Stokes equations, a wide stencil is obtained for the Laplacian operator. This wide stencil, in turn, leads to a decoupling between odd and even grid points of the pressure field that results from the pressure Poisson equation. This decoupling can lead to non-physical, spurious modes in the solution, a problem commonly known as the checkerboard problem (Ferziger et al., 2002).

Generally, this problem is avoided by using a compact stencil Laplacian. A method to do so was first developed by Rhie and Chow (Rhie and Chow, 1983). This method solves the problem of decoupled grid points and eliminates the possibility of spurious modes in the pressure field. However, this method introduces nonphysical, numerical dissipation of kinetic energy (Rhie and Chow, 1983, Felten and Lund, 2006). This dissipation disrupts the delicate interaction between convective transport and physical dissipation, especially at the smallest scales of motion. By doing so, it becomes impossible to capture the essence of turbulence, which is of high importance in accurate DNS simulations (Verstappen and Veldman, 2003). Moreover, for LES simulations this dissipation was shown to be of the same order of magnitude as the dissipation introduced by the LES model, decreasing the effectiveness of the model (Komen et al., 2021).

Many general purpose codes favour the extra stability that this method offers at the price of a lower accuracy. Unconditional stability, however, can also be achieved by mimicking the underlying symmetry properties of the continuous operators of the Navier-Stokes equations, when discretising them. A method that does this was developed for staggered Cartesian grid arrangements by Verstappen and Veldman (Verstappen and Veldman, 2003) and later extended to collocated unstructured grids by Trias *et al.* (Trias et al., 2014). Since the kinetic energy is conserved and stability is unconditional, using the method of Rhie and Chow comes at a higher price and an alternative method should be sought after. One method mentioned here is the one described by Larsson and Iaccarino (Larsson and Iaccarino, 2010), in which the kernel of the discrete Laplacian operator matrix is determined and used to eliminate the spurious

C. Marchioli et al. (Eds.): DLES 2023, ERCO 31, pp. 306–311, 2024.
https://doi.org/10.1007/978-3-031-47028-8_47

modes. However, on non-Cartesian grids, this method involves performing a singular value decomposition (SVD), for which the computational cost grows exponentially with the number of grid points, as $\mathscr{O}(N_{grid}^3)$, making this method non-viable for industrial applications (Golub and Van Loan, 1996).

In this work, properties of the discrete wide stencil Laplacian, its kernel and the relation to the connectivity of the mesh will be examined. By understanding this relation better, a prediction can be made for a set of vectors that span the nullspace, and by projecting the pressure solution field onto this nullspace, the spurious modes can be eliminated.

2 Relation Between the Mesh and the Kernel

The greatest advantage of the collocated grid formulation is the possibility to extend the solution domain to complex geometries using unstructured meshes. Therefore, the continuous operators will be discretised in a manner that is suitable for unstructured meshes. For these meshes, however, a central node will not necessarily lie between two neighbouring nodes and therefore the CDG is not defined. Here, a discretisation is used that simplifies to the CDG on uniform Cartesian meshes. Moreover, it mimics the property of CDG that gradients at a central node i only depend on values of neighbouring nodes n, i.e.:

$$[\nabla\phi]_i = \sum_n c_n \phi_n \tag{1}$$

where c_n is some coefficient.

Let the discrete wide stencil Laplacian operator be denoted by $L_c = M\Gamma_{cs}^M \Gamma_{sc}^M G$, which follows the discretisation of Trias *et al.* (Trias et al., 2014) and is a chain of operations: (1) face gradient G, (2) midpoint face-to-cell interpolation Γ_{sc}^M, (3) midpoint cell-to-face interpolation Γ_{cs}^M and (4) divergence M. The cell-centered gradient at node i is given by:

$$\begin{aligned} \left[\Gamma_{sc}^M G\phi_c\right]_i &= \frac{1}{2\left[\Omega_c\right]_i} \sum_{f\in F_f(i)} \left[\Omega_s\right]_f \frac{\phi_n - \phi_i}{\delta_{nf}} \mathbf{n}_{if} \\ &= \frac{1}{2\left[\Omega_c\right]_i} \sum_{f\in F_f(i)} A_f \phi_n \mathbf{n}_{if} \end{aligned} \tag{2}$$

where $\mathbf{n}_{if}$ is the outward-pointing face-normal vector of face f w.r.t. node i, δ_{nf} is the face-normal distance between neighbouring nodes, Ω_s denotes the face volumes, where $[\Omega_s]_f = \delta_{nf} A_f$ with face area A_f, Ω_c denotes the cell volumes and finally $\sum_{f\in F_f(i)}$ denotes a sum over the faces f that constitute cell i. The second line of Eq. 2 follows from the fact that ϕ_i can be taken out of the summation and the sum of outward-pointing surface normal vectors, $\mathbf{S}_{if} = A_f \mathbf{n}_{if}$, equals zero:

$$\sum_{f\in F_f(i)} \mathbf{S}_{if} = \mathbf{0} \tag{3}$$

Therefore, the value in the central node i does not contribute to the gradient in the central node itself and Eq. 1 holds. Similarly, the cell-centered divergence at node i is given by:

$$\begin{aligned}[M\Gamma_{cs}\psi_c]_i &= \frac{1}{2} \sum_{f \in F_f(i)} (\psi_i + \psi_n) \cdot \mathbf{n}_{if} A_f \\ &= \frac{1}{2} \sum_{f \in F_f(i)} \psi_n \cdot \mathbf{n}_{if} A_f \end{aligned} \tag{4}$$

Again, the value in the central node i does not contribute to the divergence in the central node itself. L_c is a sequence of both operators connecting nodes i to k if they share a neighbour j, its entries are given by:

$$[L_c]_{i,k} = \sum_j \frac{1}{4\,[\Omega_c]_j} \mathbf{S}_{ij} \cdot \mathbf{S}_{jk} \tag{5}$$

where $\mathbf{S}_{ij}$ is the surface-normal vector, pointing from node i to j. This implies that the Laplacian in node i does not depend on the values in its neighbours j, only on its second neighbours k.

On certain meshes, this can create checkerboard-like patterns. Non-trivial pressure modes can exist on these meshes of which the Laplacian equals zero, i.e. the pressure modes given by a vector in the nullspace of L_c. Several examples of two-dimensional meshes are shown in Fig. 1, in which cells are coloured the same way if they are connected with non-zero entries in L_c, given by Eq. 5. Meshes in which the number of faces that meet each other is always even, Fig. 1(left), show a parity. In this example the nullity of L_c is increased to 2. In the case of Cartesian meshes, Fig. 1(middle), the dot product in 5 equals zero for diagonal second neighbour pairs, giving rise to a nullity of L_c of $2^{N_{dim}}$, which equals 4 in the two-dimensional example. Finally, meshes in which an uneven number of non-orthogonal faces meet each other, Fig. 1(right), the disconnection vanishes and the nullity of L_c reduces, in this example to 1.

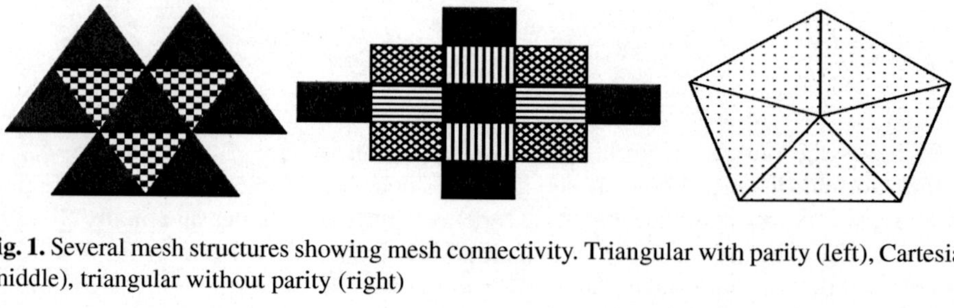

Fig. 1. Several mesh structures showing mesh connectivity. Triangular with parity (left), Cartesian (middle), triangular without parity (right)

For the current definition of L_c, the nullity can simply be seen from the connected groups in the mesh. In practice, a nullity higher than 1 will mostly occur in Cartesian meshes, therefore other mesh types are disregarded for now. What remains is to define a set of vectors that span the nullspace of L_c for Cartesian meshes.

To do so, the pressure field $\mathbf{p}_c$ is decomposed into its physical modes, $\mathbf{p}_c^{\oplus}$, and the non-physical modes, $\mathbf{p}_c^{\ominus}$, which lie in the nullspace of L_c: $\mathbf{p}_c = \mathbf{p}_c^{\oplus} + \mathbf{p}_c^{\ominus}$, such that $L_c\mathbf{p}_c = L_c\mathbf{p}_c^{\oplus}$, because $L_c\mathbf{p}_c^{\ominus} = \mathbf{0}_c$. Any $\mathbf{p}_c^{\ominus}$ will then be a linear combination of nullspace spanning vectors. For Cartesian meshes, such a set was given by (Larsson and Iaccarino, 2010), in the two-dimensional example this set is visualised in Fig. 2 and given by:

$$[\mathbf{p}_c^{\ominus(1)}]_{i,j} = 1, \quad [\mathbf{p}_c^{\ominus(2)}]_{i,j} = (-1)^i, \quad [\mathbf{p}_c^{\ominus(3)}]_{i,j} = (-1)^j, \quad [\mathbf{p}_c^{\ominus(4)}]_{i,j} = (-1)^{i+j} \tag{6}$$

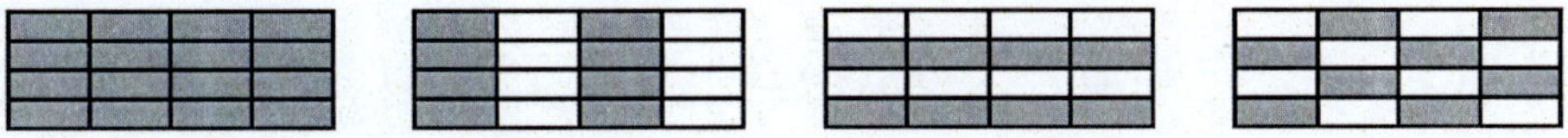

Fig. 2. Visualisation of a set of vectors spanning the nullspace of L_c

The kernel of L_c only depends on the mesh and the choice of interpolator. Therefore, all that remains is to generalise the set of Eq. 6 to other types of interpolators. Definitions of the most commonly used interpolators, linear and volumetric, are given by:

$$\left[\Gamma_{cs}^{L}\phi\right]_f = \frac{\delta_{nf,n}}{\delta_{nf}}\phi_i + \frac{\delta_{nf,i}}{\delta_{nf}}\phi_n, \quad \left[\Gamma_{cs}^{V}\phi\right]_f = \frac{\delta_{nf,i}}{\delta_{nf}}\phi_i + \frac{\delta_{nf,n}}{\delta_{nf}}\phi_n \tag{7}$$

respectively. Where $\delta_{nf,i}$ and $\delta_{nf,n}$ are the perpendicular distances between face f and centroids i and n respectively, so that:

$$\delta_{nf} = \delta_{nf,i} + \delta_{nf,n} \tag{8}$$

Now it is important to note that, in the symmetry preserving framework, the cell-to-face and face-to-cell interpolators are related to each other by (Trias et al., 2014):

$$\Gamma_{sc} = \Omega^{-1}\Gamma_{cs}^{T}\Omega_s \tag{9}$$

This leaves only one degree of freedom in L_c, namely the choice of Γ_{cs}.

To find a set of nullspace spanning vectors, the cell-centered gradient, $\Gamma_{sc}G$, is first rewritten to a Gauss gradient form, which is more commonly used in code implementations:

$$[G_G\phi]_i = \frac{1}{[\Omega_c]_i}\sum_{f\in F_f(i)} \mathbf{S}_f\phi_f \tag{10}$$

Next, as an example, the volumetric cell-centered gradient is rewritten to a Gauss Gradient form:

$$\begin{aligned}
\left[\Gamma_{sc}^{V} G\phi\right]_i &= \left[\Omega^{-1}\Gamma_{cs}^{V^T}\Omega_s G\phi\right]_i && \textit{from Eq. 9}\\
&= \frac{1}{[\Omega_c]_i}\sum_{f\in F_f(i)} \mathbf{S}_f \delta_{nf}\left(\frac{1}{\delta_{nf}}\frac{\delta_{nf,i}}{\delta_{nf}}\phi_n - \frac{1}{\delta_{nf}}\frac{\delta_{nf,i}}{\delta_{nf}}\phi_i\right) && \textit{from Eq. 7}\\
&= \frac{1}{[\Omega_c]_i}\sum_{f\in F_f(i)} \mathbf{S}_f \left(\frac{\delta_{nf,i}}{\delta_{nf}}\phi_n - \left(1-\frac{\delta_{nf,f}}{\delta_{nf}}\right)\phi_i\right) && \textit{from Eq. 8}\\
&= \frac{1}{[\Omega_c]_i}\sum_{f\in F_f(i)} \mathbf{S}_f \left(\frac{\delta_{nf,i}}{\delta_{nf}}\phi_n + \frac{\delta_{nf,f}}{\delta_{nf}}\phi_i\right) && \textit{from Eq. 3}\\
&= \left[G_G \Pi_{cs}^{L}\phi\right]_i && \textit{from Eq. 10}
\end{aligned} \tag{11}$$

where in the final step it is simply noticed that the weights are the same as for a scalar linear interpolator, given by Π_{cs}^{L}. Any cell-centered gradient can be rewritten to a Gauss gradient and a scalar interpolator, by rewriting the interpolation weight of ϕ_i as: $w_{f,i} = 1 - w_{f,n}$. The interpolators considered in this work are rewritten by:

$$\Gamma_{sc}^{M} G = G_G \Pi_{cs}^{M}, \qquad \Gamma_{sc}^{L} G = G_G \Pi_{cs}^{V}, \qquad \Gamma_{sc}^{V} G = G_G \Pi_{cs}^{L} \tag{12}$$

These relations can be used to find the nullspace spanning vectors any Laplacian operator, since we can simply find vectors for which $G_G \Pi_{cs} = \mathbf{0}_c$. Moreover, many codes use linear interpolation for the predictor velocity on the source side of the Poisson equation, combined with a linear interpolation of pressure before taking the Gauss gradient: $L_c = M\Gamma_{cs}^{L} G_G \Pi_{cs}^{L} = M\Gamma_{cs}^{L}\Gamma_{sc}^{V} G$, which is non-symmetric. This non-symmetry in the operator can disrupt the conservation of physical properties or it may even be the cause of spurious pressure modes to arise in the first place, since the image of L_c and the kernel are not orthogonal if L_c is non-symmetric. To predict a set of nullspace spanning vectors a simple adjustment can be made to the set in Eq. 6:

$$\begin{aligned}
&[\mathbf{p}_c^{\ominus(1)}]_{i,j} = 1, && [\mathbf{p}_c^{\ominus(2)}]_{i,j} = (-1)^i (\Delta x_i)^\alpha\\
&[\mathbf{p}_c^{\ominus(3)}]_{i,j} = (-1)^j (\Delta y_j)^\alpha, && [\mathbf{p}_c^{\ominus(4)}]_{i,j} = (-1)^{i+j} (\Delta x_i \Delta y_j)^\alpha
\end{aligned} \tag{13}$$

where $\alpha = [-1, 0, 1]$ for volumetric, midpoint and linear interpolations respectively. It can be easily verified that these vectors lie in the nullspace of their respective L_c. Moreover, although the vectors are not necessarily orthogonal, they are linearly independent and therefore span the nullspace of L_c. This is required since $\mathbf{p}_c$ is projected onto the nullspace and the result is subsequently subtracted to obtain $\mathbf{p}_c^{\oplus}$.

3 Conslusions

It was shown in this work that there is a strong relation between the nullity of L_c and the used mesh. Moreover, it was shown that the nullity will in practice most often be 1 for unstructured meshes, in which case the nullspace consists of the constant vector. For Cartesian meshes the nullity is known and a set of nullspace spanning vectors was derived for the most commonly used interpolators in the Laplace operator. Thereby in a strict mathematical sense, the spurious modes can effectively be filtered in almost any case. This leads to the question if the wide stenciled Laplacian can also give rise to decoupled pressure modes that do not strictly lie in its nullspace, and if they can be filtered easily as well. These questions will be addressed in future work.

Acknowledgements. This work is supported by the FusionCAT project (001-P-001722) co-financed by the European Union Regional Development Fund within the framework of the ERDF Operational Program of Catalonia 2014–2020 with a grant of 50% of total cost eligible. The work is part of the RETOtwin project (PDC2021-120970-I00) of *Ministerio de Economía y Competitividad*, Spain. J.A.H. is supported by the predoctoral grant FI 2022 (2022 FI_B1 00204) of the *Catalan Agency for Management of University and Research Grants (AGAUR).*

References

Felten, F.N., Lund, T.S.: Kinetic energy conservation issues associated with the collocated mesh scheme for incompressible flow. J. Comput. Phys. **215**(2), 465–484 (2006)

Ferziger, J.H., Perić, M., Street, R.L.: Computational Methods for Fluid Dynamics, vol. 3, pp. 196–200. Springer, Berlin (2002)

Golub, G.H., Van Loan, C.F.: Matrix Computations. Mathematical Sciences. Johns Hopkins University Press, Baltimore (1996)

Komen, E.M., Hopman, J.A., Frederix, E.M.A., Trias, F.X., Verstappen, R.W.: A symmetry-preserving second-order time-accurate PISO-based method. Comput. Fluids **225**, 104979 (2021)

Larsson, J., Iaccarino, G.: A co-located incompressible Navier-Stokes solver with exact mass, momentum and kinetic energy conservation in the inviscid limit. J. Comput. Phys. **229**(12), 4425–4430 (2010)

Rhie, C.M., Chow, W.L.: Numerical study of the turbulent flow past an airfoil with trailing edge separation. AIAA J. **21**(11), 1525–1532 (1983)

Trias, F.X., Lehmkuhl, O., Oliva, A., Pérez-Segarra, C.D., Verstappen, R.W.C.P.: Symmetry-preserving discretization of Navier-Stokes equations on collocated unstructured grids. J. Comput. Phys. **258**, 246–267 (2014)

Verstappen, R.W.C.P., Veldman, A.E.P.: Symmetry-preserving discretization of turbulent flow. J. Comput. Phys. **187**(1), 343–368 (2003)

Finite-Difference Viscous Filtering for Non-regular Meshes

R. Perrin[1(✉)] and E. Lamballais[2]

[1] Department of Mechanical Engineering, Faculty of Engineering at Sriracha, Kasetsart University Sriracha Campus, Sriracha, Chonburi, Thailand
rodolphe@eng.src.ku.ac.th

[2] Curiosity Group, Pprime Institute, CNRS - Univ-Poitiers - ISAE/ENSMA, Poitiers, France
eric.lamballais@univ-poitiers.fr

1 Introduction

In a recent study [1], the authors have proposed a new solution filtering technique for direct and large-eddy simulation (DNS/LES). This method can represent both the molecular and artificial viscosity. It is particularly easy to code in a conventional finite-difference framework where only the scheme coefficients have to be designed to ensure the expected dissipation. The main strength of this *viscous filtering* technique lies in its numerical stability features. In this sense, it can be considered as a simple and computationally efficient alternative to implicit time integration of the viscous term.

The concept of viscous filtering and its advantages in the context of DNS/LES are presented in [1]. However, as a limitation of the technique, only the case of a regular mesh has been addressed. It was a way to design the coefficients for ensuring favourable spectral features defined in the Fourier space. In particular, the accuracy conditions are derived through Taylor's expansions expressed in wavenumbers.

The purpose of this contribution is to extend this viscous filtering technique to non-regular meshes. This generalization requires to express the expected features in the physical space in order to obtain the various relations to reach a given order of accuracy. The principles of these developments are first presented. Then, a preliminary validation is shown for the Burgers equation. Finally, to assess the method in a DNS/LES context, the case of a transitional boundary layer is considered with the use of a highly stretched mesh in the near-wall region.

2 Methodology

To present the general ideas of the approach, let us consider the simple one-dimensional equation

$$\frac{\partial u}{\partial t} = F(u) + \nu \frac{\partial^2 u}{\partial x^2} \tag{1}$$

C. Marchioli et al. (Eds.): DLES 2023, ERCO 31, pp. 312–317, 2024.
https://doi.org/10.1007/978-3-031-47028-8_48

where $u(x,t)$ is the solution and ν the constant molecular viscosity. The basic principle of the method is to split the time advancement into two steps with

$$u^* = u^n + \int_{t_n}^{t_n+\Delta t} F(u)\,dt \qquad (2)$$

$$u^{n+1} = \widetilde{u^*} \qquad (3)$$

with $u^n = u(x,t_n)$ and $u^{n+1} = u(x,t_n+\Delta t)$. Step (2) consists in the time advancement of the inviscid version of Eq. (1) where the integral can be evaluated for instance by an Adams-Bashforth scheme. Step (3) corresponds to the application of a spatial filter denoted $\tilde{\cdot}$ to restore the influence of viscosity. For simplicity, the splitting error is ignored here, but it has been shown in [1] that this error is negligible for typical DNS/LES while being removable if necessary.

The spatial discretization is based on a stretched mesh $x_i = h(s_i)$ where $s_i = (i-1)\Delta s$ corresponds to regularly distributed nodes on the computational coordinate s. This coordinate transformation enables to write successive derivatives as

$$\frac{\partial u}{\partial x} = g\frac{\partial u}{\partial s} \quad , \quad \frac{\partial^2 u}{\partial x^2} = g^2\frac{\partial^2 u}{\partial s^2} + gg'\frac{\partial u}{\partial s}$$
$$\frac{\partial^3 u}{\partial x^3} = g^3\frac{\partial^3 u}{\partial s^3} + 3g^2g'\frac{\partial^2 u}{\partial s^2} + \left(g^2g^{(2)} + gg'^2\right)\frac{\partial u}{\partial s} \quad , \quad \frac{\partial^n u}{\partial x^n} = g^n\frac{\partial^n u}{\partial s^n} + \ldots \qquad (4)$$

where $g = 1/h'$. To mimic the contribution of this viscous term in step (3), a finite-difference filter scheme of the form

$$\alpha_p\tilde{u}_{i+1} + \tilde{u}_i + \alpha_m\tilde{u}_{i-1} = au_i + \frac{b_p}{2}u_{i+1} + \frac{b_m}{2}u_{i-1} + \frac{c_p}{2}u_{i+2} + \frac{c_m}{2}u_{i-2} \qquad (5)$$

is used where $u_i = u(x_i,t)$. As it is usual in a finite-difference framework, Taylor's expansion in space

$$u_{i+k} = u_i + \sum_{p=1}^{n}\frac{(k\Delta s)^p}{p!}\left.\frac{\partial^p u}{\partial s^p}\right|_i + O(\Delta s^{n+1})$$
$$\left.\frac{\partial^q u}{\partial s^q}\right|_{i+k} = \left.\frac{\partial^q u}{\partial s^q}\right|_i + \sum_{p=1}^{n}\frac{(k\Delta s)^p}{p!}\left.\frac{\partial^{p+q} u}{\partial s^{p+q}}\right|_i + O(\Delta s^{n+1}) \qquad (6)$$

provides relation orders in Δs^n between the coefficients $(\alpha_p,\alpha_m,a,b_p,b_m,c_p,c_m)$ in (5). Then, by Taylor's expansion in time

$$\tilde{u}_i = u_i + \Delta t\left.\frac{\partial u}{\partial t}\right|_i + \frac{\Delta t^2}{2}\left.\frac{\partial^2 u}{\partial t^2}\right|_i + \frac{\Delta t^3}{3!}\left.\frac{\partial^3 u}{\partial t^3}\right|_i + O(\Delta t^4) \qquad (7)$$

while switching the time derivative by their spatial counterparts, requiring that step (3) has to match a purely diffusive equation with

$$\frac{\partial u}{\partial t} = \nu\frac{\partial^2 u}{\partial x^2} \quad , \quad \frac{\partial^2 u}{\partial t^2} = \nu^2\frac{\partial^4 u}{\partial x^4} \quad , \quad \frac{\partial^3 u}{\partial t^3} = \nu^3\frac{\partial^6 u}{\partial x^6}, \qquad (8)$$

expansion (7) can also be written as

$$\tilde{u}_i = u_i + F\Delta s^2 \left.\frac{\partial^2 u}{\partial x^2}\right|_i + \frac{F^2\Delta s^4}{2} \left.\frac{\partial^4 u}{\partial x^4}\right|_i + \frac{F^3\Delta s^6}{3!} \left.\frac{\partial^6 u}{\partial x^6}\right|_i + O(F^4\Delta x^8) \tag{9}$$

where $F = \frac{\nu\Delta t}{\Delta s^2}$ is the Fourier number based on the constant cell Δs. To enable an identification between the relation orders given by (6) and (9) in the scheme (5), a conversion of the spatial derivatives from x to s is required. This can be done easily through relations (4) so that a 7×7 system $MX = N$ can be obtained where the component of $X = (\alpha_p, \alpha_m, a, b_p, b_m, c_p, c_m)^T$ are the unknown scheme coefficients of (5). This system can be solved analytically (using a symbolic calculation tool) leading to extremely long expressions for each coefficient which is a function of F, g, g', g'', $g^{(3)}$, $g^{(4)}$ and $g^{(5)}$. More conveniently, the same can be done numerically up to the machine accuracy before starting the simulation.

Specific versions of this new scheme have also been developed for performing implicit LES. In that case, an artificial dissipation can be imposed by checking

$$a - \frac{b_p}{2} - \frac{b_m}{2} + \frac{c_p}{2} + \frac{c_m}{2} - d = (1 - \alpha_p - \alpha_m)\exp\left[-\pi^2\left(\frac{\nu_0}{\nu} + 1\right)g^2 F\right] \tag{10}$$

where $d = d_p = d_m$ is a stencil extension of the initial scheme (5) while ν_0 is a numerical viscosity chosen to ensure regularization as a substitute of subgrid-scale modelling. The resulting 8×8 system can be solved as previously, leading to scheme coefficients with an extra dependency on ν_0.

Finally, the same methodology can be straightforwardly followed for the development of one-sided schemes at the boundaries.

3 Validation for Burger's Equation

To validate this new type of filter scheme, Burgers' equation has been solved, namely $F(u) = u\partial u/\partial x$ in (1). As explained in the previous section, every time step, the inviscid Burger equation is solved and then, the viscous filter scheme (5) is applied on the discrete solution to restore the influence of the viscosity. A time interval after the shock formation is computed and the solution is compared to its exact counterpart. The mesh, composed of N nodes, is regular when using the initial scheme given in [1] whereas the mesh is stretched in the shock region when using the new scheme developed here. The sixth-order version is assessed, namely the highest-order enabled by scheme (5) because of its compact stencil.

As expected, at low resolution using a regular mesh, the solution is subjected to Gibbs' phenomenon, leading to unphysical grid-to-grid oscillations in the near-shock region as exhibited in Fig. 1-left. This problem can be fixed through the increase of the resolution everywhere in the domain, which is computationally inefficient. Alternatively, the use of a stretched grid enables us to remove the unphysical oscillations without any increase of the computational cost (see Fig. 1-left for $N = 64$).

The numerical convergence analysis is based on the computation of L_∞ to estimate the numerical error defined by reference to the exact solution. Figure 1-right shows

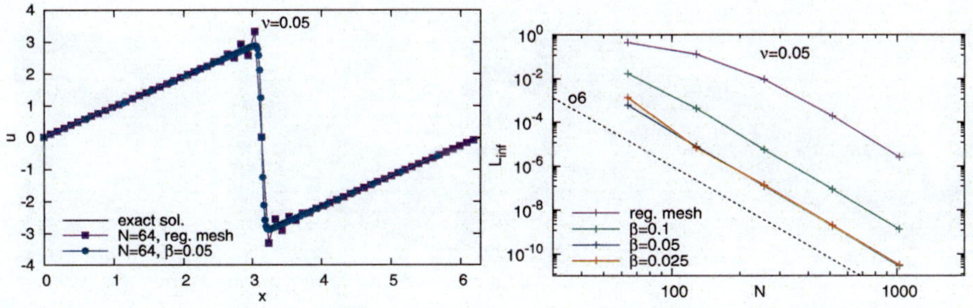

Fig. 1. Computation of a shocked Burgers' solution. Top: solution at $t = 0.05$. Bottom: evolution of L_∞ with the number N of mesh nodes.

that the expected sixth-order is correctly recovered for both the regular and non-regular meshes, but only at high resolution for the former as the signature of the Gibbs phenomenon at low resolution. The spectacular increase of the accuracy provided by the use of a stretched mesh can be observed, with for the two largest stretching parameters $\beta = (0.05, 0.025)$, a reduction by more than 2 orders of magnitude of the maximum error located near the shock. For a constant error, the computational saving corresponds to a factor of about 8 by comparison to a regular mesh.

4 Assessment by Implicit LES

For further validation, implicit LES of a spatially-evolving boundary layer subjected to bypass transition has been carried out as in [2]. To handle the present flow configuration, a Cartesian mesh of $(n_x \times n_y \times n_z) = (961, 411, 156)$ nodes is used to cover a computational domain $(L_x \times L_y \times L_z) = (1500\delta_1, 145\delta_1, 125\delta_1)$ where δ_1 is the displacement thickness of the laminar boundary layer at the inlet of the computational domain and $Re = U_0\delta_1/\nu = 300$. The inlet boundary conditions are provided by precursor simulations in which the flow in the leading edge region of the flat plate is computed together with realistic free-stream turbulence outside the boundary layer in order to trigger physically relevant bypass transition (see [2] for more details). Based on the maximum friction velocity u_τ, the resulting cell sizes in wall-units are $(\Delta x^+, \Delta y^+, \Delta z^+) \approx (23, 1-27, 12)$ through the use of a highly stretched mesh in the wall-normal direction y. The near-wall cell-size $\Delta y^+ \approx 1$ is favourable for accuracy but penalizing for numerical stability when using an explicit time advancement, in particular for the high value of numerical viscosity used here with $\nu_0/\nu = 25$. For the present second-order Adams-Bashforth (AB2) scheme, numerical stability is ensured only if the maximum Fourier number $F_{\max} = \nu\Delta t/\Delta_{\min}^2$ is less than $3.9\,10^{-3}$, imposing the use of a time step Δt reduced by a factor 18 by comparison to its maximum value allowed by the CFL condition. To alleviate this problem, the traditional approach is to make implicit the time integration of the viscous term using for instance a Crank-Nicolson (CN) scheme, leading to an hybrid AB2-CN time advancement. Here, the present viscous filtering (VF) technique is proposed as a simpler alternative designated as AB2-VF.

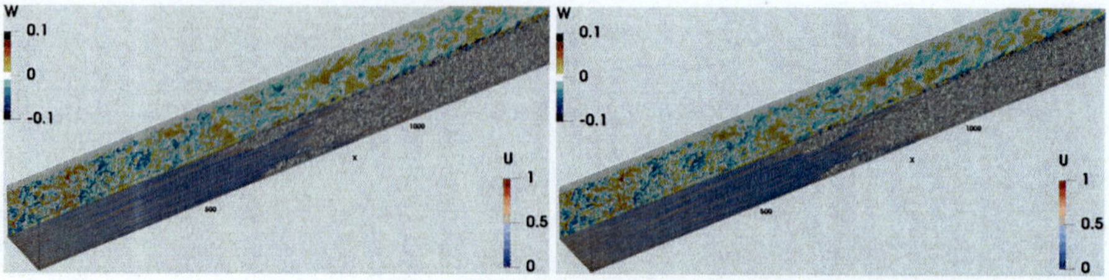

Fig. 2. Implicit LES of bypass transition of a boundary layer [2]. Maps of longitudinal and spanwise velocities at $y = \delta_1$ and $z = \delta_1$ respectively. Q-criterion isosurface in grey.

It has been observed that both AB2-CN and AB2-VF strategies enable the use of large time steps only restricted by the CFL condition while leading to virtually identical instantaneous solutions as illustrated in Fig. 2 where significant differences are only observed in flow regions where the dynamics becomes chaotic. This point can be confirmed by plotting time signals of velocity as shown in Fig. 3 for deterministic (top-left) and intermitently chaotic (top-right) boundary-layer regions. In the free-stream turbulence zone, the agreement is almost perfect up to the most downstream region because of the low level of velocity fluctuations with respect to the convection velocity U_0. Similar agreements are observed for other quantities (not shown for brevity) leading to the conclusion that AB2-CN and AB2-VF strategies are equivalent in terms of numerical accuracy and stability.

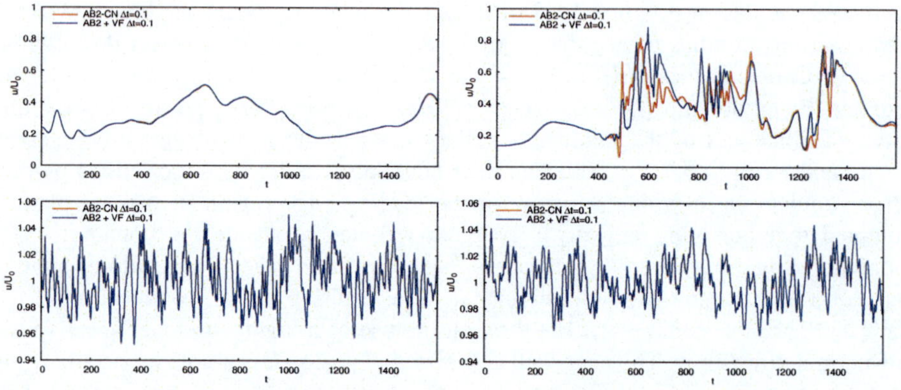

Fig. 3. Time signals of the longitudinal velocity at $(x,y) = (400\delta_1, 0.8\delta_1, 50\delta_1)$ left-top, $(x,y) = (700\delta_1, 0.8\delta_1)$ right-top, $(x,y) = (400\delta_1)$ left-bottom, $(x,y) = (700\delta_1)$ right-bottom with $z = 50\delta_1$.

5 Conclusion

The main conclusion of these one-dimensional and full-scale tests is that viscous filtering can be extended to non-regular meshes while ensuring accuracy and quasi-unconditional stability, as a valuable alternative to an implicit treatment of the viscous term in time. By comparison to the latter, viscous filtering can easily handle multidimensional problems thanks to its essentially explicit nature based on operator splitting. In particular, it does not require the use of sophisticated and iterative methods to deal with large-size and sparse matrices while enabling the use of large time steps not restricted by viscous numerical stability conditions. Among the perspectives of the present approach, one may refer to an adaptation for flows with variable viscosity in space. The extension of the present concept of viscous filtering to other frameworks of numerical methods such as finite volume/element would also deserve further developements.

Acknowledgements. This work was granted access to the HPC resources of IDRIS under the allocation A0072A07624 made by GENCI.

References

1. Lamballais, E., Vicente Cruz, R., Perrin, R.: Viscous and hyperviscous filtering for direct and large-eddy simulation. J. Comp. Phys. **431**, 110115 (2021)
2. Perrin, R., Lamballais, E.: Assessment of implicit LES modelling for bypass transition of a boundary layer. Comput. Fluids **251**, 105728 (2022)

An Energy-Preserving Unconditionally Stable Fractional Step Method for DNS/LES on Collocated Unstructured Grids

D. Santos(✉), F. X. Trias, G. Colomer, and A. Oliva

Heat and Mass Transfer Technological Center (CTTC)-Universitat Politècnica de Catalunya BARCELONA TECH (UPC), Terrassa, Spain
{daniel.santos.serrano,francesc.xavier.trias,guillem.colomer, asensio.oliva}@upc.edu

1 Introduction

A finite-volume discretization over unstructured meshes is the most used formulation to solve Navier-Stokes equations by many general purpose CFD packages codes such as OpenFOAM or ANSYS-Fluent. These codes work with collocated stencil formulations, that is, once the equations are discretized, an algorithm goes cell by cell computing the required quantities.

On the other hand, algebraic formulations maintain the discrete equations in matrix-vector form, and compute the required quantities by using matrices and vectors. A collocated fully-conservative algebraic symmetry-preserving formulation of incompressible Navier-Stokes equations was proposed by Trias et al. [1]. Assuming n control volumes and m faces:

$$\Omega \frac{d\mathbf{u}_c}{dt} + \mathbf{C}(\mathbf{u}_s)\mathbf{u}_c = \mathbf{D}\mathbf{u}_c - \Omega \mathbf{G}_c p_c, \tag{1}$$

$$\mathbf{M}\mathbf{u}_s = \mathbf{0}_c, \tag{2}$$

where $\mathbf{u}_c \in \mathbb{R}^{3n}$ and $\mathbf{p}_c \in \mathbb{R}^n$ are the cell-centered velocity and the cell-centered pressure, respectively. The staggered quantities, such as $\mathbf{u}_s \in \mathbb{R}^m$ are related to the cell-centered quantities via an interpolation operator $\Gamma_{c\to s} \in \mathbb{R}^{m\times 3n}$:

$$\mathbf{u}_s = \Gamma_{c\to s}\mathbf{u}_c. \tag{3}$$

Finally, $\Omega \in \mathbb{R}^{3n\times 3n}$ is a diagonal matrix containing the cell volumes, $\mathbf{C}(\mathbf{u}_s) \in \mathbb{R}^{3n\times 3n}$ is the discrete convective operator, $\mathbf{D} \in \mathbb{R}^{3n\times 3n}$ is the discrete diffusive operator, $\mathbf{G}_c \in \mathbb{R}^{3n\times n}$ is the cell-to-cell discrete gradient operator and $\mathbf{M} \in \mathbb{R}^{n\times m}$ is the face-to-cell discrete divergence operator. The velocity correction after applying the Fractional Step Method (FSM) to the Navier-Stokes equations reads:

$$\mathbf{u}_c^{n+1} = \mathbf{u}_c^p - \Gamma_{s\to c}\mathbf{G}p^{n+1}, \tag{4}$$

where $\Gamma_{s\to c} \in \mathbb{R}^{3n\times m}$ is the face-to-cell interpolator, which is related to the cell-to-face interpolator via the volume matrices $\Gamma_{s\to c} = \Omega^{-1}\Gamma_{c\to s}\Omega_s$, and $\mathbf{G} \in \mathbb{R}^{m\times n}$ is the cell-to-face gradient operator [1].

C. Marchioli et al. (Eds.): DLES 2023, ERCO 31, pp. 318–323, 2024.
https://doi.org/10.1007/978-3-031-47028-8_49

All the operators needed to formulate the equations can be constructed using only five discrete ones: the cell-centered and staggered control volumes (diagonal matrices), Ω_c and Ω_s, the face normal vectors, N_s, the scalar cell-to-face interpolation, $\Pi_{c\to s}$ and the cell-to-face divergence operator, $\mathbf{M}$. For more details of these operators and its construction see [1]. Due to its simplicity, these operators can be easily implemented in existing codes, such as OpenFOAM [2], which is used for the simulations of this work. However, as it was shown in [3], this code introduces a large amount of numerical dissipation. In our opinion, this is not an appropiate approach for DNS and LES simulations since this artificial dissipation interferes with the subtle balance between convective transport and physical dissipation. Hence, reliable numerical methods for DNS/LES must be free of numerical dissipation (or, at least, have a small amount), and unconditionally stable, i.e. simulations must be stable regardless of the mesh quality and resolution.

2 An Energy-Preserving Unconditionally Stable FSM

2.1 Conservation of Energy

Left-multiplying Eq. (1) by $\mathbf{u}_c^T$ and summing the result with its transpose we obtain the global discrete kinetic energy equation:

$$\frac{d}{dt}||\mathbf{u}_c||^2 = -\mathbf{u}_c^T(\mathbf{C}(\mathbf{u}_s)+\mathbf{C}(\mathbf{u}_s)^T)\mathbf{u}_c - \mathbf{u}_c^T(\mathbf{D}+\mathbf{D}^T)\mathbf{u}_c$$
$$- \mathbf{u}_c^T\Omega\mathbf{G}_c p_c - p_c^T\mathbf{G}_c^T\Omega\mathbf{u}_c. \tag{5}$$

Respecting the symmetries of these differential operators is crutial in order to respect the physical structure of the equations. In absence of diffusion, that is $\mathbf{D}=\mathbf{0}$, energy must be preserved. This will happen, if and only if, the convective operator is skew-symmetric and $\mathbf{G}=-\Omega_s^{-1}\mathbf{M}^T$ [1]. These two conditions are also mimicking the symmetries of the continuous level operators [4]. So, we do not only have physical arguments to do so, but also mathematical ones. In fact, mathematical arguments (continuous symmetry-preserving), directly imply discrete symmetry-preserving and automatically conservation of energy, but the converse is not true.

The turbulence phenomenon arises from a balance between convective transport and diffusive dissipation. These two physical processes are described (in its discrete form) by $\mathbf{C}(\mathbf{u}_s)$ and $\mathbf{D}$, respectively. At continuous level, the convective operator is skew-symmetric, and the diffusive operator is symmetric and negative-definite. If we retain these properties at the discrete level (namely $\mathbf{C}(\mathbf{u}_s)$ being a skew-symmetric matrix, $\mathbf{D}$ being a symmetric negative-definite matrix and $\mathbf{G}=-\Omega_s^{-1}\mathbf{M}^T$), the discrete convective operator is going to transport energy from resolved scales of motion to others without dissipating energy, as one should expect.

2.2 Stability of the Pressure Gradient Interpolation

Due to the fact that the pressure gradient is computed at the faces and we need to interpolate it back to the cells in order to correct our (collocated) velocity (see Eq. 4), this interpolation can lead us to some instability issues (see Eq. 5). This problem was studied in [5], thus showing the utility of an algebraic formulation. In that work, the matrix-vector formulation is used in order to study the stability of the solution in terms of the pressure gradient interpolation. To do so, the eigenvalues of $\mathbf{L} - \mathbf{L}_c$ were deeply studied ($\mathbf{L} = \mathbf{MG} \in \mathbb{R}^{n \times n}$ is the compact Laplacian operator whereas $\mathbf{L}_c = \mathbf{M}\Gamma_{c \to s}\Gamma_{s \to c}\mathbf{G} \in \mathbb{R}^{n \times n}$ is the so-called collocated wide-stencil Laplacian operator), and the cell-to-face interpolation that leads to an unconditionally stable FSM turned out to be the volume weighted:

$$\Pi_{c \to s} = \Delta_s^{-1} \Delta_{sc}^T \in \mathbb{R}^{m \times n}, \tag{6}$$

where $\Delta_s \in \mathbb{R}^{m \times m}$ is a diagonal matrix containing the projected distances between two adjacent control volumes, and $\Delta_{sc} \in \mathbb{R}^{n \times m}$ is a matrix containing the projected distance between a cell node and its corresponding face [5].

3 Test Case: Turbulent Channel Flow at $Re_\tau = 395$

In this section, the robustness and the accuracy of the method is tested in a channel flow at $Re_\tau = 395$, using different meshes. The domain chosen to carry out the simulations is the same as in Moser et al. [6]: $2\pi x 2 x \pi$. All the meshes had a y^+ around 1. Other interpolations used in our work are the midpoint interpolation ($\frac{1}{2}$ interpolation coefficients) and the linear interpolation. An example of the robustness of the method for highly-distorted unstructured meshes can be found in [7].

3.1 Accuracy of the Interpolator in High Quality Meshes

The mesh shown in Fig. 1 was chosen to test the accuracy. It is stretched towards y-direction. Figures 2 and 3 show the normalised mean velocity (in wall units) and the $< u'v' >$ component of the Reynolds stress tensor. We compare the results obtained against the DNS data obtained by Moser et al. [6]. The results are very similar for the three interpolators. We expected this because for regular Cartesian meshes all three interpolators collapse to the same one (midpoint), and if the stretching is very soft, the mesh is almost regular and Cartesian (locally).

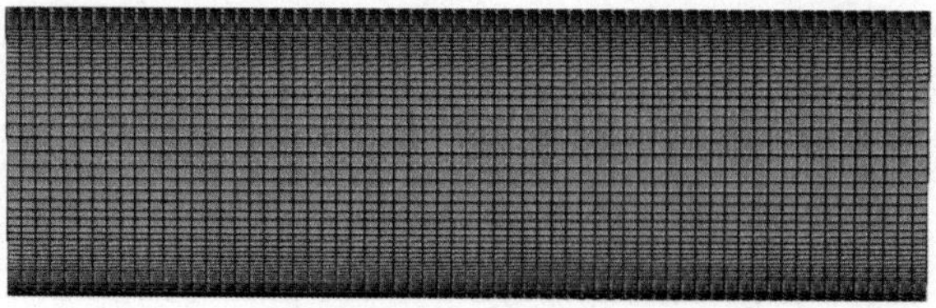

Fig. 1. 64 × 64 × 64 mesh used to test our method with different interpolators.

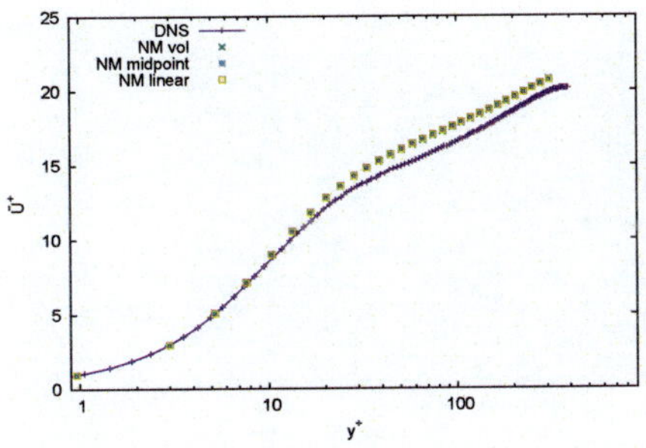

Fig. 2. Normalised mean velocity profile in wall units for channel flow at $Re_\tau = 395$.

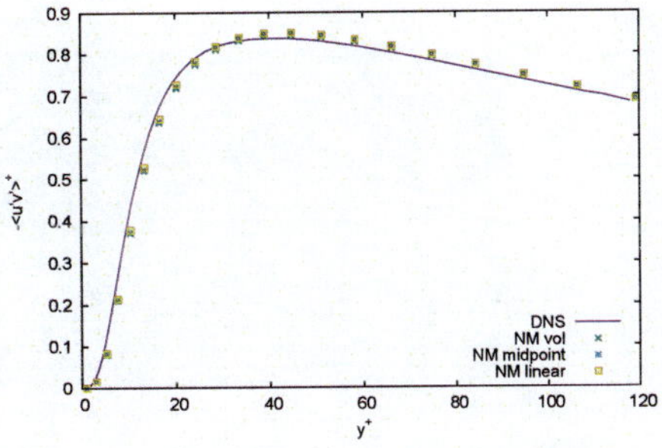

Fig. 3. Normalised turbulent shear stress profile $- < u'v' >$ in wall units at $Re_\tau =$ 395.

3.2 Robustness in Distorted Meshes

Figure 4 shows the mesh chosen to test the robustness of the method. It is stretched towards the beginning and the end, while having very long control volumes in the center. As expected with this kind of meshes, the results are not even qualitatively good (turbulence is not even triggered, the control volumes in the center filter any kind of eddy), but the only one that converges to a solution is the volume weighted, while the others were unstable.

Figure 5 shows a less distorted mesh, that could be used in daily simulations. Even for this (not so bad) mesh, the linear interpolation is not able to trigger turbulence. For the midpoint and the volume weighted interpolations, the results are very similar, as shown in Figs. 6 and 7 (results obtained with linear interpolation are so inaccurate that are not shown here):

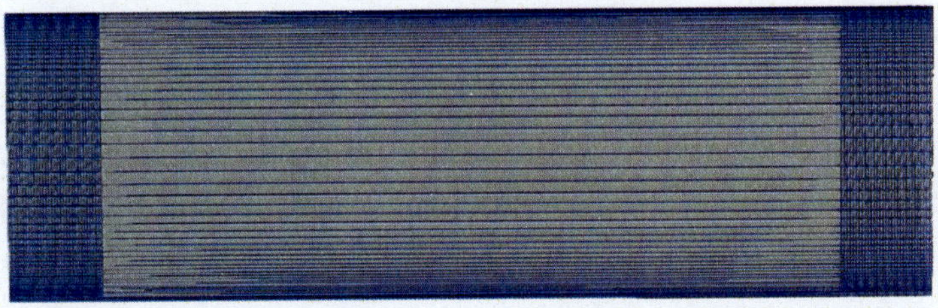

Fig. 4. Highly distorted mesh used to test the robustness of the method. Maximum aspect ratio is 250.

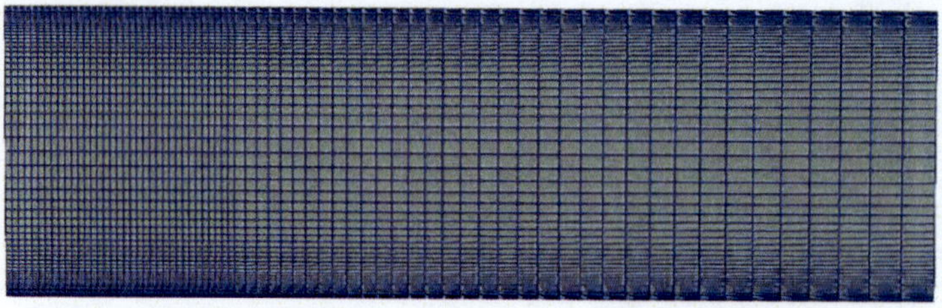

Fig. 5. Mesh which starts growing the control volumes at 1/3 of total length. Maximum aspect ratio is 5.

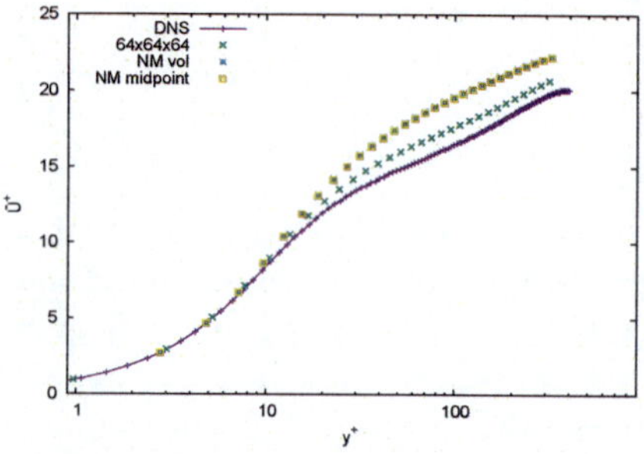

Fig. 6. Normalised mean velocity profile in wall units for channel flow at $Re_\tau = 395$.

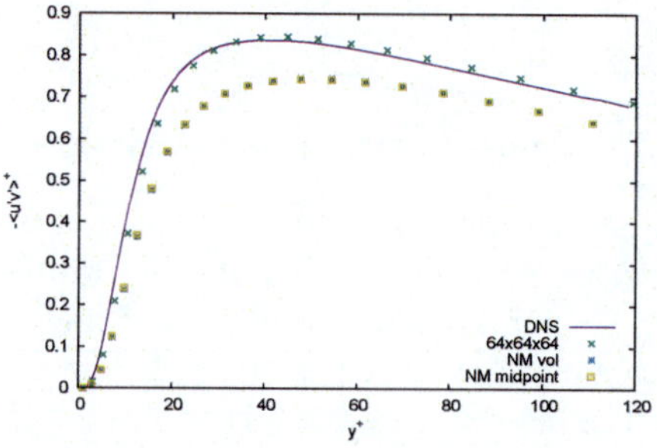

Fig. 7. Normalised turbulent shear stress profile $- < u'v' >$ in wall units at $Re_\tau = 395$.

As expected, results are worse than those obtained with a regular Cartesian mesh, but they are still reasonable. What can be surprising is that the midpoint and the volume weighted interpolators seem to give the same results, while the linear interpolation, although it is converging, gives very bad results. A possible explanation of this fact, taking into account also the results of Sect. 3.2, is that while we start stretching the mesh, the results are similar, until some point where the linear interpolation starts failing, but the other two still give similar results. Then, under more stretching, the midpoint fails while the volume weighted is unconditionally stable.

4 Conclusions and Future Work

An energy-preserving unconditionally stable fractional step method on collocated grids has been presented. Three interpolators have been tested for the pressure gradient: volume weighted, linear and midpoint. All three seem to have the same accuracy in high quality meshes. When distorting the mesh, the first one that loses accuracy and eventually blows up is the linear. The midpoint seems to be more stable than the linear, but it is still blowing up in highly distorted meshes, and the volume weighted is unconditionally stable.

Future and ongoing work related to our method would be to test the accuracy of the solution when varying progressively the distortion of the mesh, and to test it on unstructured meshes. Furthermore, it would be also interesting to test the preservation of energy using different interpolators.

Acknowledgements. This work is supported by the *Ministerio de Economía y Competitividad*, Spain, RETOtwin project (PDC2021-120970-I00), and the FusionCAT project (001-P-001722) co-financed by the European Union Regional Development Fund within the framework of the ERDF Operational Program of Catalonia 2014-2020 with a grant of 50% of total cost eligible. D. Santos acknowledges a *FI AGAUR-Generalitat de Catalunya* fellowship (2020FI_B_00839).

References

1. Trias, F.X., Lehmkuhl, O., Oliva, A., Pérez-Segarra, C.D., Verstappen, R.W.C.P.: Symmetry-preserving discretization of Navier-Stokes equations on collocated unstructured meshes. J. Comput. Phys. **258**, 246–267 (2014)
2. Komen, E., Hopman, J.A., Frederix, E.M.A., Trias, F.X., Verstappen, R.W.C.P.: A symmetry-preserving second-order time-accurate PISO-based method. Computers Fluids **225**, 104979 (2021)
3. Komen, E.M.J., Camilo, L.H., Shams, A., Geurts, B.J., Koren, B.: A quantification method for numerical dissipation in quasi-DNS and under-resolved DNS, and effects of numerical dissipation in quasi-DNS and under-resolved DNS of turbulent channel flows. J. Comput. Phys. **345**, 565–595 (2017)
4. Kreeft, J.: Mimetic spectral element method; a discretization of geometry and physics. Doctoral thesis, Fac. of Aerospace Eng., TU Delft, ISBN 9789461919281 (2013)
5. Santos, D., Muela J., Valle, N., Trias, F.X.: On the Interpolation Problem for the Poisson Equation on Collocated Meshes. 14th WCCM-ECCOMAS Congress 2020, July 2020 (postponed to January 2021 due to COVID19 pandemic)
6. Moser, R., Kim, J., Mansour, N.: Direct numerical simulation of turbulent channel flow up to $Re_\tau = 590$. Phys. Fluids **11–4**, 943–945 (1998)
7. Santos, D., Trias, F.X., Colomer, G., Pérez-Sergarra, C.D.: An energy-preserving unconditionally stable fractional step method on collocated grids. In: 8th European Congress on Computational Methods in Applied Sciences and Engineering, ECCOMAS 2022, Oslo, June 2022

DNS and LES of Buoyancy-Driven Turbulence at High Rayleigh Numbers: Numerical Methods and Subgrid-Scale Models

F. X. Trias[1(✉)], X. Álvarez-Farré[1], D. Santos[1], A. Gorobets[2], and A. Oliva[1]

[1] Heat and Mass Transfer Technological Center, Technical University of Catalonia C/Colom 11, 08222 Terrassa, Barcelona, Spain
{francesc.xavier.trias,xavier.alvarez.farre,daniel.santos.serrano, asensio.oliva}@upc.edu
[2] Keldysh Institute of Applied Mathematics, 4A, Miusskaya Sq., Moscow 125047, Russia

1 Introduction

Buoyancy-driven flows have always been an important subject of scientific studies with numerous applications in environment and technology. The most famous example thereof is the thermally driven flow developed in a fluid layer heated from below and cooled from above, *i.e.* the Rayleigh-Bénard convection (RBC). It constitutes a canonical flow configuration that resembles many natural and industrial processes, such as solar thermal power plants, indoor space heating and cooling, flows in nuclear reactors, electronic devices, and convection in the atmosphere, oceans and the deep mantle.

In the last decades significant efforts, both numerically and experimentally, have been directed at investigating the mechanisms and the detailed scaling behavior of the Nusselt number as a function of Rayleigh and Prandtl numbers in the general form $Nu \propto Ra^{\gamma} Pr^{\beta}$ [1]. In this regard, Fig. 1 shows the predictions of the Nu-number based on the classical Grossmann-Lohse (GL) theory [2] and its subsequent corrections [3, 4] where different scaling regimes, characterized by their corresponding exponents γ and β, are identified. Assuming this power-law scalings and following the same reasoning as in Ref. [5] leads to the estimations for the number of grid points shown in Fig. 2 (top). This corresponds to mesh resolution requirements for DNS and it clearly explains why nowadays DNS of RBC is still limited to relatively low Ra-numbers [1]. However, many of the above-mentioned applications are governed by much higher Ra numbers, located in the region of the $\{Ra, Pr\}$ phase space where the thermal boundary layer becomes turbulent (*i.e.* below the black dash-dotted line in Fig. 2). This region corresponds to the so-called asymptotic Kraichnan or ultimate regime of turbulence [6], with $\gamma = 1/2$. On the other hand, reaching such Ra-numbers experimentally while keeping the basic assumptions (Boussinesq approximation, adiabaticity of the closing walls, isothermal horizontal walls, perfectly smooth

C. Marchioli et al. (Eds.): DLES 2023, ERCO 31, pp. 324–330, 2024.
https://doi.org/10.1007/978-3-031-47028-8_50

surfaces...) is a very hard task; therefore, the observation of the Kraichnan regime also remains elusive [3,4].

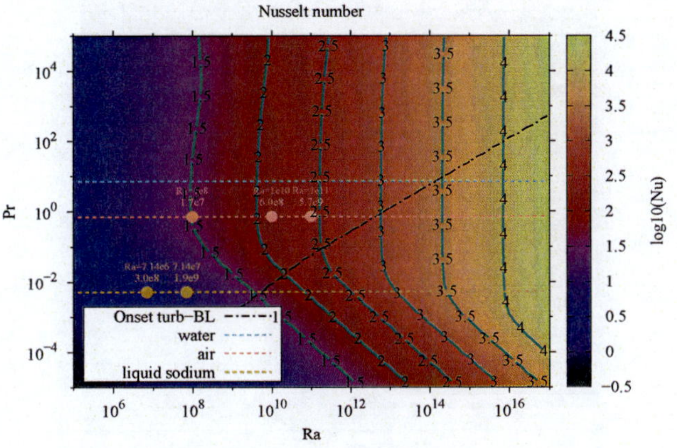

Fig. 1. Estimation of the Nusselt number of a RBC in the $\{Ra, Pr\}$ phase space given by the classical GL theory [2] and its subsequent corrections [3]. Green solid isolines represent the $log10$ of the Nusselt. Three dashed horizontal lines correspond to three different working fluids: water ($Pr = 7$), air ($Pr = 0.7$) and liquid sodium ($Pr = 0.005$). Dots displayed in the top figure correspond to the DNS simulations carried out in previous studies [5,7,8]. Black dash-dotted line is an estimation for the onset of turbulence in the thermal boundary layer.

2 LES of Buoyancy-Driven Turbulence

In this context, we may turn to LES to predict the large-scale behavior of incompressible turbulent flows driven by buoyancy at very high Ra-numbers. In LES, the large-scale motions are explicitly computed, whereas the effects of small-scale motions are modeled. Since the advent of CFD, many subgrid-scale (SGS) models have been proposed and successfully applied to a wide range of flows. However, there still exist inherent difficulties in the proper modelization of the SGS heat flux. This was analyzed in detail in the PRACE project entitled "Exploring new frontiers in Rayleigh-Bénard convection" (33.1 millions of CPU hours on MareNostrum 4 in 2018–2019), where DNS simulations of air-filled ($Pr = 0.7$) RBC up to $Ra = 10^{11}$ were carried out using meshes up to 5600M grid points (see dots displayed in Figs. 1 and 2, top). These results shed light into the flow topology and the small-scale dynamics, which are crucial in constructing the turbulent wind and energy budgets [7]. Moreover, it also provided new insights into the preferential alignments of the SGS and its dependence with the Ra-numbers [8], highlighting that the modelization of the SGS heat flux is the main difficulty that (still) precludes reliable LES of buoyancy-driven flows at (very)

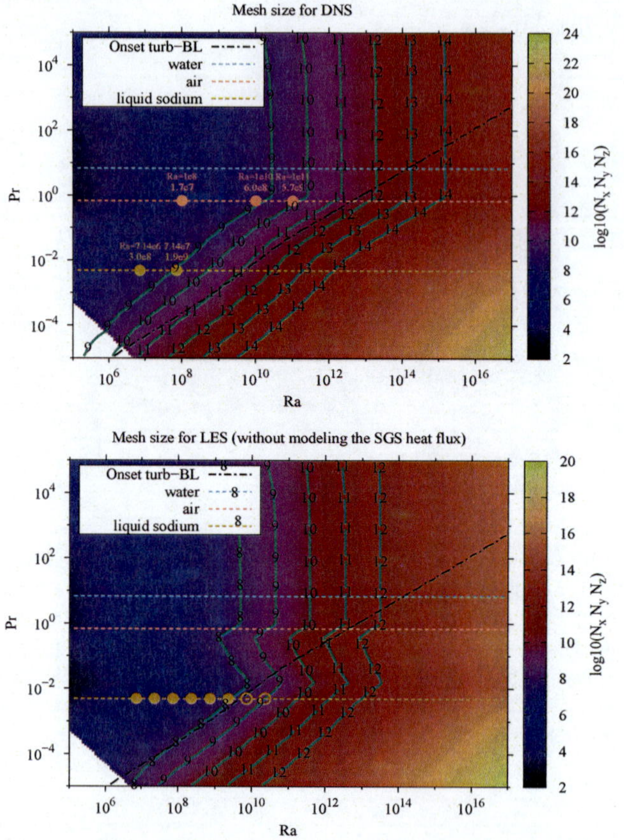

Fig. 2. Estimation of the mesh sizes for DNS (top) and LES (bottom) simulations of RBC in the $\{Ra, Pr\}$ phase space. LES estimations assume that thermal scales are fully resolved, *i.e.* no SGS heat flux model is needed. Green solid isolines represent the $log10$ of the total number of grid points. Three dashed horizontal lines correspond to three different working fluids: water ($Pr = 7$), air ($Pr = 0.7$) and liquid sodium ($Pr = 0.005$). Dots displayed in the top figure correspond to the DNS simulations carried out in previous studies [5,7,8] whereas the dots shown in the bottom figure are the set of LES simulations (being) carried out in the present work. Black dash-dotted line is an estimation for the onset of turbulence in the thermal boundary layer.

high Ra-numbers. This inherent difficulty can be by-passed by carrying out simulations at low-Prandtl numbers. In this case, the ratio between the Kolmogorov length scale and the Obukhov-Corrsin length scale (the smallest scale for the temperature field) is given by $Pr^{3/4}$; therefore, for instance, at $Pr = 0.005$ (liquid sodium) we have a separation of more than one decade. Hence, it is possible to combine an LES simulation for the velocity field (momentum equation) with the numerical resolution of all the thermal scales. Results obtained in Ref. [8] suggest that accurate predictions of the overall Nu can be obtained with meshes significantly coarser than those needed for a DNS (*e.g.* in practice for $Pr = 0.005$ we can expect mesh reductions in the range 10^2–10^3 for the total number of grid points leading to computational cost reductions in the range 10^3–10^4). This can be clearly observed in Fig. 2 (bottom), where estimations of the mesh size for LES are given with the assumption that thermal scales are fully resolved. This opens the possibility to reach the ultimate regime carrying out LES at low-Pr using meshes of "only" 10^{10}–10^{11} grid points. Nevertheless, to do so, we also need to combine proper numerical techniques for LES (also DNS) with an efficient use of modern high-performance computing (HPC) systems.

3 Numerical Methods and Algorithms for Large-Scale Simulations on Modern Supercomputers

The essence of turbulence are the smallest scales of motion. They result from a subtle balance between convective transport and diffusive dissipation. Mathematically, these terms are governed by two differential operators differing in symmetry: the convective operator is skew-symmetric, whereas the diffusive is symmetric and negative semi-definite. At discrete level, operator symmetries must be retained to preserve the analogous (invariant) properties of the continuous equations: namely, the convective operator is represented by a skew-symmetric matrix, the diffusive operator by a symmetric, negative semi-definite matrix and the divergence is minus the transpose of the gradient operator. In our opinion, this is a basic requirement for reliable DNS and LES simulations. Furthermore, these (large-scale) simulations should run efficiently on the variety of modern HPC systems (CPUs, GPUs, ARM,...) while keeping the code easy to port, optimize and maintain.

In this regard, a fully-conservative discretization for collocated unstructured grids was proposed [9]. It exactly preserves the symmetries of the underlying differential operators and is based on only five discrete operators (*i.e.* matrices): the cell-centered and staggered control volumes (diagonal matrices), Ω_c and Ω_s, the face normal vectors, N_S, the cell-to-face interpolation, $\Pi_{c\rightarrow s}$ and the cell-to-face divergence operator, M. Therefore, it constitutes a robust approach that can be easily implemented in existing codes such as OpenFOAM® [10]. Then, for the sake of cross-platform portability and optimization, CFD algorithms should rely on a very reduced set of (algebraic) kernels [11] (*e.g.* sparse-matrix vector product, SpMV; dot product; linear combination of vectors). In this approach, the basic kernels of the code shrink to dozens of lines; therefore, the portability becomes natural, and maintaining multiple implementations for different HPC architectures takes minor effort.

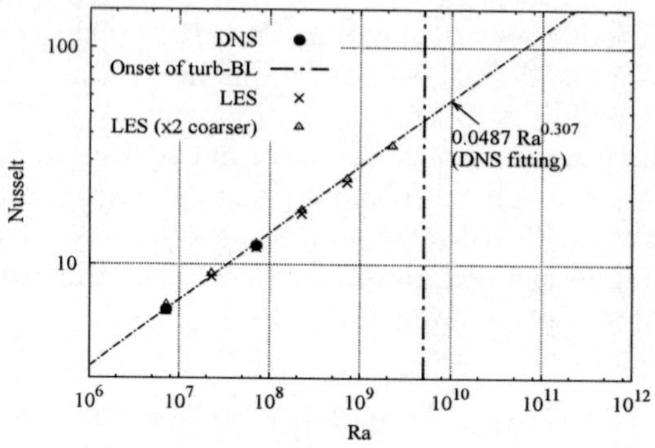

Fig. 3. Nu-vs-Ra results obtained with LES simulations at $Pr = 0.005$ using the same RBC configuration as in Ref.[8] where the two DNS results (solid black dots) were computed. The vertical dash-dotted line corresponds to the estimated Ra (for this particular Pr) where the thermal boundary layer becomes turbulent.

4 First Results and Conclusions

The above explained numerical techniques are being used to carry out a set of LES simulations of RBC at $Pr = 0.005$ for a wide range of Ra numbers (see dots in Fig. 2, bottom). The configuration is the same as in Ref.[8] where two DNS simulations (solid black dots in Fig. 3) were computed using meshes with $488 \times 488 \times 1280 \approx 305M$ ($Ra = 7.14 \times 10^6$) and $996 \times 996 \times 2048 \approx 1911M$ ($Ra = 7.14 \times 10^7$) grid points, respectively. A detailed analysis on the flow topology and how this may be affected by the discretization can be found in Ref.[5], whereas details of the parallelization strategy of the in-house STG code can be found in Ref.[12]. For the LES simulations, two levels of mesh refinement are being used: namely, a fine level that approximately corresponds to estimations shown in Fig. 2 (bottom) and a coarse level which is approximately twice coarser in each spatial direction. For instance, LES meshes at $Ra = 7.14 \times 10^7$ have respectively $44 \times 44 \times 96 \approx 0.19M$ and $90 \times 90 \times 160 \approx 1.3M$ grid points, *i.e.* ≈ 10000 and ≈ 1500 coarser compared with the DNS mesh. Meshes are designed to properly resolve the boundary layer whereas the much coarser bulk region is fine enough to guarantee that thermal scales are fully resolved, *i.e.* no SGS heat flux model is needed. Then, the SGS stress tensor is modelled using the S3PQ model [13] which was already tested for this RBC configuration in Ref. [8].

Results of the overall Nusselt number are displayed in Fig. 3. LES simulations up to $Ra = 7.14 \times 10^{10}$ (for the coarse level) and $Ra = 2.26 \times 10^{10}$ (for the fine level) are being computed using 3872 and 6272 CPU-cores on the MareNostrum 4 supercomputer. These points are located beyond the transition point for this Pr-number (see Fig. 2, bottom). Nevertheless, these simulations are not statistically converged yet and, therefore, results are not shown here. At first sight, we can observe an accurate agreement with previous DNS results. Furthermore, there is a rather good agreement with the Nu-vs-Ra scaling predicted using the DNS data. In any case, these preliminary results show the capability to obtain accurate predictions of the Nu-number using LES simulations. Accordingly to the GL theory, on-going LES simulations at higher Ra-number may show a change in the Nu-vs-Ra scaling indicating that we are finally hitting the ultimate regime of thermal turbulence.

Acknowledgements. F.X.T., X.A-F., D.S. and A.O. are supported by the *Ministerio de Economía y Competitividad*, Spain, RETOtwin project (PDC2021-120970-I00). F.X.T. and A.O. are supported by the FusionCAT project (001-P-001722) co-financed by the European Union Regional Development Fund within the framework of the ERDF Operational Program of Catalonia 2014–2020 with a grant of 50% of total cost eligible. D.S. is supported by the FI predoctoral grant (2020FI B 00839) of the Catalan Agency for Management of University and Research Grants (AGAUR). Calculations were carried out on MareNostrum 4 supercomputer at BSC. The authors thankfully acknowledge these institutions.

References

1. Stevens, R.J.A.M., Lohse, D., Verzicco, R.: Toward DNS of the ultimate regime of Rayleigh–bénard convection. In: García-Villalba, M., Kuerten, H., Salvetti, M.V. (eds.) DLES 2019. ES, vol. 27, pp. 215–224. Springer, Cham (2020). https://doi.org/10.1007/978-3-030-42822-8_28
2. Grossmann, S., Lohse, D.: Scaling in thermal convection: a unifying theory. J. Fluid Mech. **407**, 27–56 (2000)
3. Stevens, R.J.A.M., van der Poel, E.P., Grossmann, S., Lohse, D.: The unifying theory of scaling in thermal convection: the updated prefactors. J. Fluid Mech. **730**, 295–308 (2013)
4. Bhattacharya, S., Verma, M.K., Samtaney, R.: Revisiting Reynolds and Nusselt numbers in turbulent thermal convection. Phys. Fluids **33**, 015113 (2021)
5. Dabbagh, F., Trias, F.X., Gorobets, A., Oliva, A.: On the evolution of flow topology in turbulent Rayleigh-Bénard convection. Phys. Fluids **28**, 115105 (2016)
6. Kraichnan, R.H.: Turbulent thermal convection at arbitrary Prandtl number. Phys. Fluids **5**, 1374–1389 (1962)
7. Dabbagh, F., Trias, F.X., Gorobets, A., Oliva, A.: Flow topology dynamics in a three-dimensional phase space for turbulent Rayleigh-Bénard convection. Phys. Rev. Fluids **5**, 024603 (2020)
8. Trias, F.X., Dabbagh, F., Gorobets, A., Oliet, C.: On a proper tensor-diffusivity model for large-eddy simulation of buoyancy-driven turbulence. Flow Turbul. Combust. **105**, 393–414 (2020)
9. Trias, F.X., Lehmkuhl, O., Oliva, A., Pérez-Segarra, C.D., Verstappen, R.W.C.P.: Symmetry-preserving discretization of Navier-Stokes equations on collocated unstructured meshes. J. Comput. Phys. **258**, 246–267 (2014)
10. Komen, E., Hopman, J.A., Frederix, E.M.A., Trias, F.X., Verstappen, R.W.C.P.: A symmetry-preserving second-order time-accurate PISO-based method. Comput. Fluids **225**, 104979 (2021)

11. Álvarez, X., Gorobets, A., Trias, F.X.: A hierarchical parallel implementation for heterogeneous computing. Application to algebra-based CFD simulations on hybrid supercomputers. Comput. Fluids **214**, 104768 (2021)
12. Gorobets, A., Trias, F.X., Oliva, A.: A parallel MPI+OpenMP+OpenCL algorithm for hybrid supercomputations of incompressible flows. Comput. Fluids **88**, 764–772 (2013)
13. Trias, F.X., Folch, D., Gorobets, A., Oliva, A.: Building proper invariants for eddy-viscosity subgrid-scale models. Phys. Fluids **27**(6), 065103 (2015)

Turbulent Flows

New Insights on Buoyancy-Driven Turbulent Flows

Kiran Bhaganagar(✉)

Laboratory of Turbulence, Sensing and Intelligence Systems: Department of Mechanical Engineering, University of Texas, San Antonio, USA
kiran.bhaganagar@utsa.edu

1 Introduction

Naturally occurring plumes such as wildland fires [17], smoke plumes [18], volcanic plumes [16], and dust storms are all examples of buoyancy-driven turbulent flows at geophysical scales in which the active scalar, such as temperature, density, or concentration [8] dominates the flow physics. Turbulence kinetic energy (TKE) production from the buoyancy generated turbulence and the correlation between the scalar (e.g. density) and the velocity contribute to the flow physics. Our understanding of buoyancy-generated turbulent flows with active scalar contributions is still not clear. The focus of the present work is to present a high-fidelity large-eddy simulation (LES) to simulate turbulent buoyant plumes released from a circular heat source with an active scalar transport. A new numerical model - bplume-WRF-LES -to simulate buoyant plumes in the atmospheric boundary layer to account for the interactions between the macro- and micro-scale turbulence was developed. The focus of the present study is to understand the influence of atmospheric stratification on plume-generated turbulence and to quantify the TKE processes that influence plume development using the modified WRF-LES-bplume tool that has been developed. The effect of the wind is ignored in this study and the thermally stratified convective boundary layer is simulated for the case where the geostrophic forcing is set to zero. Convective rolls form in the convective boundary layer (CBL), which is referred to it as the roll-dominated CBL. The study will provide insights into the role of the relative strength of the atmospheric stratification to the source buoyancy forcings of the plume on the plume structure.

2 Numerical Methodology

In the WRF-ARM model, the governing equations are the fully-compressible Euler non-hydrostatic equations in flux form. To introduce the buoyant gas into the domain, an additional equation for the gas-mixture ratio(q_p, units: kg/kg) is introduced as follows,

C. Marchioli et al. (Eds.): DLES 2023, ERCO 31, pp. 333–343, 2024.
https://doi.org/10.1007/978-3-031-47028-8_51

$$\frac{\partial(\mu q_p)}{\partial t} + \nabla \cdot \mathbf{V} q_p = \mu \frac{\partial Q_p}{\partial z}, \tag{1}$$

where, Q_p is the gas-mixture surface-flux,$\mathbf{V} = (U,V,W) = \mu \cdot (u,v,w)$ represents the horizontal (u,v) and vertical (w) velocities

The density (ρ_m) and the pressure of the gas mixture (P_m) are modified as follows, $1/\rho_m = 1/(\rho_d + \frac{p_0}{R_p T_0} q_p)$, and $P_m = P_d + q_p P_p$

where, ρ_d and P_d are the density and pressure of dry air, respectively. ρ_p and P_p are the density and pressure of the released gas.

Within the WRF-LES framework, the unresolved turbulence is approximated using the Eddy viscosity boundary layer parameterization. The existing WRF-LES has an option for using the 1.5 order Turbulence Kinetic Energy (TKE) scheme, in which, the TKE is predicted using a prognostic energy equation. The buoyancy production term in the equation of the 1.5 TKE scheme is modified by adding the vertical gradient of the newly added gas-mixture mixing ratio. Further details and validation of the scheme are given in BhimiReddy and Bhaganagar 2020 [7].

3 Background

The background atmospheric boundary layer (ABL) is characterized by Buoyancy frequency or the Brunt-Vaisala frequency, N^2, which is given as, $N^2 = -g\frac{1}{\rho}\frac{\partial \rho}{\partial z}$. The buoyancy frequency represents the local stability of the density stratification. The CBL convective velocity scale is $w^* = (\frac{g}{\Theta_o}\overline{w'\theta'}h)^{\frac{1}{3}}$.

The convective time scale is defined as Δ/w^*, where Δ is the boundary layer (inversion layer) depth and w^* is the friction velocity. The time scales associated with the plume with respect to the buoyancy are given as $t_o = \sqrt{D/g_o\prime}$. In a convective boundary layer, the positive contribution of buoyancy forcings to the turbulence kinetic energy lifts the streaky surface structures resulting in the formation of convective rolls or thermal eddies in the atmosphere [12, 13]. Buoyant plumes released into the atmosphere from a source with high source strength rise to the top of the convective boundary layer [11]. However, the interaction between the convective rolls and the plumes for different source strengths is still not clear.

Ambient fluid is entrained into the plume at the plume interface due to shear instabilities [12]. The plume consists of a central core flow with dense hot fluid which is capped by a buoyant cloud. Close to the source, the central core is surrounded by a shear layer [19] with lighter buoyant fluid [14]. As the plume ascends, the plume density and the temperature approaches that of the atmosphere.

Table 1. Details of cases simulated for the heated plumes released into the atmosphere. The value of z_i which is the inversion height estimated from temperature and velocity profiles is 2100m. Ri_B and N^2 are the Bulk Richardson number and Buoyancy frequency estimated at 50m from the ground.

Case	*HFX*	$g_{\rho'}$	B_0	Ri_B	N^2	L	w_*	t_*	u_*	z_{fmax}/z_i	f_p
1	0.24	0.043	0.58×10^4	–2206	-1.07×10^{-4}	–8.37	2.54	0.297	1.18	0.48	0.0022
2	0.24	0.086	1.07×10^4	–2206	-1.07×10^{-4}	–8.37	2.54	0.297	1.22	0.44	0.0034
3	0.24	0.17	2.14×10^4	–2206	-1.07×10^{-4}	–8.37	2.54	0.297	1.31	0.38	0.0048

The focus of this study is to quantify the plume mixing released in a roll-dominated convective boundary layer with zero mean wind and to understand the turbulence generated in plumes of varying strengths with respect to the strength of the ambient convective rolls. The CBL has been simulated using Weather Research and Forecast model [15]. The plume is simulated using the Large-Eddy simulation framework developed by Bhaganagar and BhimiReddy [5]. The tool has been validated to simulate thermal plumes [5], and accidental and man-made plumes [3,4,6].

4 Results

WRF simulations are conducted with a constant heat flux on the lower boundary and with periodic lateral boundary conditions. A zero geostrophic forcing is imposed. The domain is of size $10D \times 10D \times 7.5D$, where, D is the source diameter. The buoyant plume is released into the domain after the boundary layer is fully developed. The diameter D is 400 m. The simulation was run for 43200 time-steps with a time-step of 1/6 s and the first 28000 time-steps taken for the development of the thermal boundary layer. Table 1 gives the details of cases simulated. A total of three cases are simulated with plume source buoyancy fluxes varying from low- to high values of 0.58×10^4, 1.07×10^4, $2.14 \times 10^4\,m^4\,s^{-3}$, and $3.52 \times 10^4\,m^4\,s^{-3}$ corresponding to Cases 1–3. The sensible surface-flux value is 0.24 Kms^{-1}. The boundary layer depth (z_i) is obtained as 2100 m. The Monin-Obukhov length scale (L) estimated is –8.37 m. The convective motions in the form of rolls have formed before releasing the plume into the ambiance. The temperature jump across the inversion layer is 8 K.

4.1 Plume Dynamics

The release of heated buoyant plumes into roll-dominated thermally stratified boundary layer is discussed. The stratification of the CBL is kept constant and the effect of increasing source buoyancy flux is analyzed. The details of the three cases discussed in this paper are given in Table 1.

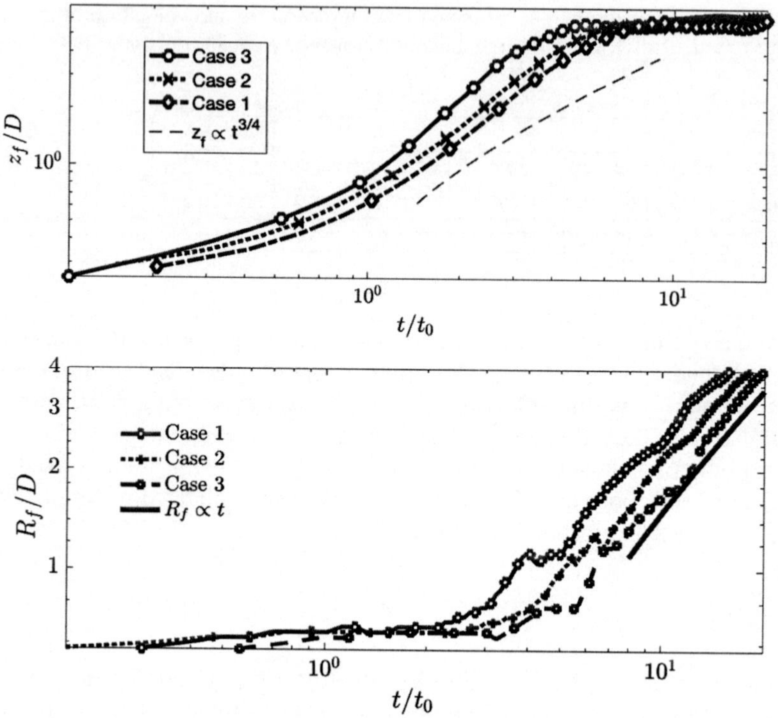

Fig. 1. For the three cases, the variation of the (a) plume rise in time. The dashed line shows the $t^{\frac{3}{4}}$ power law, (b) plume half-width in time. The solid line shows the linear curve.

Figure 1a shows the plume rise, which is the distance traveled by the leading edge of the normalized by the source diameter. The time coordinate is normalized by the plume time scale, t_o. Initially, there is a transient regime evident till $t = 1.0t_o$, beyond which the plume rise follows a power law, which varies between $t^{0.71}$–$t^{0.82}$ scaling law. The plume spread obtained from the buoyancy profiles, which is defined as the horizontal distance from the centerline, plotted versus time is shown in Fig. 1b. After traveling a distance of around $5D$ from the source, the plume expands linearly with time, before reaching the boundary layer top.

Figure 2 a and Fig. 2 b shows plume temperature in the $x-z$ (left column) and $y-z$ (right column) plane of the plume at t = 6300 s after the release. Case 1 is shown in the top panel. The forcings due to the atmospheric stratification are stronger than the plume buoyancy resulting in a significant portion of the plume overturning along with the ambient atmospheric convective rolls. The consequence of this is the plumes spread radially to a height of $2.5D$ from the source and bend towards the left side of the centerline. This is due to the clockwise rotation ($x-z$ plane) of the background convective rolls. This results in a loss of symmetry about the centerline. Initially, the plume rises vertically and it is axisymmetric as it is released from a circular area source. However, the plume is subsequently influenced by ambient forcings. As the plume reaches the

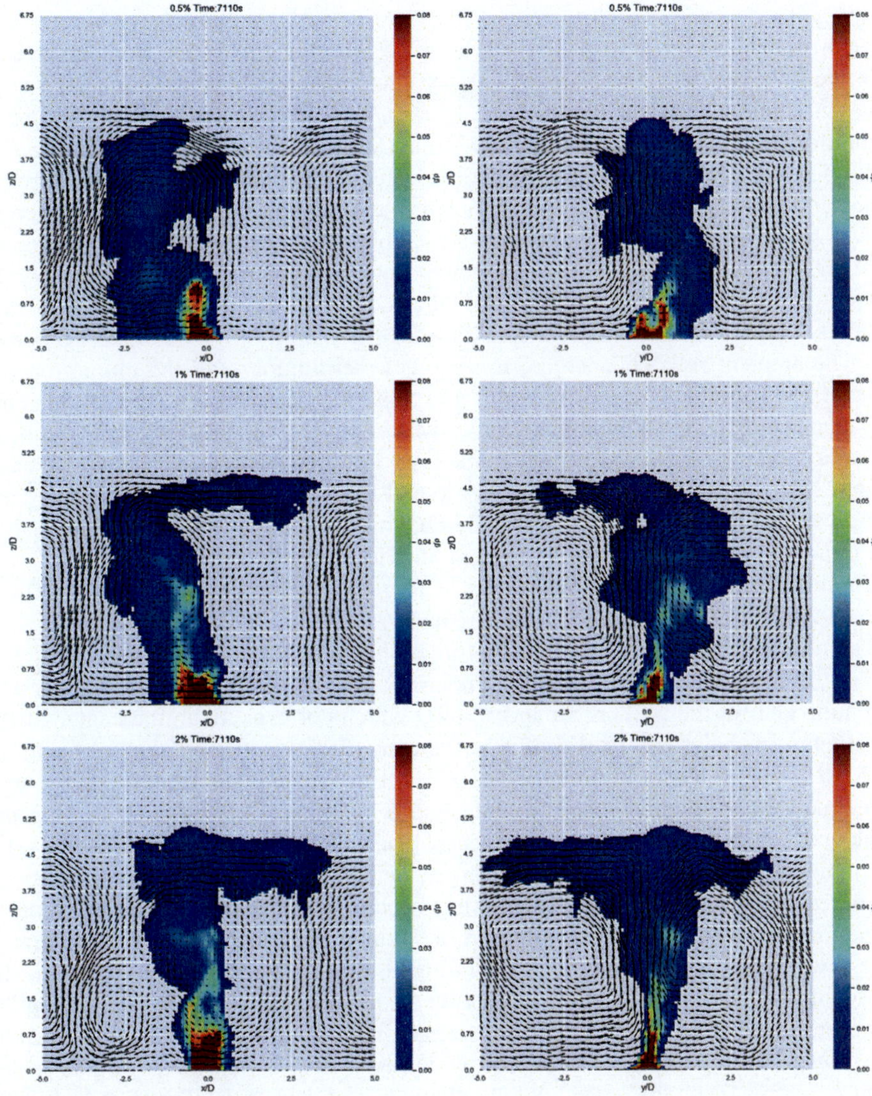

Fig. 2. Plume temperature in the $x-z$ plane (left) and $y-z$ (right) of Case 1 (top panel), Case 2 (middle panel) and Case 3 (top panel)

inversion height due to the strong mixing with the ambient rolls the plume is diluted very quickly. We refer to this as regime 1 in which the ambient convective rolls are dominant compared to the buoyancy forcings resulting in the bending of the plume and a rapid dilution before the plume reaches the inversion height. Case 2 is shown in the middle panel. The buoyancy forcings of Case 2 are higher than Case 1. Due to the initial buoyancy of the plume, it moves vertically without being influenced by the ambient rolls. However, beyond the height of $2D$ the plume loses the symmetry about the centerline. Beyond the height of $3D$, the plume bends clockwise along the $x-z$ plane (or counterclockwise along the $y-z$ plane). The plume reaches the inversion height and spreads laterally, however the strong ambient circulation dilutes the lateral spread quickly. We refer to it as regime 2 in which the ambient convective rolls result in the bending of the plume. Case 3 is shown in the top panel. The buoyancy forcings is the highest of the three cases considered. Unlike the other two regimes, the plume is not significantly influenced by the ambient convective rolls. The plume rises vertically entraining and mixing with the ambiance and at the inversion height, it spreads laterally. A similar form of lateral spreading of the plume has been observed before [18]. The ambient rolls enhance the mixing and the dilution after the plume reaches the inversion height. We refer to regime 3 in which the plume buoyancy forcings are dominant compared to the ambient convective rolls which mainly enhances the dilution rather than the direction of the plume rise. Regime 1 is the low buoyancy case, Regime 2 is the intermediate buoyancy case and Regime 3 is the high buoyancy case for a given stratification. The ratio of the ambient stratification to the plume buoyancy is an important metric to characterize the plume rise.

To understand the physical processes of Regime 1, Regime 2, and Regime 3 the mean and the turbulence characteristics are analyzed next.

The turbulent kinetic energy (TKE) profiles shown in Fig. 3 are analyzed next. The TKE in the background ABL before the release of the plume is shown in Fig. 3d. The maximum TKE in the ABL is around $z = 2D$ which corresponds to the center of the rolls. Figure 3a shows the TKE profile at the centerline ($r = 0$) and at axial locations $r = 1.0D$ (right of the centerline) and at $r = -1.0D$ (left of the centerline) for case 1 (Regime 1). For Regime 1, at the centerline ($r = 0$) after the initial transience, the turbulence is well-mixed as the TKE is nearly constant. On the other hand, at $r = -1D$ (left of the centerline) two distinct peaks are evident, at $z = 1.7D$ and at $z = 4.5D$ above the source. A significant amount of mixing occurs in the near-source region due to interactions between the ambient rolls and the plume. There is high TKE concentrated in the lower portion of the plume at off-radius locations. At the inversion height, the intense interaction of the circulating rolls with the plume results in high TKE. In regime 2, as seen in Fig. 3b, due to the strong interaction between the ambient rolls and the plume there is a high TKE is generated which results in a well-mixed plume in the region $z = 1.5-4D$ on the left side of the plume (i.e. at $r = -1.D$). For all the regimes, the source generates TKE which is of high value in the near-source region, $z = 0.5D-1D$ at the centerline. For regime 3 as seen in Fig. 3c the ambient rolls do not influence the bending of the plume as observed in the previous cases. The high TKE is mainly generated at the centerline. At off-radius locations a well-mixed region is evident.

The turbulence production due to the buoyancy (temperature and density differences) is discussed next. As shown in Fig. 4, for all the three regimes, the buoyancy production from the density source ($\overline{w\prime g\prime}$) is generated at the centerline, and higher the source buoyancy flux, the further away from the source is the peak buoyancy production. On the other hand, significant differences in the TKE production from the thermal source ($\overline{\Theta\prime g\prime}$) exist between the three regimes. TKE production dominates off-radius

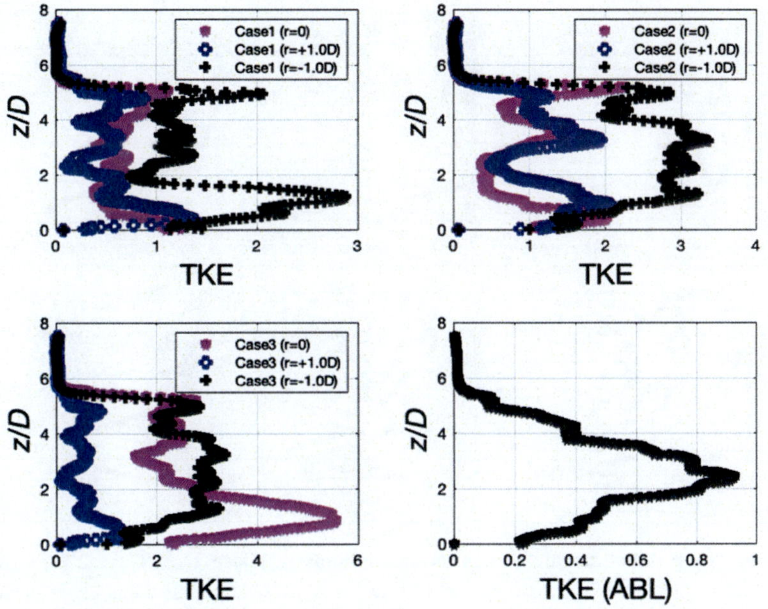

Fig. 3. TKE (a) Case 1 (b) Case 2 (c) Case 3 at r = 0, r = +1D (right of the centerline), r = −1D (left of the centerline), (d) TKE in the background ABL

locations for regime-1 and regime-2, whereas high TKE production occurs at the centerline for regime-3.

The TKE production from the shear component is given in Fig. 5. The shear stress contributes to the TKE production near the inversion layer and the higher the buoyancy flux, the larger the TKE production. As seen in Fig. 5a-c, in the region between 2.5 – 4.0D a peak value of TKE is evident, the value of which increases from regime 1 to regime 3. However, in regime 1, the interaction of the convective rolls with the plume results in additional TKE shear production in the lower portion of the plume at the off-radius locations. Similarly, the higher shear production correlates with the regions with active mixing in regime 2.

The time-averaged root-mean-square (r.m.s) of g'_ρ at different radial locations is shown in Fig. 6. At the centerline, g'_ρ is high at the source and due to dilution, the value decays with height. The higher the source buoyancy flux, the further the mixing region before the fluctuations decay to zero. Both regime 1 and regime 2 are diluted quickly and the r.m.s. of g'_ρ vanishes by $z = 1D$, whereas for regime 3, it decays and vanishes only beyond $z = 4D$. However, the trend of the fluctuation is different at the off-radius locations: At $r = -1.0D$, At $r = 2D$, the lateral spreading in regime 2 and regime 3 is associated with high r.m.s. g'_ρ near the inversion height as seen in Fig. 6d

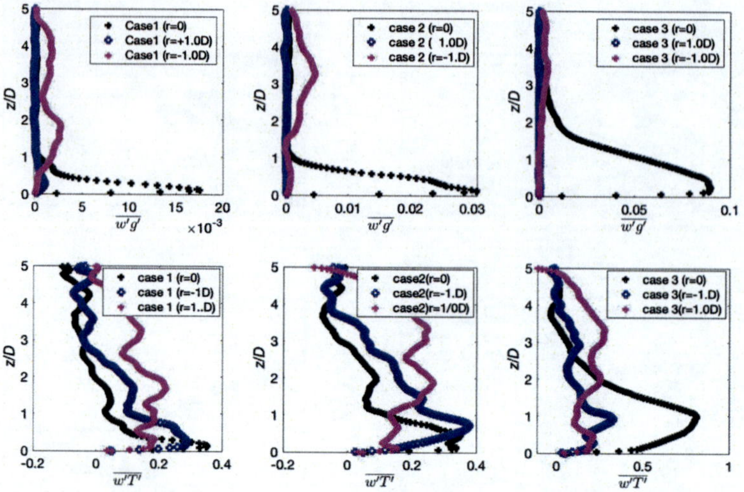

Fig. 4. Buoyancy production due to density fluctuations (a) Case 1 (b) Case 2 (c) Case 3 at r = 0 (centerline), r = +1D (right of the centerline), r = –1D (left of the centerline), Buoyancy production due to temperature fluctuations (d) Case 1, (e) Case 2, (f) Case 3

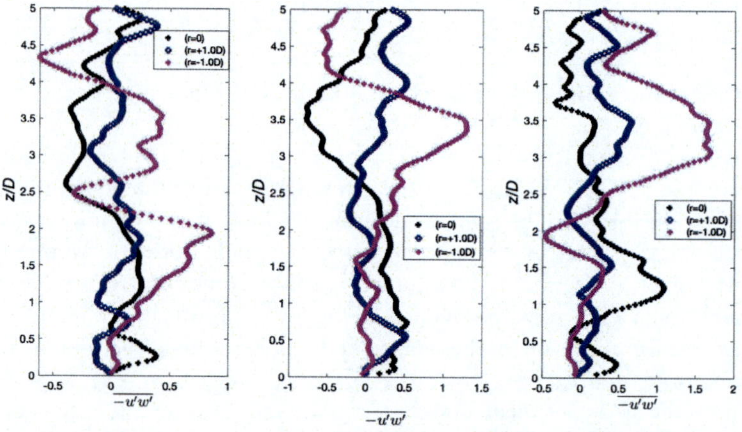

Fig. 5. Shear stress component that contributed to shear production of TKE (a) Case 1 (b) Case 2 (c) Case 3 at r = 0 (centerline), r = +1D (right of the centerline), r = –1D (left of the centerline),

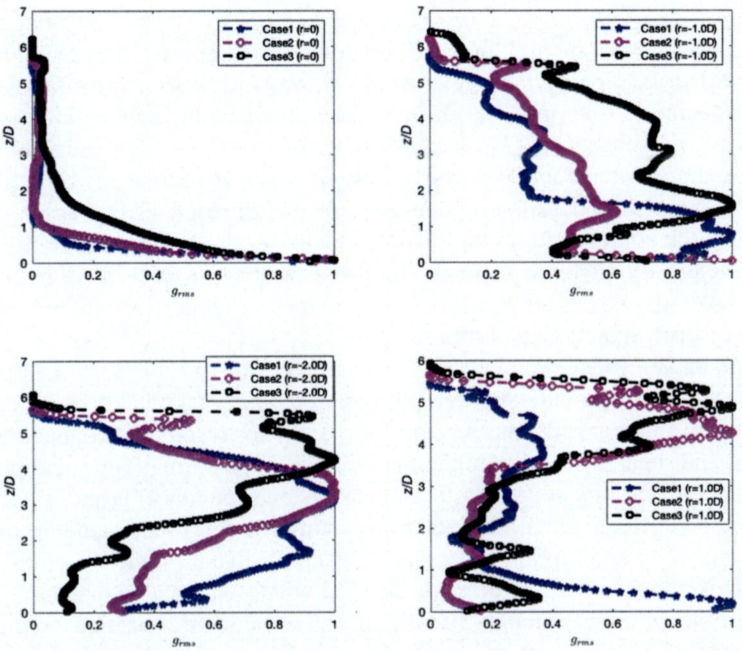

Fig. 6. The time-averaged profiles of r.m.s. of $g_{\Theta'}$ fluctuations, r.m.s. of buoyancy fluctuations at (a) Centerline ($r = 0$) (b) $r = -1.0D$ (left of the centerline) (c) $r = -2.0D$ (left of the centerline), (d) $r = 1.0D$ (right of centerline)

5 Conclusions

Quantifying the mixing and entrainment of buoyant plumes released into the atmosphere is extremely complicated due to the dynamic interactions between the buoyant plumes and the background atmospheric boundary layer[9, 10]. Towards this direction, the effect of the atmospheric stratification of the CBL with no winds on the plume dynamics has been investigated. In a convective atmospheric boundary layer (CBL), the convective rolls are the characteristic features of the unstably stratified atmosphere. WRF-LES has been used to simulate the background atmospheric state with convective rolls. Three cases of plumes with increasing buoyancy strength have been analyzed by releasing heated buoyant plumes into the ambient stratified CBL. Strong plumes (released with higher source boundary flux) rise vertically through the convective rolls and laterally expand at the inversion height. The buoyancy forcings generated are higher than the convective forcings of the atmosphere and the frontal part of the plume follows the convective rolls and spreads radially. On the other hand, for the weak plumes, the convective forcings are stronger causing the plume body and the plume head to move along with the rolls and resulting in radial bending of the plumes. Case 2 is the intermediate case where the buoyancy forcings are strong near the source and beyond which the convective forcings tilt the plume in the direction of the rolls.

In regime 1 (weak plume), due to the bending of the lower part of the plume, it results in enhanced TKE in this region (i.e. $z = 2D$) and hence causes fast dilution or mixing. In regime 2 (intermediate plume), due to the bending of the upper part of the plume, there is an enhanced TKE in this region (i.e. $z - 2D - 6D$), which restricts the lateral spreading of the plume near the inversion height. In regime 3 (strong plume), as the ambient circulation does not influence the plume, the high TKE is at the centerline and close to the source. The plume rises vertically up and spreads laterally near the inversion height. In all three regimes, the shear production of TKE is high near the inversion layer, moreover, the bending of the plume in regime 1 results in higher shear production values in the lower part of the plume.

Overall, the buoyancy production terms dominate the TKE production in plumes of all regimes. Depending on the relative strength of the buoyancy flux with respect to the strength of the ambient rolls, the bending of the plumes results in enhanced buoyancy production and shear production in the bent region of the plume. The shear production of TKE is mainly dominant at the top of the inversion (boundary) layer.

The work is offered as an important contribution to our understanding of the turbulence production and mixing processes in heated buoyant plumes, such as wildland fire and smoke plumes. As the effect of the crosswinds and mean winds have not been included in the numerical model, the role of atmospheric stratification on the plumes has been isolated. The concept of classifying the plumes into regime 1 (weak), regime 2 (intermediate), and regime 3 (strong) based on the strength of the convective rolls with respect to buoyancy forcings of the plume offer new insights into the plume dynamics in the convective boundary layer. Future work will focus on estimating the entrainment using first-principles [1,2].

Acknowledgements. Authors acknowledge Texas Advanced Computing Center (TACC) for the computational resources. Funding for the study was obtained from NASA Center for Advanced Measurements in Extreme Environments, Grant number: 80NSSC19M0194. Graphics of Fig. 2 were done by Thanh Tran.

References

1. Bhaganagar, K., Pillalamarri, N.: Lock-exchange release density currents over three-dimensional regular roughness elements. Bhaganagar J. Fluid Mech. **832**, 793–824 (2017)
2. Bhaganagar, K.: Role of head of turbulent 3-D density currents in mixing during slumping regime. Phys. Fluids **29**, 2 (2017)
3. Bhimireddy, S.R., Bhaganagar, K.: Performance assessment of dynamic downscaling of WRF to simulate convective conditions during sagebrush phase 1 tracer experiments. Atmosphere **9**(12), 505 (2018)
4. Bhaganagar, K., BhimiReddy, S.R.: Assessment of the plume dispersion due to chemical attack on April 4, 2017 in Syria. Nat. Hazards **88**(3), 1893–1901 (2017)
5. Bhaganagar, K., BhimiReddy, R.: Numerical investigation of starting turbulent buoyant plumes released in neutral atmosphere. J. Fluid Mech. **900** (2020)
6. Bhaganagar, K., Bhimireddy, S.R.: Local atmospheric factors that enhance air-borne dispersion of coronavirus-High-fidelity numerical simulation of COVID19 case study in real-time. Environ. Res. 11017 (2020)

7. BhimiReddy, S.R., Bhaganagar, K.: Implementing a new formulation in WRF-LES for buoyant plume simulations: bPlume-WRF-LES model. Mon. Weather Rev. **149**(7), 2299–2319 (2021)
8. Bonadonna, C., Pistolesi, M., Cioni, R., Degruyter, W., Elissondo, M., Baumann, V.: Dynamics of wind-affected volcanic plumes: the example of the 2011 Cordón Caulle eruption, Chile. J. Geophys. Res. Solid Earth 2242–2261 (2015)
9. Briggs, G.: Plume rise and buoyancy effects. Atmos. Sci. Power Prod. **850**, 327–366 (1984)
10. Briggs, G.: A plume rise model compared with observations. J. Air Pollut. Control Assoc. **15**(9), 433–438 (1965)
11. Bierly, E., Wendell, H.: Some restrictive meteorological conditions to be considered in the design of stacks. J. Appl. Meteorol. **1**(3), 383–390 (1962)
12. Chang, C., Bhaganagar, K.: New findings in vorticity dynamics of turbulent buoyant plumes. Phy. Fluids **33**(11) (2021)
13. Moeng, C.S., Sullivan, P.: A comparison of shear-and buoyancy-driven planetary boundary layer flows. J. Atmos. Sci. **51**(7), 999–1022 (1994)
14. Sullivan, P.P., McWilliams, J.C., Moeng, C.: A, subgrid-scale model for large-eddy simulation of planetary boundary-layer flows. Boundary-Layer Meteorol. **71**, 247–276 (1994)
15. Skamarock, W., et al.: A description of the advanced research WRF version 2, National Center For Atmospheric Research Boulder Co Mesoscale and Microscale (2005)
16. Sparks, R.S.J.: The dimensions and dynamics of volcanic eruption columns. B. Volcanol. **48**, 3–15 (1986)
17. Simpson, J.: Gravity currents in the laboratory, atmosphere, and ocean, Simpson. Annu. Rev. Fluid Mech. **14**(1), 213–234 (1982). Annual Reviews
18. Snyder, W., Lawson, R., Shipman, M., Lu, J.: Fluid modeling of atmospheric dispersion in the convective boundary layer. Bound.-Layer Meteorol. **102**(3), 335–366 (2002)
19. Weil, J.C., Snyder, W.H., Lawson, R.E., Shipman, M.S.: Experiments on buoyant plume dispersion in a laboratory convection tank. Boundary-Layer Meteorol. **102**(3), 367–414 (2002)

Assessment of the Effect of the Surface Tension Contribution on the Emulsification in Linearly Forced Turbulence

A. Begemann, T. Trummler, E. Trautner(✉), J. Hasslberger, and M. Klein

Institute of Applied Mathematics and Scientific Computing,
Department of Aerospace Engineering, University of the Bundeswehr Munich,
Werner-Heisenberg-Weg 39, 85577 Neubiberg, Germany
elias.trautner@unibw.de

1 Introduction

Emulsions are suspensions of immiscible liquids, in which the *dispersed* liquid is present in the *carrier* liquid in the form of poly-disperse droplets. In order to obtain and maintain an emulsion, a constant energy input in the form of kinetic energy is required to deform and break up droplets. Both processes increase the total droplet surface area, and thus the surface energy. Dodd and Ferrante [4] show that the power of the surface tension is either a source or a sink of turbulent kinetic energy (TKE), depending on whether the rate of change of the interfacial area between the two fluids is positive or negative. Consequently, the surface energy, or surface tension contribution, plays a central role for the emulsification process and the required energy input. In the present study, we focus on direct numerical simulations (DNS) of emulsions in homogeneous isotropic turbulence (HIT), and especially on the temporal evolution of global and local surface tension contributions during emulsification.

For the forcing, we employ a physical space forcing method introduced by Lundgren [6]. We have extended this linear forcing for two-phase flows with a PID controller to ensure a constant and prescribed TKE k and to analyze the emulsification process [2].

2 Governing Equations and Computational Method

The forced momentum equations for a two-phase flow read

$$\rho\left(\frac{\partial u_i}{\partial t}+\frac{\partial u_i u_j}{\partial x_j}\right)=-\frac{\partial p}{\partial x_i}+\frac{\partial}{\partial x_j}\left[\mu\left(\frac{\partial u_i}{\partial x_j}+\frac{\partial u_j}{\partial x_i}\right)\right]+\sigma n_i \kappa \delta_s+\mathrm{F}^* u_i \quad (1)$$

C. Marchioli et al. (Eds.): DLES 2023, ERCO 31, pp. 344–349, 2024.
https://doi.org/10.1007/978-3-031-47028-8_52

with the density ρ, the dynamic viscosity μ, the ith velocity component u_i and the pressure p. F^* is the pseudo shear term for the forcing. For the forcing augmented by a PID controller this term is $\mathrm{F}^* = \beta_{\mathrm{PID}} \cdot \mathrm{F}$, where β_{PID} denotes the controller output and F denotes the forcing parameter determined by k and the domain size.

Assuming incompressibility and homogeneity, the TKE equation derived from the momentum equation reads [4]

$$\frac{dk}{dt} = -\varepsilon + 2\,\mathrm{F}^*\,k + \Psi_\sigma. \tag{2}$$

Here, Ψ_σ denotes the surface tension contribution defined as [4]

$$\Psi_\sigma = \frac{-\sigma}{V_d \rho_d} \frac{\partial A}{\partial t}, \tag{3}$$

where V_d, ρ_d and A are the volume and the density of the dispersed phase, and the interface area.

All simulations are conducted with the state-of-the-art open-source code "PArallel Robust Interface Simulator" (PARIS) [1], which is based on the finite-volume method. It employs a red-black Gauss-Seidel solver with overrelaxation to solve the Poisson pressure equation in the framework of the projection method. A cubic staggered grid is used. The velocity components are computed at the cell faces, whereas the pressure, the density, the viscosity and the VOF marker function are stored in the cell centers. Time integration is realized with a second-order predictor-corrector method. The convective term of the momentum equation is discretized using the "Quadratic Upstream Interpolation for Convective Kinematics" (QUICK) scheme, and the viscous term is computed using central differences.

3 Numerical Setup and Configuration

We consider a cubic box with an edge length of $L = 2\pi$ and periodic boundary conditions in each direction. First, we perform single-phase simulations to obtain a fully developed single-phase HIT, and subsequently we initialize the dispersed phase as spherical droplets. This procedure is described in detail in [2].

To demonstrate the effect of Ψ_σ, we compare and discuss the emulsification of two configurations with different surface tension coefficients adapted from Begemann et al. [2]. In both configurations, the density ratio of the *dispersed* and the *carrier* phase ρ_d/ρ_c is 0.9, resembling an oil-in-water emulsion, and the global volume fraction of the dispersed fluid is 12.5%. The targeted turbulent kinetic energy is $k_0 = 0.5\,\mathrm{m}^2/\mathrm{s}^2$, and it is immediately reached and then maintained by our linear forcing approach, see also [2].

Emulsions can be characterized by the Hinze scale d_H [5], which is expected to be the most stable maximum droplet diameter in emulsions for HIT. It is given as

$$d_H = (We_{d,crit}/2)^{3/5} (\rho_c/\sigma)^{-3/5} \varepsilon^{-2/5}, \tag{4}$$

where $We_{d,crit}$ denotes the critical droplet Weber number, for which we assume $We_{d,crit} = 1.17$ following Hinze [5] and recent numerical studies [3,7].

For emulsions, a Weber number We_l using a characteristic length scale l can be defined as

$$We_l = \frac{\rho_c u'^2 l}{\sigma}, \tag{5}$$

taking into account the effect of surface tension. For HIT, l can be approximated with the integral length scale of the turbulent flow field $l = ((3/2)u'^2)^{3/2}/\varepsilon$. Here, u' can be determined from the turbulent kinetic energy $k = (3/2)u'^2$. Hence, emulsions at stationary state can be characterized by We_l, as also demonstrated and discussed in Begemann et al. [2].

We here consider two different configurations obtained by altering the surface tension coefficient σ, and therewith We_l . The case with the high σ value exhibits $We_l = 21$, and the one with a low σ value leads to $We_l = 70$.

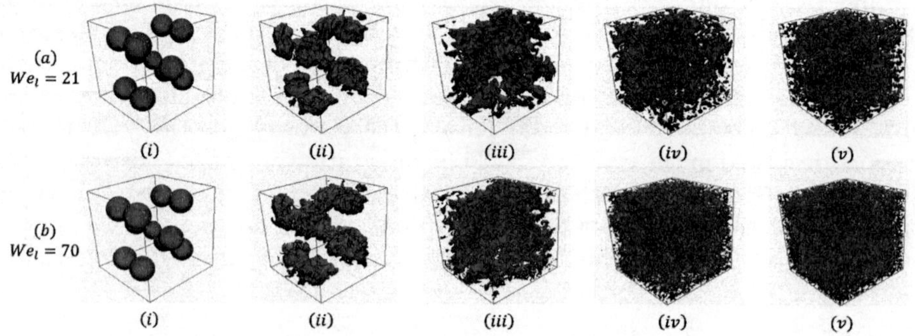

Fig. 1. Time series visualizing the emulsification process. (a) $We_l = 21$ (high σ value), (b) $We_l = 70$ (low σ value) at $t/\tau = 0$ (i), $t/\tau = 0.5$ (ii), $t/\tau = 1$ (iii), $t/\tau = 2$ (iv), and $t/\tau = 5$ (v).

4 Results

Figure 1 visualizes the emulsification of both configurations at $t/\tau = 0$ (i), $t/\tau = 0.5$ (ii), $t/\tau = 1$ (iii), $t/\tau = 2$ (iv), and $t/\tau = 5$ (v). Figure 1(i) shows the initialized droplets of the dispersed phase, which are identical for both configurations. The forcing leads to a deformation of the droplets and already at $t/\tau = 0.5$ (Fig. 1(ii)), the initialized droplets are significantly deformed. As expected, with decreasing σ (increasing We_l), the interfaces become more deformed and wrinkled. With increasing time, the dispersed structures are increasingly deformed and break up into smaller structures.

Figures 2 to 4 show the temporal evolution of the interface Area A/A_∞, the normalized interface area $(A - A_{init}) / (A_0 - A_{init})$, and the surface tension contribution Ψ_σ. The time is normalized by the eddy turn-over time $\tau = k/\varepsilon$ evaluated with k_0 and ε_0 in the statistically stationary state. A_0 and $A_\infty = L^2$ refer to the interface area at stationary state and at fully segregated state, respectively. A_{init} is the initial interface area of the inserted droplets.

Figure 2 shows that the current forcing method reaches a statistically stationary state of the normalized interface area A/A_∞ within about 4τ. Further, Fig. 3 demonstrates that the slope of the normalized change of the interface area is nearly the same for both configurations. Figure 4 plots the temporal evolution of the surface tension contribution Ψ_σ, which is not directly considered by the linear forcing. During the emulsification process, droplet break-up is the prevailing mechanism. Since this sink of TKE is not compensated, the time to reach the stationary state increases with decreasing surface tension due to more frequent break-ups. To compensate this effect, a higher energy input is required which leads to a different forcing parameter during emulsification. At statistically stationary state, break-up and coalescence are more or less balanced, and thus the net contribution of Ψ_σ vanishes.

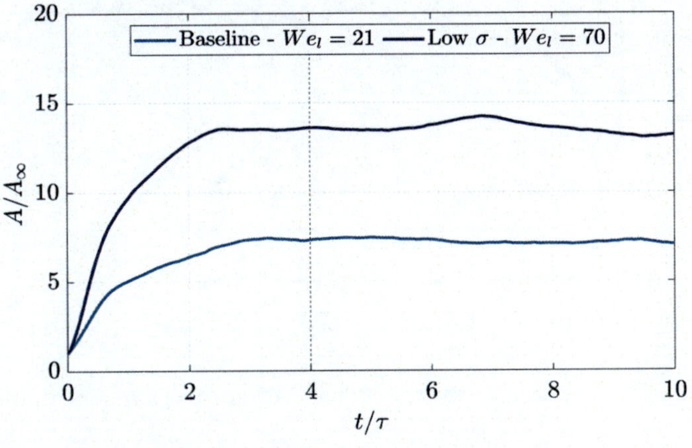

Fig. 2. Temporal evolution of the normalized interface area A/A_∞.

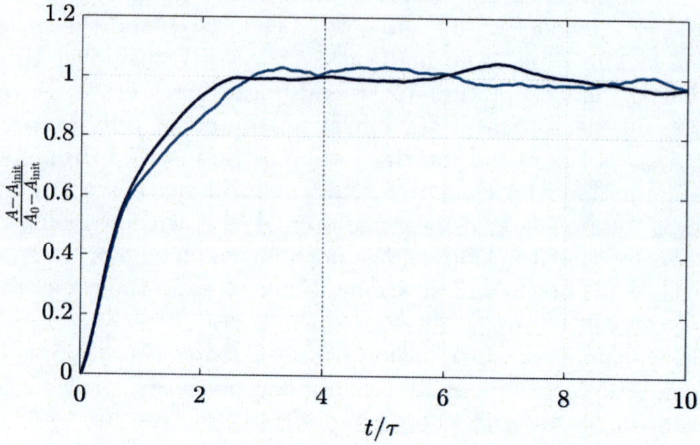

Fig. 3. Temporal evolution of the normalized interface area $(A - A_{init}) / (A_0 - A_{init})$.

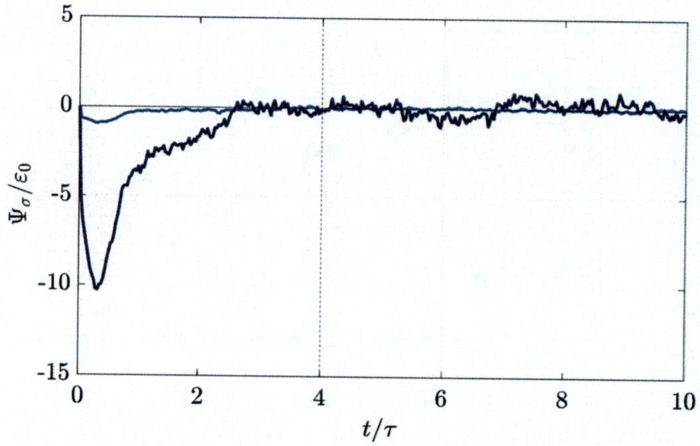

Fig. 4. Temporal evolution of the normalized surface tension contribution $\Psi_\sigma/\varepsilon_0$.

5 Conclusion and Outlook

With our study, we have investigated the effect of the surface tension contribution on the emulsification in linearly forced turbulence. Although configurations with a higher surface tension coefficient require more turbulent kinetic energy input for the emulsification, a stationary state is reached after around four eddy turnover times. Building on these observations, we plan to study the local effect of breakup and coalescence of single droplets to the net contribution in future studies.

The surface tension contribution, i.e., the surface tension coefficient, also affects the segregation of emulsions. A higher surface tension coefficient promotes the coalescence and therewith the segregation (or demixing) of an emulsion. Additionally, the surface tension coefficient also plays a role for gravity-driven segregation since it affects the droplet size distribution. Larger droplets are subject to a higher buoyancy force and rise faster, while smaller droplets have a stronger mutual interaction and rise slower. The effect of the surface tension coefficient on the segregation process and the energy release rates of the surface tension energy during the segregation process are studied in detail in our recent work [8].

Acknowledgment. This project received funding by *dtec.bw* - Digitalization and Technology Research Center of the Bundeswehr - under the project *MORE*, which is gratefully acknowledged. Further, the authors thank the Gauss Centre for Supercomputing e.V. (www.gauss-centre.eu) for providing computing time on the GCS Supercomputer SuperMUC-NG at Leibniz Supercomputing Centre (www.lrz.de).

References

1. Aniszewski, W., et al.: PArallel, robust, interface simulator (PARIS). Comput. Phys. Commun. **263**, 107849 (2021)
2. Begemann, A., Trummler, T., Trautner, E., Hasslberger, J., Klein, M.: Effect of turbulence intensity and surface tension on the emulsification process and its stationary state - a numerical study. Can. J. Chem. Eng. **100**(12), 3548–3561 (2022)
3. Crialesi-Esposito, M., Rosti, M.E., Chibbaro, S., Brandt, L.: Modulation of homogeneous and isotropic turbulence in emulsions. J. Fluid Mech. **940**, A19 (2021)
4. Dodd, M.S., Ferrante, A.: On the interaction of Taylor length scale size droplets and isotropic turbulence. J. Fluid Mech. **806**, 356–412 (2016)
5. Hinze, J.O.: Fundamentals of the hydrodynamic mechanism of splitting in dispersion processes. AIChE **1**(3), 289–295 (1955)
6. Lundgren, T.S.: Linearly forced isotropic turbulence. Center Turbul. Res. (Stanford) Annu. Res. Briefs **2003**, 461–473 (2003)
7. Mukherjee, S., Safdari, A., Shardt, O., Kenjeres, S., Van den Akker, H.E.A.: Droplet–turbulence interactions and quasi-equilibrium dynamics in turbulent emulsions. J. Fluid Mech. **878**, 221–276 (2019)
8. Trummler, T., Begemann, A., Trautner, E., Klein, M.: Numerical investigation of the segregation of turbulent emulsions. Phys. Fluids **34**(11), 113324 (2022)

Reynolds Number Dependency of Turbulent Flow over a Surface Fouled by Barnacles

A. Busse(✉) and S. Sarakinos

James Watt School of Engineering, University of Glasgow, Glasgow G12 8QQ, UK
{angela.busse,sotirios.sarakinos}@glasgow.ac.uk

1 Introduction

The shipping sector contributes a rising share to anthropogenic greenhouse gas emissions. The International Maritime Organisation reported that in 2018 total shipping, i.e., international and domestic shipping and fishing, contributed a 2.89% share (1,076 million tonnes) to global anthropogenic emission, a significant increase over its 2.76% share (977 million tonnes) in 2012 [1].

Friction drag makes the dominant contribution to the total resistance of typical merchant ships [2]. The drag of a ship hull significantly increases if the ship is affected by biofouling. Of the different forms of biofouling, calcareous macrofouling, i.e., fouling by organisms such as barnacles, tubeworms, or mussels, which have a calcareous outer shell, is considered one of the most severe forms of biofouling. Schultz [3] predicted for a typical naval surface ship of the US Navy (Oliver Hazard Perry class frigate) an increase of the total resistance by approximately 36% for medium calcareous fouling and 55% for heavy calcareous fouling. Therefore, understanding the effects of different forms of biofouling on frictional drag is of importance for the assessment of resistance of biofouled ship hulls and the cost-effective scheduling of maintenance intervals [4].

In the present study, direct numerical simulations are used to investigate the Reynolds number dependency of turbulent flow over a surface affected by a common form of calcareous macrofouling, namely fouling by acorn barnacles (order *Sessilia*). A 10% coverage state is considered, which falls into the 'sparse' roughness regime [5]. The results are compared to the flow statistics of a generic Gaussian rough surface [6].

2 Methodology

A barnacle-fouled surface was generated using the BaRGE algorithm which mimics the settlement behaviour of barnacles [7]. Barnacles are represented in simplified form as conical frustra following the approach of Sadique [8]. 10% of the surface is covered by barnacle features, while the rest of the surface remains smooth. This surface can also serve as an example for heterogeneous roughness, a class of roughness that has received to date far less attention than homogeneous rough surfaces which are statistically uniformly covered by roughness features [9]. Data from an earlier study [6] on an homogeneous irregular rough surface fully covered by roughness features with approximately Gaussian height distribution is used for comparison. This surface serves as a

C. Marchioli et al. (Eds.): DLES 2023, ERCO 31, pp. 350–355, 2024.
https://doi.org/10.1007/978-3-031-47028-8_53

representative example for many forms of engineering roughness which mostly show moderate skewness and kurtosis values [6]. Key surface parameters for both surfaces are summarized in Table 1, where δ is the mean channel half-height. For the barnacle surface, the geometric roughness height k is defined as the maximum barnacle height. In the case of the Gaussian roughness, k is set to the mean peak-to-valley height.

Table 1. Overview of key surface topographical parameters: *Sa*: mean roughness height; *Sq*: rms roughness height; *k*: geometric roughness height; *Ssk*, *Sku*: skewness and kurtosis of height distribution; *ES*: effective slope, λ_p: planform solidity (fraction of surface covered by roughness features). Note: height and length scales are given for the full-scale roughness pattern. For case 180hs height measures need to be multiplied by a factor of 1/2.

Surface	Sa/δ	Sq/δ	k/δ	Ssk	Sku	ES	λ_p
barnacle 10%	0.0089	0.018	0.1267	4.06	19.5	0.067	0.10
Gaussian	0.023	0.029	0.1667	−0.10	2.99	0.37	1.0

Table 2. Overview of simulation parameters: k geometric roughness height; N_x, N_y number of grid points in streamwise and spanwise direction. $\Delta z^+_{\min}$, $\Delta z^+_{\max}$: minimum and maximum wall-normal grid spacing. All cases expect for 180hs use the full-scale (fs) roughness pattern. Case 180hs uses 2×2 tiles of the roughness pattern scaled by a factor of 1/2 (half-scale: hs).

case	Re_τ	k/δ	k^+	N_x	N_y	$\Delta z^+_{\min}$	$\Delta z^+_{\max}$
180hs	180	0.0634	11.8	1536	768	0.667	2.58
180fs	180	0.1267	22.8	768	384	0.667	2.58
270fs	270	0.1267	34.2	768	384	0.667	2.92
395fs	395	0.1267	50.0	768	384	0.667	3.03
540fs	540	0.1267	68.4	864	432	0.667	2.58
720fs	720	0.1267	91.2	960	480	0.667	2.54

Direct numerical simulations of rough-wall turbulent channel flow driven by a constant mean streamwise pressure gradient were conducted using the in-house code iIMB [10] which employs an iterative version of the embedded boundary method of Yang & Balaras [11] to resolve the roughness features. The roughness was applied to both the lower and the upper wall of the channel (see Fig. 1); periodic boundary conditions were used in the streamwise and spanwise direction. A small wall-normal offset, $z_0/\delta = -0.0049$ was applied to the barnacle surface, so that the roughness mean plane $\langle h(x,y)\rangle$, where $h(x,y)$ is the heightmap, is located at $z = 0$.

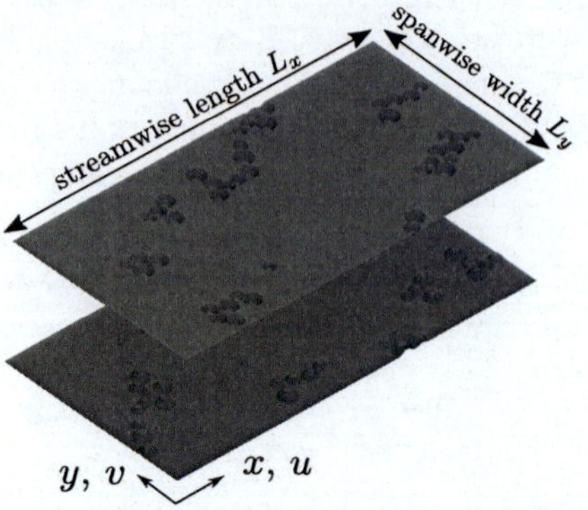

Fig. 1. Schematic illustration of the channel flow domain. The rough surface is applied to both walls of the channel. On the upper wall, the rough surface pattern is shifted to minimize local blockage effects.

Key simulation parameters are summarised in Table 2. Uniform grid spacing was used in the streamwise and spanwise direction of the flow; in the wall-normal direction, uniform spacing was set across the height of the roughness, above which the grid was stretched, reaching its maximum wall-normal spacing at the channel centre. The roughness Reynolds number $k^+ = k/\ell_d$, where k is set to the maximum barnacle height and ℓ_d is the viscous length scale of the flow, was varied using the method of Thakkar et al. [12]: For high k^+ the friction Reynolds number Re_τ of the channel flow was varied, while for low k^+ the friction Reynolds number was kept fixed and the size of the roughness features in outer units k/δ, where δ is the channel half-height, was decreased using a 'scaling and tiling' approach. In all cases, a domain length $L_x = 2\pi\delta$ and width $L_y = \pi\delta$ was used. Similar conditions were employed in the previous investigation on the Gaussian rough surface where simulations were conducted at $Re_\tau = 180, 240, 360,$ and 540 using a domain length of $L_x = 6\delta$ and width $L_y = 3\delta$ (for full details see [6]). The intrinsic average was used for the computation of the mean flow and turbulence statistics; a triple decomposition of the velocity field was applied to separate Reynolds from dispersive stresses.

3 Results

The Hama roughness function ΔU^+, i.e., the downwards shift in the mean streamwise velocity profile compared to smooth-wall conditions, is shown in Fig. 2. Fully rough behaviour is approached at the highest Reynolds number and the equivalent sand grain roughness k_s corresponds to approximately $0.5k$. Compared to Nikuradse's [13] sand-grain roughness a more gradual increase in ΔU^+ can be observed over the transitionally rough region with the barnacle surface showing a trend that is close to Colebrook formula [14] approaching the fully rough asymptote from above; in contrast, the Gaussian roughness shows a more Nikuradse like-behaviour with a steeper increase in ΔU^+ in the upper transitionally rough region. For the Gaussian surface, the streamwise Reynolds

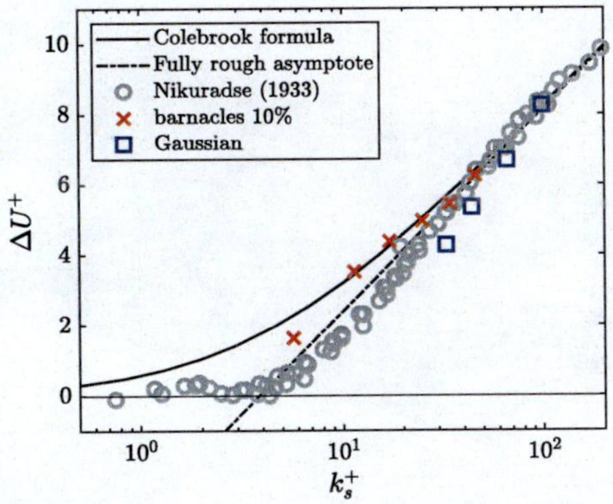

Fig. 2. Roughness function ΔU^+ versus k_s^+ for the barnacle surface with 10% coverage. For comparison, data for a Gaussian roughness [6], sand grain roughness data by Nikuradse [13] and Colebrook's empirical formula [14] are shown.

stress (shown in Fig. 3) decreases with increasing Re_τ. For the barnacle surface, a similar decrease can be observed at low Reynolds numbers, but $\langle u'u' \rangle$ levels do not reduce further for $Re_\tau = 270$ and above. For the higher Reynolds number cases, the streamwise Reynolds stress profiles display a 'roughness' peak which is associated with the crests of the barnacle features. In addition, a smooth-wall like 'inner' peak emerges, which falls into the buffer layer relative to the smooth surface. No such feature can be observed for the Gaussian rough surface.

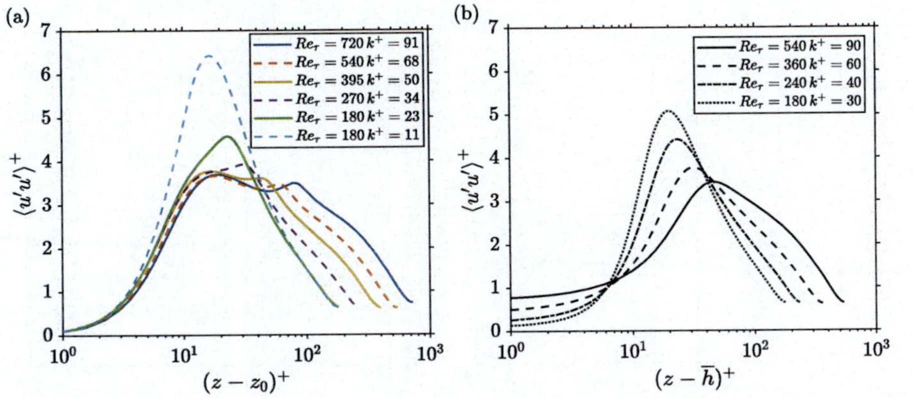

Fig. 3. Streamwise Reynolds stress profiles. (a) Barnacle 10% surface; (b) Gaussian roughness

To look into the reasons for the emergence of the inner peak, a simple geometric decomposition is applied to compute the contribution to the streamwise Reynolds stress profiles that arise above the barnacle features ('rough patches') and above the surrounding smooth surface sections ('smooth patches') by defining a corresponding

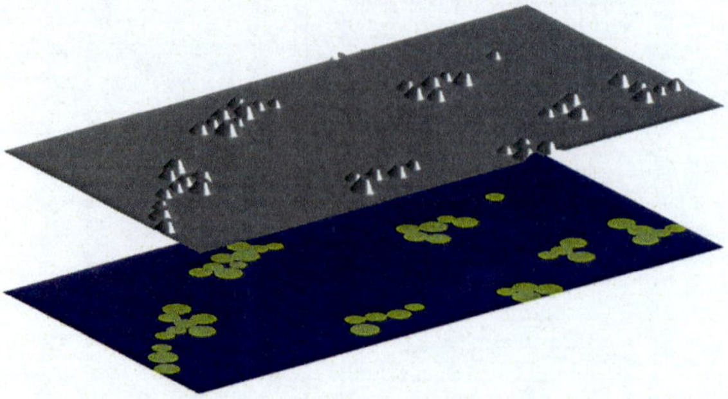

Fig. 4. Illustration of the mask used for the decomposition of the streamwise Reynolds stress profiles. Areas in blue correspond to 'smooth' patches and areas coloured in yellow to 'rough' patches.

mask $\psi(x,y)$ (see Fig. 4). Results for the decomposed profiles are shown in Fig. 5. It is evident that the inner peak arises from the smooth-wall patches, indicating a partial recovery of smooth-wall behaviour over the large connected smooth sections of this surface. This inner peak also appears to become Reynolds-number independent at the highest three Re_τ values investigated. In contrast, above the rough patches, only the outer peak can be observed, which consistently is located around the maximum barnacle height. A test of the decomposition performed on data of flow over barnacle surfaces with increasing coverage [5], shows that partial recovery of smooth-wall behaviour can only be observed at low coverage states (not shown).

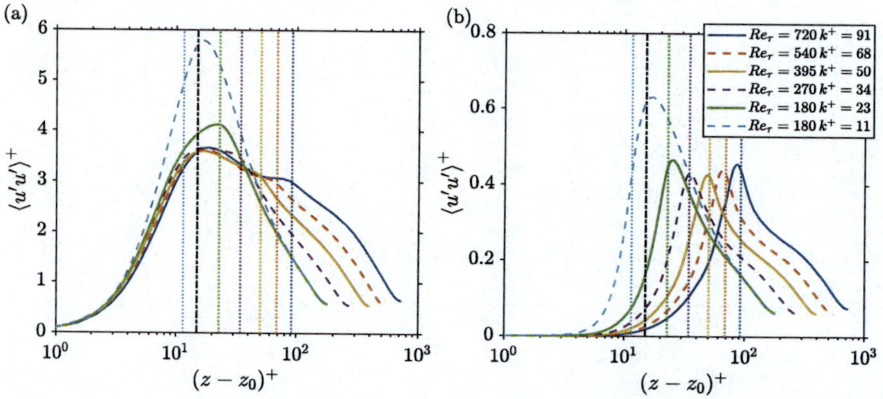

Fig. 5. Decomposed streamwise Reynolds stress profiles: (a) contribution above smooth patches; (b) contributions above rough patches. The coloured dotted vertical lines indicate the maximum barnacle height. The vertical black dash-dotted line indicates the approximate peak location of the smooth-wall Reynolds stress profile.

4 Conclusions

A surface lightly fouled by barnacles was investigated using direct numerical simulations. A comparison to data for a generic Gaussian roughness shows clear differences in the Reynolds number dependency, with the barnacle surface exhibiting a Colebrook-like behaviour whereas the Gaussian rough-surface is closer to Nikuradse's data. A characteristic feature that emerges in the $\langle u'u' \rangle$ profiles for the barnacle case is an inner peak. Using a simple geometric decomposition, this was shown to arise from a partial recovery of the smooth-wall behaviour over the blank patches. No inner-peak features were observed for the $\langle v'v' \rangle$ and $\langle w'w' \rangle$ profiles, which can be attributed to the fact that the smooth-wall peak locations of these profiles occur at significantly higher wall-normal locations.

Acknowledgements. A. B. gratefully acknowledges support via a Leverhulme Trust Research Fellowship. This work used the ARCHER UK National Supercomputing Service (http://www.archer.ac.uk).

References

1. International Maritime Organisation: Fourth IMO GHG Study 2020 - Full Report (2020)
2. Lindholdt, A., Dam-Johansen, K., Olsen, S.M., et al.: Effects of biofouling development on drag forces of hull coatings for ocean-going ships: a review. J. Coat. Technol. Res. **12**, 415–444 (2015)
3. Schultz, M.P.: Effects of coating roughness and biofouling on ship resistance and powering. Biofouling **23**(5), 331–341 (2007)
4. Schultz, M.P., Bendick, J.A., Holm, E.R., Hertel, W.M.: Economic impact of biofouling on a naval surface ship. Biofouling **27**, 89–98 (2011)
5. Sarakinos, S., Busse, A.: Investigation of rough-wall turbulence over barnacle roughness with increasing solidity using direct numerical simulations Phys. Rev. Fluids **7**(6), 064602 (2022)
6. Jelly, T.O., Busse, A.: Reynolds number dependence of Reynolds and dispersive stresses in turbulent channel flow past irregular near-Gaussian roughness. Int. J. Heat Fluid Flow **80**, 108485 (2019)
7. Sarakinos, S., Busse, A.: An algorithm for the generation of biofouled surfaces for applications in marine hydrodynamics. In: Ferrer, E., Montlaur, A. (eds.) Recent Advances in CFD for Wind and Tidal Offshore Turbines. STME, pp. 61–71. Springer, Cham (2019). https://doi.org/10.1007/978-3-030-11887-7_6
8. Sadique J.: Turbulent flows over macro-scale roughness elements - from biofouling barnacles to urban canopies, Doctoral dissertation, Johns Hopkins University, Baltimore (2016)
9. Chung, D., Hutchins, N., Schultz, M.P., Flack, K.A.: Predicting the drag of rough surfaces. Ann. Rev. Fluid Mech. **53**, 438–471 (2021)
10. Busse, A., Lützner, M., Sandham, N.D.: Direct numerical simulation of turbulent flow over a rough surface based on a surface scan. Comput. Fluids **116**, 1290147 (2015)
11. Yang, J., Balaras, E.: An embedded-boundary formulation for large-eddy simulation of turbulent flows interacting with moving boundaries. J. Comput. Phys. **215**, 12–40 (2006)
12. Thakkar, M., Busse, A., Sandham, N.D.: Direct numerical simulation of turbulent channel flow over a surrogate for Nikuradse-type roughness. J. Fluid Mech. **837**, R1 (2018)
13. Nikuradse, J.: Strömungsgesetze in rauhen rohren. VDI Forschungsheft **361**, 1–22 (1933)
14. Colebrook, C.F.: Turbulent flow in pipes with particular reference to the transition region between the smooth- and rough-pipe laws. J. Inst. Civil Eng. **11**, 133–156 (1939)

A Turbulent Plume in Crossflow

Daniel Fenton[1], Andrea Cimarelli[2], Jean-Paul Mollicone[3], Maarten van Reeuwijk[4], and Elisabetta De Angelis[1,5(✉)]

[1] Cardiff University, Cardiff CF24 3AA, UK
FentonD@cardiff.ac.uk
[2] Università di Modena e Reggio Emilia, Modena 41125, Italy
[3] University of Malta, Msida 2080, Malta
[4] Imperial College London, London SW7 2AZ, UK
[5] Università di Bologna, Forlì, Italy
e.deangelis@unibo.it

Abstract. The behaviour of a turbulent forced buoyant plume subjected to uniform crossflow is investigated utilising Direct Numerical Simulation (DNS) employing a fourth-order finite difference scheme and third-order Adams-Bashforth temporal integration. The flow features are assessed phenomenologically in the statistically steady state obtained by averaging 1,200 instantaneous 3D fields. Preliminary results on the structure of turbulent production and dissipation are shown in view of future discussions on LES modeling assumptions.

Keywords: plumes · direct numerical simulation

1 Introduction

Buoyancy driven flows are highly abundant in both industry and nature, with examples ranging from volcanic ash clouds to pollutant dispersion. Their prevalence and importance has led to great interest across many fields, from disaster management to the impact of industry on the climate. Despite decades of research, their often turbulent nature leaves their behaviour still difficult to understand and predict. In the present study, the behaviours of turbulent buoyant forced plumes subjected to uniform crossflow are investigated by means of Direct Numerical Simulation (DNS), building on the infinitely lazy plume with zero source momentum studied by Jordan et al [1], utilising the solver SPARKLE [2]. The present simulation closely resembles one of the cases studied by Jordan et al., in order that the effect of plume forcing can be inferred by comparison.

2 Statement of the Problem

The subject of the present study is that of the buoyant plume exiting from a circular source with initial velocity w_0 in the vertical z-direction subject to a uniform crossflow $\mathbf{U} = (U, 0, 0)$ in the streamwise x-direction. The governing

C. Marchioli et al. (Eds.): DLES 2023, ERCO 31, pp. 356–361, 2024.
https://doi.org/10.1007/978-3-031-47028-8_54

equations for such a flow are the Navier-Stokes equations under the Boussinesq approximation, given by

$$\boldsymbol{\nabla} \cdot \mathbf{u} = 0 \tag{1}$$

$$\frac{\partial \mathbf{u}}{\partial t} + \mathbf{u} \cdot \boldsymbol{\nabla} \mathbf{u} = -\boldsymbol{\nabla} p + \nu \nabla^2 \mathbf{u} + b\hat{\mathbf{k}} \tag{2}$$

$$\frac{\partial b}{\partial t} + \mathbf{u} \cdot \boldsymbol{\nabla} b = \kappa \nabla^2 b \tag{3}$$

where $\mathbf{u} = (u, v, w)$ is the fluid velocity and $b = g(\rho_0 - \rho)/\rho_0$ is the buoyancy field, with ρ_0 a constant reference density and g the gravitational acceleration. The kinematic pressure perturbation is given by $p = \tilde{p}/\rho_0 + gz$ with the standard pressure $\tilde{p}$. Kinematic viscosity and thermal diffusivity are given respectively by ν and κ.

3 Direct Numerical Simulation

The simulation was performed utilising the DNS code SPARKLE [2], which spatially employs a fourth-order symmetry-preserving central difference method, and integrates temporally via a third-order Adams-Bashforth scheme. The domain dimensions $L_x \times L_y \times L_z$ were chosen to be large enough for the plume to fully evolve to turbulence inside the domain with no boundary interactions [3]. The boundary conditions were chosen to be periodic in the horizontal x and y directions, and free-slip on the vertical boundaries $z = 0$ and $z = L_z$, with the exception of a constant top-hat velocity function $\mathbf{u}_0 = (0, 0, w_0)$ imposed on the plume source $\mathbf{x_0}$ with radius r_0. Neumann boundary conditions were imposed on the buoyancy at $z = 0$ and $z = L_z$, again with the exception of the plume core, which possessed a constant top-hat buoyancy b_0 as per the inflow velocity.

In order to enforce the uniform inflow in the x-direction that serves as the crossflow source under the periodic boundary conditions, a nudging region of length $L_n = 4r_0$ was introduced at the end of the domain from $x = L_x - L_n$. In this region, the velocity was gradually reduced to that of the ambient field by setting the velocity to $\mathbf{u}^* = (1 - x^*/L_n)\mathbf{u} + (x^*/L_n)\mathbf{U}$, where x^* is the distance downstream from the beginning of the nudging region - similar treatment was performed on the buoyancy to reduce it to a null field at the inflow. It is evident that therefore any dynamics in this region do not represent the physical flow and are omitted from the analysis of the results. The flow fields were initialised such that they were those of the ambient crossflow $\mathbf{U}$.

The non-dimensional parameters that define the plume in crossflow naturally must be formed by the plume source and crossflow velocities w_0 and U, the source buoyancy b_0 and the viscous and thermal diffusivities ν and κ. The five relevant non-dimensional parameters are

$$R_0 = \frac{w_0}{U},\ Ri_0 = \frac{b_0 r_0}{w_0^2},\ Ri_U = Ri_0 R_0^3,\ Re_0 = \frac{2r_0 w_0}{\nu},\ Pr = \frac{\nu}{\kappa} \tag{4}$$

where R_0 is the source-to-crossflow ratio of velocities, Ri_0 the source Richardson number, Ri_U the crossflow Richardson number, Re_0 the source Reynolds number and Pr the Prandtl number. These are reported for the present simulation in Table 1, along with the domain and grid dimensions. The plume source was centred at $\mathbf{x}_0/r_0 = (5, 12, 0)$, to fall in the middle of the domain in the y-direction, and was far enough downstream from the periodic boundary in the x-direction to avoid boundary interactions. Time averaging was performed in post-processing, utilising approximately 1,200 fully 3D snapshots of the entire domain taken once the initial transient behaviour had ended.

Table 1. Simulation parameters.

$(L_x \times L_y \times L_z)/r_0^3$	$N_x \times N_y \times N_z$	R_0	Ri_0	Ri_U	Re_0	Pr
$28 \times 24 \times 24$	$1350 \times 744 \times 744$	1.0	1.0	1.0	1000	1.0

4 Instantaneous and Mean Flow

Figure 1 is a representative snapshot of the instantaneous pressure p and enstrophy fields ω^2, where $\boldsymbol{\omega} = \nabla \times \mathbf{u}$. In the left plot, the iso-surface at a very low value of enstrophy demonstrates the characteristic topology of the Turbulent/Non-Turbulent Interface (TNTI) showing the plume expansion as it is swept downstream by the crossflow. In the right plot, tube-like Intense Vorticity Structures

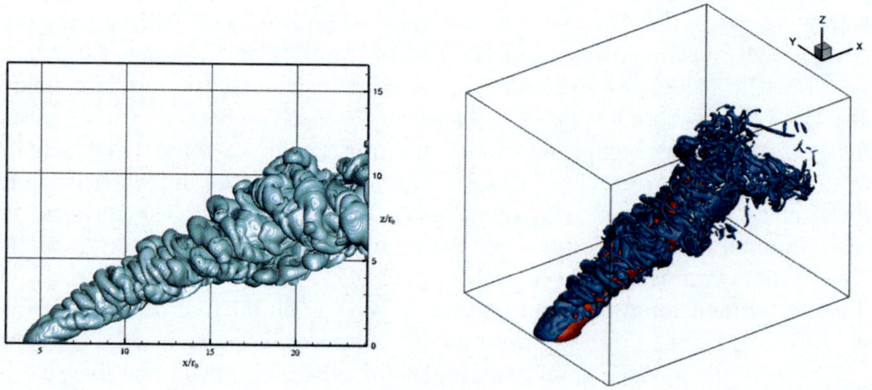

Fig. 1. 3D snapshots of the instantaneous flow fields. Left: 1% enstrophy threshold, demonstrating the turbulent non-turbulent interface. Right: enstrophy ω^2 (blue) overlaid with pressure (red), demonstrating the difference between IVSs and LVSs, respectively.

(IVSs) in blue, visualized by the iso-surface of a large enstrophy value, demonstrate the tell-tale signs of turbulent flow developing, almost immediately out of the plume source. Two Large Vorticity Structures (LVSs) are indeed shown by a iso-surface of the pressure (red).

In order to study the general features of the flow, as customary, the Reynolds decomposition of the velocity is introduced, $\mathbf{u} = \bar{\mathbf{u}} + \mathbf{u}'$, pressure $p = \bar{p} + p$ and buoyancy $b = \bar{b} + b'$, in which a prime denotes the fluctuating component and an overbar the temporal average over the available three-dimensional fields.

As the transported property, the buoyancy field allows one to easily identify the fluid within the plume. Cross sections in the y-z plane are shown in Fig. 2 for two positions along the plume. At the downstream edge of the plume source $x/r_0 = 6$, one can see the double roll structures already begin to form. In the turbulent region at $x/r_0 = 20$ we see these have concentrated into a pair of core structures symmetrical about the centreline, connected by a weak buoyancy layer across the top of the plume, which is bounded by the pressure induced by the interaction of the crossflow with the plume as it is bent over.

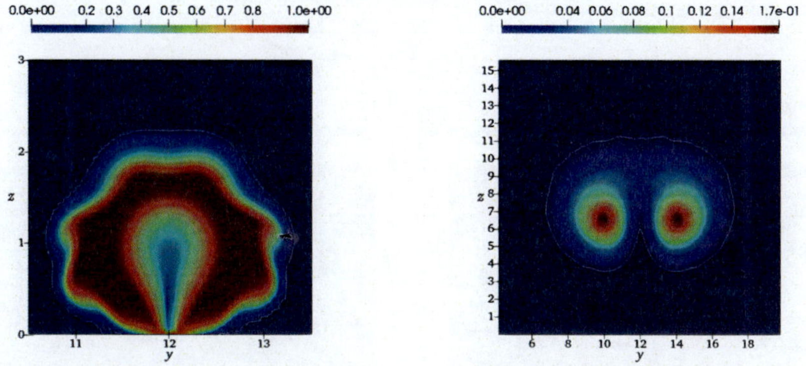

(a) $x/r_0 = 6$.

(b) $x/r_0 = 20$.

Fig. 2. Average buoyancy $\bar{b}$ in the y-z plane (a) on the downstream source edge, (b) within the fully turbulent region, outlined by the 1% threshold of the maximum buoyancy on each slice.

5 Turbulent Energy

Turbulence kinetic energy $k = u_i' u_i'/2$ measures the kinetic energy associated with the fluctuating part of the flow. The balance equation reads

$$\frac{\overline{D}}{Dt}\overline{k} = -\frac{1}{\rho}\frac{\partial \overline{u_i' p'}}{\partial x_i} - \frac{\partial \overline{k u_i'}}{\partial x_i} + \nu \frac{\partial^2 \overline{k}}{\partial x_j^2} + \overline{u_i' b'}\delta_{i3} - \overline{u_i' u_j'}\frac{\partial \overline{u}_i}{\partial x_j} - \nu \overline{\frac{\partial u_i'}{\partial x_k}\frac{\partial u_i'}{\partial x_k}} \tag{5}$$

where $\overline{D}/Dt$ represents the mean-flow material derivative, where the convective term is calculated using the mean velocity $\overline{u}_i$. The first term on the right hand side represents spatial transport due to pressure, the second and third terms represent the turbulent and viscous transport respectively. The fourth accounts for turbulent kinetic energy production due to buoyancy. The final two quantities fundamental for understanding the transfer of turbulent kinetic energy within the flow are the turbulence production,

$$\Pi = \overline{u_i' u_j'} \frac{\partial \bar{u}_i}{\partial x_j}, \tag{6}$$

and the viscous dissipation rate of k,

$$\epsilon = \nu \overline{\frac{\partial u_i'}{\partial x_j} \frac{\partial u_i'}{\partial x_j}}. \tag{7}$$

The turbulence production Π acts as a source of k when negative and a sink when positive, and describes the transfer between mean and turbulent energy, whereas ϵ is by definition positive in sign, and thus always acts as a sink of k.

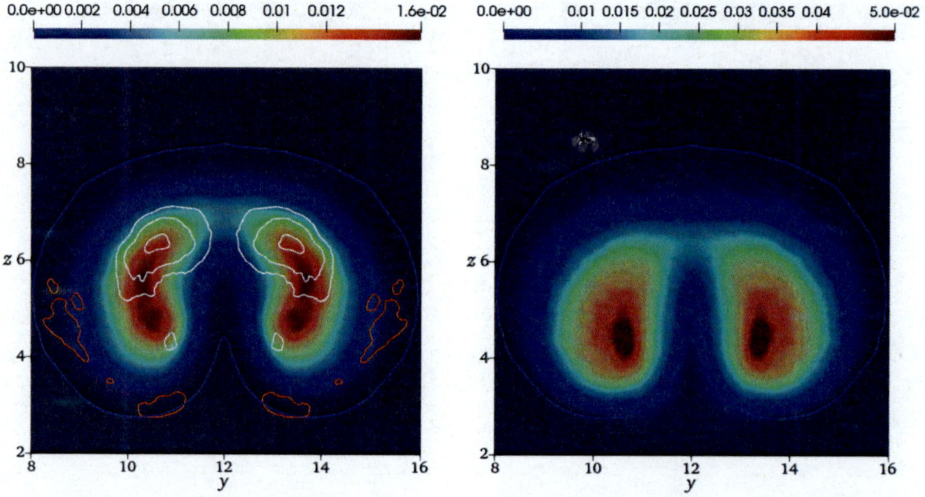

(a) k and Π (b) ϵ.

Fig. 3. Turbulence kinetic energy and contributions around the plume source in the y-z plane in the turbulent region at $x = 15$ outlined by the 1% $\bar{b}$ threshold in this plane. (a) Iso-contour of total k overlaid with contours of source turbulence production (white) and sink turbulence production (red). (b) Dissipation rate ϵ

From Fig. 3a, it is shown that the turbulent kinetic energy peaks remain confined to the centre of the two LVSs that have been identified by the pressure in Fig. 1. It is also possible to observe regions of turbulence production acting as a source of k (white contours) towards the upper edge of the plume, where it interacts most strongly with crossflow. Furthermore, Fig. 3b shows that there is considerable dissipation in the lower regions of the LVSs, where, as shown Fig. 1, the eddies break down to form smaller and smaller structures.

Figure 3a also demonstrates sinks of turbulence kinetic energy, represented by the regions outlined in red. They indicate that energy is being transferred from the turbulence back into the mean flow in these regions. While these sinks are smaller than the sources in magnitude they are not entirely negligible, with a maximum value of approximately 4% of the maximum source. It is worth mentioning that the statistical analysis of the this flow presents a number of challenges owing to the lack of homogeneous geometrical directions. Nonetheless, as shown by the quality of the plots in Fig. 3, the large number of fully 3D fields collected and used for temporal averaging allows a statistical convergence even in second order and derivative quantities.

6 Conclusions and Perspectives

The present work has discussed the first results obtained by the analysis of a large DNS dataset to study turbulence dynamics in a forced buoyant plume in crossflow. Instantaneous flow fields demonstrate the structure of the TNTI, the IVSs and the LVSs, and the time-averaged buoyancy field gives the mean structure of the bent over plume. Future work aims at investigating the full three-dimensional behaviour of the turbulent kinetic energy budget and at performing *a priori* analysis by filtering the DNS results for assessing Large Eddy Simulation models for buoyancy driven flows with Turbulent/Non-Turbulent interfaces.

Acknowledgements. The present research has been supported by the Engineering and Physical Sciences Research Council of the United Kingdom through the Grant *Multi-scale dynamics at the turbulent/non-turbulent interface of jets and plumes* EP/R042640/1.

References

1. Jordan, O.H., Rooney, G.G., Devenish, B.J., van Reeuwijk, M.: Under pressure: turbulent plumes in a uniform crossflow. J. Fluid Mech. **932**, A47 (2022). https://doi.org/10.1017/jfm.2021.1001
2. Craske, J., van Reeuwijk, M.: Energy dispersion in turbulent jets. Part 1. Direct simulation of steady and unsteady jets. J. Fluid Mech. **763**, 500–537 (2015). https://doi.org/10.1017/jfm.2014.640
3. Rooney, G.G.: Merging of a row of plumes or jets with an application to plume rise in a channel. J. Fluid Mech. **771**, R1 (2015). https://doi.org/10.1017/jfm.2015.169

Direct Numerical Simulations of Turbulent Flow in Helical Pipes

V. Lupi[1(✉)], R. Örlü[1,2], and P. Schlatter[1,3]

[1] FLOW/SimEx, KTH Engineering Mechanics, SE-100 44 Stockholm, Sweden
lupi@mech.kth.se

[2] Dept. Mechanical, Electrical and Chemical Engineering, OsloMet – Oslo Metropolitan University, 0166 Oslo, Norway

[3] Institute of Fluid Mechanics (LSTM), Friedrich-Alexander-Universität (FAU) Erlangen-Nürnberg, 91058 Erlangen, Germany

Abstract. Direct numerical simulations of the fully developed turbulent flow through helical pipes are performed. The numerical procedure is described, and a validation of the volume force driving the flow is presented. A comparison of the turbulence statistics against literature data is also reported.

1 Introduction

The flow in straight pipes has been studied extensively since the 19th century [9], and therefore it is commonly referred to as a *canonical* flow. It is both axisymmetric and homogeneous in the streamwise direction. Note that these properties should be interpreted in a statistical sense for turbulent flows. For the flow in toroidal pipes, the axial symmetry does not hold anymore, although symmetry with respect to the equatorial plane can be observed. A secondary motion of Prandtl's first kind arises because of the curvature. It can be observed for any non-zero curvature [1], and it increases the cross-plane mixing and heat transfer. On the other hand, no symmetry can be retrieved if the pipe has non-zero torsion, *i.e.* its centreline is not a planar curve, as in the case of the flow in a helical pipe.

Curved pipe flows are ubiquitous in several industrial applications and biological systems. Amongst them, toroidal pipes serve as a more idealised case to study the effects of curvature alone [1,2,8], whereas helical pipes represent a more realistic geometry for applications. The parameters governing the flow in helical pipes are the curvature δ, the torsion τ and the bulk Reynolds number Re_b, based on the bulk velocity U_b, the diameter of the pipe cross-section $D = 2R_p$ and the kinematic viscosity ν. The non-dimensional geometrical parameters are defined as

$$\delta = \frac{R_c R_p}{R_c^2 + p_s^2}, \qquad \tau = \frac{p_s R_p}{R_c^2 + p_s^2}, \tag{1}$$

where R_p, R_c and p_s are the quantities shown in Fig. 1.

To the best of the authors' knowledge, the only direct numerical simulations of turbulent helical pipe flows present in the literature are those performed by Hüttl &

C. Marchioli et al. (Eds.): DLES 2023, ERCO 31, pp. 362–367, 2024.
https://doi.org/10.1007/978-3-031-47028-8_55

Friedrich [5,6], who solved numerically the Navier–Stokes equations in the orthogonal helical coordinate system introduced by Germano [4], driving the flow with a mean pressure gradient in the streamwise direction. In the present work, the flow is driven by a volume force that mimics the effect of a pressure gradient in the streamwise direction, in the same fashion as Canton et al. [1]. A validation of the numerical implementation of the forcing is proposed, together with a comparison of the turbulence statistics obtained with the current methodology against the results present in the literature.

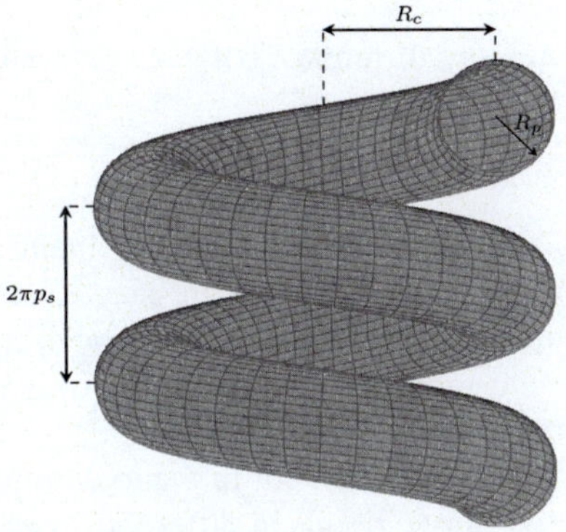

Fig. 1. Sketch of the helical pipe indicating the relevant geometrical quantities.

2 Numerical Method

The governing equations for the viscous, incompressible flow of a Newtonian fluid in a helical pipe are the incompressible Navier–Stokes equations, which read

$$\frac{\partial \mathbf{u}}{\partial t} + (\mathbf{u} \cdot \nabla)\mathbf{u} = -\nabla p + \frac{1}{Re_b} \nabla^2 \mathbf{u} + \mathbf{f}, \tag{2a}$$

$$\nabla \cdot \mathbf{u} = 0, \tag{2b}$$

where **f** is the volume force employed to drive the flow, as described in Sect. 2.1. The equations are formulated in Cartesian coordinates to avoid the treatment of the singularity at the pipe centreline, and the spatial discretisation is carried out using the spectral element method (SEM) implemented in the open source code `Nek5000` [3]. Within each element, Lagrangian interpolants of order N built on Gauss–Lobatto–Legendre (GLL) nodes are used to represent the velocity, whereas the pressure is expressed as a linear combination of Lagrangian basis functions of order $N-2$ based on Gauss–Legendre (GL) grid points, following the so-called $\mathbb{P}_N - \mathbb{P}_{N-2}$ formulation. The semi-implicit BDF3/EXT3 scheme is employed for the integration in time. A third-order backward differentiation formula (BDF3) is used to discretise the equations, and an extrapolation scheme of order three (EXT3) is then applied to express the non-linear term at the new time step.

2.1 Driving Force

The most practical way to drive the flow in a pipe or channel, whether straight or bent, is by applying a pressure gradient between the two far ends. Indeed, this approach is mainly used in experimental investigations [11]. Another valuable method is to prescribe a volume force that mimics the effect of the pressure gradient. This procedure has been used by Canton et al. [1] for the flow in a toroidal pipe. Indeed, in this case, a prescribed pressure gradient would cause a pressure discontinuity after a complete revolution.

In a straight pipe, the pressure difference between two points is

$$\Delta p = \frac{\mathrm{d}P}{\mathrm{d}s}\Delta s, \tag{3}$$

since the pressure depends only on the streamwise coordinate s. For a curved geometry, *e.g.* a bent channel, it can be shown that the pressure also varies with the distance from the centre of curvature [1]. However, when considering two points at different streamwise stations but at the same distance from the axis of revolution of the bent channel, the pressure difference can be computed using Eq. 3. Therefore, a volume force resembling the effect of the pressure gradient in a curved pipe needs to be formulated such that the difference between two points at different streamwise locations but at the same position in the cross-section depends only on Δs. For a toroidal pipe, this leads to a volume force that acts only in the streamwise direction and is inversely proportional to the distance from the axis of the torus [1]. For a helical pipe, following the same reasoning, one can find the following expression for the volume force

$$\mathbf{f} = \frac{F}{h_s}\mathbf{e}_s, \tag{4}$$

where F is a scalar constant and h_s is the scale factor for the streamwise coordinate s of the orthogonal helical reference system $\{s, r, \theta\}$. This formulation is also employed in the analytical work by Kumar [7].

2.2 Validation

In order to validate the proposed forcing, non-linear direct numerical simulations (DNS) of the flow in helical pipes with the same governing parameters Re_b, δ and τ, but with different boundary conditions at the two far ends, are performed. In one case, an inflow velocity profile is prescribed at the inlet, and the stress-free boundary condition is imposed at the outlet. This configuration is referred to as *inflow-outflow*. In the other case, periodic boundary conditions are applied between the two far ends, taking care of the rotation of the velocity components if the outward normal has a different direction on the inlet and outlet planes, and the volume forcing described above is used to drive the flow. For both cases, no-slip and impermeability conditions are applied at the wall. The validation is carried out for $Re_b = 100$, $\delta = 0.698$, $\tau = 0.267$. In the *inflow-outflow* case, the computational domain is a helical pipe with a centreline $32R_p$ long, whereas the length of the helix centreline is only $0.2R_p$ in the periodic case. Indeed, a development length needs to be considered in the *inflow-outflow* simulation to let the

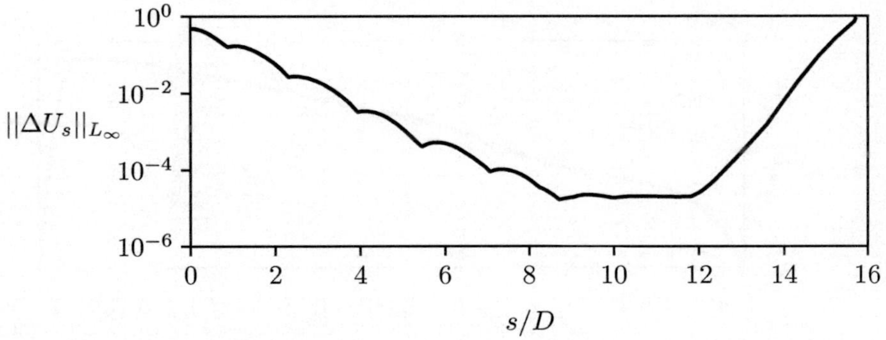

Fig. 2. L_∞-norm of the difference in the streamwise velocity component between the *inflow-outflow* case and the periodic one.

flow become homogeneous in the streamwise direction. It needs to be remarked that the flow is homogeneous along the s_w-coordinate lines of the non-orthogonal reference system proposed by Wang [10].

The velocity components at different streamwise locations of the *inflow-outflow* case are compared with those at the final cross-section of the periodic case. The L_∞-norm of the difference in the streamwise velocity component U_s between the two cases is shown in Fig. 2 as a function of the streamwise coordinate. At $s/D \approx 9$, the difference reaches its minimum, suggesting that the value of the error is sufficiently low to indicate that a fully developed state has been reached. The difference increases for $s/D \geq 12$ because of the influence of the outflow boundary condition.

3 Fully Developed Turbulent Flow

Validation of the forcing procedure for a fully developed turbulent flow is carried out by performing a DNS at $Re_b = 5613$, $\delta = 0.1$, $\tau = 0.11$ and comparing statistical quantities averaged both in time and along the s_w-coordinate lines with those obtained by Hüttl & Friedrich [6] for the same governing parameters. The no-slip and no-penetration conditions are prescribed at the wall, whereas periodic boundary conditions are imposed in the streamwise direction since the flow is driven by the volume force described in Sect. 2.1. A helical pipe with a centreline $25R_p$ long is used as the computational domain, more than 50% longer than the one used by Hüttl & Friedrich [6]. Moreover, the first off-wall point is at $r^+ \approx 0.7$, and the grid spacing in the streamwise direction is almost half of that in the reference.

The mean velocity profiles are in good agreement, as shown in Fig. 3. However, a poorer agreement is observed for the second-order moments, as displayed in Fig. 4. In particular, the present study finds a peak of the turbulent kinetic energy at the outer wall of the pipe, which is much less pronounced in the work by Hüttl & Friedrich [6]. It is worth noting that the outer wall of the pipe is the least resolved region since the mesh is conforming and the viscous length scale is smaller because of the higher friction velocity. For this reason, it is argued that the disagreement in the results is linked to a

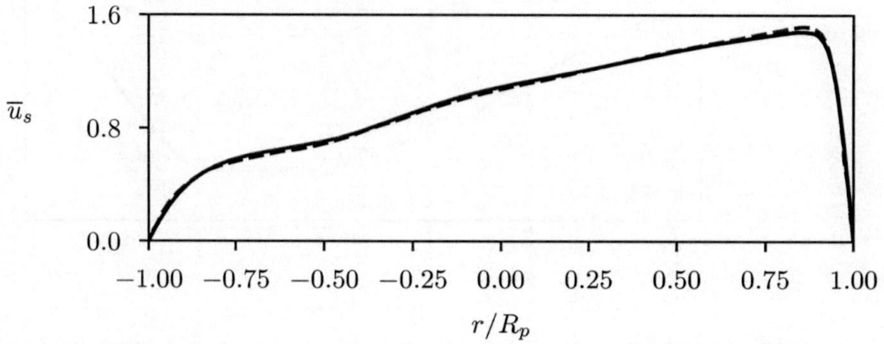

Fig. 3. Profile of the mean streamwise velocity component $\overline{u}_s$. (——) Present study, (- - - -) Hüttl & Friedrich [6].

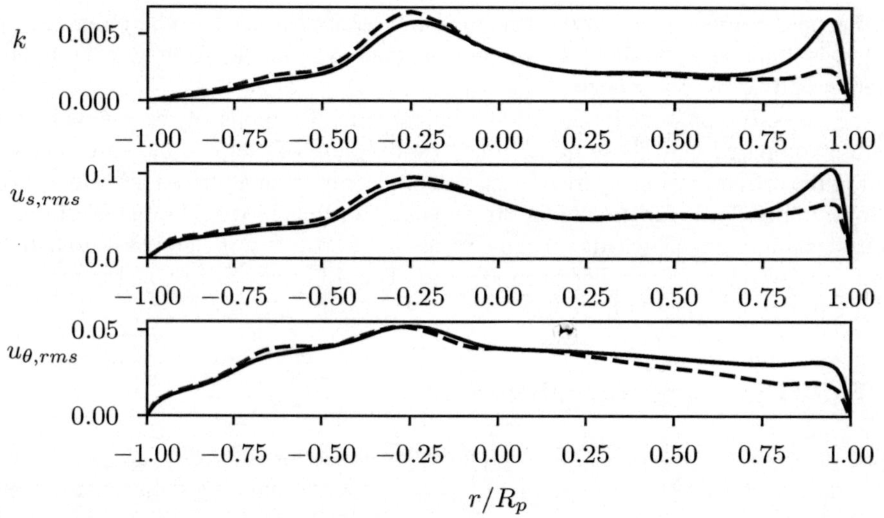

Fig. 4. Profiles of (*top*) turbulent kinetic energy and root-mean-square (rms) values of (*middle*) streamwise and (*bottom*) azimuthal velocity components. (——) Present study, (- - - -) Hüttl & Friedrich [6].

lack of resolution at the outer wall of the pipe in the literature data, most likely due to a limitation in the available computational power. Note also that Hüttl & Friedrich [6] discretised the equations using a second-order finite volume method, whereas spectral accuracy is obtained in the present study.

For very high values of the Reynolds number, the resolution requirements at the outer wall might prohibitively increase the computational cost. A possible strategy to tackle this issue is the usage of a non-conforming mesh with a higher resolution where required.

4 Conclusions and Outlook

Validation of a DNS of the turbulent flow in a helical pipe is presented and compared to the only available previous study. Good agreement is generally found while discrepancies are observed in high-order moments, ascribed to the better resolution of the present simulation. Future work aims at investigating turbulence characteristics at higher Reynolds numbers and their variation with the torsion τ. Coherent structures will also be extracted through modal decomposition techniques.

References

1. Canton, J., Örlü, R., Schlatter, P.: Characterisation of the steady, laminar incompressible flow in toroidal pipes covering the entire curvature range. Int. J. Heat Fluid Flow **66**, 95–107 (2017)
2. Canton, J., Schlatter, P., Örlü, R.: Modal instability of the flow in a toroidal pipe. J. Fluid Mech. **792**, 894–909 (2016)
3. Fischer, P.F., Lottes, J.W., Kerkemeier, S.G.: `Nek5000`: open source spectral element CFD solver (2008). https://nek5000.mcs.anl.gov/
4. Germano, M.: On the effect of torsion on a helical pipe flow. J. Fluid Mech. **125**, 1–8 (1982)
5. Hüttl, T.J., Friedrich, R.: Influence of curvature and torsion on turbulent flow in helically coiled pipes. Int. J. Heat Fluid Flow **21**(3), 345–353 (2000)
6. Hüttl, T.J., Friedrich, R.: Direct numerical simulation of turbulent flows in curved and helically coiled pipes. Comput. Fluids **30**(5), 591–605 (2001)
7. Kumar, A.: Pressure-driven flows in helical pipes: bounds on flow rate and friction factor. J. Fluid Mech. **904**, A5 (2020)
8. Noorani, A., El Khoury, G.K., Schlatter, P.: Evolution of turbulence characteristics from straight to curved pipes. Int. J. Heat Fluid Flow **41**, 16–26 (2013)
9. Reynolds, O.: An experimental investigation of the circumstances which determine whether the motion of water shall be direct or sinuous, and of the law of resistance in parallel channels. Philos. Trans. R. Soc. **174**, 935–982 (1883)
10. Wang, C.Y.: On the low-Reynolds-number flow in a helical pipe. J. Fluid Mech. **108**, 185–194 (1981)
11. Yamamoto, K., Akita, T., Ikeuchi, H., Kita, Y.: Experimental study of the flow in a helical circular tube. Fluid Dyn. Res. **16**(4), 237–249 (1995)

Check for updates

Wall-Modelled Large-Eddy Simulations of Flows with Non-uniform Roughness

T. Salomone[1,2](✉), U. Piomelli[1], and G. De Stefano[2]

[1] Queen's University, Kingston, ON, Canada
{teresa.salomone,ugo}@queensu.ca
[2] University of Campania Luigi Vanvitelli, Aversa, Italy
{teresa.salomone,giuliano.destefano}@unicampania.it

1 Introduction

Flows with non-uniform roughness, in which regions of varying roughness-height are adjacent to each other, occur often in engineering and in the natural sciences. The simplest prototype of heterogeneous roughness is the abrupt transition from a rough patch oriented normal to the flow to a smooth one; this configuration can be found very often in atmospheric boundary layers [1]. Downstream of the transition, an internal boundary layer is formed that can be separated into two regions: an equilibrium inner-layer, in which the flow has adjusted to the change of the wall boundary-condition, and an outer layer, where the flow still has memory of the upstream condition.

Several investigations of this flow have been conducted, either experimentally [1–4] or numerically [5,6]. These studies have demonstrated that, after the transition, the flow variables return to equilibrium values at different rates, with the skin-friction adapting more rapidly to the new condition than the mean velocity and Reynolds shear stresses. Direct numerical simulations (DNS) or wall-resolved large-eddy simulations (WRLES) may provide more insights than experiments, since the flow inside the roughness sublayer is resolved. Unfortunately, the prediction of high Reynolds-number flows cannot be obtained by means of these computationally expensive techniques, while Reynolds-Averaged Navier-Stokes (RANS) simulations have difficulties in predicting flows that are strongly out of equilibrium, as in the case for heterogeneous roughness. Furthermore, typical corrections used to include the effects of roughness [7] are developed using equilibrium assumptions, and are, therefore, not very accurate in non-equilibrium rough-wall boundary layers.

Methods that combine the LES approach in the outer region with a simpler methodology, on the other hand, have the ability to capture the outer-layer non-equilibrium naturally, with much lower computational costs. The two most common techniques of this type are Wall-Modelled Large-Eddy Simulation (WMLES) and hybrid RANS/LES [8,9]. In the former case, the outer-layer is obtained from the solution of the LES equations, and approximate methods are used to derive the wall stress from the outer-layer data. In hybrid RANS/LES methods, the character of the turbulence model changes in such a way that the RANS approach followed near the wall switches to LES at some distance from the surface. Corrections for modelling roughness have

C. Marchioli et al. (Eds.): DLES 2023, ERCO 31, pp. 368–373, 2024.
https://doi.org/10.1007/978-3-031-47028-8_56

been developed for both of these approaches. WMLES of the flow over heterogeneous roughness, for strips normal to the flow direction, for instance, was performed in [5], where it was pointed out that surface heterogeneity leads to sharp variations in the velocity profiles and discontinuities in the shear-stress.

To predict the flow behavior caused by step changes in roughness, identifying the limitations of the different methods is crucial. The purpose of the present work is to complement existing investigations by assessing the accuracy of two typical techniques, WMLES and hybrid RANS/LES, in predicting the sudden transition from a smooth surface to a rough one and vice-versa. Roughness is included using two different methods: the modification of either the log-law or the model equations, and the use of a localized drag force.

2 Methodology

The governing equations can be formally written the same way for both LES and unsteady RANS approaches, namely,

$$\frac{\partial \overline{u}_i}{\partial x_i} = 0, \qquad \frac{\partial \overline{u}_i}{\partial t} + \frac{\partial}{\partial x_j}(\overline{u}_i \overline{u}_j) = -\frac{\partial \overline{P}}{\partial x_i} + \nu \frac{\partial^2 \overline{u}_i}{\partial x_j \partial x_j} - \frac{\partial \tau_{ij}}{\partial x_j}, \tag{1}$$

where the overbar represents either the LES spatial filtering or the Reynolds-averaging operation. In the momentum equation, $\tau_{ij} = \overline{u_i u_j} - \overline{u}_i \overline{u}_j$ are the unresolved stresses, which have to be considered as either the subfilter-scale (SFS) stresses in LES or the Reynolds stresses in unsteady RANS. The unknown stresses are approximated using the eddy-viscosity model, namely,

$$\tau_{ij} - \frac{1}{3}\delta_{ij}\tau_{kk} = -2\nu_T \overline{S}_{ij}, \tag{2}$$

where S_{ij} is the resolved strain-rate tensor. In this work, WMLES uses the eddy-viscosity SFS model proposed in [11], and hybrid RANS/LES follows the Improved Delayed Detached-Eddy Simulation (IDDES) approach, based on the Spalart-Allmaras turbulence model. The RANS and LES fields are coupled by introducing a hybrid turbulent length-scale. Near the wall, the distance from the wall is used, whereas, in the outer layer, it is proportional to the grid size.

The present simulations were conducted for open channel flow, in the computational domain $L_x \times L_y \times L_z = 56\delta \times \delta \times 14\delta$, where δ is the open-channel height. The rough patch occupied two thirds of the domain, resembling the experimental setup in [4]. The Reynolds number based on bulk velocity was $Re_b = U_b\delta/\nu = 1.21 \times 10^5$, roughly corresponding to the experimental value at the rough-to-smooth interface. A fractional-step method was used for time integration, with the Crank-Nicolson scheme for the wall-normal diffusion, and a third-order Runge-Kutta scheme for the remaining terms. Spatial discretization was performed using central second-order finite differences on a staggered grid. In the WMLES, the spatial grid was uniform and isotropic, while, for the IDDES, it was suitably stretched in the wall-normal direction (with $y^+ < 1$ for the first grid point). Periodic boundary conditions were applied in the homogeneous directions, while a symmetry condition was imposed at the top boundary. For IDDES, no-slip conditions were applied at the wall. For WMLES, the wall-stress was obtained, given the velocity at the interface, by satisfying the log-law of the wall. Two different methods were used to incorporate the effect of roughness. One possibility was to modify either the model for the unresolved scales (for IDDES) or the wall boundary condition (for WMLES). The other one was based on the addition of a drag force to the resolved momentum equations [10]. In the WMLES the wall stress was obtained by imposing either the smooth- or rough-wall form of the logarithmic law-of-the-wall:

$$u_{if}^+ = \frac{1}{\kappa}\log y_{if}^+ + B \quad (\text{for } x < x_o), \tag{3}$$

$$u_{if}^+ = \frac{1}{\kappa}\log y_{if}^+ + B - \Delta U^+ \quad (\text{for } x > x_o) \tag{4}$$

where the subscript if denotes quantities evaluated at the inner-outer layer interface, located at $y = 0.05\delta$, $\kappa = 0.41$ is the von Kàrmàn constant, and $B = 5.0$. The roughness function ΔU^+ was chosen to match the experimental value [4]. Following [7], the Spalart-Allmaras model was modified to include the effect of roughness, changing the boundary condition for the eddy-viscosity, using a Neumann condition at the wall, instead of the traditional Dirichlet condition. Alternatively, the roughness effects are included by adding a forcing term to the right-hand-side of the momentum equation. The drag force, which is active below the roughness crest, is defined by:

$$f_i = \alpha_{ij}|\mathbf{u}_{rz}|u_{rz,j} \tag{5}$$

where $\alpha_{ij} = \text{diag}\{\alpha_t, \alpha_t, \alpha_n\}$ determines the force intensity, with the subscripts t and n denoting wall-tangential and wall-normal components, respectively. In this study, the normal component was set to zero, and the tangential component was chosen to match the equivalent sand-grain roughness-height of the experiment. Combining the two turbulence models and the two roughness models, four different methods were tested.

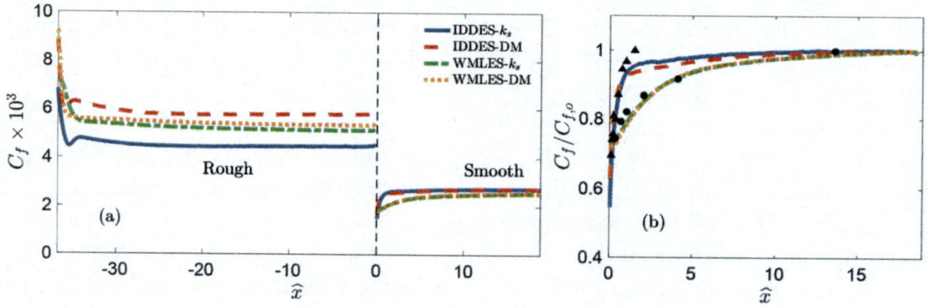

Fig. 1. (a) Skin-friction coefficient over rough and smooth strips, and (b) normalized skin-friction coefficient over the smooth strip. Triangles and circles correspond to the experimental data in [1] and [4], respectively.

3 Results

A preliminary grid-convergence study was conducted for the WMLES, by using the mean velocity profiles as diagnostics. The present results were obtained by employing $1792 \times 60 \times 448$ points, corresponding to the converged grid. For IDDES, the same wall-parallel resolution was employed, but the number of points in the wall-normal direction was increased, leading to $1792 \times 121 \times 448$ points.

The skin-friction coefficient predicted by the four models is shown in Fig. 1(a). In all cases the sudden variation of C_f is captured. At the end of the rough strip, the WMLES are in good agreement with each other, reflecting the fact that both the wall modified boundary conditions and the drag model were calibrated using fully developed rough-channel data. However, there is a significant difference between the IDDES results due to the sensitivity of the drag-model coefficient to the grid resolution. Apparently, the recovery length on the smooth strip predicted by the IDDES models is much shorter than that obtained from the WMLES.

In Fig. 1(b), the skin-friction coefficient, normalized by its value at 95% of the smooth strip, is compared to experimental data [1,4]; the position is normalized by either the open-channel height or the boundary-layer thickness at the transition location. It is worth noting that the experiments in [1] were carried out at a much higher Reynolds number, $Re_\infty = U_\infty \delta / \nu \simeq 2.73 \times 10^5$ (where U_∞ is the freestream velocity and δ the boundary-layer thickness), while the equivalent sandgrain roughness height was $k_s^+ \simeq 479$. The discrepancy between the experimental datasets can be attributed to this difference [4].

The type of roughness modelling (equation modification or drag model) does not affect the recovery length significantly; the WMLES agrees better with the experimental data at the same *Re* [4], while the IDDES follows the higher-*Re* results. An important distinction between IDDES and WMLES is the fact that, for WMLES, the wall stress depends on the velocity at the interface; the boundary-condition change does not affect the velocity at the interface immediately, but requires some downstream distance to penetrate to the height of the interface; practically, for WMLES, the wall stress has memory of the upstream flow. For IDDES, on the other hand, the different boundary condition is immediately felt throughout the layer because of the model modifications. Furthermore, the Reynolds shear-stress predicted by IDDES is higher than that obtained by WMLES, and the enhanced mixing causes the mean-velocity profile to be flatter. The mean velocity profiles are shown in Fig. 2, for four streamwise locations after the transition. The velocity is normalized by the centerline velocity $\langle U \rangle_{cl}$, wall distance by the displacement thickness δ^*. Again, the WMLES is in better agreement with the experimental data, and little difference is observed between the two roughness models. The IDDES gives a flatter velocity profile, corresponding to the above mentioned shear-stress prediction.

Figure 3 shows the streamwise development of various quantities of interest, across the rough-to-smooth interface. Three y positions are chosen for the IDDES solutions: one very close to the wall, one near the peak of ν_T in the RANS region, and another one in the outer LES zone. It is worth noting the jump in ν_T, which becomes less significant away from the wall, Fig. 3(b), and the smoother behaviour of the drag model; the two IDDES solutions collapse at a distance of less than 2δ downstream of the interface.

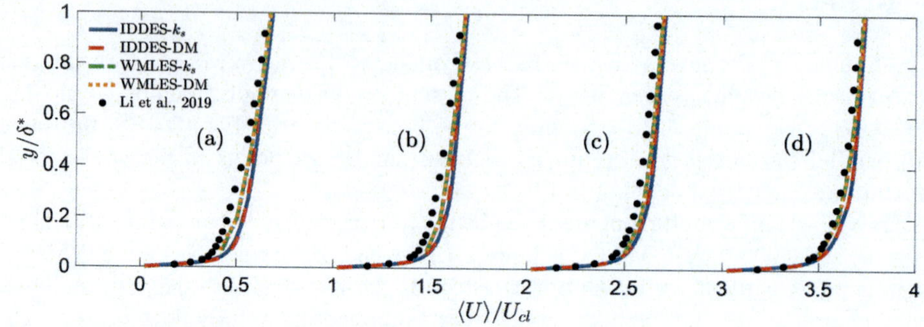

Fig. 2. Mean velocity profiles in outer units at (a) 3%, (b) 10%, (c) 19%, and (d) 63% of the length of the smooth strip, compared to experimental data [4].

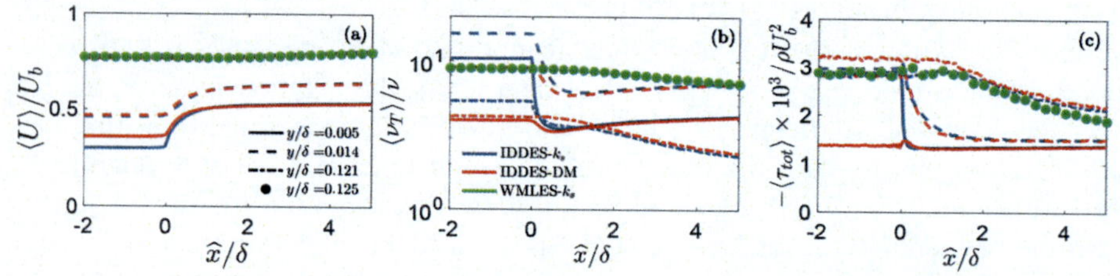

Fig. 3. Streamwise development of velocity, eddy-viscosity, and total Reynolds shear-stress.

The total stress reaches the values expected for a smooth wall very quickly, Fig. 3(c). This decrease causes the flow acceleration downstream of the interface, Fig. 3(a), also reflected in the skin-friction coefficient and mean-velocity profiles shown above. In the WMLES case, on the other hand, the eddy-viscosity and the total stress are continuous across the interface, and decrease quite slowly downstream, reflecting the lower turbulence level over the smooth wall. As a consequence, the flow acceleration is milder, as illustrated in Fig. 3(a).

4 Conclusions

Different modelling approaches were tested to predict the flow over alternating rough and smooth strips, oriented normal to the mean stream. Comparison with experimental data highlighted that WMLES captures the skin-friction recovery, and predicts fairly accurately the mean-velocity profile. In IDDES the abrupt change of the model and boundary conditions results in the excessively rapid variation of the Reynolds shear-stress, and a disproportionate flow acceleration. In the WMLES, on the other hand, the wall shear-stress is calculated based on outer-layer information and, thus, the perturbation introduced by the change in boundary condition must propagate away from the wall before the model is able to react. WMLES effectively retains memory of the upstream conditions, producing a smooth recovery of the modelled shear-stress and skin-friction coefficient. In this particular arrangement of roughness strips, WMLES are found to give better prediction of the flow field than IDDES. Also, IDDES has a much higher computational cost, and the model length-scale appears to be extremely sensitive to flow conditions.

Acknowledgments. The authors acknowledge the computational support by CINECA under the ISCRA initiative (Project HP10BSBCBZ), and by the Digital Research Alliance-Compute Canada.

References

1. Chamorro, L.P., Porté-Agel, F.: Velocity and surface shear stress distributions behind a rough-to-smooth surface transition: a simple new model. Bound. Layer Meteorol. **130**, 29–41 (2009)
2. Antonia, R.A., Luxton, R.E.: The response of a turbulent boundary layer to a step change in surface roughness. Part 1. Smooth to rough. J. Fluid Mech. **48**, 721–761 (1971)
3. Antonia, R.A., Luxton, R.E.: The response of a turbulent boundary layer to a step change in surface roughness. Part 2. Rough to smooth. J. Fluid Mech. **53**, 737–757 (1972)
4. Li, M., et al.: Recovery of wall-shear stress to equilibrium flow conditions after a rough-to-smooth step change in turbulent boundary layers. J. Fluid Mech. **872**, 472–491 (2019)
5. Bou-Zeid, E., Meneveau, C., Parlange, M.B.: Large-eddy simulation of neutral atmospheric boundary layer flow over heterogeneous surfaces: blending height and effective surface roughness. Water Resour. Res. **40**, W02505 (2004)
6. Saito, N., Pullin, D.I.: Large eddy simulation of smooth–rough–smooth transitions in turbulent channel flows. Int. J. Heat Mass Trans. **78**, 707–720 (2014)
7. Aupoix, B., Spalart, P.R.: Extensions of the Spalart-Allmaras turbulence model to account for wall roughness. Int. J. Heat Fluid Flow **434**, 454–462 (2003)
8. Piomelli, U., Balaras, E.: Wall-layer models for large-eddy simulations. Annu. Rev. Fluid Mech. **34**, 349–374 (2002)
9. Heinz, S.: A review of hybrid RANS-LES methods for turbulent flows: concepts and applications. Prog. Aerosp. Sci. **114**, 100597 (2020)
10. Varghese, J., Durbin, P.A.: Representing surface roughness in eddy resolving simulation. J. Fluid Mech. **897**, A10 (2020)
11. Vreman, A.W.: An eddy-viscosity subgrid-scale model for turbulent shear flow: Algebraic theory and applications. Phys. Fluids **451**, 3670–3681 (2004)

Direct Numerical Simulation of Scalar Transport Across the Interface Between a Porous Medium and a Turbulent Flow

S. v. Wenczowski and M. Manhart(✉)

Technical University of Munich (TUM), Acrisstraße 21, 80333 Munich, Germany
{simon.wenczowski,michael.manhart}@tum.de

1 Introduction

Mass transport processes across the interface between a porous medium and a turbulent flow field are relevant in a wide range of natural and industrial systems: Prominent examples for technical applications range from food drying up to processes within fuel cells. In the environment, the exchange of substances within the hyporheic zone is vital for the health of aquatic ecosystems, whereas the evaporation from soils must be considered for sustainable land use. Despite the apparent heterogeneity of these fields, mass transport processes within the interface region are driven by a common set of mechanisms. In this contribution, we will focus on hyporheic mass exchange to gain a better understanding of these fundamental processes of interdisciplinary interest.

Richardson and Parr (1988) reported experiments in a laboratory flume and suggested an effective diffusivity model for the mass transport over a porous media-free flow interface. Later, O'Connor and Harvey (2008) and Grant et al. (2012) combined different data sets to refine the effective diffusivity model. Voermans et al. (2018) concluded that dominant interfacial mass transport processes are determined by the permeability Reynolds number Re_K. A double-averaging framework using horizontal planes was proposed and applied by atmospheric scientists (e.g. [5]). Later, the framework was generalized for three-dimensional roughness [1] and gained popularity in environmental hydraulics (e.g. [3]).

The objective of our research is to contribute to a more comprehensive mechanistic understanding of scalar transport in the interface region. Beyond (i) distinguishing different scalar transport processes, we (ii) identify the regions of influence of individual processes and shed a light on (iii) the interaction between processes. A brief introduction of the methods and the parameter space is followed by a validation of the flow field. Mass transport processes are analysed, and their relative influence is assessed. In conclusion, the critical role of the permeability Reynolds number Re_K is confirmed, whereas its effect is shown to be highly depth-dependent.

C. Marchioli et al. (Eds.): DLES 2023, ERCO 31, pp. 374–379, 2024.
https://doi.org/10.1007/978-3-031-47028-8_57

2 Methods

For the numerical simulations, the porous medium was represented by a static random sphere pack with uniform sphere diameter D. We used LAMMPS [6] to simulate the process of pouring the spheres into an x-y-periodic domain. Thus, the resulting sediment bed showed no macro-topography like dunes or riffles, which was confirmed by the auto-correlation of the bed elevation. The position of the interface was defined by the inflection point of the porosity profile, i.e. where $\partial^2\theta/\partial z^2 = 0$.

Our MPI-parallel in-house code MGLET [2] uses a Finite-Volume discretization to solve the incompressible Navier-Stokes equations. The complex geometry of the sphere pack is included by an Immersed Boundary Method of second order. Based on the flow field, the advection-diffusion equation for a passive scalar with Schmidt number $Sc = 1$ is solved. Local grid refinement near the interface ensures that the Kolmogorov and Batchelor scales are resolved within the complete domain. All simulation cases were configured with periodic boundary conditions in the bed-parallel x- and y-direction. By means of a volume force, the flow was driven into the positive x-direction. A slip boundary condition at the top boundary of the domain was applied to approximate a free water surface. As scalar boundary conditions, fixed concentration values are imposed at the bottom and top of the domain to induce a mass flux in bed-normal direction. The surface of the spheres in the sediment bed was defined as no-slip and as impermeable to scalar fluxes. By means of a single-domain Direct Numerical Simulation (DNS), all temporal and spatial scales were resolved both in the free flow region and in the pore space of the porous medium. Thus, no a-priori assumptions are made, that may distort the observations.

In total, eight cases were simulated representing sample points within a parameter space spanned by $Re_\tau = u_\tau h/\nu$ and $Re_K = u_\tau\sqrt{K}/\nu$. As shown in Fig. 1, we considered shear Reynolds numbers in the range of Re_τ = 150–500 and permeability Reynolds numbers of Re_K = 0.4–2.8. For a monodisperse sphere pack, Re_K is tightly, yet not linearly, linked to the roughness Reynolds number k_s^+. Accordingly, cases S-150, M-150, and S-300 can be categorized as transitionally rough, whereas the remaining cases ($k_s^+ > 90$) lie within the hydraulically fully rough regime.

To discuss the strongly three-dimensional flow situation, we resort to a double-averaging technique [5]. An arbitrary quantity ϕ is first averaged in time and, consecutively, the temporal mean is averaged within a bed-parallel horizontal plane:

$$\phi_{(\mathbf{x},t)} = \overline{\phi}_{(\mathbf{x})} + \phi'_{(\mathbf{x},t)}, \qquad \overline{\phi}_{(\mathbf{x})} = \frac{1}{T}\int_0^T \phi_{(\mathbf{x},t)}\,dt \tag{1}$$

$$\overline{\phi}_{(\mathbf{x})} = \langle\overline{\phi}\rangle_{(z)} + \tilde{\overline{\phi}}_{(\mathbf{x})}, \qquad \langle\overline{\phi}\rangle_{(z)} = \frac{1}{A_f}\iint_{A_f} \overline{\phi}_{(\mathbf{x})}\,dx\,dy \tag{2}$$

In the chosen notation, $\langle\overline{\phi}\rangle$ symbolizes the intrinsic average within a horizontal plane, while spatial deviations therefrom are indicated by a tilde. Via the relation $\langle\phi\rangle^s_{(z)} = \theta_{(z)}\langle\phi\rangle_{(z)}$, a connection to the superficial horizontal average $\langle\phi\rangle^s_{(z)}$ is established, where $\theta_{(z)} = A_{f(z)}/A_0$ represents the in-plane porosity.

The simulated flow field of case M-150 was validated against experimental data from [8]. We estimated the sampling uncertainty in the experiment by reproducing the same sampling technique on the numerical data at different locations. Under consideration of the spatial heterogeneity, the plots in Fig. 2 demonstrate good agreement for the horizontally averaged streamwise velocity and the Reynolds stresses.

3 Results and Discussion

The double-averaging framework allows a decomposition of the superficially double-averaged scalar flux $\langle\overline{J_c}\rangle^s$ through any horizontal plane at a height z within the domain: Eq. 3 shows that turbulent transport, dispersive transport, and transport due to molecular diffusion contribute to $\langle\overline{J_c}\rangle^s$, which is constant over z.

$$\langle\overline{J_c}\rangle^s_{(z)} = \underbrace{\theta\left\langle\overline{w'c'}\right\rangle}_{\text{turb.}} + \underbrace{\theta\left\langle\widetilde{\overline{w}}\widetilde{\overline{c}}\right\rangle}_{\text{disp.}} - \underbrace{\theta\left\langle\Gamma_c\frac{\partial\overline{c}}{\partial z}\right\rangle}_{\text{diff.}} = \text{const}\ \forall z \tag{3}$$

To get an impression of the different processes, we consider the spatial distribution of the effective contributions to dispersive transport turbulent transport transport due to molecular diffusion. For this purpose, Fig. 3 shows an arbitrarily chosen vertical x-z-slice through simulation case M-500. Hotspots of dispersive transport in vertical direction coincide with regions of upwelling flow in the sediment bed. Directly above and slightly downstream of these hotspots, increased activity of the turbulent transport in bed-normal direction is observed within the free flow region. Diffusive transport is predominantly observed in deeper regions of the sediment where steeper gradients in the scalar concentration field are present.

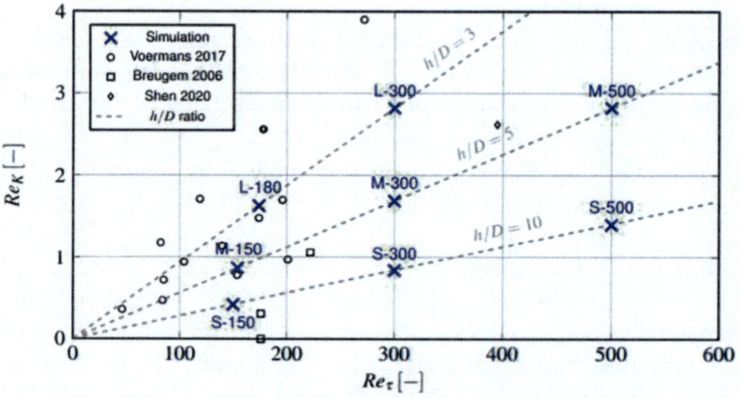

Fig. 1. Dimensionless parameter space, including data points from literature. Gray dashed lines represent fixed ratios of flow depth h (= boundary layer thickness) and sphere diameter D.

We further exploit the double-averaging framework to identify the dominant scalar transport processes in dependency of the vertical position z. For this purpose, the contribution of individual transport processes is set in relation to the total superficially averaged scalar $\langle \overline{J_c} \rangle^s$ in bed-normal direction. Unsurprisingly, turbulent scalar transport dominates within the free flow region above the interface. In the interface region around $z = 0$, the impact of turbulent transport declines rapidly with increasing depth. Instead, dispersive transport realizes most of the scalar flux below the interface. Without limitation by the bottom boundary of the domain, the influence of dispersive transport can even extend further into the sediment bed (not shown here). Only at greater depth,

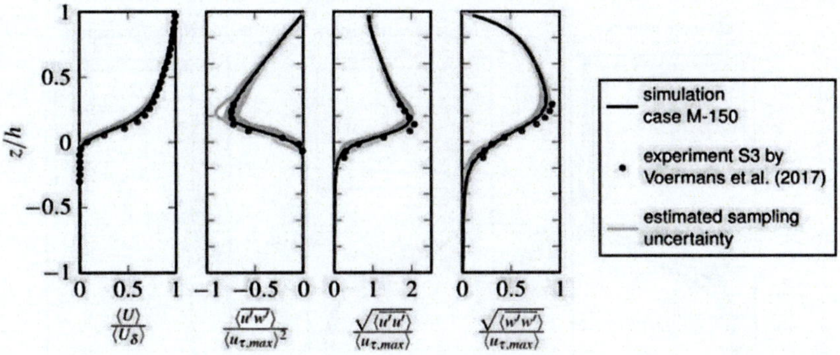

Fig. 2. Validation of the flow field of case M-150 against experiment S3 of [7]. The uncertainty was estimated by reproducing the experimental sampling approach on the simulation results.

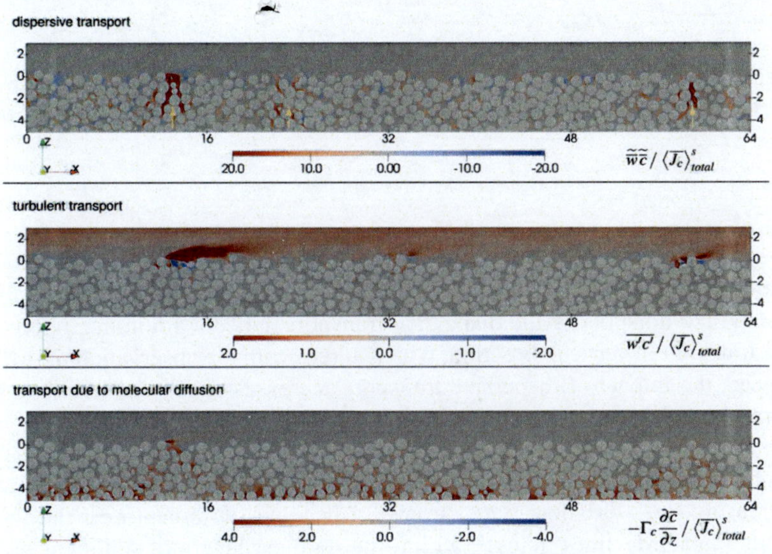

Fig. 3. Spatial distribution of the effective contributions to dispersive, turbulent, and diffusive transport within an arbitrarily chosen vertical x-z-slice through simulation case M-500.

scalar transport due to molecular diffusion becomes the most important process. As shown in Fig. 4, the relative importance of individual processes mainly depends on the permeability Reynolds number Re_K. In contrast, curves representing cases that only differ in the friction Reynolds number Re_τ seem to collapse near the interface. With increasing Re_K, the relative influence of turbulent transport increases at the interface and reaches into greater depth. For all cases, however, dispersive transport remains the dominant process below the interface, as it also gains influence with increasing Re_K. In contrast, the relative importance of scalar transport due to molecular decreases with higher Re_K.

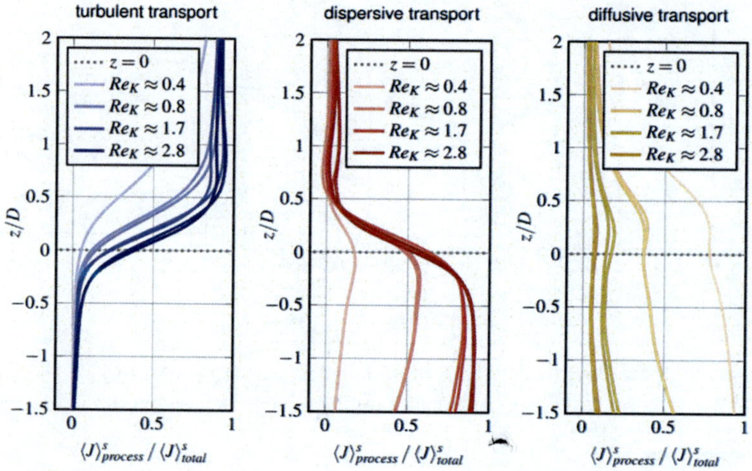

Fig. 4. Relative contribution of different scalar transport processes to the total scalar flux $\langle J \rangle^s_{total}$ in bed-normal direction. Darker curves represent a higher permeability Reynolds number Re_K.

4 Conclusion

Though the sediment bed does not exhibit any macro-topography like dunes or riffles, the chimney-like hotspots of the dispersive transport indicate a non-negligible role of advective transport in vertical direction, which agrees with the observation of [4]. Above the interface, the impact of dispersive transport decreases rapidly as turbulent mixing reduces in-plane fluctuations of the scalar concentration. The results confirm that the relative importance of the individual scalar transport processes in the interface region is predominantly determined by the permeability Reynolds number Re_K, as stated by [8]: With increasing Re_K, turbulent scalar transport becomes dominant near the interface, whereas its impact declines quickly within the topmost layers of sediment grains. In contrast, dispersive scalar transport also affects deeper regions within the sediment bed, while its influence similarly grows with Re_K. As different processes are dominant at and slightly below the interface, a comprehensive model approach would have to consider the interaction between these processes to predict hyporheic scalar transport accurately.

Acknowledgements. We thank Dr. Joey Voermans for valuable background information on his experiments. Also, we gratefully acknowledge the financial support by DFG (grant no. MA2062/15-1) as well as the computational resources provided by the Leibniz Supercomputing Center.

References

1. Giménez-Curto, L.A., Lera, M.A.C.: Oscillating turbulent flow over very rough surfaces. J. Geophys. Res. **101**, 20745–20758 (1996)
2. Manhart, M., Tremblay, F., Friedrich, R.: MGLET: a parallel code for efficient DNS and LES of complex geometries. In: Parallel Computational Fluid Dynamics 2000, pp. 449–456. Elsevier, North-Holland (2001)
3. Nikora, V.I., Goring, D.G., McEwan, I., Griffiths, G.: Spatially averaged open-channel flow over rough bed. J. Hydraul. Eng. **127**, 123–133 (2001)
4. Packman, A., Salehin, M., Zaramella, M.: Hyporheic exchange with gravel beds: basic hydrodynamic interactions and bedform-induced advective flows. J. Hydraul. Eng. **130**(7), 647–656 (2004)
5. Raupach, M.R., Shaw, R.H.: Averaging procedures for flow within vegetation canopies. Bound.-Layer Meteorol. **22**, 79–90 (1982). https://doi.org/10.1007/BF00128057
6. Thompson, A.P., et al.: LAMMPS - a flexible simulation tool for particle-based materials modeling at the atomic, meso, and continuum scales. Comput. Phys. Comm. **271**, 10817 (2022)
7. Voermans, J.J., Ghisalberti, M., Ivey, G.N.: The variation of flow and turbulence across the sediment-water interface. J. Fluid Mech. **824**, 413–437 (2017)
8. Voermans, J.J., Ghisalberti, M., Ivey, G.N.: A model for mass transport across the sediment-water interface. Water Resour. Res. **54**, 2799–2812 (2018)

Influence of Ridge Spacing on Secondary Currents in Turbulent Channel Flow over Triangular Ridges

O. Zhdanov(✉) and A. Busse

James Watt School of Engineering, University of Glasgow, Glasgow, UK
{oleksandr.zhdanov,angela.busse}@glasgow.ac.uk

1 Introduction

Surfaces with spanwise variation of topology, such as ridge-type roughness, are known to generate secondary currents, which originate from turbulence anisotropy and spatial gradients of Reynolds stresses and represent Prandtl's secondary flows of the second kind [1]. The presence of secondary currents is reflected in spanwise heterogeneity of the time-averaged streamwise velocity field in the form of alternating high- and low-momentum pathways (HMPs and LMPs), which correspond to downwash and upwash flow regions, respectively. In surfaces with ridge-type roughness, LMPs are typically observed above the ridges and HMPs occur in between them [2,3]. However, a swap in their locations can be observed when ridge spacing exceeds channel half-height [4].

Spacing between adjacent ridges is one of the primary parameters that characterises spanwise heterogeneity of ridge-type rough surfaces and consequently influences the size and strength of secondary currents [2,4]. Most previous computational investigations on spacing dependency effects considered ridges with a rectangular cross-section. Thus, due to the low number of systematic studies on other ridge shapes, it is unclear whether results obtained for rectangular ridges, for example the appearance of tertiary flows for very wide ridges [3,5], can be extrapolated to other ridge shapes. In the present ongoing study, flow over triangular ridges, which are expected to produce strong secondary currents [5], is studied in a closed channel configuration using direct numerical simulations (DNS). In the current paper, the effects of ridge spacing and Reynolds number on secondary currents and mean flow and turbulence statistics are investigated, and the results are compared to the data of Zampiron et al. [6], who conducted experiments with ridges with equilateral triangular cross-section, but placed on a rough bed in an open channel configuration.

2 Methodology

Direct numerical simulations of turbulent channel flow over equilateral triangular ridges (see Fig. 1) are conducted using the code iIMB [7]. The simulations are performed for four ridge spacings, s, namely, $\pi\delta/8$, $\pi\delta/4$, $\pi\delta/2$, and $\pi\delta$, where δ is the channel

C. Marchioli et al. (Eds.): DLES 2023, ERCO 31, pp. 380–385, 2024.
https://doi.org/10.1007/978-3-031-47028-8_58

half-height, to study the influence of this parameter on secondary currents and turbulence statistics. To investigate the Reynolds number dependency of the results, DNS are conducted at two friction Reynolds numbers, $Re_\tau = \delta u_\tau / \nu = 550$ and 1000, for each spacing. Reference smooth wall simulations are also performed at both considered values of Re_τ.

In all ridge simulations, the channel size is $3\pi\delta \times \pi\delta \times 2\delta$ in streamwise, spanwise, and wall-normal directions, respectively. Periodic boundary conditions are applied in the streamwise and spanwise directions. The flow is driven by a constant mean streamwise pressure gradient. In all considered cases, the ridges have a triangular cross-section of constant height, $h/\delta = 0.08$, and are placed on both walls of the channel, which are mirrored with respect to the channel centreline (Fig. 1). The ridge geometry is resolved using an iterative version of the embedded boundary method by Yang and Balaras [8]. The computational grid is uniform in the streamwise and spanwise directions maintaining $\Delta x^+ < 5$ and $\Delta y^+ < 5$ in all simulations. The grid spacing in spanwise direction is sufficiently fine to ensure a minimum of 16 points across ridge width, b, at both considered Re_τ. In the wall-normal direction, uniform grid spacing is applied up to the ridge height ($\Delta z^+_{min} = 2/3$); above, the grid-spacing is gradually increased up to the channel centre. The maximum wall-normal grid spacing is limited to $\Delta z^+_{max} < 5$ in all cases.

In all ridge simulations, statistical data are acquired for a minimum of 140 flow through times. For the computation of statistical quantities, such as Reynolds and dispersive stresses, the double-averaging approach [9] is used combined with intrinsic averaging [10], i.e., only the fluid occupied region is considered when computing averages below the ridge height.

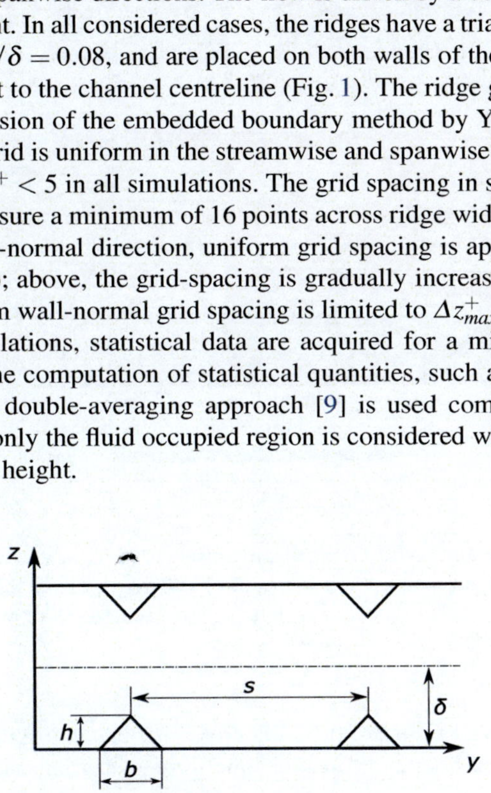

Fig. 1. Schematic of a channel with triangular ridges. δ is the channel half-height, s the spacing between ridges in spanwise direction, h the height of the ridge, and b the width of the ridge.

3 Results

The time and phase averaged streamwise velocity fields for all studied ridge spacings are presented in Fig. 2 for both considered Re_τ. In each case the presence of secondary currents can be observed with the upwash flow regions above the ridges and downwash flow regions in between them. No swap in their location occurs even at the largest ridge spacing $s = \pi\delta$, which exceeds channel half-height by more than 3 times. The present results for a closed channel configuration are consistent with previous studies

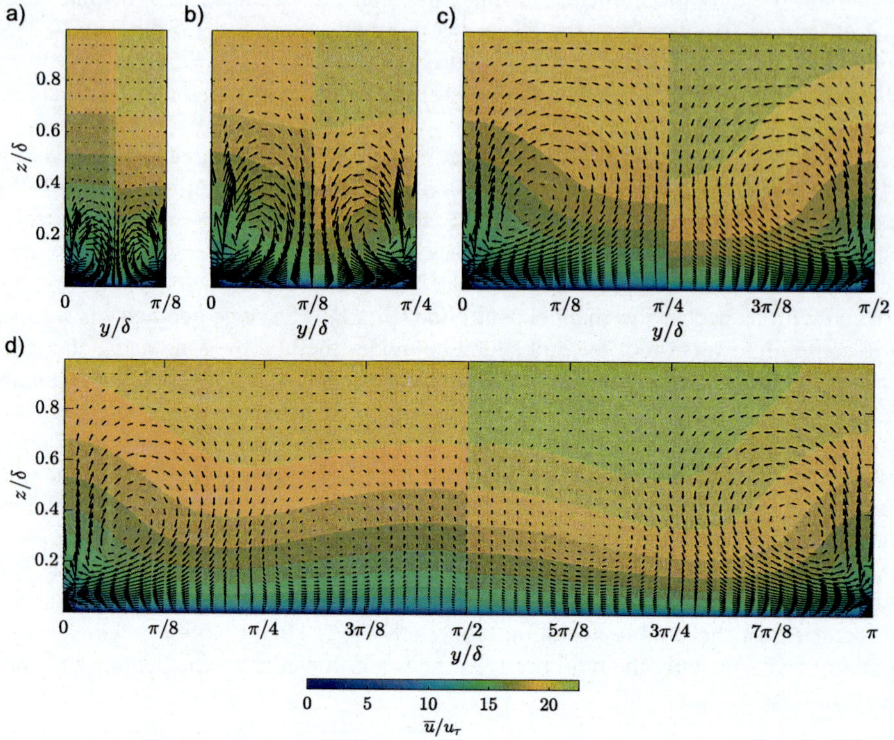

Fig. 2. Contours of time and phase averaged streamwise velocity $\bar{u}/u_\tau$ with vectors of $\bar{v}/u_\tau$ and $\bar{w}/u_\tau$ for all studied ridge spacings: a) $s = \pi\delta/8$, b) $s = \pi\delta/4$, c) $s = \pi\delta/2$, d) $s = \pi\delta$ at both considered friction Reynolds number. The left half of each panel shows the velocity field at $Re_\tau = 550$ and the right half at $Re_\tau = 1000$.

on surfaces covered with ridges of different cross-sections in turbulent boundary layers [2,3] and experimental results of open channel flow over triangular ridges on a rough bed [6].

At the lowest spacing $s = \pi\delta/8$ the secondary currents are confined to the near-wall region and for higher y/δ the mean flow field appears homogeneous. As the ridges are placed further apart, secondary currents increase in size until they extend up to the channel half-height at $s = \pi\delta/2$. As expected from the experimental study by Zampiron et al. [6] on triangular ridges, the degree of spanwise flow heterogeneity increases as ridges are placed further apart due to increase in the size of secondary currents.

This trend is reversed at the largest considered spacing $s = \pi\delta$. For this case, in addition to secondary currents, which are still present on both sides of each ridge, tertiary flows are observed in between ridges in the form of a pair of counter-rotating streamwise vortices. The emergence of tertiary flows can be explained by the full development of secondary currents, which cannot grow in size beyond δ. With the appearance of tertiary flows, the flow field in between ridges becomes more homogeneous in spanwise direction compared to the $s = \pi\delta/2$ case, where only secondary currents are present.

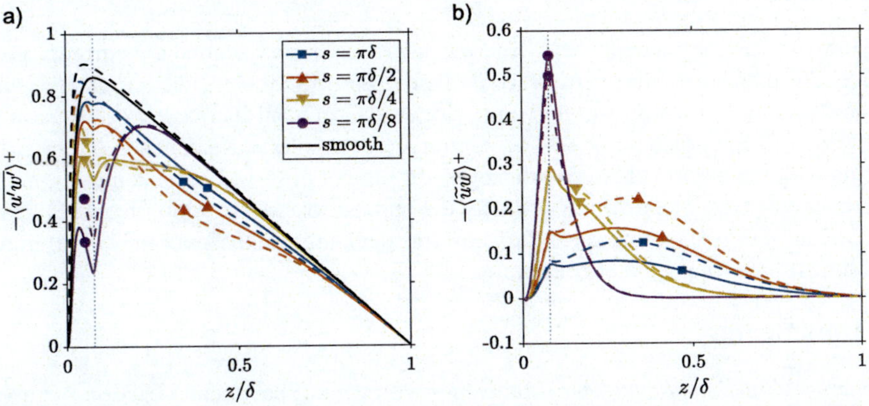

Fig. 3. Profiles of (a) Reynolds shear stress and (b) dispersive shear stress for all studied ridge spacings at both considered friction Reynolds numbers. Reference smooth-wall data is also shown. Solid lines show data for $Re_\tau = 550$; dashed lines show data for $Re_\tau = 1000$. The ridge height is marked with a thin dotted vertical line.

However, above the ridges upwash regions reach up the channel half-height. Reynolds number does not affect the topology of the flow including locations and size of secondary currents, and the mean flow fields at $Re_\tau = 550$ and $Re_\tau = 1000$ look qualitatively the same, except for the higher mean velocities at the higher Re_τ.

The Reynolds $\langle\overline{u'w'}\rangle^+$ and dispersive $\langle\widetilde{u}\widetilde{w}\rangle^+$ shear stress profiles from all simulations are presented in Fig. 3. As the ridge spacing is increased from $\pi\delta/8$ to $\pi\delta/2$ the levels of Reynolds shear stress in the outer layer drop and this drop is compensated by an increase in the dispersive shear stress. However, at the maximum ridge spacing $\pi\delta$, where tertiary flows are present, some recovery in the Reynolds shear stress profile is observed, and the level of dispersive shear stress drops compared to spacing of $\pi\delta/2$. These results are in line with the experimental study of Zampiron et al. [6], where results obtained above ridge height were reported. In addition, the present DNS provide insight into the behaviour of shear stresses below the ridge crests and their overall behaviour with Reynolds number. At all considered ridge spacings, the Reynolds shear stress profiles exhibit two peaks, one above the ridge crest and one below. At the narrowest ridge spacing, these two peaks are clearly separated and the magnitude of the peak below the roughness height is significantly lower compared to the one above. As the ridges are placed further apart, the two peaks move closer to each other and the difference in their magnitude reduces. These trends hold at both considered friction Reynolds numbers.

Looking at the influence of Re_τ on the individual ridge spacing cases, it can be observed that for the smallest ridge spacing $s = \pi\delta/8$ the effect of increasing Reynolds number is evident only below and immediately above the ridge crest, where $\langle\overline{u'w'}\rangle^+$ levels increase with Re_τ, whereas the dispersive shear stress profiles follow an opposite trend. At both friction Reynolds numbers the profiles collapse onto the smooth wall profile for $y/\delta \approx 0.3$ and above.

However, as the spacing between ridges increases and secondary currents grow in size, the dispersive shear stress levels below the ridge height do not change with Reynolds number. In contrast, above the ridges a significant increase in $\langle \widetilde{u}\widetilde{w} \rangle^+$ can be observed with Reynolds number. The magnitude of this increase depends on the ridge spacing with the highest change with Reynolds number observed for spacing $s = \pi\delta/2$. The Reynolds shear stress follows exactly the opposite trends. In addition, $\langle \overline{u'w'} \rangle^+$ profiles for surfaces with strong secondary currents deviate from the smooth wall reference data almost up to the channel centreline.

4 Conclusions

Turbulent channel flow over triangular ridges with varied spacing and friction Reynolds number was investigated numerically using direct numerical simulations. Variation in ridge spacing was found to have significant influence on both size and strength of secondary currents generated by triangular ridges. The secondary currents grow in size as spacing between ridges is increased until they occupy all available space in the channel. With the further increase in s, secondary currents cannot grow larger as they already reach δ scale, and tertiary flows emerge. At all considered ridge spacings, high- and low-momentum pathways are observed in between and above ridges, respectively, and no swap in their location is recorded at any investigated s. These findings are consistent with the experimental results of Zampiron et al. [6].

Overall, secondary currents over triangular ridges are found to exhibit similar behaviour as secondary flows over rectangular ridges, where the size of the secondary currents increases with spanwise spacing and their strength is maximised when s reaches the outer length scale of the flow [2]. The present study shows that while variation of Reynolds number has no effect on topology of the flow at any ridge spacing, Re_τ influences the levels of Reynolds and dispersive shear stresses. For cases with strong secondary currents, $\langle \widetilde{u}\widetilde{w} \rangle^+$ levels in the outer layer increase with Re_τ, while for narrow ridge spacings this increase is not observed ($s = \pi\delta/8$) or significantly smaller ($s = \pi\delta/4$).

While spanwise ridge spacing is the primary parameter that influences the level of secondary currents, other parameters, such as ridge base width, also affect secondary currents [3,5]. However, base width has mainly been investigated for rectangular ridges and far less is known for its effect in the context of other ridge shapes. Tertiary flows were observed above very wide rectangular ridges [3,5], but it is not known whether a similar effect could be observed above ridges with triangular cross-section. Although linearised models suggest that even very wide triangular ridges cannot generate tertiary flows [11], this prediction still needs to be tested either experimentally or using DNS. Furthermore, only limited shape variation is possible for a rectangular cross-section where all internal angles are fixed to 90° by definition, while other ridge cross-sections, such as triangles, can allow a wider range of modifications.

Some of these questions will be addressed in the next stage of this project, where the influence of triangular ridge configuration on secondary currents and turbulence statistics will be studied. First, the effect of ridge base width will be considered by changing equilateral triangular ridge cross-section to isosceles triangles. Next, the cross-section will be changed to scalene triangles to establish the influence of varied side length.

Acknowledgements. We gratefully acknowledge support for this work by the United Kingdom's Engineering and Physical Sciences Research Council [grant number EP/V002066/1]. This work used the ARCHER2 UK National Supercomputing Service (https://www.archer2.ac.uk).

References

1. Anderson, W., Barros, J.M., Christensen, K.T., Ankit, A.: Numerical and experimental study of mechanisms responsible for turbulent secondary flows in boundary layer flows over spanwise heterogeneous roughness. J. Fluid Mech. **768**, 316–347 (2015)
2. Vanderwel, C., Ganapathisubramani, B.: Effects of spanwise spacing on large-scale secondary flows in rough-wall turbulent boundary layers. J. Fluid Mech. **774**, R2 (2015)
3. Hwang, G.H., Lee, J.H.: Secondary flows in turbulent boundary layers over longitudinal surface roughness. Phys. Rev. Fluids **3**, 014608 (2018)
4. Yang, J., Anderson, W.: Numerical study of turbulent channel flow over surfaces with variable spanwise heterogeneities: topographically-driven secondary flows affect outer-layer similarity of turbulent length scales. Flow Turbul. Combust. **10**, 1–17 (2018)
5. Medjnoun, T., Vanderwel, C., Ganapathisubramani, B.: Effects of heterogeneous surface geometry on secondary flows in turbulent boundary layers. J. Fluid Mech. **886**, A31 (2020)
6. Zampiron, A., Cameron, S., Nikora, V.: Secondary currents and very-large-scale motions in open-channel flow over streamwise ridges. J. Fluid Mech. **887**, A17 (2020)
7. Busse, A., Lützner, M., Sandham, N.D.: Direct numerical simulation of turbulent flow over a rough surface based on a surface scan. Comput. Fluids **116**, 1290147 (2015)
8. Yang, J., Balaras, E.: An embedded-boundary formulation for large-eddy simulation of turbulent flows interacting with moving boundaries. J. Comput. Phys. **215**, 12–40 (2006)
9. Raupach, M.R., Shaw, R.H.: Averaging procedures for flow within vegetation canopies. Bound.-Layer Meteorol. **22**, 79–90 (1982)
10. Gray, W.G., Lee, P.C.Y.: On the theorems for local volume averaging of multiphase systems. Int. J. Multiph. Flow **3**, 333–340 (1977)
11. Zampino, G., Lasagna, D., Ganapathisubramani, B.: Modelling secondary structures developing over longitudinal ridges with different geometries. In: 12th International Symposium on Turbulence and Shear Flow Phenomena (TSFP12), Osaka, Japan, 19-22 July 2022 (2022)

Author Index

C. Marchioli et al. (Eds.): DLES 2023, ERCO 31, pp. 387–389, 2024.
https://doi.org/10.1007/978-3-031-47028-8